BIOFILMS IN BIOELECTROCHEMICAL SYSTEMS

BIOFILMS IN BIOELECTROCHEMICAL SYSTEMS

From Laboratory Practice to Data Interpretation

HALUK BEYENAL
JEROME T. BABAUTA

WILEY

Published by John Wiley & Sons, Inc., Hoboken, New Jersey
Published simultaneously in Canada

Library of Congress Cataloging-in-Publication Data:

Beyenal, H.
Biofilms in bioelectrochemical systems : from laboratory practice to data interpretation / Haluk Beyenal, Jerome T. Babauta.
pages cm
Includes index.
ISBN 978-1-118-41349-4 (hardback)
1. Fuel cells–Materials. 2. Energy harvesting. 3. Biofilms–Industrial applications.
4. Bioelectrochemistry–Industrial applications. I. Babauta, Jerome T. II. Title.
TK2931.B488 2015
621.31′24290284–dc23

2015010200

Typeset in 10/12pt, TimesLtStd by SPi Global, Chennai, India

10 9 8 7 6 5 4 3 2 1

1 2015

CONTENTS

CONTRIBUTORS LIST

Jerome T. Babauta, The Gene and Linda Voiland School of Chemical Engineering and Bioengineering, Washington State University, Pullman, WA, USA

Haluk Beyenal, The Gene and Linda Voiland School of Chemical Engineering and Bioengineering, Washington State University, Pullman, WA, USA

Darryl A. Boyd, National Research Council, Washington, DC, USA

Orianna Bretschger, Microbial and Environmental Genomics Group, J. Craig Venter Institute, San Diego, CA, USA

Bart Chadwick, Space and Naval Warfare Systems Center Pacific, San Diego, CA, USA

Jeffrey S. Erickson, Center for Bio/Molecular Science and Engineering, Naval Research Laboratory, Washington, DC, USA

Francisco Fabregat-Santiago, Grup de Dispositius Fotovoltaics i Optoelectrònics, Universitat Jaume I, Spain

Jim Fredrickson, Biological Sciences Division, Pacific Northwest National Laboratory, Richland, WA, USA

Sixto Giménez, Grup de Dispositius Fotovoltaics i Optoelectrònics, Universitat Jaume I, Spain

Annemiek ter Heijne, Sub-department of Environmental Technology, Wageningen University, Netherlands

Lewis Hsu, Space and Naval Warfare Systems Center Pacific, San Diego, CA, USA

Shun'ichi Ishii, Microbial and Environmental Genomics Group, J. Craig Venter Institute, San Diego, CA, USA

Cornelius Ivory, The Gene and Linda Voiland School of Chemical Engineering and Bioengineering, Washington State University, Pullman, WA, USA

Jeff Kagan, Space and Naval Warfare Systems Center Pacific, San Diego, CA, USA

Andrew Kuprat, Fundamental and Computational Sciences Directorate, Pacific Northwest National Laboratory, Richland, WA, USA

Derek R. Lovley, Department of Microbiology, University of Massachusetts, Amherst, USA

Nikhil S. Malvankar, Department of Microbiology, University of Massachusetts, Amherst, USA

Lisa McDonald, Microbial and Environmental Genomics Group, J. Craig Venter Institute, San Diego, CA, USA

Sudeep C. Popat, Swette Center for Environmental Biotechnology, Biodesign Institute, Arizona State University, USA

John Regan, Department of Civil and Environmental Engineering, The Pennsylvania State University, University Park, PA, USA

Gemma Reguera, Department of Microbiology and Molecular Genetics, Michigan State University, East Lansing, MI, USA

Clare E. Reimers, College of Earth, Ocean and Atmospheric Sciences, Oregon State University, OR, USA

Ryan Renslow, Environmental Molecular Sciences Laboratory, Pacific Northwest National Laboratory, Richland, WA, USA

Jared N. Roy, George Mason University, Manassas, VA, USA

Jim Schenk, The Department of Chemistry, Washington State University, Pullman, WA, USA

Rachel M. Snider, Center for Bio/Molecular Science and Engineering, Naval Research Laboratory, SW, Washington, DC, USA; BioTechnology Insititute, University of Minnesota, Saint Paul, MN, USA

Crystal Snowden, Microbial and Environmental Genomics Group, J. Craig Venter Institute, San Diego, CA, USA

Allison M. Speers, Department of Microbiology and Molecular Genetics, Michigan State University, East Lansing, MI, USA

Sarah M. Strycharz-Glaven, Center for Bio/Molecular Science and Engineering, Naval Research Laboratory, Washington, DC, USA; BioTechnology Insititute, University of Minnesota, Saint Paul, MN, USA

Shino Suzuki, Microbial and Environmental Genomics Group, J. Craig Venter Institute, San Diego, CA, USA

Leonard M. Tender, Center for Bio/Molecular Science and Engineering, Naval Research Laboratory, Washington, DC, USA; BioTechnology Insititute, University of Minnesota, Saint Paul, MN, USA

César I. Torres, Swette Center for Environmental Biotechnology, Biodesign Institute, Arizona State University, USA

Francisco Fabregat-Santiago, Grup de Dispositius Fotovoltaics i Optoelectrònics, Universitat Jaume I, Spain

Jerome T. Tudala, The Gene and Linda Voiland School of Chemical Engineering and Bioengineering, Washington State University, Pullman, WA, USA

Hengjing Yan, Department of Chemistry and Biochemistry, University of California Santa Barbara, Santa Barbara, CA, USA

Rachel A. Yoho, Swette Center for Environmental Biotechnology, Biodesign Institute, Arizona State University, USA

PREFACE

Almost a decade has passed since the rediscovery of the microbial fuel cell, yet the power generation of these devices has not advanced significantly. This is mainly due to the research being focused solely on the improvement of power generation rather than on a fundamental understanding of electron transfer processes at the anode and cathode. This book was written to support bioelectrochemical systems researchers wanting to discover novel information about microbial respiration on anodes and on cathodes, as well as to improve our understanding of extracellular electron transfer. The book focuses on the methods used to study electron transfer processes in biofilms growing on electrodes and presents several successful applications of microbial fuel cells. Throughout the book, we refer to electrochemically active biofilms (EABs) as biofilms that exchange electrons with conductive surfaces, that is, electrodes. The reason we call biofilms that exchange electrons with conductive surfaces electrochemically active is that we can grow them in such a way that we can quantify electron transfer rates. In particular, biofilms grown on inert electrodes allow us to quantify electron transfer processes and identify mechanisms.

We, the editors, believe that it would currently be very difficult to use microbial fuel cells as alternative energy sources that could compete with fossil fuels. Instead, we recognize that the study of the process itself, cells growing and respiring on electrodes, is novel in its own right and will allow researchers to generate new knowledge on biological electron transfer processes. Also, these processes are fundamental and can critically contribute toward many different applications. Research on EABs is new and requires multidisciplinary approaches based on integrated techniques. The potential applications will require the integration of microbiology, biochemistry, and electrochemistry, as well as engineering. We note that the authors and reviewers of this book are microbiologists, electrochemists, engineers, and physicists. The diverse authorship of this book attests to the multidisciplinary nature of EAB research.

The first chapter introduces EABs and provides an overview. The second chapter describes in detail how to grow *Geobacter* biofilms. Successful bioelectrochemical systems applications and research efforts require a comprehensive understanding of the microbial activities associated with the bioelectrocatalytic conversion of chemical and electrical inputs. Chapters 3 and 4 describe methods and protocols for microbial biomass quantification, DNA extraction from electrode surfaces, 16S rRNA gene sequencing, and sequence data analysis. Chapter 5 focuses on cyclic voltammetry of anodic and cathodic biofilms. It provides practical information and fundamental knowledge to researchers who study biofilms on electrodes. Chapter 6 describes redox conduction and experimental methods that enable researchers to perform electron transport rate measurements for their own types of biofilms. This chapter focuses on electron hopping. Chapter 7 summarizes the methods used to measure the newly discovered conductive properties of biofilms directly. It also describes the physical meaning of electronic conductivity and the various mechanisms of conductivity, then discusses in detail the experimental methods applied to measure conductivity in living biofilms directly and the results obtained using these methods. Chapter 8 focuses on the principles and theory of electrochemical impedance spectroscopy and its application to studying EABs. Chapter 9 presents a generic model that incorporates diffusion- and conduction-based mechanisms to describe EABs that can utilize conductive and mediated electron transfer processes simultaneously. This chapter shows how mathematical modeling can be used as a unifying tool for investigating EABs and electron transfer mechanisms. The last two chapters describe the power management, construction, and deployment of microbial fuel cells. They emphasize the details in the design and operation of microbial fuel cells. Innovative power management platforms are detailed in Chapter 10 along with engineering and environmental considerations that have influenced the development of microbial fuel cells powering electronic devices. Chapter 11 describes the construction, deployment, and operation of larger-scale microbial fuel cells, as well as segmented systems. Overall, we expect that this book will be an integrated reference that ties together the principles of all the disciplines used to study EABs. Our main focus is to provide hands-on procedures for researchers who are new to this area.

We hope that this book will serve as a reference for the researchers and educators who wish to use EABs in bioelectrochemical systems and in the classroom. We are especially excited regarding the use of EABs in the classroom, as they expose students to a unique form of microbial respiration. In research, we anticipate that the information presented in this book will stimulate novel ideas for improving electron transfer efficiencies and developing viable technologies in the future. We acknowledge support from the Office of Naval Research, Department of Energy, National Science Foundation, and Department of Defense, which sponsored projects that allowed us to develop the methodology and results presented in this book. Beyenal acknowledges partial support by the “Fundamental and Applied Chemical and Biological Catalysts to Minimize Climate Change, Create a Sustainable Energy Future, and Provide a Safer Food Supply” project (#WNP00807) and National Institute Of Environmental

Health Sciences (R25ES23632) for his time to edit chapters. Babauta acknowledges the NIH Biotechnology Training Program at Washington State University, which sponsored part of his PhD education on EABs, and Gamry Instruments for supporting a summer internship on electrochemical tools to study EABs. We would like to thank Sarah Hall from Wiley, who motivated us initially to organize this book. We specifically thank Dr. Edmund H. Immergut, a consulting editor at Wiley, who followed up the book idea and strongly supported our book. Finally, we thank Anita Lekhwani and Cecilia Tsai for working with us to finalize the book. We thank Dr. Linda A. Chrisey, a program manager at the Office of Naval Research, who supported fundamental research on electron transfer processes that has contributed greatly to this book. Her strong support for EAB research enabled us to write this book.

We thank Jeffrey London and Ryan Renslow from PNNL for the image on the cover. The expert comments from many reviewers have improved the chapters considerably, and we are thankful for these reviewers. Finally, we thank the authors of the various chapters in this book for their contributions. It has been a great pleasure working with such a diverse group of expert researchers.

HALUK BEYENAL
JEROME T. BABAUTA

1

INTRODUCTION TO ELECTROCHEMICALLY ACTIVE BIOFILMS

JEROME T. BABAUTA AND HALUK BEYENAL

The Gene and Linda Voiland School of Chemical Engineering and Bioengineering, Washington State University, Pullman, WA, USA

1.1 INTRODUCTION

Microbial respiration is based on electron transfer from electron donors to electron acceptors – a series of reactions facilitated by a cascade of energetic substances; these are well-known reactions described in the literature [1–4]. The donors and acceptors of electrons are typically dissolved substances; however, some microorganisms can use solid electron donors and/or solid electron acceptors, such as minerals and metals, in respiration [5]. Electron transfer by microorganisms to and from external electron acceptors or donors is termed *extracellular electron transfer*. Extracellular electron transfer is typically studied using the model organisms *Shewanella oneidensis* and *Geobacter sulfurreducens*. However, the focus on extracellular electron transfer has been extended toward interspecies electron transfer, as well as electron transfer into microbes [6–10]. The exact mechanisms of extracellular electron transfer between microorganisms and solid substances remain a matter of debate in the literature [1, 2, 11–20]. One of the goals of this book is to provide fundamental knowledge needed to study the exact nature of extracellular electron transfer processes. Two points of view are usually presented in this debate: (1) that electrons are transferred by conduction through extracellular materials or elongated appendages called nanowires [12, 14, 17, 18, 21, 22] and (2) that electrons are transferred across a redox gradient by electrochemical reactions using either freely diffusing redox mediators, also known as electron shuttles [11, 15, 16, 19, 20, 23–25], or bound redox mediators at sufficient density

Biofilms in Bioelectrochemical Systems: From Laboratory Practice to Data Interpretation, First Edition.
Edited by Haluk Beyenal and Jerome Babauta.

within the biofilm to allow for electron hopping across redox sites [26, 27]. This book covers both mechanisms and describes how to perform the measurements for each mechanism. Biofilms with microorganisms capable of electron transfer to and from solid electron acceptors have been used in microbial fuel cells (MFCs) to harvest energy from various environmental processes [28]. The biofilms grown on the electrodes of MFCs are called electrochemically active biofilms (EABs), which admittedly is a misnomer, as all microorganisms are electroactive in the respiration process. EABs are also known under several other names in the literature dedicated to MFCs, such as electricigens, electrochemically active microbes, exoelectrogenic bacteria, and anode-respiring or anodophilic species. However, because the hallmark of EABs is the ability to exchange electrons with solid surfaces such as electrodes, we believe that the term "EABs" refers to the most basic property of these biofilms. Because this book focuses on extracellular electron transfer processes between biofilms and inert electrodes, the term "EABs" is fully appropriate. The link connecting EABs, electrochemistry, and electrochemical techniques is also considerably clearer for a more general audience.

The use of EABs in MFCs is not new. For various reasons, these devices have attracted some attention in the literature recently. In particular, researchers have recognized their potential use as alternative sources of energy. The attention MFCs receive is fully justified, although some expectations of their ability to deliver large amounts of energy combined simultaneously with high power appear overly optimistic. It has been demonstrated that MFCs are successful sources of energy to power electronic devices that consume low levels of power continuously or to power electronic devices requiring higher power intermittently [29–35]. Lately, the concept of power management systems is gaining popularity [33, 36–44]. Much of the interest in using MFCs stems from the idea of harvesting energy from wastewater treatment processes, which at present are wasteful processes in which energy-rich streams are reclaimed without useful energy being obtained [45]. There has been an estimate presented at conferences referring to the amount of energy that could be harvested from all wastewater treated in the United States if the entire chemical oxygen demand were converted to the equivalent number of electrons and used to power external devices. In our opinion, this calculation resembles the computations estimating how much gold could be extracted from seawater – and how rich one could get by doing so – if we were to treat all the oceans in the world. There is some truth in these calculations, of course, but they neglect the costs and the technical problems associated with harvesting energy from wastewater or extracting gold from seawater. MFCs are fascinating devices, and they no doubt will find practical applications through integrated power management systems. In the short term, however, it is difficult to see how they can meaningfully contribute to solving the impending energy crisis. We are afraid that similar claims may be promoted on similar microbial technologies based on MFCs [46, 47]. Just how practical these applications will become remains to be seen.

Collectively, MFCs and the newer biologically catalyzed electrochemical cells have come to be known as bioelectrochemical systems (BESs) [48–52]. As BES research becomes more sophisticated, it appears that BESs can provide new insights into the fundamental mechanisms of electron transfer between microorganisms and solid substances. This application can deliver interesting results sooner than the expectations of harvesting large amounts of energy from wastewater processes can be fulfilled. There is a lot of excitement about using BESs, and understandably, all expectations may not be fulfilled.

The immediate goal of this book is to describe the state-of-the-art research techniques for studying electron transfer processes in EABs. The future of MFCs or any technology based on them will be based on fundamental electrochemical and microbial research on electron transfer processes in biofilms. This directly translates to revealing the processes governing electron transfer between a biofilm and an electrode and not the incremental improvement of power output from MFCs. Thus, we would also like to call attention to something we find promising: the use of BESs as tools of discovery in studying the process of electron transfer in EABs. For example, high-throughput, efficient BESs could be used to select for new EABs [53, 54]. Many researchers could use this technology as a tool for understanding the biochemistry of these unique microorganisms. This use of BESs may be much less glamorous than the promise of delivering power to the national grid, but at the same time, the expectations we set are more realistic than converting all or even a large part of the chemical oxygen demand in wastewater into electron equivalents and using it to power external devices. As BES researchers, we are somewhat concerned that focusing on glamorous, but, in our opinion, currently unrealistic, expectations may do damage to a legitimate and very interesting field of research. We hope that this book can help identify and evaluate the strengths and the limitations of BESs for both generating energy and studying the mechanisms of electron transfer between microorganisms in biofilms and solid substances. Electron transfer between biofilms and solid surfaces was known long ago [55]; however, MFC research and related research tools have critically contributed to new developments and mechanisms of extracellular electron transfer by combining electrochemistry with microbiology. In the following sections of this chapter, we describe EAB research from an electrochemical perspective, focusing on single electrodes rather than on reactor systems. We use the term BES to describe generic electrochemical cells utilizing microbe-based half-cells. The term "biofilm electrode" refers to the microbe-based half-cell where the EAB is grown. Finally, we note that these sections have been extended and re-written from a previously published work to serve as an introduction to this book [56].

1.2 ELECTROCHEMICALLY ACTIVE BIOFILM PREPARATION AND REACTOR CONFIGURATIONS

Although there are many techniques for quantifying EAB extracellular electron transfer mechanisms, the quality and interpretation of the results are highly dependent on the way the study is conducted. Factors that are often selected arbitrarily, such as (1) the biofilm electrode material, (2) how the EABs are grown on the biofilm electrodes, (3) the reactor configuration used to grow the EABs, and (4) the reactor configurations used to study the extracellular electron transfer processes, have a critical impact on the resulting EAB and its ability to participate in extracellular electron transfer processes. Identifying the effects of each factor on EAB performance may serve as a basis for optimizing systems toward maximizing the rate of energy conversion.

1.2.1 Electrode Materials

The electrode material used to construct a biofilm electrode affects the measured current and open circuit potential (OCP) of the EAB grown on it, and therefore, the choice of electrode material is important for the standardization of reported values. Traditionally, cheaper graphite, carbon paper, carbon granule, carbon brush, or carbon felt electrodes are used in MFC practical applications [57, 58]. These carbon materials suffer from high background currents that can mask the electrochemical responses of redox species at low concentrations. In our laboratory, we often use glassy carbon electrodes to observe electrochemical activity. One advantage of glassy carbon is that the background currents are practically zero in the potential ranges in which EABs are studied; another is that it is nonporous. In addition, there is significant literature on electrochemistry utilizing glassy carbon electrodes, potentially opening up a vast amount of literature to EAB studies. The use of glassy carbon electrodes would provide more universal current values when fundamental electron transfer of EABs is studied. For these reasons, we recommend researchers use, or at least test, their systems with glassy carbon electrodes. There are various glassy carbon types, and readers are referred to the following reference for a more thorough review [59].

When glassy carbon electrodes are not compatible with an experiment, common substitutes include gold and indium tin oxide (ITO) electrodes [60]. Gold offers the advantage of a significant literature on self-assembled monolayers and the modification of surface functional groups [61–63]. Thin gold films on glass substrate have also been used in advanced spectroscopic techniques for direct electron transfer studies [64]. ITO is used in spectroelectrochemical experiments in which an optically transparent electrode is required [65, 66]. Users should be aware of the resistivity of ITO electrodes and their durability, as the conductive film is thin compared to glassy carbon [66]. Platinum and other catalytic electrode materials could have unanticipated effects on an experiment and are best avoided. For the supporting (auxiliary) electrode that completes the electrochemical cell and functions as the electron source/sink for the electrons derived from the EAB, a cheaper carbon electrode with a larger surface area can be used. Examples of results obtained from different electrode materials are given in Chapter 5.

1.2.2 Reactors and Electrode Configurations Used to Study Electrochemically Active Biofilms

The positions of the biofilm electrode, supporting electrode, and reference electrode (RE) in a BES have direct effects on the measured current and should be accounted for beyond spatial geometric considerations [59, 67, 68]. For example, the simplest system to configure to study electron transfer in EABs may be an MFC (Fig. 1.1a), although it is generally used to quantify power production in

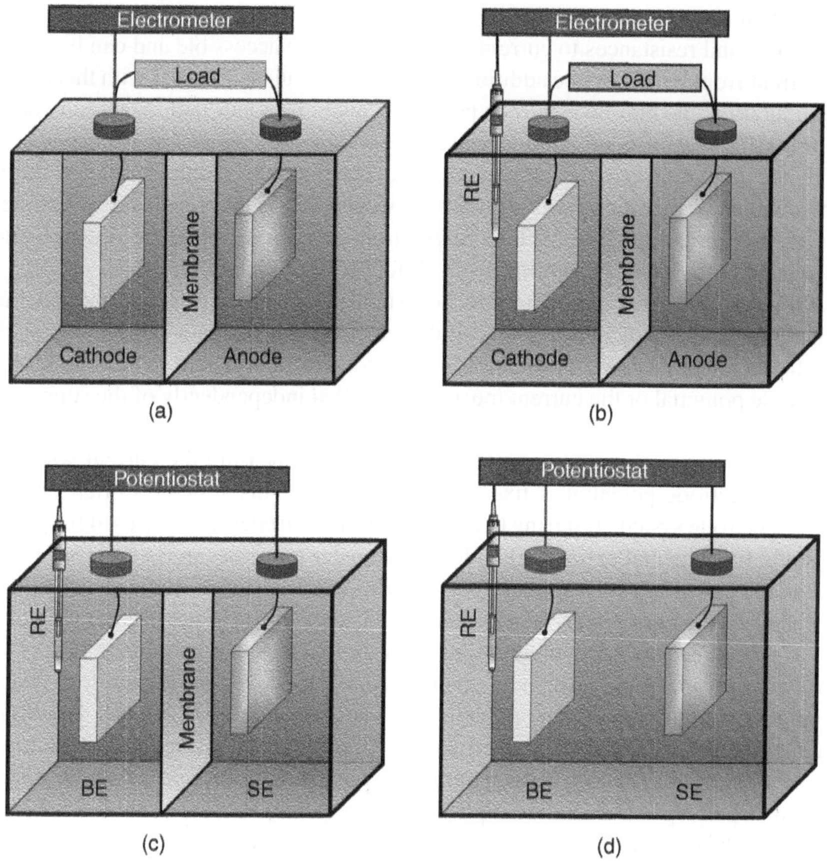

Figure 1.1 EABs can be studied using four different configurations: (a) an MFC with an anode and a cathode; (b) an MFC with an anode, a cathode, and a reference electrode (RE) used to monitor individual electrode potentials (against the RE); (c) a BES with a biofilm electrode (BE), an isolated supporting electrode (SE), and an RE connected to a potentiostat; and (d) a BES with all three electrodes immersed in the same solution.

practical research [67, 69]. The geometry or other experimental parameters of an MFC can be optimized to produce more power. However, an MFC cannot be used to obtain information about the EABs on the individual electrodes because only cell potential can be measured. In practical terms, it is very difficult to control the conditions for either electrode independently and thus form a conclusion on the role of the EAB. Without knowing the individual electrode potentials, it would be very difficult to determine the fundamental reason for an increase in power in an MFC. Thus, how the EAB responded to variation in electrode potentials could not be understood through electrochemical theory. The end result would be the inability to study electron transfer in EABs. The MFC reactor configuration can be enhanced by inserting an RE to measure individual electrode potentials and characterize overpotentials and potential losses (Fig. 1.1b) [69, 70]. The individual electrode potentials and resistances to current flow then become accessible and can be related to current from EABs. While additional information can be obtained with this reactor configuration, the biofilm electrode potential cannot be controlled. Ambiguity arises in the MFC reactor configuration when an experimental parameter is changed and both the potential and the current change. Having two variables change during operation makes comparisons difficult between the results generated in different research laboratories. The MFC configuration only serves practical research such as quantifying power generation from different wastewaters, in which capacity it is fully adequate. For the results to be extended beyond a practical purpose and related back to fundamental extracellular electron transfer processes (i.e., for the electrochemical response of the biofilm electrode to be isolated), either the biofilm electrode potential or the current must be controlled independently of the supporting electrode.

Therefore, a potentiostat is generally required to measure the current while the biofilm electrode potential is fixed (Fig. 1.1c or d). This system is often called a three-electrode system, referring to the number of electrodes, and it is used frequently to study fundamental extracellular electron transfer processes in EABs [15, 71–73]. When an experimental parameter such as the initial electron donor concentration is changed, the current can then be correlated without the effect of a varying electrode potential. Reactor configurations with an ion-selective membrane (Fig. 1.1c) and without one (Fig. 1.1d) have distinct advantages. For membrane-less reactor configurations, the membrane potential loss is eliminated [67, 74–76]. The disadvantage is that the supporting electrode reaction products are free to diffuse to the biofilm electrode. This could potentially generate uncontrolled experimental parameters. Chapter 5 gives several example cases. In membrane-less microbial electrolysis cells, the diffusion of hydrogen from the supporting electrode to the working electrode can be utilized by EABs to produce a current higher than that expected with the supplied electron donor [77, 78]. Regardless of the use of ion-selective membranes, potentiostatic systems provide more control over electron transfer than the MFC mode of operation (Fig. 1.1a and b). There is often a misconception that the use of a potentiostat damages the EABs on a biofilm electrode. Potentiostats allow users to set the biofilm electrode potential to a wide range of values, including those that would irreversibly damage the EAB. In reality, it is usually inexperienced users who damage EABs with potentiostats because such an outcome was not possible in an MFC configuration.

1.2.3 Current-Limiting Electrode

The current-limiting electrode is the electrode that cannot pass a higher current than the other electrode, either because of its small size or because of limiting electrode reaction kinetics. If the biofilm electrode (with an EAB) limits the current of the BES, this means that the performance of the BES is limited by the EAB, and electron transfer in EABs can only be studied under this condition. The knowledge of which electrode limits the current is critical when BESs are studied. In the case of

MFCs (Fig. 1.1a), because both current and potential are variable when the resistance to the current flow is changed, it is important to confirm that the EAB under investigation is limiting the current. The simplest way to determine which electrode in an MFC setup is the current-limiting electrode is to monitor the individual electrode potentials using an RE (Fig. 1.1b). When the resistor load is changed, the current-limiting electrode will undergo a significant change in electrode potential, whereas the non-current-limiting electrode will not [30]. This knowledge is critical, especially when sediment microbial fuel cells (SMFCs) and other field-deployed BESs are studied. The current-limiting electrode will always be the highest priority and the electrode in need of immediate improvement. The current-limiting electrode concept also applies to potentiostatic systems (Fig. 1.1c and d). The potentiostatic mode controls the biofilm electrode potential such that perturbations of the biofilm electrode potential cause a measurable change in the EAB under investigation. Thus, the current is controlled by the EAB. If, at any time, reactions at the supporting electrode affect the EAB under investigation, the controlled electrode cannot be called the limiting electrode, and steps must be taken to ensure that the effect of the supporting electrode can be assumed to be negligible. This is especially important in BESs that place the biofilm electrode of interest and the supporting electrode in the same solution (Fig. 1.1d). For fundamental electron transfer investigations, this concept cannot be ignored.

1.2.4 The Preferred Polarization Potential for Growing Electrochemically Active Biofilms

The effect of the polarization potential (anode potential) has been studied for various BESs and EABs [16, 54, 79–86]. There is no consensus on the exact magnitude of the potential to apply; however, there is a clear understanding that applying a polarization potential more positive than the OCP of the biofilm electrode is sufficient to drive electrons from the EAB to the biofilm electrode. The concept of an optimal polarization potential is misleading, as the polarization potential can be limited by external factors such as the energy efficiency of the BES [87]. The preferred polarization potential when electron transfer from an EAB is studied must be explored experimentally and chosen from a range of polarization potentials from near OCP to a few hundred millivolts more positive than OCP. For example, if maximum current is desired, then a polarization potential that is in the current-limiting region for EABs (current independent from polarization potential) should be used [87]. The polarization potential could also be used to select for different types of EABs, with different abilities for extracellular electron transfer [54]. However, for use in MFCs and other BESs, the polarization potential should be comparable to what is observed in the actual selection process or application. Interestingly, this is one of the more important concepts for studying extracellular electron transfer in SMFC applications. There, conditioning of anodes and cathodes would dramatically affect the performance. Therefore, techniques have been developed to adjust the potential of the electrode to a desired value incrementally, often slowly, to mimic the behavior seen during power generation.

1.2.5 Electrode Acclimatization and Growing Electrochemically Active Biofilms

Electrode acclimatization refers to the processes in which EABs are allowed time to populate an electrode surface; it is often used for multispecies EABs [54, 86, 88–97]. The purpose of acclimatization is to increase electrode performance by enhancing biofilm attachment and/or to allow the biofilm electrode to reach a steady state OCP prior to use in a BES [98]. The method of acclimatization affects the type of EAB grown on the biofilm electrode and can be focused on control of the current or of the biofilm electrode potential. Four acclimatization methods are common in the literature:

1. *Closed circuit:* The biofilm electrode and the supporting electrode are short circuited or connected across a resistor.
2. *Open circuit:* The biofilm electrode is left disconnected.
3. *Controlled cell potential:* A constant potential is applied between the biofilm electrode and the supporting electrode.
4. *Controlled electrode potential:* A constant polarization potential is placed between the biofilm electrode and the RE.

Closed circuit and open circuit acclimatization are the simplest methods to configure. Closed circuit acclimatization allows the biofilm electrode to reach a steady state cell potential and is focused on enhanced steady state electron transfer. The choice of resistor controls the amount of current allowed to pass [79, 99, 100]. Chapter 11 gives an example of this procedure being used to acclimatize the cathode and anodes of an SMFC. Open circuit acclimatization allows the biofilm electrode potential to develop a steady state OCP utilizing natural redox processes in the environment. Controlled cell potential acclimatization and controlled electrode potential acclimatization require powered external equipment irrespective of the natural redox processes in the environment. Controlled cell potential acclimatization ensures steady state electron transfer at a researcher-specified level. Controlled electrode potential acclimatization ensures steady state electron transfer irrespective of the supporting electrode. Both methods allow the user to expose the system to potentials not normally sustainable or possible. However, only controlled electrode potential acclimatization gives the researcher direct and consistent control of the biofilm electrode potential. To choose one method over the other, critical decisions must be made. Firstly, does the researcher prefer to select for biofilm processes or natural redox processes that can take advantage of an applied potential or polarization potential? Does the researcher prefer to produce electrodes that reflect only the natural redox processes? When a current is passed through an electrode to or from its surroundings, this will affect the state of redox processes around it. The choice of how electrodes are acclimatized affects the end result and should be reported clearly.

Growing EABs refers normally to pure cultures in the laboratory and can be achieved using two distinct acclimation methods. We should note that acclimatization refers to field experiments, whereas acclimation refers to laboratory experiments.

The first method is to grow EABs on an electrode in the presence of a soluble electron acceptor. This method is in line with open circuit acclimatization because no polarization potential is required for EAB growth. Once the EAB has reached a desired state (thickness, surface coverage, metabolic activity, and OCP), the soluble electron acceptor can be removed and the EAB can be switched to respiring on the biofilm electrode where a current is measured. The second method is to grow the EAB on an electrode that acts as the sole electron acceptor. This method is in line with closed circuit, controlled cell potential, and controlled electrode potential acclimatization in which the current produced reflects biofilm growth. Once an EAB reaches a desired state (thickness, surface coverage, metabolic activity, and current), it can be used for further investigation. Both methods are able to produce laboratory-scale EABs; however, the EABs resulting from these two methods have different biofilm properties and electron transfer capabilities. Most likely, this is due to the acclimation of the EAB to each electron acceptor. Nevin et al. [101] observed different 3D biofilm structures and electron transfer capabilities in *G. sulfurreducens* depending on whether the EAB was grown on fumarate as a soluble electron acceptor or on an electrode as a solid state electron acceptor [101, 102]. We have also observed structural differences between EABs grown using the two methods in our laboratory.

1.3 ELECTROCHEMICAL TECHNIQUES FOR STUDYING EXTRACELLULAR ELECTRON TRANSFER OF ELECTROCHEMICALLY ACTIVE BIOFILMS

The majority of this book is dedicated to describing how to study extracellular electron transfer of EABs. Once the correct reactor configuration has been chosen and the EAB has been successfully grown or acclimatized on an electrode, the next step is to study the electron transfer properties of the biofilm electrode using electrochemical techniques. There are many electrochemical techniques available; however, we introduce only those frequently used to study electron transfer processes in EAB literature.

1.3.1 Long-Term Electrode Polarization of Electrochemically Active Biofilms

In a long-term electrode polarization experiment, a selected polarization potential is applied to an electrode with an EAB grown on it and the current is measured. In this way, the total charge transferred in a batch system or the steady state current produced by the EABs in a continuous system can be measured. A long-term electrode polarization experiment identifies sustainable current generation that can be systematically related to controlled parameters such as polarization potential. While the technique appears similar to controlled electrode potential acclimatization, in which EABs are grown on polarized electrodes, the intents of the two are distinct and should be distinguished. For example, *G. sulfurreducens* DL-1 was allowed to acclimatize on an electrode for 5 months and resulted in selection for a new strain, *G. sulfurreducens*

KN400 [97]. Subsequently, the sustainable current productions of the two strains were measured using long-term electrode polarization experiments. The key value of performing electrode polarization experiments is to observe the electrochemical response of an EAB to a change in a controlled experimental parameter or the effects of acclimatization methods.

When a significant current (above background and noise levels) is generated by an EAB, such as that shown in Figure 1.2, the EAB is thought to exchange electrons with the biofilm electrode. To confirm that the current generated is related to the metabolism of the EAB or its metabolic by-products, certain long-term electrode polarization experiments can be performed. In batch experiments, the total consumption of the electron donor can be correlated to the total charge transferred. If the current trends toward zero when the electron donor concentration goes to zero, then the oxidation of the electron donor is the source of electrons and the coulombic efficiency can be calculated [67]. For example, the current production by *G. sulfurreducens* biofilms is directly related to the consumption of acetate in batch mode [103]. In continuous experiments, the electron donor feed concentration can be altered and subsequent steady state current values measured. Replacing the bulk solution with fresh growth medium during electrode polarization experiments has been done to probe soluble extracellular electron transfer mechanisms. For example, the spent solution in a *S. oneidensis* MR-1 biofilm reactor was replaced to show that soluble redox mediators were responsible for the steady state current generated [15]. The bulk solution of *G. sulfurreducens* biofilms was replaced with acetate-free growth medium to show acetate dependence [103]. Other sophisticated experiments can be done to isolate controlled parameters in EAB experiments that affect extracellular electron

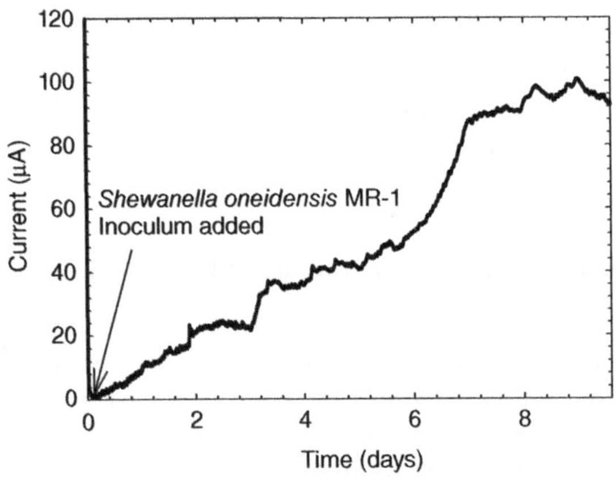

Figure 1.2 Current generation by *Shewanella oneidensis* MR-1 biofilm on a graphite electrode under anaerobic conditions in the reactor configuration shown in Figure 1.1c. The current increased steadily over a period of 9 days. The polarization potential was 0 $mV_{Ag/AgCl}$. Reproduced with permission from Babauta et al. [56]. Copyright 2012 Taylor & Francis. http://www.informaworld.com.

transfer mechanisms. EABs can be genetically engineered to enhance/inhibit current generation. For example, it was observed that no current could be produced by a ΔpilA *G. sulfurreducens* mutant on gold electrodes [60]. The pH can be adjusted to correlate proton transfer with current generation [104, 105]. Chapter 2 demonstrates similar procedures for *G. sulfurreducens* biofilms, and Chapter 5 discusses in detail the importance of background current.

1.3.2 Cyclic Voltammetry

When a steady state current or a current higher than the background is identified as being the result of EABs, cyclic voltammetry (CV) can be used to identify the biofilm electrode potential at which active redox couples related to the EAB are oxidized or reduced. CV is an electrochemical technique that applies a linear polarization potential scan from an initial polarization potential to a final polarization potential and measures the current. Because redox couples can only be reduced or oxidized at certain electrode potentials, CV can determine the biofilm electrode potential range in which extracellular electron transfer can occur in EABs [72, 106]. Under well-controlled conditions, CV can be used to determine whether EABs have the capability for electron transfer, whether freely diffusing species or surface-adsorbed species contribute to electron transfer, and whether EABs engage in catalytic activity toward specific substrates [15, 72, 73, 106–113]. CV studies, however, do not reflect the ability of EABs to produce long-term, sustainable current; this issue is reserved for long-term electrode polarization experiments. Often, CV is coupled to long-term electrode polarization experiments in which CV can explain how the active redox couples are affected by systematic changes in controlled parameters.

There is an implicit assumption, however, that EABs growing on electrode surfaces can be described as a "well-controlled" condition in which CV can be applied to study reaction mechanisms as in pure electrochemical systems. Beyond reproducibility of the biofilm electrode surface, simply characterizing biofilm structure itself has historically been difficult [114–116]. Furthermore, the result of biofilm heterogeneity is local variation of not only diffusion coefficients, but also flow velocities [117–120]. The unknown mass transfer conditions suggest that not all cells in the EAB contribute equally to current production. Several chapters in this book are dedicated to the use of CV to characterize electron transfer processes in EABs.

1.3.3 Limitations of Electrochemical Techniques

Electrochemical investigations in complex systems such as EABs require more physical and chemical evidence to determine whether an observed electrochemical response was caused by a change in an experimental parameter. Figure 1.3 shows a scanning electron microscope (SEM) image of a biofilm growing on an electrode. The electrode was deployed in the environment and polarized for more than 3 months. The cells on the electrode were then imaged. The biofilm was grown on a graphite electrode that had micropores. The cells had many appendages attached to

Figure 1.3 An SEM image of cells growing on an electrode.

the electrode. One could simply hypothesize that these were nanowires facilitating electron transfer to the electrodes. For a scientific study, such a hypothesis needs to be further tested: the conductivity of these "nanowires" must be measured. However, the critical question is: "is it enough to demonstrate that these wires are conductive?" The answer is no. That the cells have conductive nanostructures does not mean that this is the primary mechanism through which the cells transfer electrons to the electrode. In this case, genetic engineering can provide an excellent tool! Through overexpression of these nanostructures, genetic engineering can allow us to understand their role. Let us assume that when the nanostructures were overexpressed, it was found experimentally that more current was generated. Does this really mean that the current increase was due to the nanostructures? The answer depends critically on whether genetic engineering changed only the nanostructures and nothing else changed in the cell or the resulting biofilm. This is the biggest challenge and limitation in electrochemical techniques. There is no single technique that can address this question. Many control experiments should be designed to quantify the changes in biomass production in both cases. The increase in nanostructure production could cause better cell growth, and the increased cell growth could be responsible for the increased current. Moreover, what about charge transfer limitations? Is it possible that genetic engineering also changes protein expression and one of these proteins is involved in charge transfer at the electrode/biofilm interface? Electrochemical impedance spectroscopy (EIS) can be helpful in addressing this question. What about the possibility of redox proteins being overexpressed? CV or square wave voltammetry (SWV) or EIS possibly could provide additional information about limiting steps and redox couples involved in the electron transfer processes. The most critical limitation of electrochemical techniques discussed in this chapter is that by themselves they are of limited use for understanding electron transfer processes. Coupled or even multiple techniques are required to understand electron transfer processes in EABs. If the research were concerned with the electron transfer processes for a single redox couple with only one or two coupled chemical reactions, our job would be easy. However, in EABs, an entire metabolic process of cascading biochemical reactions is coupled to electron transfer, and it may be impossible to change only one variable at a time. Therefore, individual electrochemical techniques will have limited use unless they are coupled with other techniques.

In the literature, researchers have used additional molecular techniques, such as the generation of mutants with different gene/protein expression levels, to provide a physiological link to electrochemical investigations. Electrochemical techniques such as CV and SWV accurately describe the nature of the electron transfer event in the presence of EABs (reversibility, mass transfer limitations, properties of redox couples, and reaction steps); however, they do not give any evidence on how EABs participate in the electron transfer or what aspect of EABs promotes electron transfer. Thus, the presence of redox couples in EABs does not necessarily mean that they participate in electron transfer. More importantly, the presence or absence of electrochemical activity (current peaks observed in CV or SWV) does not necessarily mean that it is or that it is not the source of long-term, sustainable current in EABs. An example of this is the ability of certain microorganisms to utilize soluble exogenous electron shuttles in their surroundings for extracellular electron transfer. For example, it was shown that *Desulfitobacterium hafniense* strain DCB2 could utilize exogenous quinone-like mediators to produce sustainable current in MFCs [121]. In our laboratory, we found that an iron-reducing, biofilm-forming *Paenibacillus* sp. could use exogenous flavins and anthraquinone-2,6-disulfonate to reduce ferrihydrite [122]. The electrochemical activity of these types of EABs would not be observable with CV in pure cultures in the laboratory, which highlights the importance of syntrophic interactions in mixed species EABs [123]. In fact, the interpretation of electrochemical investigations with CV has been a topic of intense debate in EAB research [14, 124].

In summary, the main limitation of electrochemical techniques is that no single technique is adequate for studying electron transfer processes in EABs. However, the use of multiple or coupled techniques as summarized in the following sections and detailed further in this book brings about other limitations, such as the extent to which one can master all these techniques. This is a common issue in interdisciplinary fields such as this. In practical terms, there are many different research groups who have expertise in different techniques, and all these techniques may be required. Therefore, it is critical for each group to detail their protocol so that other researchers can apply them for their research. This has been a major limitation in EAB research so far and is responsible for the writing of this book. One of the goals of this book is to overcome this limitation.

1.4 COUPLED TECHNIQUES

The coupling of electrode polarization, CV, and SWV with methods that directly measure physical or chemical parameters varying inside EABs in situ addresses issues with the interpretation of electrochemical data. The goal of this coupling is to correlate the properties that vary within the EAB under electrode-respiring conditions. Specifically, future EAB research and advanced techniques should focus on variation that occurs within the biofilm and not just in the bulk solution. In fact, the correlation of variation in biofilms and bulk solution will advance the field significantly. This correlation can be done by: (1) directly measuring the kinetics of redox mediators inside biofilms, (2) resolving local concentrations of chemical species inside biofilms, (3) resolving the physical location of electrochemically active species in situ, and (4) correlating limiting current with local biofilm parameters. There are several available tools and techniques that can be used, such as microelectrodes, spectroelectrochemical methods, and microscopy.

1.4.1 Microelectrodes

Microelectrodes have been used to study the stratification of various biofilm properties in biofilms and water-sediment interfaces since the early 1990s [125–128]. For example, microelectrodes have been used to measure concentrations of oxygen, hydrogen, hydrogen sulfide, and carbon dioxide, as well as pH, redox potential, and local flow velocities [127–135]. Voltammetric microelectrodes that use techniques such as CV and pulse voltammetry have also been used to detect chemicals such as metal cations and flavins [133, 136]. Because microelectrodes have dimensions as small as 1–5 µm and are minimally invasive, they are well suited to studying changes in both the EAB and the bulk solution above the EAB during electrode respiration. A distinct advantage of microelectrodes is that they can correlate local biofilm properties with the bulk biofilm properties, electron transfer rates, and electron transfer mechanisms. On the other hand, the data should be critically interpreted when a highly heterogeneous EAB is investigated because the results can vary between locations [137].

Microelectrodes can be used in conjunction with voltammetric techniques in two modes. An example of the first, in which depth profiles are taken during constant polarization, is shown in Figure 1.4. A redox microelectrode was stepped down toward the bottom of a *S. oneidensis* biofilm. At each step, redox potential was measured, producing a redox potential depth profile inside the biofilm. In the presence of soluble redox mediators, redox potential is expected to increase toward the biofilm electrode. However, the redox potential profiles in Figure 1.4 show a decreasing redox potential profile, indicating that the *S. oneidensis* biofilm was not redox controlled [138]. An example of the second mode is shown in Figure 1.5. A hydrogen peroxide microelectrode (Fig. 1.5a) was placed approximately 100 µm above a glassy carbon electrode in river water. The glassy carbon electrode simulated a cathode in an SMFC in which oxygen reduction is the cathodic reaction. One by-product of oxygen reduction is hydrogen peroxide, and a hydrogen peroxide

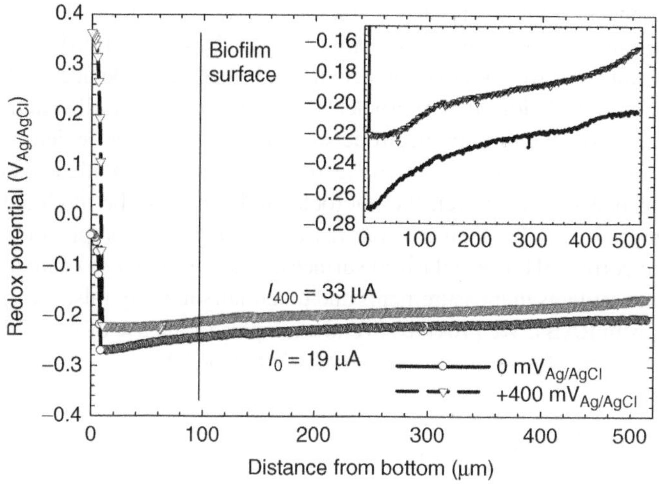

Figure 1.4 Redox potential inside a *Shewanella oneidensis* MR-1 biofilm grown on a graphite electrode. Reprinted (adapted) with permission from Ref. [138]. Copyright (2011) American Chemical Society.

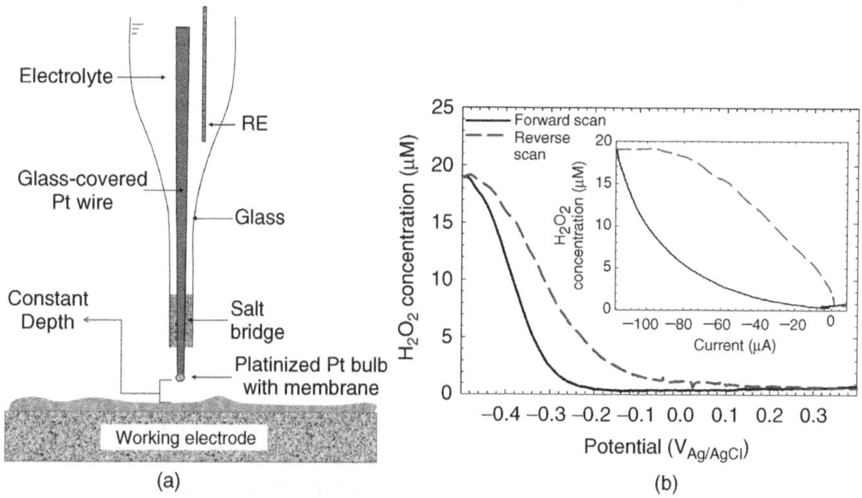

Figure 1.5 (a) Diagram of a hydrogen peroxide microelectrode. (b) Hydrogen peroxide concentration measured approximately 100 μm above a glassy carbon electrode during a CV scan. The inset shows current versus hydrogen peroxide concentration. Reproduced with permission from Ref. [56]. Copyright 2012 Taylor & Francis. http://www.informaworld.com.

microelectrode was used to detect hydrogen peroxide generation during a CV scan. In this case, the microelectrode was held at a constant distance from the biofilm electrode, and the hydrogen peroxide concentration was monitored as a function of potential and current. Figure 1.5b shows that hydrogen peroxide was produced at potentials below $-200\,mV_{Ag/AgCl}$. The inset in Figure 1.5b shows how the hydrogen peroxide concentration correlated with the measured current. The forward and reverse currents are well separated, demonstrating the effect of diffusion processes. We should note that we could not detect any hydrogen peroxide accumulating in the bulk solution. It is obvious that microelectrode techniques are well suited to studying the chemistry near the biofilm electrode surface, which cannot be determined using other techniques. The use of microelectrodes is critical to identifying electron transfer mechanisms. However, the microelectrodes need to be modified to operate on polarized surfaces. For example, our research group modified pH microelectrodes to show the correct pH near polarized surfaces. The improper use of microelectrodes could lead to artifacts in measurements and potentially identify false electron transfer mechanisms in EABs. We provide several examples of the use of microelectrodes to study electron transfer mechanisms in EABs in Chapter 5.

1.4.2 Spectroelectrochemistry

Spectroelectrochemistry combines electrochemistry with spectroscopy; it correlates the change in the spectral signature, usually in the UV/vis/IR range, of electrochemically active compounds above an electrode set at a chosen polarization potential [139]. Spectroelectrochemical techniques have been used to identify the redox state of metallo-proteins such as cytochromes containing heme in purified protein preparations [140]. The large literature covering redox titrations of cytochromes using absorbance measurements makes it possible to monitor the redox state of these proteins in EABs. For example, evanescent wave UV/vis spectroscopy was used to compare the in vivo oxidation/reduction of c-type cytochromes attached to whole cells of *Shewanella loihica* PV-4 with the oxidation/reduction of purified c-type cytochromes [141]. Attenuated total reflection surface-enhanced infrared absorption spectroscopy (ATR-SEIS) was used to measure the redox state of c-type cytochromes on *G. sulfurreducens* cells attached to gold electrodes and were able to determine that c-type cytochromes were directly oxidized/reduced by the biofilm electrode [64]. These spectroelectrochemical experiments provided detailed knowledge of the bacterial outer membrane/electrode interaction. A mini-review covering these techniques has been written for biological systems [142].

Only recently, however, has spectroelectrochemistry been used for in situ measurements of the redox state of cytochromes in thick, pregrown EABs in which extracellular electron transfer through the biofilm matrix was studied. It was shown that the c-type cytochromes inside thick *G. sulfurreducens* biofilms probed under nonturnover conditions were completely reduced at polarization potentials below $-350\,mV_{SHE}$ and completely oxidized at potentials above $+100\,mV_{SHE}$, demonstrating long-range extracellular electron transfer through the cytochrome network [66].

1.4.3 Microscopy

The in situ imaging of biofilms using fluorescent microscopes is a standard technique in biofilm research [143]. The commercial availability of optically transparent electrodes has allowed the use of in situ confocal laser scanning microscopy (CLSM) in EABs. The development of current with the biofilm has been observed for thick biofilms, as well as for monolayers of cells [144]. This coupling provides the ability to monitor biofilm parameters such as surface coverage, biovolume, and biofilm roughness with current. For example, Nevin et al. [101] used CLSM to monitor the difference in biofilm structure between *G. sulfurreducens* biofilms grown on fumarate and *G. sulfurreducens* biofilms grown on an electrode. McLean et al. [145] used CLSM to locate individual strains of *S. oneidensis* MR-1 in a mixed culture biofilm. In addition, microscopic techniques allow us to determine single-cell electron transfer rates [144, 146]. We should note that single-cell electron transfer rates may be critical when electron transfer mechanisms are studied using mutants. Mutants often exhibit a different growth rate than their wild-type counterparts. Typically, mutants grow slower than the wild type. Therefore, when current generation is compared between a mutant and its wild type, current should be quantified on a per-cell basis. This minimizes the effect of the growth rate on experimental observations. When chemical indicator probes are added to the influent growth medium, CLSM can correlate the chemical properties of EABs as well. CLSM was successfully used in conjunction with a pH indicator probe, C-SNARF-4, to measure pH changes inside *G. sulfurreducens* biofilm in a flow cell MFC [147]. The measured pH profiles that developed within the biofilm during electrode respiration provided evidence that pH varied appreciably inside EABs. The use of microscopy would critically allow us to correlate biofilm structure with its function related to electron transfer.

1.4.4 Future Techniques Coupled to Study Electrochemically Active Biofilms

Any number of combinations of electrochemical techniques with analytical experiments can be designed, and these composite techniques will likely guide future EAB research because of the complexity of electron transfer inside EABs. Two potential examples include quartz crystal microbalance (QCM) and nuclear magnetic resonance (NMR). In QCM, a quartz crystal is oscillated at selected frequencies, and variations in mass on a surface attached to the quartz crystal are measured [139]. The technique can easily detect the deposition of monolayers of material and can reach an atomic resolution of mass. When QCM is associated with an electrode and the biofilm electrode potential is controlled with a potentiostat, it becomes electrochemical quartz crystal microbalance (e-QCM), which has the capability of correlating mass change with current and could determine whether any adsorption reaction steps precede or follow electron transfer steps [148, 149]. However, e-QCM is typically used with solid surfaces; its use for thick biofilms could be limited because of their viscoelastic nature. It may find its use in probing thin biofilms. For example, e-QCM was used to monitor the effect of electrode potential on *Escherichia coli* biofilm development [149]. We provide detailed examples of e-QCM in Chapter 5. NMR uses high-strength magnetic fields and electromagnetic pulses to detect spin-active nuclei inside molecules. Some of the NMR biofilm reactors that are currently used for monitoring metabolites and diffusion coefficients could be modified to incorporate a three-electrode system [120, 145, 150, 151]. Because of the electromagnetic shielding effects of conductive materials, the three electrodes would need to be oriented appropriately in order not to block the radio frequency pulse. Simultaneously performing electrochemical experiments and monitoring in situ metabolic reactions in situ could play an important role in understanding the fundamental processes occurring in EABs and would allow for both mass and electron balances on the system.

1.5 MODELING ELECTROCHEMICALLY ACTIVE BIOFILMS

Along with the coupled techniques that could make critical contributions to our understanding of EAB extracellular electron transfer, mathematical modeling is expected to provide a theoretical basis for studying extracellular electron transfer. Originally, EAB modeling had its roots in models developed specifically to describe the operation of MFCs and BESs: the goal was to link microbial processes to the power output of MFCs and optimize these processes on the basis of model predictions [152]. Newer models now focus on the complex extracellular electron transfer mechanisms in EABs. The main goals of the models are to predict current and relate it to electron and proton transfer. Chapter 9 provides details of model development for studying extracellular electron transfer in EABs.

1.6 CURRENT STATUS OF RESEARCH ON ELECTROCHEMICALLY ACTIVE BIOFILMS

The direction in which this field of research will head is difficult to predict, as new applications of BESs are continually being published [47]. Interestingly, a majority of the researchers are still interested in demonstrating that it is possible to produce energy using various substrates and microorganisms, without realizing that this was demonstrated nearly 100 years ago! Although many publications still focus on MFC research, the current status of our knowledge of electron transfer in EABs critically demonstrates that BESs are far from solving the energy demands of the future. However, BESs can be used as tools for discovering novel capabilities of bacteria in various environments and may eventually lead to promising applications. We believe there are five main issues that need to be addressed in this process.

1. ***Scaling up of the current density generated by electrochemically active biofilms***
   One of the limitations of MFCs, and of BESs generally, is that the current density does not scale up linearly with the active surface area of the biofilm electrode to meet the demand of high-capacity processes. The scaling up of MFCs has been partially addressed by methods such as reducing anodic and cathodic overpotentials, increasing solution conductivity, decreasing mass transfer resistances, decreasing electrode spacing, using novel air cathodes, and stacking [70, 100, 153–160]. While these improvements in MFC design and scalability have increased the maximum power density, they have not addressed the

fundamental scalability of the current generation of anodes and cathodes inside these systems [155]. For example, Cheng and Logan [153] showed that the scaling up of the maximum power generation in wastewater MFCs was more closely related to the cathode surface area (62% increase in power by doubling the cathode surface area) than to the anode surface area (12% increase in power by doubling the anode surface area) [153]. The disparity in the percent increases in power highlights our incomplete fundamental knowledge of scaling up MFCs. There is room for improvement, which may involve using models specifically designed to optimize MFCs [161, 162]. Why MFC power production can only be improved at values less than 100% (increase in power per doubling of electrode area) is a question we believe to be at the heart of the scaling up of MFCs and BESs. In particular, the scalability of anodes in MFCs and BESs should be directly related to the extracellular electron transfer mechanisms that EABs utilize. Future scaling up studies need to focus on this aspect. The same outlook also applies to biocathodes. There is a critical need to address the individual contribution of each electrode to the overall energy conversion. Otherwise, the energy conversion of these devices will continue to be suboptimal. However, we note that there are electronic methods known as power management systems currently in use to alleviate some of these bottlenecks [33, 36–44].

2. ***Translation of electrochemically active biofilm extracellular electron transfer research to bioelectrochemical systems***
   Figure 1.6 shows how potential losses restrict the amount of power available to an MFC. These losses are a large factor controlling the overall potential of an MFC when a current is drawn from it. Generally, the OCP of an anode is about $-500\,mV_{SCE}$, and the OCP of a cathode is $+400\,mV_{SCE}$ [31, 98]. Therefore, the typical maximum potential we would expect is about 900 mV, although it might be possible to reach higher values. However, as soon as a load is applied to an MFC and current starts to flow, potential losses at both the anode and the

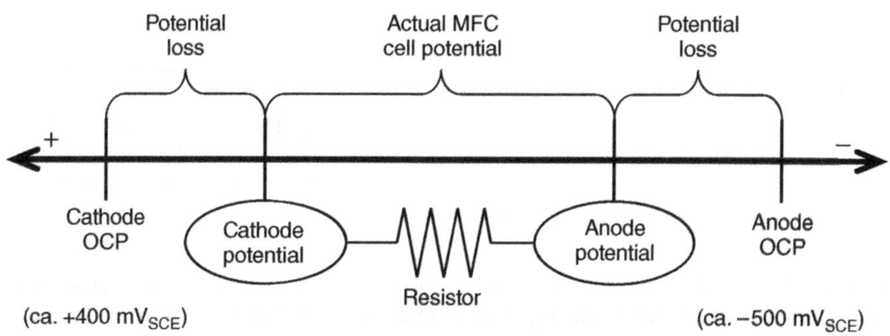

Figure 1.6 Potential losses at both the anode and the cathode restrict the amount of power that remains for the MFC when a resistor is connected. Activation, ohmic, and concentration losses reduce the anode and cathode potentials, lowering the cell potential from the maximum at OCP. The distances on the line are not drawn to scale. Reproduced with permission from Ref. [56]. Copyright 2012 Taylor & Francis. http://www.informaworld.com.

cathode affect the biofilm electrode potentials. The anode potential increases, while the cathode potential decreases, reducing the cell potential. Depending on the relative sizes of the biofilm electrodes, anode potentials fall between −100 and +100 mV_{SCE}, while cathode potentials typically stabilize between +200 and 0 mV_{SCE} [31]. Frequently, the practical values of anodes in MFCs do not align with the polarization potentials used to cultivate EABs in the laboratory, yet many published research articles describe the usefulness of the data for advancing MFCs. For example, *G. sulfurreducens* biofilms are typically grown at 0 $mV_{Ag/AgCl}$ or higher and have electrode potentials that cannot be observed in practical MFC applications. These EAB studies have significant relevance to studying electron transfer mechanisms, and we are not debating on this point. They provide a fundamental level of knowledge about EAB electrophysiology. What is missing, however, is the translation of that fundamental knowledge to practical use. Many EAB studies do not address the compatibility of their systems with what is observed in real devices, such as when we attempt to power an electronic device. Does what we do in the laboratory positively affect what we can do in the field? A critical factor could be that EAB studies focusing on MFC/BES applications should test their electrodes in real MFCs/BESs.

To demonstrate the concepts in Figure 1.6, we constructed a seawater SMFC as shown in Figure 1.7a [163]. The individual electrode potentials were monitored using a data acquisition unit. After anode burial, the anodes and cathodes were allowed to reach a steady state OCP, typically greater than 0.7 V. The steady state electrode potentials before the system was loaded with an external resistor were approximately −400 and +400 $mV_{Ag/AgCl}$ for the anode and cathode, respectively. Once these OCP values were observed, the electrodes

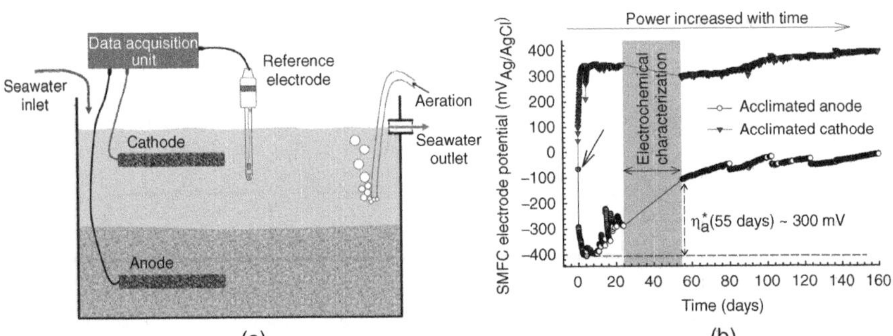

Figure 1.7 (a) Schematic of a laboratory seawater SMFC constructed using ocean sediment and seawater. (b) Anode and cathode potentials of the seawater SMFC over time. Shaded areas represent time periods during electrochemical characterization. Note that once the OCP values of the anode and the cathode had been observed, they were connected across a 108-kΩ resistor until a stable cell potential was observed. The resistor value was then systematically reduced to a final value of 180 Ω, with the goal of maintaining the operating potential of the SMFC around 0.4 V. Reproduced with permission from Ref. [163]. Copyright 2014 WILEY-VCH Verlag GmbH & Co. KGaA, Weinheim.

were connected across a 108-kΩ resistor until a stable cell potential was observed. The resistor value was then systematically reduced to a final value of 180 Ω, with the goal of maintaining the operating potential of the SMFC at about 0.4 V. The SMFC was continuously operated in this mode except during electrochemical characterization. Figure 1.7b shows the anode and cathode potentials. As time passed, the average power was 1.7 mW compared to the background at day 55. However, initially, both the cathode and the anode operated near -75 $mV_{Ag/AgCl}$ (black arrow). Within the first few days, the potential of the cathode increased to a potential of approximately +375 $mV_{Ag/AgCl}$; it then slowly increased to a final value of +395 $mV_{Ag/AgCl}$. The potential of the anode initially decreased to -400 $mV_{Ag/AgCl}$ and subsequently increased to a value slightly below 0 $mV_{Ag/AgCl}$. The shift in the anode potential as power increased is representative of the anode overpotential, η_a^*. Because the change in cathode potential was less than the change in anode potential, we can see that the seawater SMFC was anode limited. Using CV, we were able to determine the potentials at which the acclimated anode and cathode could generate current. We also show a control unacclimated cathode for comparison. Our aim was to identify the overpotentials associated with the seawater SMFC and the electrochemical differences between acclimated and unacclimated cathodes. From Figure 1.7b, we calculated that the anode and acclimated cathode of SMFC #1 passed ±2.76 mA during power generation. Superimposing this current value on the CVs in Figure 1.8, we estimated the approximate overpotentials operating in the seawater SMFC. Simply, the distances between where the anode and cathode cyclic voltammograms intersect with the x-axis to the intersection with the dotted lines represent the approximate overpotentials, η_a^* and η_c^*, necessary to drive 2.76 mA and generate 1 mW of power across a 180-Ω resistor. Qualitatively, η_a^* is larger than η_c^*, which explains why the anode

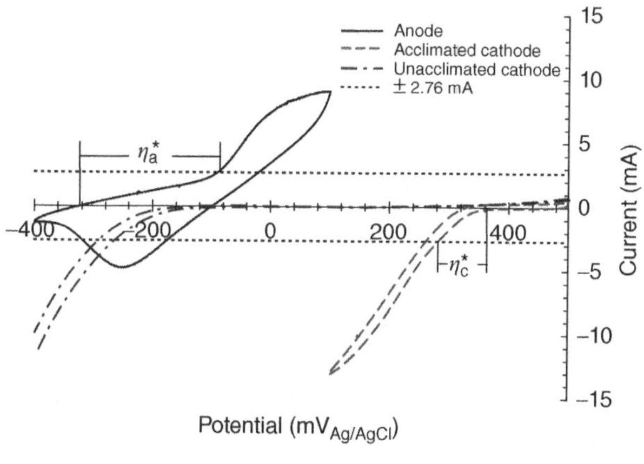

Figure 1.8 Cyclic voltammograms of the anode, acclimated cathode, and unacclimated cathode. The scan rate was 1 mV s^{-1}. Reproduced with permission from Ref. [163]. Copyright 2014 WILEY-VCH Verlag GmbH & Co. KGaA, Weinheim.

potentials increased more than the acclimated cathode potentials in Figure 1.7b. Figures 1.7 and 1.8 demonstrate the utility of electrochemical analysis in relating the results of an SMFC experiment to the expected behavior (Fig. 1.6).

3. *Extracellular electron transfer mechanisms in electrochemically active biofilms*
   To date, the questions of whether electrons are transferred through conductive nanowires, bound redox mediators, or diffusing redox mediators and of how electrons enter these pathways have yet to be unequivocally answered. Also of importance is the role of extracellular polymeric substances (EPSs) in electron transfer. For example, redox-active proteins were found in the EPSs of *Shewanella* sp. HRCR-1, which were distinct from the redox-active proteins commonly found on the cell surface [164]. EPSs are known to facilitate oxidation/reduction reactions to minerals [2, 165]. How do EPSs interact with the cells in EABs? There is a significant amount of knowledge missing on these electron transfer mechanisms that have been proposed to account directly for the path of electrons from inside cells to the biofilm electrode. While many publications in the literature push various ideas and concepts, what is needed now is direct demonstration of key findings on the electron transfer processes. Identifying the functions of key components of EABs in electron transfer could critically improve our understanding of these systems and allow us to improve the devices in which EABs are used. In addition, new methods of measuring the parameters reported in the past could add validity and create a functional base from which we can develop theory. Several chapters in this book are dedicated to addressing these measurements.
4. *Bulk measurements versus measurements inside biofilms*
   EAB processes are an interfacial phenomenon: EABs interact with the electrode inside the biofilm diffusive and reactive layers at the electrode surface. Extracellular electron transfer is directly related to these microscale layers, whereas diffusion processes above these layers are linked indirectly. Thus, we expect that the surface concentrations of the redox-active compounds and the local solution properties inside EABs are more relevant and critical than the corresponding values in the bulk. Correlating and fitting lines to bulk data may have little significance to the fundamental processes occurring inside EABs. Direct measurements inside EABs are preferred, such as measuring pH inside EABs or measuring the spectroelectrochemical properties of EABs [66, 138, 147]. This is especially important because the cell density inside some EABs is not uniformly distributed and predictions based on simple diffusion may not apply [120].
5. *How far have we really come in MFC research related to electrochemically active biofilms?*
   Rismani-Yazdi et al. [70] published a table on the MFC-related review papers from 2001 to 2007 that listed 15 reviews. Running generic keyword searches using "review" and another limiter – "microbial fuel cell," "extracellular

electron transfer," "bioelectrochemical system," "microbial electrolysis cell," or "biofuel cell," we found that the number of reviews on MFCs, BESs, and enzymatic fuel cells increased significantly from 2008 to 2011 [5, 49, 52, 57, 69, 166–215]. Most of these reviews claimed that BES devices have promising applications for wastewater treatment, biosensors, the creation of value-added products, or the production of sustainable energy, yet only one of them put forth relevant figures showing our progress toward producing useful power in MFCs [201]. We noted previously that Logan and Regan [216] put forth a similar figure. However, we feel that the figure showing an exponential increase of power density in reported MFCs over the years from 1997 to 2009 needs to be re-evaluated. There is a very clear trend across multiple experiments and MFC configurations that power densities reported as milliwatt per square meter are related to electrode size [155]. We believe that even when MFCs and BESs are optimized experimentally via design considerations, we will still lack a fundamental understanding of how we can optimize EAB extracellular electron transfer to scale linearly with electrode size. MFC stacking does not address this limitation: instead, it shows how inefficient MFCs still are as energy-converting devices. Although we are critical of what has been done to date with MFCs, we believe that MFCs and BESs have the potential to be used in high-throughput processes in the future. However, this will not occur until our fundamental understanding of the mechanisms of extracellular electron transfer in mixed species EABs has progressed substantially [123]. Rather than focusing on new ways to apply MFCs and BESs, we should focus on how to use these devices to elucidate mechanisms of extracellular electron transfer. It is to be hoped that the electrochemical techniques outlined in this book will be of some use in heading toward this goal.

1.7 FUTURE DIRECTIONS IN ELECTROCHEMICALLY ACTIVE BIOFILMS RESEARCH

EAB research is new, and there have been exciting discoveries related to electron transfer mechanisms. Many new techniques have been developed, which can be used to understand electron transfer processes in EABs. This book focuses on these techniques and demonstrates data generated in many leading laboratories. While our knowledge of EAB electron transfer mechanisms is much more advanced than it was just 5 years ago, this knowledge has yet to infiltrate the BES literature as a whole. We hope that BES research will benefit from this book describing how EAB extracellular electron transfer is studied. What we are looking for in the future are BES studies that optimize both the design of BESs and the electron transfer expected from EABs. We would also like to see BES studies address some of the questions concerning EAB extracellular electron transfer – that is, practical BES studies testing hypotheses developed in nonpractical EAB extracellular electron transfer studies. We believe that EAB research will eventually be successful if we focus on fundamental research!

ACKNOWLEDGMENTS

This research was supported by the US Office of Naval Research (ONR), grant #N00014-09-1 0090, and the National Science Foundation (NSF), grant #0954186.

REFERENCES

1. Bird LJ, Bonnefoy V, Newman DK. Bioenergetic challenges of microbial iron metabolisms. Trends Microbiol 2011;19:330–340.
2. Gralnick JA, Newman DK. Extracellular respiration. Mol Microbiol 2007;65:1–11.
3. Kraft B, Strous M, Tegetmeyer HE. Microbial nitrate respiration – genes, enzymes and environmental distribution. J Biotechnol 2011;155:104–117.
4. Stams AJM, de Bok FAM, Plugge CM, van Eekert MHA, Dolfing J, Schraa G. Exocellular electron transfer in anaerobic microbial communities. Environ Microbiol 2006;8:371–382.
5. Nealson KH, Finkel SE. Electron flow and biofilms. MRS Bull 2011;36:380–384.
6. Erable B, Feron D, Bergel A. Microbial catalysis of the oxygen reduction reaction for microbial fuel cells: a review. ChemSusChem 2012;5:975–987.
7. Kato S, Hashimoto K, Watanabe K. Microbial interspecies electron transfer via electric currents through conductive minerals. Proc Natl Acad Sci U S A 2012;109:10042–10046.
8. Kimura Z, Okabe S. Acetate oxidation by syntrophic association between *Geobacter sulfurreducens* and a hydrogen-utilizing exoelectrogen. ISME J 2013;7:1472–1482.
9. Shrestha PM, Rotaru AE. Plugging in or going wireless: strategies for interspecies electron transfer. Front Microbiol 2014;5:237.
10. Strycharz-Glaven SM, Glaven RH, Wang Z, Zhou J, Vora GJ, Tender LM. Electrochemical investigation of a microbial solar cell reveals a nonphotosynthetic biocathode catalyst. Appl Environ Microbiol 2013;79:3933–3942.
11. Coursolle D, Baron DB, Bond DR, Gralnick JA. The Mtr respiratory pathway is essential for reducing flavins and electrodes in *Shewanella oneidensis*. J Bacteriol 2010;192:467–474.
12. El-Naggar MY, Wanger G, Leung KM, Yuzvinsky TD, Southam G, Yang J, Lau WM, Nealson KH, Gorby YA. Electrical transport along bacterial nanowires from *Shewanella oneidensis* MR-1. Proc Natl Acad Sci U S A 2010;107:18127–18131.
13. Leung KM, Wanger G, El-Naggar MY, Gorby Y, Southam G, Lau WM, Yang J. *Shewanella oneidensis* MR-1 bacterial nanowires exhibit p-type, tunable electronic behavior. Nano Lett 2013;13:2407–2411.
14. Malvankar NS, Vargas M, Nevin KP, Franks AE, Leang C, Kim BC, Inoue K, Mester T, Covalla SF, Johnson JP, Rotello VM, Tuominen MT, Lovley DR. Tunable metallic-like conductivity in microbial nanowire networks. Nat Nanotechnol 2011;6:573–579.
15. Marsili E, Baron DB, Shikhare ID, Coursolle D, Gralnick JA, Bond DR. Shewanella secretes flavins that mediate extracellular electron transfer. Proc Natl Acad Sci U S A 2008;105:3968–3973.
16. Peng L, You SJ, Wang JY. Electrode potential regulates cytochrome accumulation on *Shewanella oneidensis* cell surface and the consequence to bioelectrocatalytic current generation. Biosens Bioelectron 2010;25:2530–2533.

17. Reguera G, McCarthy KD, Mehta T, Nicoll JS, Tuominen MT, Lovley DR. Extracellular electron transfer via microbial nanowires. Nature 2005;435:1098–1101.

18. Reguera G, Nevin KP, Nicoll JS, Covalla SF, Woodard TL, Lovley DR. Biofilm and nanowire production leads to increased current in *Geobacter sulfurreducens* fuel cells. Appl Environ Microbiol 2006;72:7345–7348.

19. Snider RM, Guissepi-Elie A, Strycharz-Glaven S, Tender L. 2011. On the conductive nature of biofilms of *Geobacter sulfurreducens*. Abstr Pap Am Chem Soc, 241.

20. Velasquez-Orta SB, Head IM, Curtis TP, Scott K, Lloyd JR, von Canstein H. The effect of flavin electron shuttles in microbial fuel cells current production. Appl Microbiol Biotechnol 2010;85:1373–1381.

21. Liu X, Tremblay PL, Malvankar NS, Nevin KP, Lovley DR, Vargas M. A *Geobacter sulfurreducens* strain expressing *Pseudomonas aeruginosa* type IV pili localizes OmcS on pili but is deficient in Fe(III) oxide reduction and current production. Appl Environ Microbiol 2014;80:1219–1224.

22. Vargas M, Malvankar NS, Tremblay PL, Leang C, Smith JA, Patel P, Synoeyenbos-West O, Nevin KP, Lovley DR. Aromatic amino acids required for pili conductivity and long-range extracellular electron transport in *Geobacter sulfurreducens*. mBio 2013;4:e00105–e00113.

23. Flynn TM, O'Loughlin EJ, Mishra B, DiChristina TJ, Kemner KM. Sulfur-mediated electron shuttling during bacterial iron reduction. Science 2014;344:1039–1042.

24. Roy JN, Babanova S, Garcia KE, Cornejo J, Ista LK, Atanassov P. Catalytic biofilm formation by *Shewanella oneidensis* MR-1 and anode characterization by expanded uncertainty. Electrochim Acta 2014;126:3–10.

25. TerAvest MA, Rosenbaum MA, Kotloski NJ, Gralnick JA, Angenent LT. Oxygen allows *Shewanella oneidensis* MR-1 to overcome mediator washout in a continuously fed bioelectrochemical system. Biotechnol Bioeng 2014;111:692–699.

26. Edwards MJ, Baiden NA, Johs A, Tomanicek SJ, Liang LY, Shi L, Fredrickson JK, Zachara JM, Gates AJ, Butt JN, Richardson DJ, Clarke TA. The X-ray crystal structure of *Shewanella oneidensis* OmcA reveals new insight at the microbe-mineral interface. Febs Lett 2014;588:1886–1890.

27. Snider RM, Strycharz-Glaven SM, Tsoi SD, Erickson JS, Tender LM. Long-range electron transport in *Geobacter sulfurreducens* biofilms is redox gradient-driven. Proc Natl Acad Sci U S A 2012;109:15467–15472.

28. Sharma V, Kundu PP. Biocatalysts in microbial fuel cells. Enzyme Microb Technol 2010;47:179–188.

29. Dewan A, Ay SU, Karim MN, Beyenal H. Alternative power sources for remote sensors: a review. J Power Sources 2014;245:129–143.

30. Dewan A, Beyenal H, Lewandowski Z. Intermittent energy harvesting improves the performance of microbial fuel cells. Environ Sci Technol 2009;43:4600–4605.

31. Dewan A, Donovan C, Heo D, Beyenal H. Evaluating the performance of microbial fuel cells powering electronic devices. J Power Sources 2010;195:90–96.

32. Donovan C, Dewan A, Heo D, Beyenal H. Batteryless, wireless sensor powered by a sediment microbial fuel cell. Environ Sci Technol 2008;42:8591–8596.

33. Donovan C, Dewan A, Peng HA, Heo D, Beyenal H. Power management system for a 2.5 W remote sensor powered by a sediment microbial fuel cell. J Power Sources 2011;196:1171–1177.

34. Lewandowski Z, Beyenal H, Dewan A, Gao HW, Meehan A. 2009. Microbial fuel cells to power submersed electronic devices. Abstr Pap Am Chem Soc, 238.
35. Shantaram A, Beyenal H, Raajan R, Veluchamy A, Lewandowski Z. Wireless sensors powered by microbial fuel cells. Environ Sci Technol 2005;39:5037–5042.
36. Andersen SJ, Pikaar I, Freguia S, Lovell BC, Rabaey K, Rozendal RA. Dynamically adaptive control system for bioanodes in serially stacked bioelectrochemical systems. Environ Sci Technol 2013;47:5488–5494.
37. Crepaldi M, Chiolerio A, Tommasi T, Hidalgo D, Canavese G, Stassi S, Demarchi D, Pirri FC. 2013. A low complexity wireless microbial fuel cell monitor using piezoresistive sensors and impulseradio ultra-wide-band. Smart Sensors, Actuators, and MEMS VI 8763.
38. Meehan A, Gao HW, Lewandowski Z. Energy harvesting with microbial fuel cell and power management system. IEEE Trans Power Electron 2011;26:176–181.
39. Mohan SV, Chandrasekhar K. Solid phase microbial fuel cell (SMFC) for harnessing bioelectricity from composite food waste fermentation: Influence of electrode assembly and buffering capacity. Bioresour Technol 2011;102:7077–7085.
40. Park JD, Ren ZY. Hysteresis controller based maximum power point tracking energy harvesting system for microbial fuel cells. J Power Sources 2012;205:151–156.
41. Schrader PS, Doolan C, Reimers CE, Girguis PR, Wolf M, Green D. 2013. Sensors and acoustic modems powered by benthic microbial fuel cells at the MARS observatory. Oceans-IEEE.
42. Winfield J, Chambers LD, Stinchcombe A, Rossiter J, Ieropoulos I. The power of glove: Soft microbial fuel cell for low-power electronics. J Power Sources 2014;249:327–332.
43. Yang F, Zhang DX, Shimotori T, Wang KC, Huang Y. Study of transformer-based power management system and its performance optimization for microbial fuel cells. J Power Sources 2012;205:86–92.
44. Zhang DX, Yang F, Shimotori T, Wang KC, Huang Y. Performance evaluation of power management systems in microbial fuel cell-based energy harvesting applications for driving small electronic devices. J Power Sources 2012;217:65–71.
45. Du ZW, Li HR, Gu TY. A state of the art review on microbial fuel cells: a promising technology for wastewater treatment and bioenergy. Biotechnol Adv 2007;25:464–482.
46. Rosenbaum MA, Franks AE. Microbial catalysis in bioelectrochemical technologies: status quo, challenges and perspectives. Appl Microbiol Biotechnol 2014;98:509–518.
47. Wang HM, Ren ZYJ. A comprehensive review of microbial electrochemical systems as a platform technology. Biotechnol Adv 2013;31:1796–1807.
48. Arechederra R, Minteer SD. Organelle-based biofuel cells: immobilized mitochondria on carbon paper electrodes. Electrochim Acta 2008;53:6698–6703.
49. Ivanov I, Vidakovic-Koch T, Sundmacher K. Recent advances in enzymatic fuel cells: experiments and modeling. Energies 2010;3:803–846.
50. Manohar AK, He Z, Mansfeld F. The use of electrochemical impedance spectroscopy (EIS) for the evaluation of the electrochemical properties of bioelectrochemical systems. In: Rabaey K, Angenent LT, Schröder U, Keller J, editors. Bioelectrochemical Systems: From Extracellular Electron Transfer to Biotechnological Application. London, UK: IWA Publishing; 2010. p 169–183.
51. Rabaey K, Rodriguez J, Blackall LL, Keller J, Gross P, Batstone D, Verstraete W, Nealson KH. Microbial ecology meets electrochemistry: electricity-driven and driving communities. ISME J 2007;1:9–18.

52. Rosenbaum M, Schroder U. Photomicrobial solar and fuel cells. Electroanalysis 2010;22:844–855.

53. Call DF, Logan BE. A method for high throughput bioelectrochemical research based on small scale microbial electrolysis cells. Biosens Bioelectron 2011;26:4526–4531.

54. Torres CI, Krajmalnik-Brown R, Parameswaran P, Marcus AK, Wanger G, Gorby YA, Rittmann BE. Selecting anode-respiring bacteria based on anode potential: phylogenetic, electrochemical, and microscopic characterization. Environ Sci Technol 2009;43:9519–9524.

55. Cohen B. The bacterial culture as an electrical half-cell. J Bacteriol 1931;21:18–19.

56. Babauta J, Renslow R, Lewandowski Z, Beyenal H. Electrochemically active biofilms: facts and fiction. A review. Biofouling 2012;28:789–812.

57. Wei JC, Liang P, Huang X. Recent progress in electrodes for microbial fuel cells. Bioresour Technol 2011;102:9335–9344.

58. Zhou MH, Chi ML, Luo JM, He HH, Jin T. An overview of electrode materials in microbial fuel cells. J Power Sources 2011;196:4427–4435.

59. Kissinger PT, Heineman WR, editors. Laboratory Techniques in Electroanalytical Chemistry. 2nd ed. New York, NY: Marcel Dekker, Inc.; 1996.

60. Richter H, McCarthy K, Nevin KP, Johnson JP, Rotello VM, Lovley DR. Electricity generation by *Geobacter sulfurreducens* attached to gold electrodes. Langmuir 2008;24:4376–4379.

61. Eckermann AL, Feld DJ, Shaw JA, Meade TJ. Electrochemistry of redox-active self-assembled monolayers. Coordin Chem Rev 2010;254:1769–1802.

62. Gooding J, Yang WR. The self-assembled monolayer modification of electrodes – some recent advances in biological application. Actual Chim 2008;320:85–89.

63. Mandler D, Kraus-Ophir S. Self-assembled monolayers (SAMs) for electrochemical sensing. J Solid State Electrochem 2011;15:1535–1558.

64. Busalmen JP, Esteve-Nunez A, Berna A, Feliu JM. ATR-SEIRAs characterization of surface redox processes in *G. sulfurreducens*. Bioelectrochemistry 2010;78:25–29.

65. Jain A, Gazzola G, Panzera A, Zanoni M, Marsii E. Visible spectroelectrochemical characterization of *Geobacter sulfurreducens* biofilms on optically transparent indium tin oxide electrode. Electrochim Acta 2011;56:10776–10785.

66. Liu Y, Kim H, Franklin RR, Bond DR. Linking spectral and electrochemical analysis to monitor c-type cytochrome redox status in living *Geobacter sulfurreducens* biofilms. Chemphyschem 2011;12:2235–2241.

67. Logan BE, Hamelers B, Rozendal RA, Schrorder U, Keller J, Freguia S, Aelterman P, Verstraete W, Rabaey K. Microbial fuel cells: methodology and technology. Environ Sci Technol 2006;40:5181–5192.

68. Manohar AK, Mansfeld F. The internal resistance of a microbial fuel cell and its dependence on cell design and operating conditions. Electrochim Acta 2009;54:1664–1670.

69. Clauwaert P, Aelterman P, Pham TH, De Schamphelaire L, Carballa M, Rabaey K, Verstraete W. Minimizing losses in bio-electrochemical systems: the road to applications. Appl Microbiol Biotechnol 2008;79:901–913.

70. Rismani-Yazdi H, Carver SM, Christy AD, Tuovinen IH. Cathodic limitations in microbial fuel cells: an overview. J Power Sources 2008;180:683–694.

71. Bouhenni RA, Vora GJ, Biffinger JC, Shirodkar S, Brockman K, Ray R, Wu P, Johnson BJ, Biddle EM, Marshall MJ, Fitzgerald LA, Little BJ, Fredrickson JK, Beliaev AS, Ringeisen BR, Saffarini DA. The role of *Shewanella oneidensis* MR-1 outer surface structures in extracellular electron transfer. Electroanalysis 2010;22:856–864.
72. Fricke K, Harnisch F, Schroder U. On the use of cyclic voltammetry for the study of anodic electron transfer in microbial fuel cells. Energy Environ Sci 2008;1:144–147.
73. Strycharz SM, Malanoski AP, Snider RM, Yi H, Lovley DR, Tender LM. Application of cyclic voltammetry to investigate enhanced catalytic current generation by biofilm-modified anodes of *Geobacter sulfurreducens* strain DL1 vs. variant strain KN400. Energy Environ Sci 2011;4:896–913.
74. Li FX, Sharma Y, Lei Y, Li BK, Zhou QX. Microbial fuel cells: the effects of configurations, electrolyte solutions, and electrode materials on power generation. Appl Biochem Biotechnol 2010;160:168–181.
75. Logan BE, Call D, Cheng S, Hamelers HVM, Sleutels THJA, Jeremiasse AW, Rozendal RA. Microbial electrolysis cells for high yield hydrogen gas production from organic matter. Environ Sci Technol 2008;42:8630–8640.
76. Rozendal RA, Sleutels THJA, Hamelers HVM, Buisman CJN. Effect of the type of ion exchange membrane on performance, ion transport, and pH in biocatalyzed electrolysis of wastewater. Water Sci Technol 2008;57:1757–1762.
77. Lee HS, Torres CI, Parameswaran P, Rittmann BE. Fate of H(2) in an upflow single-chamber microbial electrolysis cell using a metal-catalyst-free cathode. Environ Sci Technol 2009;43:7971–7976.
78. Parameswaran P, Torres CI, Lee HS, Rittmann BE, Krajmalnik-Brown R. Hydrogen consumption in microbial electrochemical systems (MXCs): the role of homo-acetogenic bacteria. Bioresour Technol 2011;102:263–271.
79. Aelterman P, Freguia S, Keller J, Verstraete W, Rabaey K. The anode potential regulates bacterial activity in microbial fuel cells. Appl Microbiol Biotechnol 2008;78:409–418.
80. Cao XX, Fan MZ, Liang P, Huang X. Effects of anode potential on the electricity generation performance of *Geobacter sulfurreducens*. Chem J Chin Univ 2009;30:983–987.
81. Cheng KY, Ho G, Cord-Ruwisch R. Affinity of microbial fuel cell biofilm for the anodic potential. Environ Sci Technol 2008;42:3828–3834.
82. Huang LP, Chai XL, Chen GH, Logan BE. Effect of set potential on hexavalent chromium reduction and electricity generation from biocathode microbial fuel cells. Environ Sci Technol 2011;45:5025–5031.
83. Lee HS, Torres CI, Rittmann BE. Effects of substrate diffusion and anode potential on kinetic parameters for anode-respiring bacteria. Environ Sci Technol 2009;43:7571–7577.
84. Nam JY, Tokash JC, Logan BE. Comparison of microbial electrolysis cells operated with added voltage or by setting the anode potential. Int J Hydrogen Energy 2011;36:10550–10556.
85. Srikanth S, Mohan SV, Sarma PN. Positive anodic poised potential regulates microbial fuel cell performance with the function of open and closed circuitry. Bioresour Technol 2010;101:5337–5344.
86. Wang X, Feng YJ, Ren NQ, Wang HM, Lee H, Li N, Zhao QL. Accelerated start-up of two-chambered microbial fuel cells: effect of anodic positive poised potential. Electrochim Acta 2009;54:1109–1114.

87. Lee HS, Rittmann BE. Characterization of energy losses in an upflow single-chamber microbial electrolysis cell. Int J Hydrogen Energy 2010;35:920–927.

88. Cheng SA, Kiely P, Logan BE. Pre-acclimation of a wastewater inoculum to cellulose in an aqueous-cathode MEC improves power generation in air-cathode MFCs. Bioresour Technol 2011;102:367–371.

89. Finkelstein DA, Tender LM, Zeikus JG. Effect of electrode potential on electrode-reducing microbiota. Environ Sci Technol 2006;40:6990–6995.

90. Hong YY, Call DF, Werner CM, Logan BE. Adaptation to high current using low external resistances eliminates power overshoot in microbial fuel cells. Biosens Bioelectron 2011;28:71–76.

91. Kassongo J, Togo CA. Performance improvement of whey-driven microbial fuel cells by acclimation of indigenous anodophilic microbes. Afr J Biotechnol 2011;10:7846–7852.

92. Kim JR, Min B, Logan BE. Evaluation of procedures to acclimate a microbial fuel cell for electricity production. Appl Microbiol Biotechnol 2005;68:23–30.

93. Larrosa-Guerrero A, Scott K, Katuri KP, Godinez C, Head IM, Curtis T. Open circuit versus closed circuit enrichment of anodic biofilms in MFC: effect on performance and anodic communities. Appl Microbiol Biotechnol 2010;87:1699–1713.

94. Liu JL, Lowy DA, Baumann RG, Tender LM. Influence of anode pretreatment on its microbial colonization. J Appl Microbiol 2007;102:177–183.

95. Liu Y, Harnisch F, Fricke K, Sietmann R, Schroder U. Improvement of the anodic bioelectrocatalytic activity of mixed culture biofilms by a simple consecutive electrochemical selection procedure. Biosens Bioelectron 2008;24:1006–1011.

96. Rodrigo MA, Canizares P, Garcia H, Linares JJ, Lobato J. Study of the acclimation stage and of the effect of the biodegradability on the performance of a microbial fuel cell. Bioresour Technol 2009;100:4704–4710.

97. Yi HN, Nevin KP, Kim BC, Franks AE, Klimes A, Tender LM, Lovley DR. Selection of a variant of *Geobacter sulfurreducens* with enhanced capacity for current production in microbial fuel cells. Biosens Bioelectron 2009;24:3498–3503.

98. Renslow R, Donovan C, Shim M, Babauta J, Nannapaneni S, Schenk J, Beyenal H. Oxygen reduction kinetics on graphite cathodes in sediment microbial fuel cells. Phys Chem Chem Phys 2011;13:21573–21584.

99. Jadhav GS, Ghangrekar MM. Performance of microbial fuel cell subjected to variation in pH, temperature, external load and substrate concentration. Bioresour Technol 2009;100:717–723.

100. Zhang L, Zhu X, Li J, Liao Q, Ye DD. Biofilm formation and electricity generation of a microbial fuel cell started up under different external resistances. J Power Sources 2011;196:6029–6035.

101. Nevin KP, Kim BC, Glaven RH, Johnson JP, Woodard TL, Methe BA, DiDonato RJ, Covalla SF, Franks AE, Liu A, Lovley DR. Anode biofilm transcriptomics reveals outer surface components essential for high density current production in *Geobacter sulfurreducens* fuel cells. PLoS One 2009;4:e5628.

102. Inoue K, Leang C, Franks AE, Woodard TL, Nevin KP, Lovley DR. Specific localization of the c-type cytochrome OmcZ at the anode surface in current-producing biofilms of *Geobacter sulfurreducens*. Environ Microbiol Rep 2011;3:211–217.

103. Bond DR, Lovley DR. Electricity production by *Geobacter sulfurreducens* attached to electrodes. Appl Environ Microbiol 2003;69:1548–1555.

104. Patil SA, Harnisch F, Koch C, Hubschmann T, Fetzer I, Carmona-Martinez AA, Muller S, Schroder U. Electroactive mixed culture derived biofilms in microbial bioelectrochemical systems: the role of pH on biofilm formation, performance and composition. Bioresour Technol 2011;102:9683–9690.

105. Torres CI, Marcus AK, Rittmann BE. Proton transport inside the biofilm limits electrical current generation by anode-respiring bacteria. Biotechnol Bioeng 2008;100:872–881.

106. Marsili E, Sun J, Bond DR. Voltammetry and growth physiology of *Geobacter sulfurreducens* biofilms as a function of growth stage and imposed electrode potential. Electroanalysis 2010;22:865–874.

107. Carmona-Martinez AA, Harnisch F, Fitzgerald LA, Biffinger JC, Ringeisen BR, Schroder U. Cyclic voltammetric analysis of the electron transfer of *Shewanella oneidensis* MR-1 and nanofilament and cytochrome knock-out mutants. Bioelectrochemistry 2011;81:74–80.

108. Katuri KP, Kavanagh P, Rengaraj S, Leech D. *Geobacter sulfurreducens* biofilms developed under different growth conditions on glassy carbon electrodes: insights using cyclic voltammetry. Chem Commun 2010;46:4758–4760.

109. Marsili E, Rollefson JB, Baron DB, Hozalski RM, Bond DR. Microbial biofilm voltammetry: direct electrochemical characterization of catalytic electrode-attached biofilms. Appl Environ Microbiol 2008;74:7329–7337.

110. Meitl LA, Eggleston CM, Colberg PJS, Khare N, Reardon CL, Shi L. Electrochemical interaction of *Shewanella oneidensis* MR-1 and its outer membrane cytochromes OmcA and MtrC with hematite electrodes. Geochim Cosmochim Acta 2009;73:5292–5307.

111. Rabaey K, Boon N, Siciliano SD, Verhaege M, Verstraete W. Biofuel cells select for microbial consortia that self-mediate electron transfer. Appl Environ Microbiol 2004;70:5373–5382.

112. Richter H, Nevin KP, Jia HF, Lowy DA, Lovley DR, Tender LM. Cyclic voltammetry of biofilms of wild type and mutant *Geobacter sulfurreducens* on fuel cell anodes indicates possible roles of OmcB, OmcZ, type IV pili, and protons in extracellular electron transfer. Energy Environ Sci 2009;2:506–516.

113. Tender LM, Strycharz S, Yi H, Nevin KP, Kim BC, Franks A, Klimes A, Lovley DR. 2009. Cyclic voltammetry of electrode-bound biofilms of *Geobacter sulfurreducens*. Abstr Pap Am Chem Soc, 238.

114. Renslow R, Lewandowski Z, Beyenal H. Biofilm image reconstruction for assessing structural parameters. Biotechnol Bioeng 2011;108:1383–1394.

115. Yang XM, Beyenal H, Harkin G, Lewandowski Z. Quantifying biofilm structure using image analysis. J Microbiol Methods 2000;39:109–119.

116. Yang XM, Beyenal H, Harkin G, Lewandowski Z. Evaluation of biofilm image thresholding methods. Water Res 2001;35:1149–1158.

117. Beyenal H, Lewandowski Z. Mass-transport dynamics, activity, and structure of sulfate-reducing biofilms. AIChE J 2001;47:1689–1697.

118. Beyenal H, Lewandowski Z. Internal and external mass transfer in biofilms grown at various flow velocities. Biotechnol Prog 2002;18:55–61.

119. Beyenal H, Tanyolac A, Lewandowski Z. Measurement of local effective diffusivity in heterogeneous biofilms. Water Sci Technol 1998;38:171–178.

120. Renslow RS, Majors PD, McLean JS, Fredrickson JK, Ahmed B, Beyenal H. In situ effective diffusion coefficient profiles in live biofilms using pulsed-field gradient nuclear magnetic resonance. Biotechnol Bioeng 2010;106:928–937.

121. Milliken CE, May HD. Sustained generation of electricity by the spore-forming, Gram-positive, *Desulfitobacterium hafniense* strain DCB2. Appl Microbiol Biotechnol 2007;73:1180–1189.

122. Ahmed B. 2011. Uranium immobilization in subsurface sediments [dissertation]. Pullman (WA): Washington State University. 225 p.

123. Kiely PD, Regan JM, Logan BE. The electric picnic: synergistic requirements for exoelectrogenic microbial communities. Curr Opin Biotechnol 2011;22:378–385.

124. Strycharz-Glaven SM, Snider RM, Guiseppi-Elie A, Tender LM. On the electrical conductivity of microbial nanowires and biofilms. Energy Environ Sci 2011;4:4366–4379.

125. Cronenberg CCH, Vandenheuvel JC. Determination of glucose diffusion-coefficients in biofilms with microelectrodes. Biosens Bioelectron 1991;6:255–262.

126. Glud RN, Ramsing NB, Revsbech NP. Photosynthesis and photosynthesis-coupled respiration in natural biofilms quantified with oxygen microsensors. J Phycol 1992;28:51–60.

127. Vanhoudt P, Lewandowski Z, Little B. Iridium oxide pH microelectrode. Biotechnol Bioeng 1992;40:601–608.

128. Zhang TC, Bishop PL. Experimental-determination of the dissolved-oxygen boundary-layer and mass-transfer resistance near the fluid-biofilm interface. Water Sci Technol 1994;30:47–58.

129. Beyenal H, Davis CC, Lewandowski Z. An improved Severinghaus-type carbon dioxide microelectrode for use in biofilms. Sens Actuators BChem 2004;97:202–210.

130. Beyenal H, Lewandowski Z. Combined effect of substrate concentration and flow velocity on effective diffusivity in biofilms. Water Res 2000;34:528–538.

131. Lee WC, Debeer D. Oxygen and pH microprofiles above corroding mild-steel covered with a biofilm. Biofouling 1995;8:273–280.

132. Xia FH, Beyenal H, Lewandowski Z. An electrochemical technique to measure local flow velocity in biofilms. Water Res 1998;32:3631–3636.

133. Xu K, Dexter SC, Luther GW. Voltammetric microelectrodes for biocorrosion studies. Corrosion 1998;54:814–823.

134. Yang SN, Lewandowski Z. Measurement of local mass-transfer coefficient in biofilms. Biotechnol Bioeng 1995;48:737–744.

135. Yu T, Bishop PL. Stratification of microbial metabolic processes and redox potential change in an aerobic biofilm studied using microelectrodes. Water Sci Technol 1998;37:195–198.

136. Nguyen HD, Renslow R, Babauta J, Ahmed B, Beyenal H. A voltammetric flavin microelectrode for use in biofilms. Sens Actuators, B 2012;161:929–937.

137. Nguyen HD, Cao B, Mishra B, Boyanou MI, Kemner KM, Fredrickson JK, Beyenal H. Microscale geochemical gradients in Hanford 300 Area sediment biofilms and influence of uranium. Water Res 2012;46:227–234.

138. Babauta JT, Nguyen HD, Beyenal H. Redox and pH microenvironments within *Shewanella oneidensis* MR-1 biofilms reveal an electron transfer mechanism. Environ Sci Technol 2011;45:6654–6660.

139. Bard AJ, Faulkner LR, editors. Electrochemical Methods: Fundamentals and Applications. 2nd ed. New York, NY: John Wiley & Sons, Inc.; 2001.

140. Bowden EF, Hawkridge FM, Chlebowski JF, Bancroft EE, Thorpe C, Blount HN. Cyclic voltammetry and derivative cyclic voltabsorptometry of purified horse heart cytochrome-C at tin-doped indium oxide optically transparent electrodes. J Am Chem Soc 1982;104:7641–7644.

141. Nakamura R, Ishii K, Hashimoto K. Electronic absorption spectra and redox properties of C type cytochromes in living microbes. Angew Chem Int Ed Engl 2009;48:1606–1608.

142. Ataka K, Kottke T, Heberle J. Thinner, smaller, faster: IR techniques to probe the functionality of biological and biomimetic systems. Angew Chem Int Ed Engl 2010;49:5416–5424.

143. Lewandowski Z, Beyenal H. Fundamentals of Biofilm Research. Boca Raton (FL): CRC Press; 2007.

144. McLean JS, Wanger G, Gorby YA, Wainstein M, McQuaid J, Ishii SI, Bretschger O, Beyenal H, Nealson KH. Quantification of electron transfer rates to a solid phase electron acceptor through the stages of biofilm formation from single cells to multicellular communities. Environ Sci Technol 2010;44:2721–2727.

145. McLean JS, Majors PD, Reardon CL, Bilskis CL, Reed SB, Romine MF, Fredrickson JK. Investigations of structure and metabolism within *Shewanella oneidensis* MR-1 biofilms. J Microbiol Methods 2008;74:47–56.

146. Jiang X, Hu J, Petersen ER, Fitzgerald LA, Jackan CS, Lieber AM, Ringeisen BR, Lieber CM, Biffinger JC. Probing single- to multi-cell level charge transport in *Geobacter sulfurreducens* DL-1. Nat Commun 2013;4:2751.

147. Franks AE, Nevin KP, Jia HF, Izallalen M, Woodard TL, Lovley DR. Novel strategy for three-dimensional real-time imaging of microbial fuel cell communities: monitoring the inhibitory effects of proton accumulation within the anode biofilm. Energy Environ Sci 2009;2:113–119.

148. Varela H, Malta M, Torresi RM. Low cost in situ techniques in electrochemistry: the quartz crystal microbalance. Quim Nova 2000;23:664–679.

149. Xie XH, Li EL, Tang ZK. EQCM and EIS study of the effect of potential of zero charge on *Escherichia coli* biofilm development. Int J Electrochem Sci 2010;5:1018–1025.

150. Majors PD, McLean JS, Pinchuk GE, Fredrickson JK, Gorby YA, Minard KR, Wind RA. NMR methods for in situ biofilm metabolism studies. J Microbiol Methods 2005;62:337–344.

151. Majors PD, McLean JS, Scholten JCM. NMR bioreactor development for live in-situ microbial functional analysis. J Magn Reson 2008;192:159–166.

152. Zhang XC, Halme A. Modeling of a microbial fuel-cell process. Biotechnol Lett 1995;17:809–814.

153. Cheng SA, Logan BE. Increasing power generation for scaling up single-chamber air cathode microbial fuel cells. Bioresour Technol 2011;102:4468–4473.

154. Dekker A, Ter Heijne A, Saakes M, Hamelers HVM, Buisman CJN. Analysis and improvement of a scaled-up and stacked microbial fuel cell. Environ Sci Technol 2009;43:9038–9042.

155. Dewan A, Beyenal H, Lewandowski Z. Scaling up microbial fuel cells. Environ Sci Technol 2008;42:7643–7648.

156. Ieropoulos I, Greenman J, Melhuish C. Microbial fuel cells based on carbon veil electrodes: stack configuration and scalability. Int J Energy Res 2008;32:1228–1240.

157. Jiang DQ, Li XA, Raymond D, Mooradain J, Li BK. Power recovery with multi-anode/cathode microbial fuel cells suitable for future large-scale applications. Int J Hydrogen Energy 2010;35:8683–8689.

158. Logan BE. Scaling up microbial fuel cells and other bioelectrochemical systems. Appl Microbiol Biotechnol 2010;85:1665–1671.

159. Zhuang L, Feng CH, Zhou SG, Li YT, Wang YQ. Comparison of membrane- and cloth-cathode assembly for scalable microbial fuel cells: construction, performance and cost. Process Biochem 2010;45:929–934.

160. Zhuang L, Zheng Y, Zhou S, Yuan Y, Yuan H, Chen Y. Scalable microbial fuel cell (MFC) stack for continuous real wastewater treatment. Bioresour Technol 2012;106:82–88.

161. Marcus AK, Torres CI, Rittmann BE. Conduction-based modeling of the biofilm anode of a microbial fuel cell. Biotechnol Bioeng 2007;98:1171–1182.

162. Picioreanu C, Head IM, Katuri KP, van Loosdrecht MCM, Scott K. A computational model for biofilm-based microbial fuel cells. Water Res 2007;41:2921–2940.

163. Babauta JT, Hsu L, Atci E, Kagan J, Chadwick B, Beyenal H. Multiple cathodic reaction mechanisms in seawater cathodic biofilms operating in sediment microbial fuel cells. ChemSusChem 2014;7:2898–2906.

164. Cao B, Shi LA, Brown RN, Xiong YJ, Fredrickson JK, Romine MF, Marshall MJ, Lipton MS, Beyenal H. Extracellular polymeric substances from Shewanella sp. HRCR-1 biofilms: characterization by infrared spectroscopy and proteomics. Environ Microbiol 2011;13:1018–1031.

165. Sand W, Gehrke T. Extracellular polymeric substances mediate bioleaching/biocorrosion via interfacial processes involving iron(III) ions and acidophilic bacteria. Res Microbiol 2006;157:49–56.

166. Arechederra RL, Minteer SD. Self-powered sensors. Anal Bioanal Chem 2011;400: 1605–1611.

167. Debabov VG. Electricity from microorganisms. Microbiology 2008;77:123–131.

168. Fleming JT. Electronic interfacing with living cells. Adv Biochem Eng Biotechnol 2010;117:155–178.

169. Franks AE, Nevin KP. Microbial fuel cells, a current review. Energies 2010;3:899–919.

170. Guven G, Prodanovic R, Schwaneberg U. Protein engineering – an option for enzymatic biofuel cell design. Electroanalysis 2010;22:765–775.

171. Hallenbeck PC, Ghosh D, Skonieczny MT, Yargeau V. Microbiological and engineering aspects of biohydrogen production. Indian J Microbiol 2009;49:48–59.

172. Hamelers HV, Ter Heijne A, Sleutels TH, Jeremiasse AW, Strik DP, Buisman CJ. New applications and performance of bioelectrochemical systems. Appl Microbiol Biotechnol 2010;85:1673–1685.

173. Harnisch F, Schroder U. From MFC to MXC: chemical and biological cathodes and their potential for microbial bioelectrochemical systems. Chem Soc Rev 2010;39:4433–4448.

174. Haruyama T. Design and fabrication of a molecular interface on an electrode with functional protein molecules for bioelectronic properties. Electrochemistry 2010;78:888–895.

175. Huang LP, Cheng SA, Chen GH. Bioelectrochemical systems for efficient recalcitrant wastes treatment. J Chem Technol Biotechnol 2011;86:481–491.

176. Huang LP, Regan JM, Quan X. Electron transfer mechanisms, new applications, and performance of biocathode microbial fuel cells. Bioresour Technol 2011;102:316–323.

177. Ito J, Petzold CJ, Mukhopadhyay A, Heazlewood JL. The role of proteomics in the development of cellulosic biofuels. Curr Proteomics 2010;7:121–134.

178. Kannan AM, Renugopalakrishnan V, Filipek S, Li P, Audette GF, Munukutla L. Bio-batteries and bio-fuel cells: leveraging on electronic charge transfer proteins. J Nanosci Nanotechnol 2009;9:1665–1678.
179. Landoulsi J, Cooksey KE, Dupres V. Review – interactions between diatoms and stainless steel: focus on biofouling and biocorrosion. Biofouling 2011;27:1105–1124.
180. Lee HS, Vermaas WF, Rittmann BE. Biological hydrogen production: prospects and challenges. Trends Biotechnol 2010;28:262–271.
181. Lee JW, Kjeang E. A perspective on microfluidic biofuel cells. Biomicrofluidics 2010;4:041301.
182. Lefebvre O, Uzabiaga A, Chang IS, Kim BH, Ng HY. Microbial fuel cells for energy self-sufficient domestic wastewater treatment – a review and discussion from energetic consideration. Appl Microbiol Biotechnol 2011;89:259–270.
183. Li WW, Sheng GP, Liu XW, Yu HQ. Recent advances in the separators for microbial fuel cells. Bioresour Technol 2011;102:244–252.
184. Liu Y, Kim E, Ghodssi R, Rubloff GW, Culver JN, Bentley WE, Payne GF. Biofabrication to build the biology-device interface. Biofabrication 2010;2:022002.
185. Mahadevan R, Palsson BO, Lovley DR. In situ to in silico and back: elucidating the physiology and ecology of Geobacter spp. using genome-scale modelling. Nat Rev Microbiol 2011;9:39–50.
186. Meunier CF, Yang XY, Rooke JC, Su BL. Biofuel cells based on the immobilization of photosynthetically active bioentities. ChemCatChem 2011;3:476–488.
187. Moehlenbrock MJ, Minteer SD. Extended lifetime biofuel cells. Chem Soc Rev 2008;37:1188–1196.
188. Mohapatra BR, Dinardo O, Gould WD, Koren DW. Biochemical and genomic facets on the dissimilatory reduction of radionuclides by microorganisms – a review. Mineral Eng 2010;23:591–599.
189. Namour P, Jaffrezic-Renault N. Sensors for measuring biodegradable and total organic matter in water. TrAC Trend Anal Chem 2010;29:848–857.
190. Ong YT, Ahmad AL, Zein SHS, Tan SH. A review on carbon nanotubes in an environmental protection and green engineering perspective. Braz J Chem Eng 2010;27:227–242.
191. Osman MH, Shah AA, Walsh FC. Recent progress and continuing challenges in bio-fuel cells. Part II: Microbial. Biosens Bioelectron 2010;26:953–963.
192. Osman MH, Shah AA, Walsh FC. Recent progress and continuing challenges in bio-fuel cells. Part I: enzymatic cells. Biosens Bioelectron 2011;26:3087–3102.
193. Pant D, Van Bogaert G, Diels L, Vanbroekhoven K. A review of the substrates used in microbial fuel cells (MFCs) for sustainable energy production. Bioresour Technol 2010;101:1533–1543.
194. Pasco NF, Weld RJ, Hay JM, Gooneratne R. Development and applications of whole cell biosensors for ecotoxicity testing. Anal Bioanal Chem 2011;400:931–945.
195. Ponomareva ON, Arlyapov VA, Alferov VA, Reshetilov AN. Microbial biosensors for detection of biological oxygen demand (a review). Appl Biochem Microbiol 2011;47:1–11.
196. Pugh S, McKenna R, Moolick R, Nielsen DR. Advances and opportunities at the interface between microbial bioenergy and nanotechnology. Can J Chem Eng 2011;89:2–12.
197. Qian F, Morse DE. Miniaturizing microbial fuel cells. Trends Biotechnol 2011;29:62–69.

198. Rachinski S, Carubelli A, Mangoni AP, Mangrich AS. Microbial fuel cells used in the production of electricity from organic waste: a perspective of future. Quim Nova 2010;33:1773–1778.

199. Ramanavicius A, Ramanaviciene A. Hemoproteins in design of biofuel cells. Fuel Cells 2009;9:25–36.

200. Rasouli M, Phee LSJ. Energy sources and their development for application in medical devices. Expert Rev Med Devices 2010;7:693–709.

201. Rinaldi A, Mecheri B, Garavaglia V, Licoccia S, Di Nardo P, Traversa E. Engineering materials and biology to boost performance of microbial fuel cells: a critical review. Energy Environ Sci 2008;1:417–429.

202. Rosenbaum M, Aulenta F, Villano M, Angenent LT. Cathodes as electron donors for microbial metabolism: Which extracellular electron transfer mechanisms are involved? Bioresour Technol 2011;102:324–333.

203. Rubenwolf S, Kerzenmacher S, Zengerle R, von Stetten F. Strategies to extend the lifetime of bioelectrochemical enzyme electrodes for biosensing and biofuel cell applications. Appl Microbiol Biotechnol 2011;89:1315–1322.

204. Schaetzle O, Barriere F, Baronian K. Bacteria and yeasts as catalysts in microbial fuel cells: electron transfer from micro-organisms to electrodes for green electricity. Energy Environ Sci 2008;1:607–620.

205. Strik DP, Timmers RA, Helder M, Steinbusch KJ, Hamelers HV, Buisman CJ. Microbial solar cells: applying photosynthetic and electrochemically active organisms. Trends Biotechnol 2011;29:41–49.

206. Su LA, Jia WZ, Hou CJ, Lei Y. Microbial biosensors: a review. Biosens Bioelectron 2011;26:1788–1799.

207. Torres CI, Marcus AK, Lee HS, Parameswaran P, Krajmalnik-Brown R, Rittmann BE. A kinetic perspective on extracellular electron transfer by anode-respiring bacteria. Fems Microbiol Rev 2010;34:3–17.

208. Vaddiraju S, Tomazos I, Burgess DJ, Jain FC, Papadimitrakopoulos F. Emerging synergy between nanotechnology and implantable biosensors: a review. Biosens Bioelectron 2010;25:1553–1565.

209. Wang HY, Bernarda A, Huang CY, Lee DJ, Chang JS. Micro-sized microbial fuel cell: a mini-review. Bioresour Technol 2011;102:235–243.

210. Wesley MJ, Lerner RN, Kim ES, Islam S, Liu Y. Biological fixed film. Water Environ Res 2011;83:1150–1186.

211. Willner I, Yan YM, Willner B, Tel-Vered R. Integrated enzyme-based biofuel cells – a review. Fuel Cells 2009;9:7–24.

212. Wrana N, Sparling R, Cicek N, Levin DB. Hydrogen gas production in a microbial electrolysis cell by electrohydrogenesis. J Clean Prod 2010;18:S105–S111.

213. Xu X, Ying YB. Microbial biosensors for environmental monitoring and food analysis. Food Rev Int 2011;27:300–329.

214. Zhao F, Slade RCT, Varcoe JR. Techniques for the study and development of microbial fuel cells: an electrochemical perspective. Chem Soc Rev 2009;38:1926–1939.

215. Zhou JP, McCreanor PT, Montalto F, Erdal ZK. Sustainability. Water Environ Res 2011;83:1414–1438.

216. Logan BE, Regan JM. Electricity-producing bacterial communities in microbial fuel cells. Trends Microbiol 2006;14:512–518.

2

THEORETICAL AND PRACTICAL CONSIDERATIONS FOR CULTURING *Geobacter* BIOFILMS IN MICROBIAL FUEL CELLS AND OTHER BIOELECTROCHEMICAL SYSTEMS

ALLISON M. SPEERS AND GEMMA REGUERA

Department of Microbiology and Molecular Genetics, Michigan State University, East Lansing, MI, USA

2.1 INTRODUCTION

Bioelectrochemical systems (BESs) are devices that harness the electroactivity of microorganisms as they catalyze oxidation and/or reduction reactions using electrodes (reviewed in [1]). In the classical BES configuration, the device consists of two chambers, the anode and the cathode, separated by a proton-exchange membrane (PEM) to allow the selective passage of protons from the anode to the cathode. Each chamber houses an electrode, and the electrodes are connected to each other through external electrical circuitry that permits the passage of electrons generated at the anode electrode to the cathode electrode. The anode chamber thus functions as a reactor where a microbial culture grows by coupling the oxidization of a substrate to the reduction of the anode electrode, while in the cathode chamber, the electrons are used to reduce a catholyte. Both electrodes can also be housed in the same chamber, although reactive species present in the medium or generated in the microbial and electrochemical reactions can affect microbial growth and BES performance. In this chapter, we describe the theory and practice of BESs driven by some of the most efficient and best-studied anode-reducing microorganisms – those in the genus

Biofilms in Bioelectrochemical Systems: From Laboratory Practice to Data Interpretation, First Edition. Edited by Haluk Beyenal and Jerome Babauta.

Geobacter. We focus, in particular, on double-chambered configurations but describe the constraints associated with growing *Geobacter* bacteria in single-chambered BESs. We begin with a general description of BESs designed for power generation (microbial fuel cells, MFCs) and those originally designed for hydrogen production in the cathode (microbial electrolysis cells; MECs), as these are the BESs most often used to culture *Geobacter* bacteria.

2.1.1 Microbial Fuel Cells (MFCs)

The archetype of BES is the MFC, a device where electrical current is harnessed from electroactive microorganisms growing in the anode chamber (bioanode). The microorganisms can be planktonic (suspended in the liquid medium) and/or attached to the electrode (anode biofilm), the latter being the most efficient mode of growth to achieve high power densities [2]. The electrical potential that drives the transfer of electrons from the anode to the cathode electrode in an MFC is partially generated from the reduction in the cathode solution of substrates (the catholyte) with a thermodynamically favorable reduction potential. O_2 is often the preferred catholyte both because of its high theoretical reduction potential (E^0 = 0.805 V vs. the standard hydrogen electrode or V_{SHE} at 25 °C, pH 7, pO_2 of 0.2 [3]) and because it is inexpensive and freely available in air. However, the low solubility of O_2 in water (~8 mg l^{-1} at 1 atm and 25 °C [4]) requires the continuous sparging of air through the liquid in the cathode chamber. Alternatively, the whole cathode chamber can be replaced with an air cathode (i.e., an electrode exposed to air) often doped with a metal catalyst (e.g., platinum) to increase the kinetics of O_2 reduction [5]. Although this configuration decreases some of the costs associated with the construction of the cathode chamber [3], doping the cathode electrode can be costly. Furthermore the activation energy required to completely reduce O_2 to H_2O (a four-electron transfer reaction) is still high for these systems, and the actual cathodic potential achieved (~0.25 V_{SHE}) is much lower than the theoretical potential. This effect, known as overpotential, is especially high for the reduction of O_2, thus making this catholyte less desirable for efficient MFC operation. The incomplete reduction of O_2 can also generate H_2O_2 (hydrogen peroxide, generated through a two-electron transfer reaction), which has a lower standard potential (E^0 = 0.695 V_{SHE}) than the complete reduction of O_2, and may inhibit microbial growth and reduce cathodic current [3]. In addition, O_2 can traverse the PEM and poison strictly anaerobic bacteria in the anode chamber. Furthermore, it can be used as an electron acceptor by some anode microorganisms, thus diverting electrons from the electronic circuit and decreasing power production.

To bypass these limitations, O_2 can be replaced by liquid catholytes such as ferricyanide or permanganate, which also have a high reduction potential (E^0 = 0.361 V_{SHE} [6] and E^0 = 1.382 V_{SHE} [7], respectively). This increases the cell voltage and power density while decreasing the risk of O_2 intrusions into the anode [6]. Liquid catholytes must be continually replenished throughout the operation of the MFC, making them impractical for scaled-up applications. They are, however, useful for laboratory studies, as described later in more detail.

Microorganisms can also be cultured in the cathode chamber to catalyze the reduction of substrates using the cathode electrode as the electron donor for their metabolism. By using microorganisms as catalysts of the cathodic reactions, biocathodes bypass the need to replenish liquid catholytes or to use expensive metal catalysts, thus reducing the material cost of the material cost of MFC operation. The relatively high metabolic diversity of microorganisms that can accept electrons from the cathode also allows these systems to catalyze a wide range of reductive reactions such as O_2 to H_2O, proton to hydrogen, nitrate to molecular nitrogen, and bicarbonate to methane (reviewed in [1]). Some of the reactions catalyzed by biocathodes also generate added-value products such as hydrogen, methane, liquid fuels, and other chemicals, thus compensating for the low power densities often achieved by MFCs and making these systems more economically feasible.

2.1.2 Microbial Electrolysis Cells (MECs)

Electrical power can also be applied to the BES to increase the electrical potential between the anode and cathode electrodes and enable cathodic reactions (reviewed in [1]) not attainable in cathode-limited systems such as MFCs. Energy is added to the system with an external power source or by poising the anode potential with a potentiostat. This type of BES is often referred to as a MEC because it enables the electrochemical reduction of protons to hydrogen at the cathode (reviewed in [8]). These BESs are especially suitable for microbial studies and have been used extensively for *Geobacter* bacteria. For this reason, we describe the theory and practice of MEC operation using *Geobacter* cultures in detail in the following sections.

2.2 *Geobacter*-DRIVEN BESs

Most BESs described today rely on the ability of electroactive microorganisms to couple their oxidative metabolism to the reduction of the anode electrode. Many microorganisms can catalyze the reduction of an electrode using redox-active mediator molecules, either exogenous or endogenous, but only a few have been reported to directly transfer metabolic electrons to the electrode in the absence of mediators (reviewed in [2]). Mediator-less systems are often enriched in bacteria from the family *Geobacteraceae* (reviewed in [9]). This taxonomic group in the *∂-proteobacteria* includes genera such as *Geobacter* (type genus) and *Desulfuromonas*, which oxidize acetate to CO_2 using naturally abundant, insoluble Fe(III) and Mn(IV) oxides as electron acceptors [10]. Because of the insoluble nature of the metal oxides, these bacteria have evolved mechanisms to colonize and directly transfer electrons from the cell envelope to the electron acceptor using metalloproteins [11–13] and conductive protein filaments or pili [14] exposed on their outer membrane. As cells face similar physical limitations when they reduce metal oxides and electrodes, the mechanisms involved in the two reactions share many similarities [2].

Members of the *Geobacteraceae* are among the few well-studied electrode-reducing microorganisms available in pure culture and with established genetic systems and sequenced genomes. *Geobacter metallireducens* and *Desulfuromonas acetoxidans* were the first members to be shown to couple the oxidation of acetate and benzoate, respectively, to the reduction of a graphite electrode in BESs [15]. However, most BES studies have used the model representative *Geobacter sulfurreducens* [16], the first member of the family to have a sequenced genome [17] and a genetic system [18] and the best studied at the mechanistic level. For this reason, the focus of this chapter is on culturing approaches that enable the growth of *G. sulfurreducens* in BESs. It is our expectation that the chapter provides sufficient detail for even the most inexperienced readers to grow *G. sulfurreducens* in BECs and understand the operational and biological constraints that limit cell growth and current production in these systems. At the same time, the protocols here can be modified to suit the individual needs of the researcher and experimental objectives.

2.2.1 Operational Parameters Affecting the Growth of *Geobacter* Biofilms in BESs

In the standard, two-chambered configuration, both chambers contain liquid medium, but *G. sulfurreducens* is cultured only in the anode chamber. There, it grows by forming an electrochemically active biofilm on the surface of the anode electrode. Biofilm growth on the electrode surface is ultimately dependent on the amount of energy derived from the oxidation of the electron donor coupled to the reduction of the electrode and the amount and type of carbon source available for assimilation [19]. The electrical potential between the anode and cathode electrode determines the reduction potential of the anode electrode and, therefore, the amount of energy that the anode biofilms generate from the oxidation of the electron donor. *Geobacter* bacteria also need to assimilate carbon for gluconeogenesis, biomass synthesis, and other anabolic reactions [20], as well as for the synthesis of the structural and electronic components of the biofilm matrix, such as electrically conductive pili [14, 21, 22] and *c*-type cytochromes associated with the biofilm matrix [23]. For this reason, carbon-containing electron donors that are efficiently assimilated and oxidized by the cells result in rapid biofilm growth and current production [19]. Hence, the choice of carbon source and electron donor has a direct impact on the electrochemical performance of *Geobacter*-driven BESs as well. On the basis of these considerations, it is important to carefully select the type of electron donor and the electrical potential between the anode and cathode electrode to maximize the performance of *Geobacter*-driven BESs.

2.2.1.1 Acetate as the Preferred Electron Donor Depending on the operational parameters, the electrode-attached cells can grow and form a multilayered biofilm. The metabolism of *Geobacter* bacteria is especially suited for the oxidation and assimilation of acetate, and as a result, acetate is the preferred electron donor by *Geobacter*-driven BESs [19]. Acetate is oxidized in the tricarboxylic acid (TCA) cycle to produce energy for growth and electrons for electrode reduction while also providing carbon, via its conversion to pyruvate, for gluconeogenesis, biomass synthesis, and other anabolic reactions [20]. Coulombic efficiencies (CEs) between 80% and 85% are typical for acetate-driven BESs [19, 24]. Thus, 80–85% of the acetate is completely oxidized to CO_2 with the concomitant generation of current, whereas the remaining 15–20% is diverted to build the biofilm biomass.

Studies with the model representative *G. sulfurreducens* in acetate-fed MECs have provided evidence supporting a developmental biofilm process on the anode electrode. On inoculation in the anode chamber, a subset of planktonic cells attaches to the electrode (adhesion phase; Fig. 2.1) and current is produced. The magnitude of the initial current increase is dependent on the amount of cells inoculated into the anode chamber: higher planktonic cell densities lead to more rapid electrode colonization [25]. The rates of electron transfer (2 mA (mg protein)$^{-1}$) reported during this initial phase of colonization are low [25]. This supports the notion that the cells attached to the electrode are diverting energy from the oxidation of the electron donor toward the synthesis of cellular components required for biofilm growth and electron transfer to the electrode. Current production then increases exponentially (exponential growth phase, Fig. 2.1), as the attached cells grow and divide to form a multilayered biofilm. The rate of current increase during this exponential phase closely matches the growth rate of *G. sulfurreducens* cells grown planktonically in medium with soluble Fe(III) and is independent of the cell density of the initial inoculum [25]. Thus, planktonic

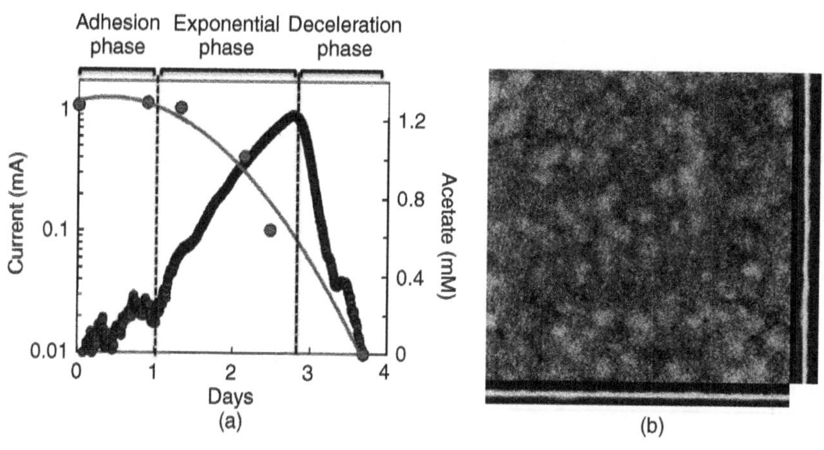

Figure 2.1 (a) Stages in the development of an anode biofilm by *G. sulfurreducens* coupled to current production (red symbols) and acetate utilization (shown as decrease in acetate concentration, blue symbols) in an MEC. After an initial adhesion phase, the attached cells grow exponentially on the anode electrode coupling the oxidation of the electron donor (acetate) to current production (exponential phase). Once the electron donor concentration decreases to growth-limiting levels, the biofilm cells enter stationary phase, the biofilm stops growing, and current production declines until all the residual acetate has been utilized (deceleration phase). (b) CLSM micrograph (200 × 200 μm^2 field) of an acetate-fed anode biofilm is ~10 μm thick examined at the end of the experiment (3.7 days). The anode biofilm was grown using the standard cultivation medium described in Table 2.1 and incubating the MEC at 30 °C. (*See insert for color representation of this figure.*)

TABLE 2.1 Stock Solutions for DB Medium.

10× DB Stock Solution	Concentration (g l^{-1})
KCl	3.80
NH_4Cl	2.00
$NaH_2PO_4 \cdot H_2O$	0.69
$CaCl_2 \cdot 2H_2O$	0.40
$MgSO_4 \cdot 7H_2O$	2.00
100× Mineral mix	g l^{-1}
Nitrilotriacetic acid (NTA)	1.500
Adjust pH of NTA solution to 8.5 with NaOH, then add	
$MnCl_2 \cdot 4H_2O$	0.100
$FeSO_4 \cdot 7H_2O$	0.300
$CoCl_2 \cdot 6H_2O$	0.170
$ZnCl_2$	0.100
$CuSO_4 \cdot 5H_2O$	0.040
$AlK(SO_4)_2 \cdot 12H_2O$	0.005
H_3BO_3	0.005
Na_2MoO_4	0.090
$NiCl_2$	0.120
$NaWO_4 \cdot 2H_2O$	0.020
Na_2SeO_4	0.100
Adjust pH of solution to 7.0 with NaOH	
100× Vitamin mix	mg l^{-1}
Biotin	2.0
Folic acid	2.0
Pyridoxine hydrochloride	10.0
Riboflavin	5.0
Thiamine hydrochloride	5.0
Nicotinic acid	5.0
DL-Calcium pantothenate	5.0
Vitamin B_{12}	0.1
p-Aminobenzoic acid	5.0
Thioctic (lipoic) acid	5.0

cell recruitment is unlikely to be significant in this phase [25]. The rate of electron transfer also increases fourfold (8 mA (mg protein)$^{-1}$) during the exponential phase of biofilm growth [25] until the electron donor becomes growth limiting. At this point, current production declines rapidly (deceleration phase, Fig. 2.1). Re-feeding the electron donor (Fig. 2.2) or initiating a continuous flow of medium with electron donor [21] when the current is declining alleviates the growth limitation imposed by electron donor availability, and biofilm growth and current production can be reinitiated and even increased.

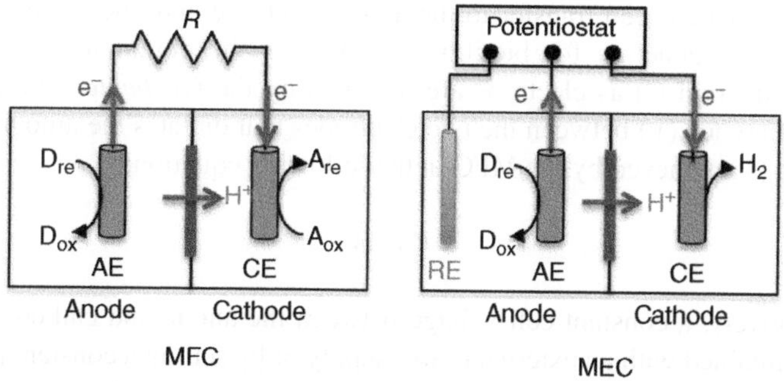

Figure 2.2 Schematic of a two-chambered MFC or an MEC configuration, both having an anode electrode (AE), where *Geobacter* biofilms catalyze the oxidation of the electron donor (D_{re} to D_{ox}). A proton-exchange membrane separates the two chambers to allow the diffusion of protons (H^+) from the anode to the cathode chamber. In the MFC, the anode electrode is wired directly to the cathode electrode (CE), and the amount of electrons (e^-) generated by the anode biofilms is dependent on the reduction potential of the electron acceptor (reaction A_{ox} to A_{re}) used as catholyte. In the MEC, the cathode limitation is bypassed using a potentiostat, which sets a constant potential of the anode electrode versus a reference electrode (RE) and allows the H^+ and the e^- to combine on the cathode electrode to generate H_2.

2.2.1.2 Alternative Electron Donors *G. sulfurreducens* can also utilize formate, lactate, or hydrogen, alone or in combination with acetate as electron donors [19]. H_2, however, does not support the growth of the anode biofilm and current generation when provided as sole electron donor [19], although it can be oxidized to support current production by biofilms pregrown with acetate [26]. This is because *G. sulfurreducens* cannot grow autotrophically with H_2 and requires a carbon source to support cell growth [27]. Formate and lactate can theoretically produce more negative cell voltages (−0.403 and −0.325 V, respectively) than acetate (−0.277 V), thereby providing more energy for growth. However, neither formate nor lactate can be used as effectively in *G. sulfurreducens*-driven BESs to produce current compared to acetate [19]. This is because although formate and lactate can be fully oxidized in a BES, metabolic cellular constraints prevent its efficient assimilation as carbon source and biofilm growth is limited [19]. However, small amounts of acetate added to formate-driven BESs are sufficient to stimulate the assimilation of formate carbon and support the growth of the anode biofilm and current production with CEs of about 75% [19]. Lactate, on the other hand, generates an excess flux of carbon and limits cell growth and current production in BESs [19]. Yet, it can be oxidized by acetate-grown biofilms yielding current levels and CEs (~ 90%) comparable to acetate [19].

2.2.1.3 MFC Versus MEC: Cathode-Limited Versus Unlimited Systems The catholyte in *Geobacter*-driven MFCs is generally selected for its high reduction potential. This maximizes the cell voltage (calculated as the potential difference between the anode and cathode chambers), as well as the amount of electrons that *Geobacter* cells can transfer to the anode electrode and, therefore, the amount of energy they generate for biofilm growth. It is the combination of electrical current production (I) as electrons are liberated by the *Geobacter* biofilm and the voltage difference (V) between the two electrodes that dictates the amount of power production (P) achieved by the MFC in the following equation:

$$P = IV$$

Alternatively, a constant cell voltage between the anode and cathode electrodes can be maintained with an external power supply or by setting a constant potential of the anode electrode versus a reference electrode with a potentiostat. Figure 2.2 shows a schematic of an MFC compared to an MEC in which a potentiostat is used to poise the anode electrode using a reference electrode placed in the anode chamber.

The choice of MFC or MEC depends on whether power or H_2 is the desired energetic output of the system, respectively. Bypassing the limitation of the cathodic reaction in MECs leads to systems whose performance is controlled by the catalytic activity of the anode biofilms. For this reason, these systems are also referred to as anode-controlled. Not surprisingly, MECs are advantageous in biological studies in which reproducibility is an important consideration, for example, when comparing the rates of oxidation of various electron donors or the effect of mutations in biofilm growth and/or current production, as explained in more detail in the following sections.

2.2.1.3.1 MFCs (Cathode-Limited Systems) In the standard MFC system with O_2 as the catholyte, the growth of *G. sulfurreducens* biofilm and the amount of electrons that it transfers to the electrode are limited by the low potential (lower than the theoretical maximum) of the cathodic reduction of O_2 to H_2O and the low solubility of O_2 in water, which reduces its availability. Furthermore, as *G. sulfurreducens* can also use O_2 as a terminal electron acceptor under microaerophilic conditions [28], O_2 intrusions from the cathode chamber into the anode chamber can divert metabolic electrons away from the anode electrode and can potentially inhibit the cell's metabolism or be lethal to the cells. The growth limitations imposed by the use of O_2 as the catholyte prevent the formation of multilayered biofilms on the anode electrode and also limit power production in these systems (Fig. 2.3). In contrast, replacing O_2 with ferricyanide increases the cell voltage and promotes more biofilm growth on the anode electrode (Fig. 2.3). This is because the actual reduction potential of ferricyanide is higher than that of O_2 under typical BES operational conditions. This, in turn, increases power density and improves the performance of the MFC. Another strategy to overcome the cathodic limitation is to increase the surface area of the cathode electrode such that it is many times larger than the anode. Thus, in addition to carefully selecting the catholyte to enable maximum biofilm growth on the anode electrode without diverting electrons away from the electrode, designers of MFCs must always examine the trade-offs between increasing the cell voltage, decreasing the internal resistance, and minimizing material costs while producing the greatest amount of power [3].

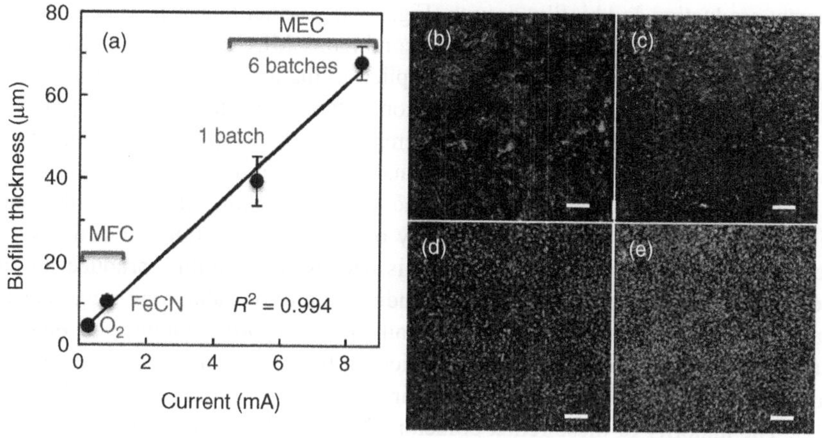

Figure 2.3 (a) Correlation between the growth of *G. sulfurreducens* on the anode electrode (quantified as biofilm thickness) and current generation (mA) in cathode-limited systems (MFCs) with oxygen (O_2) or ferricyanide (FeCN) as the catholyte and in cathode-unlimited systems (MECs) in which the anode electrode was poised at a potential of 300 mV versus an Ag/AgCl reference electrode using a potentiostat and fed with 10 mM acetate once (one batch) or six times (six batches). (b–e) CLSM micrographs of the anode biofilms from MFCs with O_2 (b) or FeCN (c) catholytes and from MECs fed once (d) or six times (e). Scale bar, 20 μm. All BES used fresh water medium and were incubated at 25 °C, as described in Reguera et al. [21]. (*See insert for color representation of this figure.*)

2.2.1.3.2 MECs (Cathode-Unlimited or Anode-Controlled Systems) The typical approach to remove the cathodic limitations on the growth of *G. sulfurreducens* biofilms is to poise the anode electrode at a constant potential. This anoxic, poised system also eliminates the possibility of O_2 intrusions in the anode chamber and maintains consistency between independent BES experiments – an important consideration when investigating mutant phenotypes [21] or metabolic activities of anode biofilms [19, 29]. When compared to cathode-limited MFCs run with O_2 or ferricyanide as the catholyte, an MEC equipped with an anode electrode poised at a constant potential promotes the growth of *G. sulfurreducens* biofilms (measured as biofilm thickness) and yields substantially more current production (Fig. 2.3). All the systems in Figure 2.3 were operated in batch mode in the same dual-chambered, H-type fuel cells equipped with graphite electrodes and same growth conditions (fresh water medium with 10 mM acetate as the electron donor and incubated at 25 °C), as previously described [21, 26]. By increasing the electrical potential between the anode and cathode electrode, the amount of electricity harvested in these systems (from MFCs with O_2 and ferricyanide to MECs) is also increased. Furthermore, the current generated was directly proportional to the thickness or height of the anode biofilms (Fig. 2.3a). Replenishing the medium of the anode chamber several times (six batches) with fresh medium supplemented with acetate as soon as the electron donor from the previous batch had been consumed led to additional increases in current production and biofilm growth (Fig. 2.3a). Further linear increases in current generation have also been reported for continuous MEC systems, in which fresh medium with acetate is continuously supplied to the anode chamber at a flow rate of 0.5 ml min^{-1} to maintain a constant concentration of electron donor [21]. This is consistent with cathode-unlimited systems in which electron donor availability is the only variable limiting biofilm growth and electrode reduction.

It is important to note that despite the vertical stacking of cells as the anode biofilms grow in thickness, the efficiency of current production per cell does not decrease as current increases [21]. This is because the biofilms produce an electroactive matrix of *c*-cytochromes and conductive pili to enable long-range electron transfer across the multilayered community [21, 23, 30]. Evidence to date also supports the notion that the electron carriers in the biofilm matrix are functionally diverse, as indicated by the fact that their oxidation potentials span a wide range [31]. The majority of these redox cofactors, however, have oxidation potentials too negative to be involved in electron transport when acetate is the electron source. Whether these electron carriers can be functional when using electron donors other than acetate or at different electrode potentials has not been empirically investigated.

2.2.1.4 BES Configuration and Operation Most *Geobacter*-driven BESs described to date are two-chambered devices operated as cathode-unlimited systems (or MECs) in order to maximize the growth and electroactivity of the *Geobacter* biofilms on the anode electrode. As mentioned earlier, a small amount of electricity from a power supply, battery, or, in some cases, an MFC [32, 33] can be used to polarize the anode and cathode electrodes and promote the growth of a multilayered anode biofilm in a cathode-unlimited system. A power supply, for example, keeps the potential difference between the anode and cathode electrodes constant. A potentiostat, on the other hand, sets the potential of the anode (or cathode) electrode with respect to a reference electrode with a known potential. Depending on whether the anode or the cathode electrode needs to be poised at a constant potential, the reference electrode is placed in the anode or cathode chamber, respectively. The reference electrode could be a saturated calomel electrode (SCE; $E = +0.244$ V) or a silver/silver chloride (Ag/AgCl; $E = +0.197$ V, saturated) electrode, the latter being reasonably inexpensive and, therefore, used most often. When placing the reference electrode in the anode chamber to poise the anode electrode, for example, the potential at the cathode automatically adjusts to the current that is being produced at the anode electrode [34]. The potentiostat also allows real-time measurement of the current produced at the anode electrode, which can be recorded with a computer connected to the potentiostat. This provides a real-time view of the metabolic activity and performance of the anode biofilm. The benefit of this setup is that the potential conditions of the anode electrode (termed working electrode) can be held constant over the duration of the experiment, as well as between experiments, thus enabling comparative studies of anode biofilm growth and current production by different mutant strains [21] or by the same strain grown under different conditions such as, for example, different electron donors [19]. Alternatively, the voltage can be changed to alter the redox environment experienced by the anode biofilm. In this way, potentiostat-driven MECs become powerful tools to study the physiology and metabolism of *Geobacter* bacteria.

Single-chambered BESs can also be used to grow *Geobacter* bacteria. They are often operated as cathode-unlimited MEC systems in much the same way as a two-chambered system [25, 35], except that the anode and cathode electrodes are placed in the same vessel, usually in close proximity to one another. Single-chambered MECs do not require the physical separation of the anode and cathode electrodes to maintain a potential difference because the potential is provided externally (by a potentiostat or power source) rather than by the electrochemical properties of the anode biofilm and catholyte. One of the potential down sides to the single-chamber MEC is that the hydrogen produced at the cathode electrode can diffuse freely to the anode, where it can be consumed as an electron donor by *Geobacter* bacteria. This can potentially prevent the accurate measurement of CE, as some of the electrons converted into current by the biofilm will also come from cathodic hydrogen. Furthermore, the presence of H_2 may also change the efficiency of acetate metabolism in *Geobacter* cells. H_2 and acetate can, for example, be simultaneously used as electron donors by *G. sulfurreducens*, but the rates of consumption are slightly decreased when the two electron donors are used simultaneously, consistent with competitive inhibition between the two metabolic pathways [36]. The ultimate decision of which configuration to use will depend on the experimental goals.

2.2.1.5 Electrode Materials The choice of anode electrode in *Geobacter*-driven BESs follows the general considerations of any other BESs. The electrode needs to be highly conductive and nonreactive but also inexpensive. For this reason, graphite felt, carbon paper, carbon cloth, reticulated vitreous carbon, graphite or carbon fiber brushes, and granular or activated carbon are commonly used in BES systems (reviewed in [37]). When operated as an MFC, where the ultimate goal is to reach the highest power production, the electrodes are also chosen so they provide a high surface area. Thus, porous electrode materials such as carbon cloth or carbon brushes are popular choices for MFCs [2, 37]. In MFCs, the connections between the electrodes and the external circuit and load must also be highly conductive to minimize the internal resistance, which limits power production.

The anode electrode material also influences the adhesion of the cells to the anode electrode and their ability to transfer electrons to it, which ultimately controls the growth and electroactivity of the anode biofilms. Interestingly, information about the effect of electrode materials and surface chemistry on *Geobacter* adhesion and growth is scarce. Thus, as with the BES configuration, the ultimate choice of electrode materials will depend on the specific research goals, and it needs to be validated empirically. Although rigorous studies on the effect of electrode materials on biofilm formation by *G. sulfurreducens* are lacking, it has been reported to grow on graphite stick [26], graphite rod [19], graphite paper [38], glassy carbon [35], carbon cloth [39], and gold [40] electrodes.

The cathode material need not be the same as the anode material. In fact, its selection is especially important in cathode-limited systems. In an MFC, the anode and cathode electrodes catalyze distinct reactions, and the choice of electrode material should take into account their specific catalytic performance. When O_2 is the catholyte, for example, the cathode electrode must contain a catalyst that promotes the complete reduction of O_2 to H_2O. These catalysts can include platinum, transition metals, nonprecious metals, or activated carbon [3]. The cathode electrode in these systems is also often constructed to be many times larger in surface area than the anode electrode to minimize the effect of the limiting cathodic reaction [3].

2.3 STANDARD PROTOCOL TO CULTURE *Geobacter* IN BESs

The following protocol has been optimized in our laboratory for growing biofilms of the model representative *G. sulfurreducens* in both single- and double-chambered BESs using acetate as the electron donor. Operation as MFC or MEC does not affect the general culturing procedure.

2.3.1 Bacterial Strains and Culture Conditions

Most studies of *Geobacter*-driven BESs use *G. sulfurreducens* strain (ATCC 51573), which is routinely grown in an anaerobic medium named DB described by Speers and Reguera [19]. We named the medium DB after Daniel Bond, at the University of Minnesota, who first described an anaerobic medium for culturing *G. sulfurreducens* in MECs [38], and which we modified to make the DB medium by supplementing it with 10 ml l^{-1} of a 100× vitamin mixture [41] to promote biofilm growth and electroactivity. The medium is prepared from three stock solutions (Table 2.1) routinely stored at 4 °C (the mineral and vitamin mixes are also stored in the dark).

The protocol for the preparation of 1 l of DB medium with 20 mM acetate and 40 mM fumarate (DB-AF) is as follows:

1. Stir in the three stock solutions (100 ml 10× DB stock, 10 ml 100× mineral mix, and 10 ml 100× vitamin mix) to mix.
2. Add the electron donor (e.g., acetate as sodium acetate, 20 mM) and fumarate (sodium fumarate, 40 mM) as the electron acceptor, from concentrated stock solutions.
3. Adjust the pH of the solution to 6.8 with NaOH. (We routinely use 2 N NaOH for pH adjustment during media preparation.)
4. Bring the solution to 1 l final volume with double distilled water.
5. Dispense the solution in a side-armed flask and stir and degas under vacuum for 30 min.
6. Add 2 g of $NaHCO_3$ and stir to mix.
7. Dispense into culture vessels (10 ml per pressure tube or 100 ml per 120 ml serum bottle).

8. Sparge with an anaerobic N_2:CO_2 (80:20) gas mix to remove O_2 from solution and headspace (30 min for tubes or 60 min for bottles) following standard protocols for making anaerobic media [42].
9. Seal with butyl rubber stoppers and aluminum crimps and autoclave in dry cycle for 30 min.

Note: Chemglass Life Sciences offers a good selection of glassware and accessories for anaerobic media preparation such as 18 mm × 150 mm glass tubes (anaerobic tubes, catalog # CLS-4209-10), 20 mm blue butyl rubber stoppers (CLS-4209-14), and 20 mm aluminum seals (CLS-4209-12).

This acetate–fumarate medium (abbreviated DB-AF) supports the planktonic growth of *G. sulfurreducens* to high cell densities at 30 °C, making it ideal to grow cultures for use as inoculum in BESs.

2.3.2 Preparing the Cell Inoculum for the BES

A standard procedure to prepare an inoculum of *G. sulfurreducens* for a BES involves inoculating an active culture of *G. sulfurreducens* into 100 ml of DB-AF medium contained in a 120-ml glass serum bottle. A culture in the bottle initiated with cells at an initial optical density at 600 nm (OD_{600}) of 0.02 and incubated at 30 °C without shaking will often reach stationary phase in 36–48 h. This stationary-phase culture in DB-AF provides the cells used to inoculate the anode chamber of the BES. A flow chart describing the major steps in this procedure is depicted in Figure 2.4.

The cells are harvested when the culture reaches early stationary phase (OD_{600} in a disposable cuvette 0.55–0.6). Visually, stationary-phase cultures have cells already settled at the bottom of the bottle, where they form a red film. The bottle is then gently inverted to re-suspend the cell film and transferred to an anaerobic enclosure (such as a glove bag, COY Labs). The anaerobic enclosure is highly recommended, as it helps maintain the anaerobic conditions throughout the procedure. Approximately 40 ml of the culture is then dispensed in 50 ml centrifuge tubes previously left overnight in the anaerobic enclosure to remove any O_2. Disposable conical centrifuge tubes (e.g., 50 ml Fisherbrand™ Easy Reader™ Plastic Centrifuge Tubes, catalog # 05-539-6) can also be used, provided the available centrifuge is compatible with this type of tubes and is able to reach the centrifugal speeds (4000–5000 × *g*) necessary to pellet down the cells from the culture. After the caps are sealed tightly, the centrifuge tubes are removed from the anaerobic enclosure and centrifuged at approximately 4600 × *g* for 8 min at room temperature. The tubes are then carefully transferred again to the anaerobic enclosure to decant the supernatant fluids, leaving a red cell pellet at the bottom. The cell pellet is then gently re-suspended in 10 ml of DB medium (with no acetate or fumarate additions), and the cell suspension is subjected to a new cycle of centrifugation. The cell pellet collected in this last centrifugation step is gently re-suspended in 10 ml of anaerobic DB medium supplemented with the electron donor of choice provided at the same concentration used in the anode chamber. We herein describe a protocol using 1 mM acetate (DB-A medium). The 10-ml cell suspension prepared in DB-A can then be dispensed into a pressure tube and sealed with

Grow *G. sulfurreducens* culture
(Inoculate cells to initial OD_{600} of 0.02 in 100 ml DB-AF medium incubated at 30 °C for 36–48 h)

↓

Pellet cells by centrifugation
(In an anaerobic enclosure, harvest cells from 40 ml of the early stationary phase culture, centrifuge at 4600 x *g*, 8 min, room temperature)

↓

Wash cells in fumarate-free medium
(In an anaerobic enclosure, decant supernatant, gently re-suspend cells with 10 ml DB medium)

↓

Pellet cells by centrifugation
(Centrifuge at 4600 x *g*, 8 min, room temperature)

↓

Re-suspend cells in BES medium
(In an anaerobic enclosure, decant supernatant, gently re-suspend cells with 10 ml DB-A medium)

↓

Inoculate BES
(Anaerobically transfer cell suspension in DB-A medium into the anode chamber of the BES)

Figure 2.4 Flowchart showing the steps in the preparation of *G. sulfurreducens* cells for inoculation into a BES.

a butyl rubber stopper and aluminum crimp before being removed from the anaerobic enclosure. We routinely use this cell suspension as the inoculum of a 150-ml BES chamber with a 100-ml reactor volume and equipped with 12 cm^2 electrode surface area.

2.3.3 Setting Up the BES and Inoculating the Anode Chamber

For a two-chambered BES, the glass chambers are assembled with the PEM membrane (e.g., a Nafion membrane; Ion Power, Inc., New Castle, DE) held together between the chambers by a metal clamp and equipped with the anode and cathode electrodes. We use custom glass chambers made in our own glass facility by retrofitting 100 ml PYREX® screw cap storage bottles (Corning, catalog # 1395-100) with a glass, threaded side-arm. Conveniently, premade two-chambered fuel cells are now commercially available from several companies. Adams & Chittenden Scientific Glass, for example, offers a good selection of H-type fuel cell vessels, which are made from standard 100- or 250-ml borosilicate media bottles and fitted with various types of glass, threaded ports, and port attachments for gas sparging, sampling, holding electrodes, and for crimp-style septum fillings using butyl rubber stoppers. The assembled device is autoclaved to sterilize all its components, and each chamber is then filled with sterile, anaerobic DB medium with the electron donor of choice (e.g., 1 mM acetate, which makes DB-A medium). In general, 150 ml BES chambers can be filled with 100 ml of DB-A medium. (Note that the reactor volume includes the volume of the chamber plus the side port sealed with the PEM.) The anode is initially filled with only 90 ml of medium, as the remaining 10 ml is provided with the cell inoculum, as described previously. The reference electrode is sterilized separately, usually with 95% ethanol for 30 s, and immersed into the anode chamber's medium using sterile techniques. This sterilization procedure is not always effective, and more stringent decontamination is sometimes necessary. We have successfully used Tyndallization in anaerobic Luria Bertani (LB) medium for this purpose (four cycles, each comprising 30 min in boiling medium and 24 h at 30 °C and then 1 min in 70% ethanol before rinsing with sterile water for the most stubborn contaminants in reference electrodes [29]).

Figure 2.5 shows a schematic of a BES prototype with two chambers similar to the one described in this protocol. The prototype includes graphite rod electrodes (Alfa Aesar, 1.27 cm diameter, 99% metals basis, 12 cm^2), inlets and outlets for

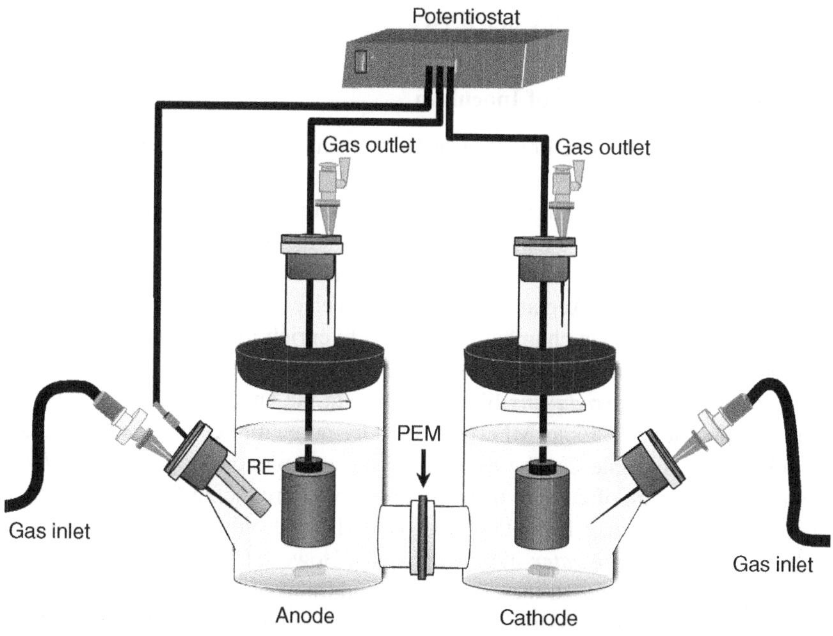

Figure 2.5 Schematic of a two-chambered BES. Shown are the anode (left) and cathode (right) chambers separated by a proton-exchange membrane (PEM). Each chamber is equipped with a graphite rod electrode, a stir bar, and inlet and outlet needles for gas circulation. The inlet needles are connected to the gas supply by tubes. The anode chamber is also equipped with a reference electrode (RE) when a potentiostat is used to poise the anode electrode.

gas sparging, and electrical connections to a potentiostat for MEC operation. Gas sparging uses an anaerobic mix of N_2:CO_2 (80:20; 99.5% purity) dispensed through needles fitted with sterile 0.2-μm filters, which filter-sterilize the incoming gas to prevent contamination of the reactor medium. To reduce operational costs, we recommend the use of heat-resistant 0.2-μm filters so they can be autoclaved and reused (such as Fisherbrand 0.22 μm, polytetrafluoroethylene PTFE syringe filters, catalog # 09-720-7, which we routinely autoclave up to five times). The chambers are also equipped with stir bars to ensure proper mixing in the anode and cathode chambers. The potential of the anode (working) electrode is often maintained at 240 mV versus an Ag/AgCl reference electrode using a potentiostat, and it is allowed to stabilize (usually 30 min) at the incubation temperature (e.g., 30 °C) until the current is stably maintained as a baseline close to 0 mA. However, these parameters can be changed to accommodate to the requirements of a particular experiment or availability of resources. Note, for example, that the MECs shown in Figure 2.3 use an anode potential of 300 mV and incubations at 25 °C. The potentiostat can also be replaced by a power source or, when operated in MFC mode, by a resistor; in these cases, the reference electrode is not needed.

Once the BES device is set up, the 10 ml cell suspension prepared in DB-A is transferred in its entirety into the BES anode chamber using strictly anaerobic techniques [42]. This is the equivalent of a 40% cell inoculum in reference to the original stationary-phase source culture in DB-AF.

2.3.4 Alternative Sources of Inoculum

Geobacter spp. can also be enriched in BESs from mixed-species sources such as domestic wastewater, industrial wastewater, or activated sludge (reviewed in [9]) using techniques described in this book. These sources of inoculum contain species of *Geobacter*, which can be enriched for in the anode biofilm with specific electron donors. As the proportion of *Geobacter* members increases, so does the electrical performance of the BES. Acetate, as the preferred electron donor for *Geobacter* spp., can be added to the anode chamber of the BES to enrich for these organisms. MFCs powered by mixed-species consortia grown with acetate, for example, are primarily enriched with *G. sulfurreducens*-like sequences in the anode biofilm and have similar power output and CEs as MFCs powered by pure cultures of *G. sulfurreducens* [39, 43]. While formate can be used as an electron donor for *G. sulfurreducens*, formate-fed BESs are often dominated by low-power-producing *Paracoccus* species, which efficiently oxidize formate to CO_2 and H_2, and only contain a smaller percentage (~14%) of *Geobacter*-like species [44]. However, the utilization of formate by *G. sulfurreducens* can be increased by the addition of small amounts of acetate, which facilitates formate assimilation as a carbon source [19]. Lactate, on the other hand, is a poor carbon and electron donor for *G. sulfurreducens* [19]. As a result, lactate enriches for closely related organisms such as *Pelobacter propionicus* and *Desulfuromonas* sp. [44].

Interestingly, BESs fed with substrates such as glucose and cellulose, which cannot be directly utilized by *Geobacter* spp. as electron donors or carbon sources, also enrich for *G. sulfurreducens*-like organisms as the predominant members of the anode biofilms [9]. This is due to the syntrophic relationship of the complex BES community in which organisms ferment complex substrates into acetate, formate, hydrogen, and/or lactate, which can then be used as electron donors to support the growth of *G. sulfurreducens*-like organisms [29, 45]. BESs-fed complex substrates such as wastewater also often enrich for *Geobacter* spp. [9]. However, CEs with these substrates are generally lower than in BESs fed with acetate because fewer *Geobacter* spp. are present in the anode biofilms and because some of the carbon and electrons in the substrates are diverted to support the growth of the other organisms [9]. While these complex substrates are not converted to current as efficiently as simple substrates, the action of the mixed-species biofilms enables reactions of industrial and biotechnological importance such as wastewater treatment, bioremediation of toxic chemicals, and processing of complex substrates for the synthesis of added-value chemicals and fuels. For example, the synergistic activities of the cellulolytic bacterium *Cellulomonas uda* and *G. sulfurreducens* in an MEC resulted in substantial energy recoveries (~73%) from the degradation and fermentation of chemically pretreated corn stover into ethanol with the cogeneration of cathodic hydrogen [29].

2.3.5 Methods to Monitor the Growth of *Geobacter* Biofilms

2.3.5.1 Current Production Current production can be used to monitor the growth of *Geobacter* biofilms during early stages of growth, as the exponential phase of biofilm growth matches the exponential rates of current production measured in the BES [25]. When a *Geobacter*-driven BES is operated in MFC mode, the cell voltage is measured across the resistor with the use of a multimeter. The current can then be calculated with Ohms law as $I = V/R$, where current (I) is in amps, voltage (V) is in volts, and resistance (R) is in ohms. In *Geobacter*-driven MECs, current production at the anode electrode can be directly measured with the use of a potentiostat (as in the example shown in Fig. 2.1). This enables a real-time view of the growth of the biofilm. The thickness of the anode biofilm is controlled by the amount of energy available for cell growth, which, in turn, depends on the voltage between the anode and cathode (and, therefore, the operational mode: MFC vs. MEC, as shown in Fig. 2.3), the concentration and type of electron donor, mass transport limitations, and incubation time [19]. The current yield is proportional to the biofilm thickness [21], which in turn depends on the concentration and type of electron donor [19]. Thus, changes in the concentration or type of electron donor will affect the current production very rapidly, and these changes can be detected with the potentiostat almost instantly [19].

2.3.5.2 Coulombic Efficiency The electrical conversion efficiency is an important parameter to assess the performance of the BES, and it also provides valuable insights into the growth and electroactivity of the anode biofilms. The CE allows to estimate how efficiently the electron donor is oxidized to generate current. CE is calculated by dividing the number of electrons recovered as current (I) over time (t) by the number of electrons available in the amount of acetate (or other electron donor) consumed by the biofilms [3], using the following equation:

$$\mathrm{CE} = \frac{\int_0^t I\, dt}{8F\,\Delta c}$$

where F is Faraday's constant, Δc represents the moles of acetate consumed over the duration of the experiment (t), and 8 is the number of moles of electrons per mole of acetate.

The concentration of the electron donor in the anode chamber thus needs to be monitored throughout the BES experiment in order to calculate CEs. Acetate, the preferred electron donor of *G. sulfurreducens*, can be easily measured by high-performance liquid chromatography (HPLC) using a 300 × 7.8 mm Aminex HPX-87H column (Biorad, catalog # 125-0140) at 30 °C with 5 mM H_2SO_4 as the eluent at a flow rate of 0.6 ml min^{-1}. Organic acids such as acetate, formate, and lactate can be detected with a UV detector at 210 nm [46]. Samples can be removed from the anode chamber periodically to monitor the consumption of electron donor and accumulation of metabolic intermediates or products. The cells from the sample are first removed by centrifugation, and the supernatant is then filtered through 0.22 μm filters before injection into the HPLC. In addition, a guard column (e.g., Biorad's Micro-Guard IG Cation H Cartridge, # 125-0502) can be used to ensure that no residual cells or other particulates make their way to the main HPLC column. Under optimal growth conditions, the *Geobacter* biofilm will consume most of the available acetate, and then the current will decrease rapidly until all the acetate is consumed (Fig. 2.1). Other electron donors such as ethanol and methanol can also be detected by HPLC using a differential refractometer [46] or by using a gas chromatography system with flame ignition to vaporize the sample. H_2 can be detected with a gas chromatograph (GC) equipped with a molecular sieve column.

From a biological point of view, the CE allows the researcher to estimate the efficiency of electron donor oxidation by *Geobacter* cells and, indirectly, the electrochemical activity of the biofilms. CEs of approximately 80% (±3%) have been reported for *G. sulfurreducens* anode biofilms growing in MECs with 1 mM acetate at a set anode potential of 0.24 V versus a 3 M Ag/AgCl reference electrode [19]. This indicates that, on average, 80% of the acetate was converted into current, while the remaining substrate was used as source of carbon and energy to build cell biomass. In contrast, a pilin-deficient mutant strain, which cannot produce *Geobacter* pilus nanowires to build thick biofilms and to promote their electrochemical activity, produces low levels of current over long periods of time and forms thinner biofilms on the anode electrode [21]. However, the mutant's CE is similar to the wild-type strain, indicating that, on average, the mutant converted as much acetate into current as the wild type, just over a prolonged period of time. Thus, the inability of the mutant to produce pili to grow and wire thick, electroactive biofilms, rather than the rates of acetate oxidation, limited current production.

CEs also provide important information about the metabolic constraints that may limit current production and biofilm growth in MECs. As we described earlier,

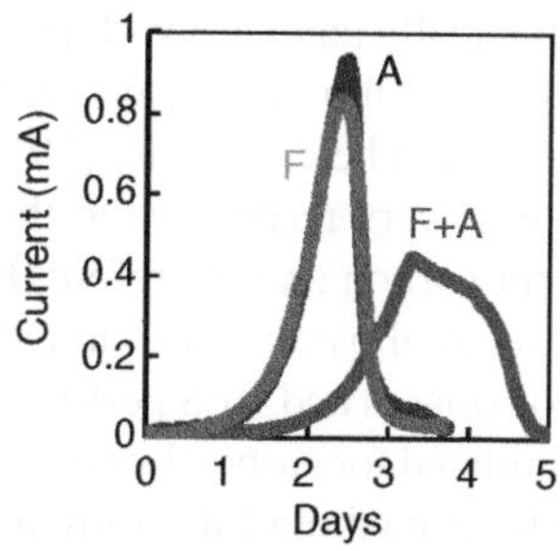

	CE (%)	I_{max}(mA)
A	83	0.9
F	93	0.4
F+A	75	0.8

Figure 2.6 Current production in *G. sulfurreducens* MECs with 1.1 mM acetate (A, black), 4 mM formate (F, medium gray), and 4 mM formate with 0.1 mM acetate (F+A, light gray). Coulombic efficiencies (CEs) and current maxima (I_{max}) for each MEC plot are shown in the table on the right side.

formate can also sustain current production in *G. sulfurreducens* MECs, but formate carbon assimilation limits current production in these systems. Figure 2.6 shows, for example, a formate-fed MEC in which formate was added at concentrations (4 mM) that provide the same electron equivalents as in 1-mM acetate. Initially, the anode biofilms grew and produced current as when acetate is the electron donor but only reached about half the current maxima before entering the deceleration phase [19]. The deceleration phase observed in formate-fed MECs is also longer than with acetate, with current declining slowly for 1–2 days until all the formate had been used. Despite the low performance of formate-fed MECs, CEs are higher (93 ± 1%) than with acetate. Thus, over time, the anode biofilms converted more formate to electricity, yet at the expense of limiting the availability of formate carbon for biofilm biomass. Not surprisingly, formate-fed anode biofilms are half as thick as acetate-fed biofilms [19]. Hence, while the CE shows that formate oxidation is very efficient in *G. sulfurreducens*, carbon assimilation through gluconeogenesis, biomass synthesis, and other anabolic activities, which is needed to build and sustain the electroactivity of the anode biofilms, is not. To alleviate this metabolic constraint, acetate can be added to the MEC in low (0.1 mM) concentrations to generate acetyl-CoA, which is used for the assimilation of formate carbon as pyruvate in the reaction of the pyruvate formate lyase [19]. The small amount of acetate is sufficient to promote the assimilation of formate carbon and supported maximum current as in acetate-fed MECs (Fig. 2.6). As more formate is diverted for carbon assimilation than for oxidation, CEs in formate-fed MECs supplemented with acetate decreased to approximately 75%.

2.3.5.3 Quantifying Biofilm Parameters As the amount of biofilm biomass is proportional to current production, quantifying biomass is important to assess BES performance. The anode biofilms can be detached from the electrode with a spatula or razor, and the biofilm biomass can be calculated from the total cell protein [21]. Alternatively, the cells can be stained with fluorescence dyes such as the BacLight viability kit (Invitrogen) and examined by Confocal Laser Scanning Microscopy (CLSM). The biofilm CLSM micrographs can then be statistically analyzed with programs such as COMSTAT [47]. This enables the analysis of biofilm parameters such as biomass, height (thickness), surface coverage, and structural heterogeneity [19].

The structural organization of the biofilm is very responsive to the efficiency of electron donor oxidation and assimilation for carbon and, therefore, to current production [19]. Hence, quantifying biofilm structural parameters can provide valuable information about the metabolism of electron donors and BES performance. Acetate, for example, is efficiently oxidized and assimilated for carbon by *Geobacter* bacteria, thus supporting the highest yields of biofilm growth and the formation of thicker, relatively uniform, confluent biofilms. This also maximizes electrode coverage and cell stacking per electrode surface area and increases the performance of the BES. In contrast, formate or lactate cannot be efficiently assimilated as carbon sources to support cell growth, and the biofilm structure changes to increase nutrient flow and minimize diffusion limitations in the anode biofilms. As a result, formate- and lactate-fed biofilms are more porous and heterogeneous and have reduced electrode coverage [19].

In summary, the practical considerations described in this chapter provide researchers with knowledge necessary to operate *Geobacter*-driven BESs while ensuring optimal growth of the electroactive biofilms and maximum system performance. When coupled with the theoretical knowledge of how *Geobacter* cells grow as biofilms, oxidize electron donors, and sustain current production, which is also discussed in this chapter, researchers have a powerful tool to troubleshoot any shortcomings arousing during BES operation and to explore new applications for these systems.

ACKNOWLEDGMENTS

The authors acknowledge support from grant MCB-1021948 from the National Science Foundation.

REFERENCES

1. Logan BE, Rabaey K. Conversion of wastes into bioelectricity and chemicals by using microbial electrochemical technologies. Science 2012;337:686–690.
2. Borole AP, Reguera G, Ringeisen B, Wang Z-W, Feng Y, Kim BH. Electroactive biofilms: current status and future research needs. Energy Environ Sci 2011;4:4813–4834.
3. Logan BE. Microbial Fuel Cells. Wiley; 2008.
4. Kim BH, Chang IS, Gadd GM. Challenges in microbial fuel cell development and operation. Appl Microbiol Biotechnol 2007;76:485–494.
5. Fan YZ, Hu HQ, Liu H. Enhanced Coulombic efficiency and power density of air-cathode microbial fuel cells with an improved cell configuration. J Power Sources 2007;171:348–354.

6. Logan BE, Hamelers B, Rozendal R, Schröder U, Keller J, Freguia S, Aelterman P, Verstraete W, Rabaey K. Microbial fuel cells: methodology and technology. Environ Sci Technol 2006;40:5181–5192.

7. You SJ, Zhao QL, Zhang JN, Jiang JQ, Zhao SQ. A microbial fuel cell using permanganate as the cathodic electron acceptor. J Power Sources 2006;162:1409–1415.

8. Logan BE, Call D, Cheng S, Hamelers HVM, Sleutels THJA, Jeremiasse AW, Rozendal RA. Microbial electrolysis cells for high yield hydrogen gas production from organic matter. Environ Sci Technol 2008;42:8630–8640.

9. Kiely PD, Regan JM, Logan BE. The electric picnic: synergistic requirements for exoelectrogenic microbial communities. Curr Opin Biotechnol 2011;22:378–385.

10. Lovley DR. Dissimilatory Fe(III) and Mn(IV)-reducing prokaryotes. In: Dworkin M, Falkow S, Rosenberg E, Schleifer K-H, editors. The Prokaryotes, 3rd ed., Volume 2. Springer; 2006. p 635–658.

11. Leang C, Coppi MV, Lovley DR. OmcB, a c-type polyheme cytochrome, involved in Fe(III) reduction in *Geobacter sulfurreducens*. J Bacteriol 2003;185:2096–2103.

12. Mehta T, Childers SE, Glaven R, Lovley DR, Mester T. A putative multicopper protein secreted by an atypical type II secretion system involved in the reduction of insoluble electron acceptors in *Geobacter sulfurreducens*. Microbiology 2006;152:2257–2264.

13. Mehta T, Coppi MV, Childers SE, Lovley DR. Outer membrane *c*-type cytochromes required for Fe(III) and Mn(IV) oxide reduction in *Geobacter sulfurreducens*. Appl Environ Microbiol 2005;71:8634–8641.

14. Reguera G, McCarthy KD, Mehta T, Nicoll JS, Tuominen MT, Lovley DR. Extracellular electron transfer via microbial nanowires. Nature 2005;435:1098–1101.

15. Bond DR, Holmes DE, Tender LM, Lovley DR. Electrode-reducing microorganisms that harvest energy from marine sediments. Science 2002;295:483–485.

16. Caccavo F Jr, Lonergan DJ, Lovley DR, Davis M, Stolz JF, McInerney MJ. *Geobacter sulfurreducens* sp. nov., a hydrogen- and acetate-oxidizing dissimilatory metal-reducing microorganism. Appl Environ Microbiol 1994;60:3752–3759.

17. Methé BA, Nelson KE, Eisen JA, Paulsen IT, Nelson W, Heidelberg JF, Wu D, Wu M, Ward N, Beanan MJ, Dodson RJ, Madupu R, Brinkac LM, Daugherty SC, DeBoy RT, Durkin AS, Gwinn M, Kolonay JF, Sullivan SA, Haft DH, Selengut J, Davidsen TM, Zafar N, White O, Tran B, Romero C, Forberger HA, Weidman J, Khouri H, Feldblyum TV, Utterback TR, Van Aken SE, Lovley DR, Fraser CM. Genome of *Geobacter sulfurreducens*: metal reduction in subsurface environments. Science 2003;302:1967–1969.

18. Coppi MV, Leang C, Sandler SJ, Lovley DR. Development of a genetic system for *Geobacter sulfurreducens*. Appl Environ Microbiol 2001;67:3180–3187.

19. Speers AM, Reguera G. Electron donors supporting growth and electroactivity of *Geobacter sulfurreducens* anode biofilms. Appl Environ Microbiol 2012;78:437–444.

20. Yang TH, Coppi MV, Lovley DR, Sun J. Metabolic response of *Geobacter sulfurreducens* towards electron donor/acceptor variation. Microb Cell Fact 2010;9:90.

21. Reguera G, Nevin KP, Nicoll JS, Covalla SF, Woodard TL, Lovley DR. Biofilm and nanowire production lead to increased current in microbial fuel cells. Appl Environ Microbiol 2006;72:7345–7348.

22. Reguera G, Pollina RB, Nicoll JS, Lovley DR. Possible nonconductive role of *Geobacter sulfurreducens* pilus nanowires in biofilm formation. J Bacteriol 2007;189:2125–2127.

23. Rollefson JB, Stephen CS, Tien M, Bond DR. Identification of an extracellular polysaccharide network essential for cytochrome anchoring and biofilm formation in *Geobacter sulfurreducens*. J Bacteriol 2011;193:1023–1033.

24. Call DF, Wagner RC, Logan BE. Hydrogen production by *Geobacter* species and a mixed consortium in a microbial electrolysis cell. Appl Environ Microbiol 2009;75:7579–7587.

25. Marsili E, Sun J, Bond DR. Voltammetry and growth physiology of *Geobacter sulfurreducens* biofilms as a function of growth stage and imposed electrode potential. Electroanalysis 2010;22:865–874.

26. Bond DR, Lovley DR. Electricity production by *Geobacter sulfurreducens* attached to electrodes. Appl Environ Microbiol 2003;69:1548–1555.

27. Coppi MV, O'Neil RA, Lovley DR. Identification of an uptake hydrogenase required for hydrogen-dependent reduction of Fe(III) and other electron acceptors by *Geobacter sulfurreducens*. J Bacteriol 2004;186:3022–3028.

28. Lin WC, Coppi MV, Lovley DR. *Geobacter sulfurreducens* can grow with oxygen as a terminal electron acceptor. Appl Environ Microbiol 2004;70:2525–2528.

29. Speers AM, Reguera G. Consolidated bioprocessing of AFEX-pretreated corn stover to ethanol and hydrogen in a microbial electrolysis cell. Environ Sci Technol 2012;46:7875–7881.

30. Nevin KP, Kim BC, Glaven RH, Johnson JP, Woodard TL, Methé BA, Didonato RJ, Covalla SF, Franks AE, Liu A, Lovley DR. Anode biofilm transcriptomics reveals outer surface components essential for high density current production in *Geobacter sulfurreducens* fuel cells. PLoS One 2009;4:e5628.

31. Snider RM, Strycharz-Glaven SM, Tsoi SD, Erickson JS, Tender LM. Long-range electron transport in *Geobacter sulfurreducens* biofilms is redox gradient-driven. Proc Natl Acad Sci U S A 2012;109:15467–15472.

32. Sun M, Sheng GP, Zhang L, Xia CR, Mu ZX, Liu XW, Wang HL, Yu HQ, Qi R, Yu T, Yang M. An MEC–MFC-coupled system for biohydrogen production from acetate. Environ Sci Technol 2008;42:8095–8100.

33. Wang A, Sun D, Cao G, Wang H, Ren N, Wu WM, Logan BE. Integrated hydrogen production process from cellulose by combining dark fermentation, microbial fuel cells, and a microbial electrolysis cell. Bioresour Technol 2011;102:4137–4143.

34. Nam JY, Tokash JC, Logan BE. Comparison of microbial electrolysis cells operated with added voltage or by setting the anode potential. Int J Hydrogen Energy 2011;36:10550–10556.

35. Marsili E, Rollefson JB, Baron DB, Hozalski RM, Bond DR. Microbial biofilm voltammetry: direct electrochemical characterization of catalytic electrode-attached biofilms. Appl Environ Microbiol 2008;74:7329–7337.

36. Brown DG, Komlos J, Jaffé PR. Simultaneous utilization of acetate and hydrogen by *Geobacter sulfurreducens* and implications for use of hydrogen as an indicator of redox conditions. Environ Sci Technol 2005;39:3069–3076.

37. Logan BE. Scaling up microbial fuel cells and other bioelectrochemical systems. Appl Microbiol Biotechnol 2010;85:1665–1671.

38. Srikanth S, Marsili E, Flickinger MC, Bond DR. Electrochemical characterization of *Geobacter sulfurreducens* cells immobilized on graphite paper electrodes. Biotechnol Bioeng 2008;99:1065–1073.

39. Ishii S, Watanabe K, Yabuki S, Logan BE, Sekiguchi Y. Comparison of electrode reduction activities of *Geobacter sulfurreducens* and an enriched consortium in an air-cathode microbial fuel cell. Appl Environ Microbiol 2008;74:7348–7355.
40. Richter H, McCarthy K, Nevin KP, Johnson JP, Rotello VM, Lovley DR. Electricity generation by *Geobacter sulfurreducens* attached to gold electrodes. Langmuir 2008;24:4376–4379.
41. Balch WE, Fox GE, Magrum LJ, Woese CR, Wolfe RS. Methanogens: reevaluation of a unique biological group. Microbiol Rev 1979;43:260–296.
42. Speers AM, Cologgi DL, Reguera G. Anaerobic cell culture. Curr Protoc Microbiol 2009;Appendix 4:Appendix 4F.
43. Nevin KP, Richter H, Covalla SF, Johnson JP, Woodard TL, Orloff AL, Jia H, Zhang M, Lovley DR. Power output and columbic efficiencies from biofilms of *Geobacter sulfurreducens* comparable to mixed community microbial fuel cells. Environ Microbiol 2008;10:2505–2514.
44. Kiely PD, Rader G, Regan JM, Logan BE. Long-term cathode performance and the microbial communities that develop in microbial fuel cells fed different fermentation endproducts. Bioresour Technol 2011;102:361–366.
45. Ren Z, Ward TE, Regan JM. Electricity production from cellulose in a microbial fuel cell using a defined binary culture. Environ Sci Technol 2007;41:4781–4786.
46. McKinlay JB, Zeikus JG, Vieille C. Insights into *Actinobacillus succinogenes* fermentative metabolism in a chemically defined growth medium. Appl Environ Microbiol 2005;71:6651–6666.
47. Heydorn A, Nielsen AT, Hentzer M, Sternberg C, Givskov M, Ersboll BK, Molin S. Quantification of biofilm structures by the novel computer program COMSTAT. Microbiology 2000;146:2395–2407.

3

MICROBIAL COMMUNITY CHARACTERIZATION ON POLARIZED ELECTRODE SURFACES

John M. Regan

Department of Civil and Environmental Engineering, The Pennsylvania State University, University Park, PA, USA

Hengjing Yan

Department of Chemistry & Biochemistry, University of California Santa Barbara, Santa Barbara, CA, USA

3.1 INTRODUCTION

An essential component of every bioelectrochemical system (BES) is the anode-reducing (exoelectrogenic) and/or cathode-oxidizing (exoelectrotrophic) activity of microorganisms. BESs are often operated with pure cultures for the purposes of studying the physiology of exoelectrogenic and exoelectrotrophic bacteria. However, applications of BESs in nonsterile environments such as wastewater and sediments necessarily entail the colonization of electrodes and other system surfaces with mixed-culture microbial communities. While nonmicrobial system components such as membranes and metal cathode catalysts often constrain performance, the microbial assemblage composition and structure can influence and explain electrochemical productivity, stability, and losses. These communities can involve cooperation and competition among microbial populations that contribute to (e.g., via substrate conditioning) or impair (e.g., through methanogenic losses) electrochemical performance, or even add nonelectrochemical functionality such as nitrogen removal (Fig. 3.1). Moreover, there is often an interdependence between the enriched microbial community and design and operating parameters such as

Biofilms in Bioelectrochemical Systems: From Laboratory Practice to Data Interpretation, First Edition.
Edited by Haluk Beyenal and Jerome Babauta.

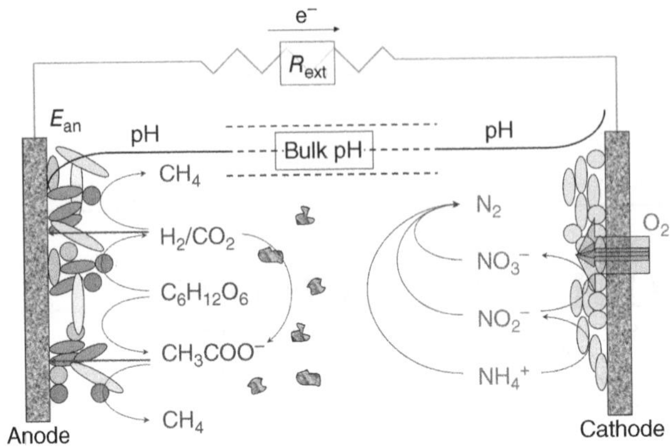

Figure 3.1 Ecology in mixed-culture air-cathode microbial fuel cell and several influencing factors.

inoculum, substrate [1], anode material [2], anode potential [3, 4], and operational mode [5]. Therefore, the characterization of these microbial communities can assist in establishing these relationships to promote the rational engineering of these communities and for monitoring and problem diagnosis.

This chapter provides information on methods that can be used to characterize microbial consortia in BESs. The emphasis is on several nucleic acid-based techniques whose use has been prevalent in the BES literature. There is also a brief description of several bulk biomass characterizations that do not differentiate community members, but rather serve a supporting function in community characterization efforts.

3.2 NUCLEIC ACID-BASED ANALYSES

Most BES studies that have included microbial community characterization have used 16S rRNA and its gene as the target molecules. The results from such phylogenetic characterizations must be appropriately interpreted, as there are some well-characterized biases that can emerge. First of all, as a nonfunctional genetic marker, one must be cautious about inferring function from ribosomal RNA (rRNA) identification due to strain variability and facultative phenotypes. Secondly, the polymerase chain reaction (PCR) amplification of the 16S rRNA gene can cause quantitative misrepresentation of community composition due to different rRNA operon copy numbers among different bacteria [6] and primer bias [7]. There can also be differential DNA extraction efficiencies among populations [8], which would be problematic for any analysis method using extracted nucleic acids. However, with these qualifiers, phylogenetic characterization is still the most common strategy due to the diversity of known mechanisms involved in microbially catalyzed electrode redox reactions, the lack of universal molecular targets to capture these mechanisms, the presence and involvement of nonexoelectrogenic or nonexoelectrotrophic populations in the community, and the likely presence of unknown exoelectrogens or exoelectrotrophs in these communities. This section focuses on 16S rRNA-based techniques but could be adapted to focus on other gene targets.

3.2.1 Denaturing Gradient Gel Electrophoresis

Denaturing gradient gel electrophoresis (DGGE) is most appropriately used for the screening of community differences over time [9, 10] or from different treatments [4, 11]. The technique involves PCR amplification of a target gene using a primer with a GC-rich extension (GC clamp) from the 5′ end and the subsequent separation of these PCR products on a gel cast with a gradient of denaturant. As the PCR products migrate in the direction of increasing denaturant, the double-stranded DNA begins to melt as a function of the sequence (higher GC content requires more denaturant), eventually leaving only the GC clamp intact. This predominantly single-stranded configuration hinders the continued migration of the products, which allows a sequence-dependent separation of the PCR products [12, 13]. The resultant DGGE patterns, or "fingerprints," can then be compared among different samples, and individual bands of interest (e.g., predominant bands or bands that are notably present in some samples but absent in others) can be excised and sequenced for specific identification.

Several factors should be considered in determining whether DGGE is an appropriate technique for a given community analysis effort and for getting the most out of DGGE data. Common to all post-PCR methods, DGGE includes the potential introduction of biased community characterizations for the aforementioned reasons, and discussions of DGGE-inferred community compositions should acknowledge this potential. Often, the comparative interpretation of DGGE profiles from different samples is handled somewhat arbitrarily and subjectively. This can be addressed through methods such as principal component analysis, which can be automated and also informative about the conditions that correlate with observed community differences [3]. Shortcomings associated with DGGE include the comigration of distinct bands, thereby masking diversity, and the need for band excision to ascertain identity, which involves the additional effort of cloning and sequencing. If the primary interest is in identifying only those populations that change with different treatments, DGGE might be appropriate. However, if broad community characterization is the goal [1, 5], the analysis of clone libraries or another approach makes more sense because the use of DGGE for this purpose requires clone libraries anyway and constrains the sequences that can be recovered to those that are in the visible and distinct DGGE bands.

3.2.1.1 PCR for DGGE Analysis The first step in DGGE analysis is DNA extraction followed by PCR amplification of a target gene. There are a number of commercially available kits for DNA extraction from environmental samples. We routinely use the bead-beating-based PowerSoil™ DNA Isolation Kit (MO BIO Laboratories,

TABLE 3.1 16S rRNA Gene-Targeted PCR Primer Pairs Commonly Used for DGGE Analysis.

Primer	Sequence (5′–3′)	Variable Region (s)	Annealing Conditions (°C)	Approximate Length (bp)	References
341F	CCT ACG GGA GGC AGC AG	V3	TD[a] 65–55	194	[17]
534R	ATT ACC GCG GCT GCT GG				
341R	CCT ACG GGA GGC AGC AG	V3–V5		586	[18]
926R	CCG TCA ATT CMT TTG AGT TT				
968F	AAC GCG AAG ACC CTT AC	V6	TD 57–54	91	[19]
1041R	GCG TGT GTA CAA GAC CC				

[a]TD – touchdown.

Inc., Carlsbad, CA), but there are alternative vendors and products that may be used with equivalent results and are often cited in the literature.

Primer selection is critical to the success of any PCR-based method. There are several 16S rRNA gene-targeted PCR primer pairs (Table 3.1) commonly used to give a product length that is appropriate for DGGE discrimination [14]. These primers are designed to target conserved sequence regions (e.g., all *Bacteria*) and flank one or more hypervariable sequence regions [15] that allow significant discrimination of fragments from different microbial templates. It should be noted that the 968F/1041R primer pair has nucleotide mismatches with commonly recovered sequences from BES communities [16] and should not be used for community analyses from these systems.

For PCR amplification prior to DGGE, one of the primers should be ordered with a GC clamp on the 5′ end. The following is a suitable GC clamp sequence and the modification of primer 968F with this clamp sequence:

- GC clamp: 5′-CGC CCG CCG CGC CCC GCG CCC GGC CCG CCG CCC CCG CCC C-3′.
- GC968F: 5′-CGC CCG CCG CGC CCC GCG CCC GGC CCG CCG CCC CCG CCC CAA CGC GAA GAA CCT TAC-3′.

The thermal profile used in PCR amplification is a function of the specific primers due to their annealing temperatures, and sometimes the use of touchdown PCR to recover products from templates that are not a perfect match with the primers (Table 3.1). For example, with the 341F/534R primer pair, we use the following profile: initial denaturation at 95 °C for 10 min; 10 touchdown cycles of denaturation at 94 °C for 1 min, annealing at 65 to 55 °C for 1 min (decreasing 1 °C each cycle), and extension at 72 °C for 2 min; 25 cycles of denaturation at 94 °C for 1 min, annealing at 55 °C for 1 min, and extension at 72 °C for 2 min; and a final extension at 72 °C for 30 min [3].

We routinely perform PCR in 25-μl volumes, which provides sufficient sample for subsequent processing while conserving reagents. The reaction mixture is composed of the following components:

- 12.5-μl Master Mix (2×) (The master mix composition for QIAGEN's *Taq* PCR Master Mix Kit includes 0.05 U μl^{-1} *Taq* DNA polymerase, 2× PCR buffer, 3 mM $MgCl_2$, and 400 μM of each dNTP, but there are numerous equivalent products and vendors.)
- 0.25–0.5 μM (final concentration) of each primer
- 10–50 ng extracted DNA
- balance water.

Several of these constituents can be varied to optimize PCR and overcome problems with inhibition. For example, the final $MgCl_2$ concentration with the aforementioned kit is 1.5 mM, but this can easily be increased by addition from a 25-mM $MgCl_2$ stock solution. The DNA concentration can also be varied if inhibition is a problem due to carryover of something in the sample matrix. This can be tested by inspecting PCR product concentration with different DNA template additions and spiking the sample DNA extraction with a positive-control template.

3.2.1.2 DGGE Protocol Preparation of the gel for DGGE is considerably more extensive than for typical gel electrophoresis. The instructions that follow are for the Bio-Rad DCode system (Fig. 3.2). The glass plates that will sandwich the gel during casting should be in good condition without scratches and chipped edges. The plates along with a set of 1-mm spacers should be washed with detergent, rinsed with tap water, and dried with laboratory wipes. Then place the spacers on both side edges of the larger glass plate and the smaller glass plate on top of them and screw the sandwich clamps to the sides of the cassette. Make sure that the spacers and both glass plates are aligned along the bottom edge. Place the sandwich on top of the rubber gasket of the casting stand and press down the handles to lock it in place. Add deionized (DI) water into the space between the glass plates to confirm that the sandwich assembly does not leak, reassembling if necessary. Pour out the water and use filter paper to blot the inner sides of the two glass plates. Then allow the glass plates to air-dry completely.

Once the gel casting cassette is ready, the denaturant gradient gel can be prepared. This involves making two solutions, one for each limit of the denaturant range. The description in this case assumes a denaturant range of 30–60% and a 7% gel but can be modified as necessary to achieve optimum separation of GC-clamped PCR products. The solutions are prepared in 50-ml centrifuge tubes according to the recipes in Table 3.2 and then filtered with a 0.45-μm filter. To 20 ml of each filtered solution,

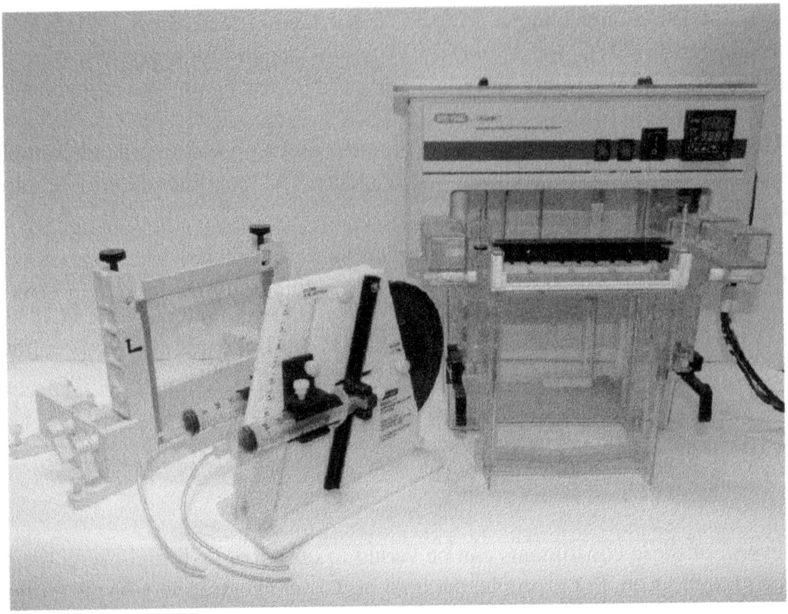

Figure 3.2 Bio-Rad D-Code DGGE components. Photo courtesy of Hiroyuki Kashima. (*See insert for color representation of this figure.*)

TABLE 3.2 Recipes for Low (30%) and High (60%) Denaturant Solutions.

Denaturant Condition (%)	40% Acrylamide/ Bis (ml)	50× TAE Buffer (ml)	Deionized Formamide (ml)	Urea (g)	Distilled H_2O (ml)
30	7	0.8	4.8	5.04	27.4
60	7	0.8	9.6	10.08	22.6

add 180 μl of 10% APS (ammonium persulfate solution; 0.3 g ammonium persulfate in 30 ml DI water) and 18 μl TEMED (*N,N,N′,N′*-Tetramethylethylenediamine) and stir this mixture gently by hand. TEMED promotes polymerization of the acrylamide, so the following processes need to be performed quickly. Using two 30-ml syringes, draw up each of these solutions, remove air from the syringes, and connect the syringes to a three-way tube assembled with a Y-fitting. Attach a needle to the other end of the three-way tube, and place this end between the glass plates with the needle pointing directly down. Put the two syringes in the syringe holders on the correct sides of the gradient maker, and turn the wheel slowly with a constant speed to dispense the solutions. Do not allow air bubbles to be trapped in the gel. Stop turning the wheel when the solutions have filled the space between the two glass plates, and then insert the comb on top of the glass plates. (Wash the tubes and the needle with water immediately to prevent the residual solution from polymerizing in the tube and

needle.) Leave the gel for 1–2 h to polymerize. Then carefully remove the comb and rinse the slots with DI water or 0.5× Tris–acetate–EDTA (TAE) buffer to remove nonpolymerized gel. Use filter paper to absorb the residual water or TAE buffer from the slots.

After the gel has been cast, the next steps of DGGE involve loading and running the gel. Fresh 0.5× TAE buffer should be added up to the fill line of the buffer tank, and the system should be turned on for at least an hour prior to sample loading to allow the buffer temperature to reach 60 °C. Prepare the sandwich core by wiping the gaskets with water, to clean them and provide a good seal with the gel sandwich. Then remove the gel sandwich from the casting stand and snap the sandwich clamps into the sandwich core (which can accommodate two gel sandwiches, one on either side) with the smaller glass plate facing the gasket of the core. Turn off the preheated unit, remove the lid, and lower the sandwich core into the buffer tank with the red dot on the right-hand side. Replace the lid and turn the unit on with stirring to keep the buffer warm until the samples are ready to load. Samples and markers should be mixed with a loading buffer; we use a glycerol-based buffer with bromophenol blue. We also avoid loading the outer four wells, as the band patterns are typically poor on the edges. Our typical electrophoresis protocol is 12 h at 70 V, but there are many different parameters in the literature, and they should be adjusted as necessary to suit a particular primer pair and denaturing gradient.

Following electrophoresis, the gel is stained and visualized. There are several common methods of staining gels, including the use of intercalating fluorochromes such as ethidium bromide and variants of the SYBR dye. We have found the best sensitivity using silver staining [20], and this approach is detailed as follows. Firstly, the system is turned off and the lid removed. Then the sandwich core is removed from the buffer, and the gel sandwich is removed from the core. The sandwich clamps can then be removed from the edges of the sandwich assembly, and the smaller glass plate and spacers can be removed. With the gel supported on the larger glass plate, the gel can be immersed in a bath of 10% acetic acid (prepared using DI water and 100% acetic acid) for 30 min. The plate-supported gel is then washed three times with water for 5 min each wash. The gel is then stained for 30 min by immersing in 1 l of solution containing 1.5-ml formaldehyde and 1-g $AgNO_3$. After silver staining, the gel is washed with water for 5 s and then developed in 2 l of 3% sodium carbonate (w/v) solution containing 450 μl of 10 mg ml^{-1} $Na_2S_2O_3 \cdot 5H_2O$ and 3-ml formaldehyde. The plate and gel are then transferred to 10% acetic acid bath for 5 min and finally washed with water for 5 min. The gel is now ready for visualization, which can be done using a high-resolution scanner to generate publication-quality images.

While DGGE is most appropriately a community-screening method, there is often interest in specifically identifying pronounced bands or bands that are differentially present under variable conditions. This can be done by excising the band from the gel using a sterile razor blade, eluting the DNA in 50 μl of sterile Tris–EDTA buffer, and then cloning and sequencing the eluted DNA. It is preferable to clone the DNA to assure the band is from a single template sequence and not multiple sequences that comigrated. The details of cloning and sequencing are presented in the following section.

3.2.2 Clone Libraries

The construction of clone libraries has also been a common method of microbial community analysis in BESs, providing individual sequences of a selected number of PCR-amplified target gene fragments that can be used for phylogenetic identification. The basic steps of cloning involve PCR amplification of a target gene, ligation of the PCR amplicons into a cloning vector, transformation of host cells with the modified vector, selection and screening of transformants, extraction and purification of the plasmid from selected transformants, and sequencing of the PCR product insert in each plasmid. Each of these steps is described in more detail in the following section. As a post-PCR method, cloning also incurs the shortcomings mentioned earlier, and the interpretation of cloning results should account for these issues. One of the main concerns with clone libraries is the number of cloned sequences necessary to adequately characterize the community. Rarefaction curves [21] and statistical metrics [22] are standard tools to estimate the extent of community coverage, but there appears to be no general agreement in molecular microbial ecology on what constitutes sufficient community characterization. Early cloning efforts involved very small libraries, and this expanded as sequencing became less expensive. We often use a 96-well plate for sequencing clones, which could yield 96 clone sequences or 48 sequences from longer inserts requiring sequencing from both directions. Nonetheless, with the emergence of affordable pyrosequencing and the high number of sequences available using this approach, cloning should be considered a shallow characterization of community diversity that is unlikely to retrieve community members that are low in abundance.

As with DGGE, PCR is the first step in clone library development. This requires the selection of a primer pair that provides the desired level of specificity for the target gene. Typically, the 16S rRNA gene is the target, and Table 3.1 includes information on some commonly used PCR primers for this target. However, other genes can be targeted to characterize the diversity of some functional trait of interest. Unlike with DGGE, cloning does not require the electrophoretic separation and distinction of PCR amplicons, which enables the use of other 16S rRNA gene-targeted primers that would not work well with DGGE. One such primer pair that is often preferred for clone library construction, because it provides nearly full-length 16S rRNA gene sequencing, is 8F (5′-AGAGTTTGATCCTGGCTCAG-3′) and 1492R (5′-GGYTACCTTGTTACGACTT-3′). Following PCR, it is common to purify the PCR products using a spin column or gel extraction kit to remove unincorporated primers and dNTPs, and nonspecific products if they are present (gel extraction kit only). There are many suppliers of these PCR product cleanup kits. Another strategy that eliminates the need for a cleanup kit involves using a 0.1–0.2% low melting point (LMP) agarose inlay within a standard agarose gel of whatever concentration achieves sufficient separation of the PCR products from any nonspecific products [23]. After electrophoretically separating the PCR products into the LMP agarose inlay, the desired band(s) can be directly pipetted from the low-concentration agarose and used in subsequent reactions.

The simplest way to ligate PCR products into a vector for clone library development is with the use of a TA cloning vector. These vectors have 3′ thymine overhangs on each end of the linearized plasmid to take advantage of the deoxynucleotidyl transferase activity of *Taq* DNA polymerase, which adds an extra adenine on the 3′ ends of the PCR products. This built-in complementarity with 3′ adenylated PCR products eliminates the need for restriction digestions to create compatible ends between the vector and the PCR products. There are numerous TA cloning vectors available. For shorter PCR products, we routinely use Life Technologies' TOPO cloning kit (Fig. 3.3), which has topoisomerase I on each end of the linearized plasmid to catalyze the ligation of PCR products in as little as 5 min at room temperature, as opposed to as long as overnight at 4 °C for other TA cloning vectors that require ligase addition. For the cloning of longer PCR products, such as nearly full-length 16S rRNA gene fragments, we have noticed declined ligation efficiency with the topoisomerase-based vector, yielding few correct transformants. In this case, we have observed improved success with the Promega pGEM®-T vector [24], which offers the same TA cloning, restriction, and selection features. This product also requires the addition of DNA ligase, but a rapid ligation buffer shortens incubation times to 1 h. There are also addition DNA ligase-based products that have an accelerated ligation strategy.

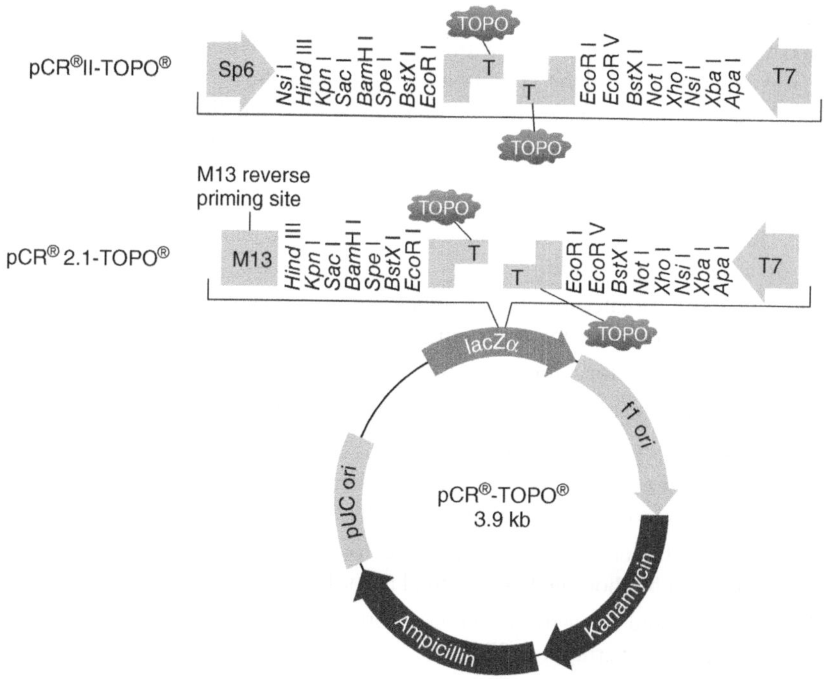

Figure 3.3 Map of Life Technologies' TOPO cloning vector. Reproduced by permission of Thermo Fisher Scientific Inc.

Regardless of the specific product being used, the manufacturer will provide simple instructions for ligation that can be followed directly. The vector can be stretched to twice as many reactions as specified by using half the volume, which still provides sufficient volume for the subsequent steps. A typical reaction mixture (for the TOPO vector) would include the following components:

- approximately 1-μl PCR products;
- 0.5-μl salt solution;
- 0.5-μl TOPO vector;
- balance sterile water to 3 μl.

Due to the very small volumes of reaction components, it is best to prepare a master mix for the total number of samples being ligated, and then dispense this mix into individual tubes. We also routinely perform the ligation for at least 30 min with the addition of salt solution, and not the minimum 5-min ligation, as recommended by the manufacturer for longer and mixed PCR products.

The next step in cloning is the transformation of competent *Escherichia coli* host cells. These can be prepared in-house through electroporation or chemical treatments (e.g., calcium chloride or rubidium chloride) or ordered from a vendor separately or as part of a cloning kit. The least expensive approach is to prepare the competent cells in-house, but this involves a trade-off in time. The expense of vendor-purchased cells can be reduced by using half the recommended volume per transformation, which works fine unless there are too few transformants recovered. (For example, we have had this experience using the TOPO-based vector and long PCR product inserts.) The manufacturer's transformation instructions can be followed directly. For the transformation of chemically competent One Shot® *E. coli*, this entails the following steps, for a half-volume transformation:

- Thaw on ice one vial of One Shot cells. Dispense half the volume (i.e., 25 μl) to a sterile 1.5-ml tube.
- Add 1 μl of ligated reaction product to the cells in one of the tubes and mix gently. Competent cells are fragile and can be damaged by vigorous mixing.
- Incubate on ice for 5 min.
- Heat-shock the cells for 30 s at 42 °C using a water bath.
- Immediately transfer the tubes to ice.
- Add 125-μl room temperature Super optimal broth (SOC) medium, which promotes the recovery of the competent cells.
- Cap the tubes and shake them at 37 °C for 1 h. (This time has worked well for us, but can be shortened or extended accordingly.)
- Spread 25 and 50 μl from each transformation on prewarmed selective plates, and incubate overnight at 37 °C.

We routinely use ampicillin selection of transformants and blue-white screening for transformants with an insert in the vector. We then confirm that the insert is the desired PCR product by doing colony PCR in which we prepare PCR tubes with primers that target the vector sites flanking the insert location (e.g., SP6 and T7, or M13 and M13 reverse), touch a white colony with a sterile toothpick, and swirl the toothpick end in the PCR tube to introduce cells from the colony. We also use this tooth pick to inoculate both an indexed plate of clones and an Lysogeny Broth (LB) tube for the subsequent step, which are incubated overnight at 37 °C. For those colonies that show the expected PCR product length (which should account for a slight increase in length due to the short fragments of the plasmid that are incorporated in the products), the plasmid is extracted from cells in the Lysogeny Broth (LB) tube and purified using a plasmid miniprep kit. These come in many formats, including spin columns, vacuum tubes, and 96-well plates.

The last step of clone library development involves sequencing and sequence analysis. There are numerous nucleic acid sequencing facilities that conduct sequencing reactions from cloned fragments, using standard primers that target common vector sequences (e.g., SP6 and T7, or M13 and M13 reverse) or user-supplied primers to target the insert fragment. These facilities generally provide instructions for sample preparation, which may include specifications for nucleic acid concentration and purity. A typical Sanger sequencing reaction can provide 700 bases or more from a high-quality template at the optimal concentration. If the insert is longer than this (e.g., when using primers that provide nearly full-length 16S rRNA gene amplicons), sequencing should be performed in both directions from the vector, and the two resultant sequences can be merged, provided the sequences are long enough to provide a region of overlap at the ends. When sequences are retrieved from the facility, they need to be inspected for quality control prior to comparative sequence analysis. There are many programs that can be used for this function. We typically use Molecular Evolutionary Genetics Analysis(MEGA, Fig. 3.4), which is a free analysis tool available for download at www.megasoft.net. Sequence editing always involves trimming the ends, as the initial 30–40 bases are never good quality and the 3′ end of the sequence will show diminished quality. Moreover, the 5′ end should be trimmed to remove any vector or primer sequence, if that is detected beyond the poor-quality 5′ sequence. In addition to trimming the ends, examine the quality of the remaining sequence data and make any edits that are clearly justifiable. Obvious errors in base calling are often due to a high background signal. Also, look closely at the sequence electropherogram for any crowded or separated bases, as this is often an indication of a problematic location. For fragments that were sequenced in both directions, the edited reverse sequence should be converted to its reverse complement, and then an alignment between the forward and reversed sequence pair will show whether they overlap and can be merged into one full-length sequence.

Following appropriate editing, each sequence can be phylogenetically characterized, and the entire clone library can be analyzed and often graphically depicted in a phylogenetic tree. MEGA can perform all of these functions, but there are many alternate analysis tools for this purpose. Phylogenetic identification involves the comparison of each retrieved sequence with a database of sequences. Two commonly used databases include GenBank, which is administered by the National

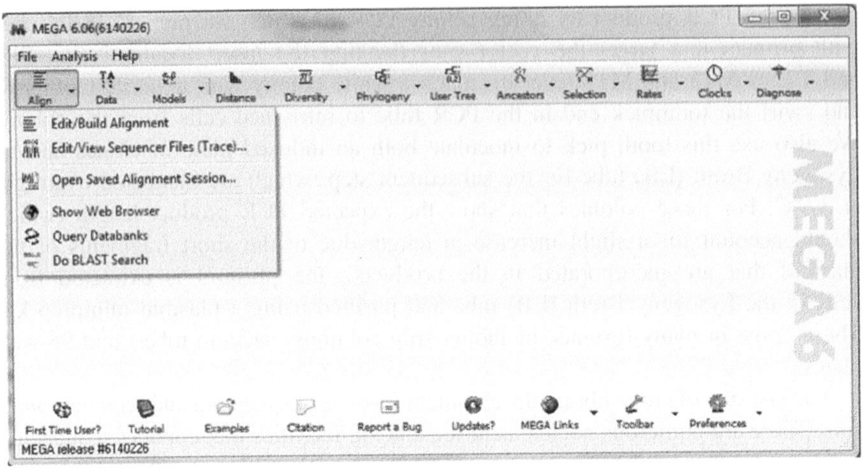

Figure 3.4 Screen capture of the MEGA program (version 5.2.1).

Center for Biotechnology Information (www.ncbi.nlm.nih.gov), and the Ribosomal Database Project (RDP), which is administered by the Center for Microbial Ecology at Michigan State University (http://rdp.cme.msu.edu/index.jsp) and specifically contains rRNA gene sequences. Both sites provide sequence analysis tools; the NCBI has a search algorithm called BLAST (Basic Local Alignment Search Tool), and the RDP has SeqMatch or Classifier. We generally restrict our reporting of phylogenetic inferences to sequences from isolates, which can be specified as a restriction in the RDP tools, unless the accession information for a cloned sequence appears particularly relevant (e.g., from an iron-reducing enrichment). To analyze the entire clone library, the sequences need to be aligned with each other and relevant reference strains. Again, this can be performed directly in MEGA, which has embedded the ClustalW and Muscle alignment tools, or many other packages. Phylogenetic trees should include bootstrapping analysis, which indicates the reliability of a tree topology. These bootstrap values should be reported at the nodes of tree branches, showing the percentage of times that particular topology was generated in the replicated random sampling of sequence fragments.

3.2.3 Pyrosequencing

Pyrosequencing is a technique that addresses the issue of shallow clone library coverage of community diversity. Depending on the degree of sample multiplexing, which is the generation of PCR products from each sample that are labeled with a unique nucleotide barcode and then the running of multiple samples together, this technique generates thousands to tens of thousands of sequences per sample. Pyrosequencing is beginning to make inroads in community characterization of BES systems [16, 25–28]. The initial introduction of this sequencing technology permitted only very short sequence reads that did not provide sufficient resolution in phylogenetic identification. This has been overcome with technology advancements, which now enable approximately 500 bases of quality sequence per read. Cost has also been a limiting factor, but given the sheer number of sequences generated and considering the time and materials required for multiple cloning steps that are not included in pyrosequencing, it is significantly less expensive on a per-sequence basis than cloning. Also, there are now a number of sequencing facilities that provide pyrosequencing on a per-sample basis and pool samples from multiple customers, which greatly reduces the cost per sample. Pyrosequencing offers a much deeper level of community coverage than cloning, allowing the recovery of sequences from rare populations in the community that would likely go undetected in a clone library or DGGE analysis. However, it is unknown whether those rare populations play a significant role in the community ecology, and it is interesting to note that pyrosequencing-derived rarefaction curves generally do not show improved asymptotic coverage of the community diversity.

Pyrosequencing is often performed on PCR products amplified from community 16S rRNA genes. As with all PCR-involving methods, there is the potential for bias in sequence representation associated with factors such as differential cell lysis and DNA recovery, as well as primer selection, and these should be acknowledged in the discussion of the resultant data. Sample DNA can be sent directly to facilities that perform pyrosequencing, often with the user-specified primers and depth of sequencing desired from each sample. These facilities can then perform PCR using the desired primers modified with adapters and barcodes, carry out the pyrosequencing, and provide the raw sequencing data, as well as often some level of bioinformatics analysis of the data. An explanation of the pyrosequencing methodology is beyond the scope of this review, but the reader is directed to existing reviews on this subject [29–31].

Given the abundance of sequences provided with this technique, the manual inspection and analysis of sequences are not feasible. Rather, the data are handled through sequence analysis pipelines that perform quality control functions, separate sequences into their respective samples from which they were derived on the basis of unique sample barcodes, phylogenetically identify sequences, and analyze the collection of sequences from each sample and between samples. We typically make use of the mothur software package (available at www.mothur.org), but there are other pipelines available for this purpose such as the RDP [32] (accessible at http://rdp.cme.msu.edu/index.jsp) and QIIME [33] (available at http://qiime.sourceforge.net/). Quality control involves eliminating low-quality sequences that are identified on the basis of considerations such as mismatches with the primers or barcodes, homopolymers of a certain length (because pyrosequencing error rates increase when quantifying a number of specific nucleotides in a run [34]), short length, and chimeric origin [35]. Similarly to the removal of primer and vector sequences in cloning and sequencing, the primer and barcode sequences must be trimmed from pyrosequencing data. The remaining high-quality sequences are then classified by comparison to some reference database of sequences, such as the SILVA reference database (http://www.arb-silva.de/). All of these functions can be programmed for batch processing, which allows the rapid processing of the thousands of sequences that result from this sequencing technology.

3.2.4 Fluorescent In Situ Hybridization

Fluorescent in situ hybridization (FISH) is a complementary technique that allows examination of specific subsets of a community, often selected on the basis of some metabolic function of interest (e.g., methanogens) or guided by the resulting community profile from one of the preceding community analysis techniques (e.g., confirming a *Geobacter*-dominated community). However, FISH by itself is not suitable for high-resolution community characterization. FISH has been used somewhat sparingly in BES studies [5, 36–38]. It is particularly useful for examining issues related to localization within an aggregate or biofilm, which might be relevant in BESs to investigate whether exoelectrogenic populations are stratified with respect to distance from the anode or uniformly distributed across the biofilm thickness. Moreover, FISH does not introduce the quantitative bias associated with PCR-based techniques, making it a good technique for confirming PCR-derived quantitation. FISH is not without its own biases, however. Even with thorough probe optimization to define the hybridization conditions that provide the intended probe specificity using pure cultures of target and nontarget cells, there is no way of being certain in a mixed culture that there are no false positive responses from populations that were not included in the optimization process. Conversely, as FISH (at least the conventional technique) targets rRNA, which has an intracellular abundance that is a function of growth rate, it is possible to have false negatives if a target population is not sufficiently active to yield above-background fluorescence. Given these concerns, the results derived from FISH should be reported as probe-positive cell numbers.

Just as the specificity of PCR-based techniques is in the selection of primers, FISH specificity is determined by the probe sequence. The literature has many probes designed and optimized for a range of taxonomic levels, from the EUB mix probe set designed to target all bacteria [39] to species-level probes. Some probes relevant for many BES studies are listed in Table 3.3. For the design of new probes, there are a number of tools available to assist in selecting candidate probe sequences, such as PRIMROSE [40]. However, there is no substitute for testing the probes with target and mismatch control cultures because even a perfectly matched probe sequence may yield dim probe-positive cells due to the variable rRNA accessibility within the ribosome complex [41, 42]. There are approaches to increase accessibility, such as the use of unlabeled helper probes [43], but positive-control verification is still necessary. The design and optimization of FISH probes is beyond the scope of this chapter.

The FISH protocol consists of sample fixation, probe hybridization, probe washing, and visualization. Fixation is typically performed with a 3% paraformaldehyde final concentration, which involves adding a 3:1 volume ratio of 4% paraformaldehyde to a suspended sample or immersing a biofilm sample (e.g., fibers from a graphite fiber brush anode) in a 3% paraformaldehyde solution. Alternatively, if the goal is to determine relative or absolute abundance of a target organism and not the intact biofilm structure, biofilms can be detached from the substratum and processed in suspension. The dissolution of paraformaldehyde during preparation of this fixative solution is enhanced at high pH, low salt, and elevated temperature

TABLE 3.3 Probes Commonly Used in FISH Analysis.

Probe	Sequence (5′–3′)	Specificity	References
EUB Mix[a]		Bacteria	[39]
EUB338-I	GCTGCCTCCCGTAGGAGT		
EUB338-II	GCAGCCACCCGTAGGTGT		
EUB338-III	GCTGCCACCCGTAGGTGT		
NON338	ACTCCTACGGGAGGCAGC	EUB338-I reverse complement, nonspecific binding	
GEO2	GAAGACAGGAGGCCCGAAA	*G. sulfurreducens*	[44]
HGEO2-1	GTCCCCCCCTTTTCCCGCAAGA	Helper probes for GEO2[b]	[44]
HGEO2-2	CTAATGGTACGCGGACTCATCC		
Arch915	GTGCTCCCCCGCCAATTCCT	Archaea	[45]

[a]Equimolar EUB338-I, EUB338-II, and EUB338-III.
[b]Unlabeled.

conditions. For a 50 ml preparation of 4% paraformaldehyde, we would use the following steps:

- Add 20 µl of 10 M NaOH and 2 g of paraformaldehyde to 33 ml of water.
- Heat at 60 °C under hood until paraformaldehyde dissolves.
- Add 16.5 ml of 3× phosphate buffered saline PBS, cool to 4 °C, and adjust pH to 7.2 with 1 N HCl.
- Filter through a 0.2-µm filter, keep refrigerated, and use within 48 h.

There are additional fixatives that have been used to permeabilize Gram-positive cells, such as ethanol or mutanolysin [46]. One can also find a large range of fixation times reported in the literature, but a 2-h fixation at 4 °C is typically used. Once the cells are fixed, the suspended cells should be washed in PBS and can then be used directly or stored at −20 °C in 1:1 PBS:ethanol for subsequent use.

The hybridization and washing steps can be performed in suspension, by immersion of biofilms in the hybridization and wash buffers, or by detachment of the biofilm and subsequent processing of cells in suspension. The following description is for the processing of samples on Teflon-coated microscope slides with 7–10-mm wells, but the same principles would apply for suspended or intact biofilm samples. The Teflon provides a hydrophobic barrier that prevents cross contamination of samples and reagents across wells. Small volumes (3–10 µl) of fixed cells can be spotted on the wells, air-dried, and then dehydrated by immersing slides in a 50%, 80%, and 96% ethanol gradient for 2 min each. If we are using FISH to enumerate cell density in a sample, and not just relative cell density, we avoid spotting the sample on the slides and allowing them to air-dry because this creates a drying front in the wells that leads to a nonuniform distribution of cells in the well. A remedy to this concern involves filtering the cells and transferring the uniformly distributed cells from the filter to a gelatin-coated slide [47]. Routine controls that we run during each FISH analysis include a no-probe well for autofluorescence, a positive-control well that has fixed cells of a target pure culture, a positive-control well that uses the EUB mix probe set, and a negative-control well that uses the nonEUB338 probe (the reverse complement of EUB338) to characterize nonspecific binding fluorescence.

Given that FISH is performed at a fixed temperature, the critical features that control hybridization and washing stringency are the formamide concentration in the hybridization buffer and the salt concentration in the wash buffer. Hybridization buffers are composed of variable %formamide, 0.9 M NaCl, 20 mM Tris–HCl (pH 7.2), and 0.01% sodium dodecyl sulfate (SDS). Wash buffer composition is variable molarity NaCl, 20 mM Tris–HCl (pH 8.0), and 0.01% SDS. The formamide and salt concentrations can be found in the literature for probes that have already been optimized. For the design of new probes, this requires experiments with variable formamide concentration buffers and the determination of target cell and nontarget cell fluorescence intensity. The wash buffer salt concentration for a given probe is designed to provide equivalent stringency as the corresponding hybridization buffer, where stringency is determined by melting temperature. This can be estimated with the following equation:

$$T_m\,(^\circ\mathrm{C}) = 16.6\log[\mathrm{NaCl}] + 0.41\,(\%\mathrm{GC}) + 81.5 - 0.61\,(\%\mathrm{formamide}) - \frac{500}{\#\mathrm{bases}}$$

Probes with equivalent optimum formamide concentrations can be hybridized simultaneously. For probes with different buffer stringencies, the probe with greater stringency conditions (i.e., higher formamide concentration in the hybridization buffer, lower salt concentration in the wash buffer) should be hybridized and washed first, followed by the probe with less stringent conditions.

Hybridization and washing can be performed in a 50-ml polypropylene centrifuge tube for each slide. A moist hybridization chamber can be established in the tube by adding a paper towel saturated in 0.9 M NaCl and preincubating at 46 °C for at least 15 min. While this is equilibrating, the hybridization buffers and probes can be added to each well of each slide. When the slides are prepared, they can each be inserted horizontally in a preincubated tube and incubated at 46 °C for 1.5 h. During hybridization, the corresponding wash buffers can be prepared for each hybridization condition (~50 ml for each slide) and put in the incubator to prewarm them. After hybridization is complete, the slides are rinsed with 2 ml of prewarmed wash buffer, and then each slide is immersed in the corresponding buffer and incubated at 48 °C for 20 min. Slides are then rinsed with 2-ml cold water and air-dried. It is common to counterstain the cells with DAPI, to allow total cell counts. This can be done directly in the slide wells for 5 min. Then the unbound DAPI is washed, and the slides are mounted with an antifadent that scavenges oxygen to reduce photobleaching of fluorochromes. There are several commercial products available, some including the DAPI counterstain, or a DABCO solution can be prepared with 25 mg DABCO/ml of PBS:glycerol at 1:9 (v/v) and pH adjusted to 8.6. After adding the antifadent, a large coverslip is sealed to the slide by painting nail polish along each edge. Note that all of these steps should be performed in limited light conditions, as the fluorochrome are light sensitive.

The visualization of FISH preparations can be done with wide-field epifluorescence microscopy for optically thin samples ($\lesssim$10 μm) or using confocal microscopy to optically section thicker samples such as intact biofilms. It should be noted though that confocal microscopy has sample thickness constraints as well. These include shadowing that can be observed in dense biofilm samples as thin as 30 μm due to the absorption of excitation and emission light from biofilm material overlying the focal plane. There are also focal depth constraints for the objective lenses. For thicker samples, an alternative approach is to perform cryosectioning on fixed samples [48], and then perform the hybridization and washing steps on cryosections. We use the water-soluble Optimum Cutting Temperature (OCT) medium to cryoembed biofilm samples, freeze the medium on dry ice, and detach the embedded biofilm from the substratum with a sterile razor blade. The detached biofilm can then be fixed to the cryostat chuck with a drop of OCT, and sections just under 10 μm can be reliably prepared. The sections can be placed on a well slide, allowed to air-dry, and then the hybridization and wash steps can be performed as described previously.

3.3 ANALYSIS OF BIOFILM BIOMASS

In addition to community composition, it is often necessary to measure total biomass concentration in the biofilm. Without total biomass measurements, changes in the relative distributions of populations that are observed from community analyses may not correspond to changes in total abundance of these populations. For example, an increase in the percent of *Geobacter* spp. in an anode community may not correspond to an increase in total *Geobacter* spp. abundance. It has been demonstrated that biofilm architecture develops over the first several weeks of operation, with a corresponding improvement in system performance [10]. Therefore, total biomass is necessary to properly characterize these transient acclimation periods. Biofilm architecture has also been shown to depend in part on external resistance and anode potential [4], and a direct correlation between biomass concentration and current production in *Geobacter sulfurreducens* biofilms has been demonstrated [49]. These observations imply that biomass-normalized specific rates of substrate utilization and current generation are essential for a fair comparison of different communities or pure cultures, where unmeasured differences in biomass density might give different apparent rates.

Several biomass measurement techniques have been used in BES research, including total protein, dry weight, quantitative real-time PCR [3, 50], and total and viable cell counts following nucleic acid staining [10]. It is important to remember when using microscopy-based techniques that porous anode materials can harbor cells within the anode matrix that will not be visualized, but that might be accounted for using extracted assays such as protein or qPCR, or by intact dry weight. Another method of estimating total biomass changes is the measurement of biofilm thickness. As mentioned previously for the visualization of FISH preparations, this could include nonspecific staining and confocal microscopy for moderate biofilm thickness, or it may require cryosectioning of the biofilm if the biofilm thickness exceeds the constraints of confocal microscopy.

ACKNOWLEDGMENTS

Funding for this chapter was provided by the Army Research Office and the US Army Engineer Research and Development Center under Award W911NF-11-10148.

REFERENCES

1. Kiely PD, Rader G, Regan JM, Logan BE. Long-term cathode performance and the microbial communities that develop in microbial fuel cells fed different fermentation endproducts. Bioresour Technol 2011;102(1):361–366.
2. Larrosa-Guerrero A, Scott K, Katuri KP, Godinez C, Head IM, Curtis T. Open circuit versus closed circuit enrichment of anodic biofilms in MFC: effect on performance and anodic communities. Appl Microbiol Biotechnol 2010;87(5):1699–1713.
3. Jung S, Regan JM. Influence of external resistance on electrogenesis, methanogenesis, and anode prokaryotic communities in microbial fuel cells. Appl Environ Microbiol 2011;77(2):564–571.
4. Ren ZY, Yan H, Wang W, Mench MM, Regan JM. Characterization of microbial fuel cells at microbially and electrochemically meaningful time scales. Environ Sci Technol 2011;45(6):2435–2441.
5. Kiely PD, Cusick R, Call DF, Selembo PA, Regan JM, Logan BE. Anode microbial communities produced by changing from microbial fuel cell to microbial electrolysis cell operation using two different wastewaters. Bioresour Technol 2011;102(1):388–394.
6. Crosby LD, Criddle CS. Understanding bias in microbial community analysis techniques due to rrn operon copy number heterogeneity. Biotechniques 2003;34(4):790–794.
7. Fredriksson NJ, Hermansson M, Wilen BM. The choice of PCR primers has great impact on assessments of bacterial community diversity and dynamics in a wastewater treatment plant. PLoS One 2013;8(10):e76431.
8. Vishnivetskaya TA, Layton AC, Lau MC, Chauhan A, Cheng KR, Meyers AJ, Murphy JR, Rogers AW, Saarunya GS, Williams DE, Pfiffner SM, Biggerstaff JP, Stackhouse BT, Phelps TJ, Whyte L, Sayler GS, Onstott TC. Commercial DNA extraction kits impact observed microbial community composition in permafrost samples. FEMS Microbiol Ecol 2014;87:217–230.
9. Jung SH, Regan JM. Comparison of anode bacterial communities and performance in microbial fuel cells with different electron donors. Appl Microbiol Biotechnol 2007;77(2):393–402.
10. Ramasamy RP, Cloud-Owen SR, Yan H, Mench MM, Regan JM. Time-course correlation of biofilm properties and electrochemical performance in single-chamber microbial fuel cells. Bioresour Technol 2011;102(1):416–421.
11. Xing DF, Cheng S, Regan JM, Logan BE. Change in microbial communities in acetate- and glucose-fed microbial fuel cells in the presence of light. Biosens Bioelectron 2009;25(1):105–111.

12. Díez B, Pedrós-Alió C, Marsh TL, Massana R. Application of denaturing gradient gel electrophoresis (DGGE) to study the diversity of marine picoeukaryotic assemblages and comparison of DGGE with other molecular techniques. Appl Environ Microbiol 2001;67(7):2942–2951.
13. Muyzer G, Smalla K. Application of denaturing gradient gel electrophoresis (DGGE) and temperature gradient gel electrophoresis (TGGE) in microbial ecology. Antonie van Leeuwenhoek 1998;73(1):127–141.
14. Wu Y, Hayes VM, Osinga J, Mulder IM, Looman MW, Buys CH, Hofstra RM. Improvement of fragment and primer selection for mutation detection by denaturing gradient gel electrophoresis. Nucleic Acids Res 1998;26(23):5432–5440.
15. VandePeer Y, Chapelle S, DeWachter R. A quantitative map of nucleotide substitution rates in bacterial rRNA. Nucleic Acids Res 1996;24(17):3381–3391.
16. Yates MD, Kiely PD, Call DF, Rismani-Yazdi H, Bibby K, Peccia J, Regan JM, Logan BE. Convergent development of anodic bacterial communities in microbial fuel cells. ISME J 2012;6(11):2002–2013.
17. Watanabe K, Kodama Y, Harayama S. Design and evaluation of PCR primers to amplify bacterial 16S ribosomal DNA fragments used for community fingerprinting. J Microbiol Methods 2001;44(3):253–262.
18. Muyzer G, Brinkhoff T, Nübel U, Santegoeds C, Schäfer H, Wawer C. Denaturing gradient gel electrophoresis (DGGE) in microbial ecology. In: Kowalchuk GA, de Bruijn FJ, Head IM, Akkermans AD, van Elsas JD, editors. Molecular Microbial Ecology Manual. The Netherlands: Kluwer; 2004. p 743–770.
19. Ren NQ, Xing D, Rittmann BE, Zhao L, Xie T, Zhao X. Microbial community structure of ethanol type fermentation in bio-hydrogen production. Environ Microbiol 2007;9(5):1112–1125.
20. Radojkovic D, Kusic J. Silver staining of denaturing gradient gel electrophoresis gels. Clin Chem 2000;46(6):883–884.
21. Hughes JB, Hellmann JJ. The application of rarefaction techniques to molecular inventories of microbial diversity. Methods Enzymol 2005;397:292–308.
22. Kemp PF, Aller JY. Estimating prokaryotic diversity: when are 16S rDNA libraries large enough? Limnol OceanogrMethods 2004;2:114–125.
23. Hao M, Difazio S. An efficient method for purification of PCR products for sequencing. BioTechniques 2008;44(7):921–923.
24. Litterer L. Comparing Cloning Efficiency of the pGEM®-T and pGEM®-T Easy Vectors to the TOPO TA Cloning® Vectors. Promega Corporation; 2009.
25. Lee TK, Van Doan T, Yoo K, Choi S, Kim C, Park J. Discovery of commonly existing anode biofilm microbes in two different wastewater treatment MFCs using FLX Titanium pyrosequencing. Applied Microbiology and Biotechnology 2010;87(6):2335–2343.
26. Lu L, Xing D, Ren N. Pyrosequencing reveals highly diverse microbial communities in microbial electrolysis cells involved in enhanced H2 production from waste activated sludge. Water Res 2012;46:2425–2434.
27. Zhang GD, Zhao Q, Jiao Y, Wang K, Lee DJ, Ren N. Biocathode microbial fuel cell for efficient electricity recovery from dairy manure. Biosens Bioelectron 2012;31(1):537–543.
28. Vargas IT, Albert IU, Regan JM. Physical distribution of bacterial communities on volumetric and planar anodes in single-chamber air-cathode microbial fuel cells. Biotechnol Bioeng 2013;110(11):3059–3062.

29. Metzker ML. Applications of next-generation sequencing. Sequencing technologies – the next generation. Nat Rev Genet 2010;11(1):31–46.
30. Metzker ML. Emerging technologies in DNA sequencing. Genome Res 2005;15(12):1767–1776.
31. Mardis ER. Next-generation DNA sequencing methods. Annu Rev Genomics Hum Genet 2008;9:387–402.
32. Cole JR, Wang Q, Fish JA, Chai B, McGarrell DM, Sun Y, Brown CT, Porras-Alfaro A, Kuske CR, Tiedje JM. Ribosomal Database Project: data and tools for high throughput rRNA analysis. Nucleic Acids Res 2014;42(D1):D633–D642.
33. Caporaso JG, Kuczynski J, Stombaugh J, Bittinger K, Bushman FD, Costello EK, Fierer N, Peña AG, Goodrich JK, Gordon JI, Huttley GA, Kelley ST, Knights D, Koenig JE, Ley RE, Lozupone CA, McDonald D, Muegge BD, Pirrung M, Reeder J, Sevinsky JR, Turnbaugh PJ, Walters WA, Widmann J, Yatsunenko T, Zaneveld J, Knight R. QIIME allows analysis of high-throughput community sequencing data. Nate Methods 2010;7(5):335–336.
34. Balzer S, Malde K, Jonassen I. Systematic exploration of error sources in pyrosequencing flowgram data. Bioinformatics 2011;27(13):I304–I309.
35. Quince C, Lanzén A, Curtis TP, Davenport RJ, Hall N, Head IM, Read LF, Sloan WT. Accurate determination of microbial diversity from 454 pyrosequencing data. Nat Methods 2009;6(9):639–641.
36. Ren Z, Steinberg LM, Regan JM. Electricity production and microbial biofilm characterization in cellulose-fed microbial fuel cells. Water Sci Technol 2008;58(3):617–622.
37. He Z, Minteer SD, Angenent LT. Electricity generation from artificial wastewater using an upflow microbial fuel cell. Environ Sci Technol 2005;39(14):5262–5267.
38. Chung K, Okabe S. Continuous power generation and microbial community structure of the anode biofilms in a three-stage microbial fuel cell system. Appl Microbiol Biotechnol 2009;83(5):965–977.
39. Daims H, Brühl A, Amann R, Schleifer KH, Wagner M. The domain-specific probe EUB338 is insufficient for the detection of all Bacteria: Development and evaluation of a more comprehensive probe set. Syst Appl Microbiol 1999;22(3):434–444.
40. Ashelford KE, Weightman AJ, Fry JC. PRIMROSE: a computer program for generating and estimating the phylogenetic range of 16S rRNA oligonucleotide probes and primers in conjunction with the RDP-II database. Nucleic Acids Res 2002;30(15):3481–3489.
41. Fuchs BM, Syutsubo K, Ludwig W, Amann R. In situ accessibility of *Escherichia coli* 23S rRNA to fluorescently labeled oligonucleotide probes. Appl Environ Microbiol 2001;67(2):961–968.
42. Fuchs BM, Wallner G, Beisker W, Schwippl I, Ludwig W, Amann R. Flow cytometric analysis of the in situ accessibility of *Escherichia coli* 16S rRNA for fluorescently labeled oligonucleotide probes. Appl Environ Microbiol 1998;64(12):4973–4982.
43. Fuchs BM, Glöckner FO, Wulf J, Amann R. Unlabeled helper oligonucleotides increase the in situ accessibility to 16S rRNA of fluorescently labeled oligonucleotide probes. Appl Environ Microbiol 2000;66(8):3603–3607.
44. Richter H, Lanthier M, Nevin KP, Lovley DR. Lack of electricity production by *Pelobacter carbinolicus* indicates that the capacity for Fe(III) oxide reduction does not necessarily confer electron transfer ability to fuel cell anodes. Appl Environ Microbiol 2007;73(16):5347–5353.

45. Stahl DA, Amann R. Development and application of nucleic acid probes. In: Stackebrandt E, Goodfellow M, editors. Nucleic Acid Techniques in Bacterial Systematics. West Sussex, England: John Wiley & Sons; 1991. p 205–248.
46. de los Reyes FL, Raskin L. Role of filamentous microorganisms in activated sludge foaming: relationship of mycolata levels to foaming initiation and stability. Water Res 2002;36(2):445–459.
47. Zilles JL, Peccia J, Kim MW, Hung CH, Noguera DR. Involvement of Rhodocyclus-related organisms in phosphorus removal in full-scale wastewater treatment plants. Appl Environ Microbiol 2002;68(6):2763–2769.
48. Yu FP, Callis GM, Stewart PS, Griebe T, McFeters GA. Cryosectioning of biofilms for microscopic examination. Biofouling 1994;8:85–91.
49. Reguera G, Nevin KP, Nicoll JS, Covalla SF, Woodard TL, Lovley DR. Biofilm and nanowire production leads to increased current in *Geobacter sulfurreducens* fuel cells. Appl Environ Microbiol 2006;72:7345–7348.
50. Ren Z, Ward TE, Logan BE, Regan JM. Characterization of the cellulolytic and hydrogen-producing activities of six mesophilic clostridium species. J Appl Microbiol 2007;103(6):2258–2266.

4

CHARACTERIZATION OF ELECTRODE-ASSOCIATED BIOMASS AND MICROBIAL COMMUNITIES

ORIANNA BRETSCHGER, CRYSTAL SNOWDEN, AND LISA MCDONALD

Microbial and Environmental Genomics Group, J. Craig Venter Institute, San Diego, CA, USA

SHINO SUZUKI, AND SHUN'ICHI ISHII

Microbial and Environmental Genomics Group, J. Craig Venter Instituteand the Japan Agency for Marine-Earth Science and Technology (JAMSTEC), Nangoku-shi, Kochi, Japan

4.1 INTRODUCTION

The successful application of microbial fuel cells (MFCs) and bioelectrochemical systems (BESs) requires an understanding, and ultimately the optimization, of microbial activities associated with the bioelectrocatalytic conversion of chemical and electrical inputs. Researchers must always consider that MFC/BES reactors utilize *living* microorganisms to drive catalytic activity, and these microbes will respond to system changes in different ways than abiotic catalysts.

When using microbial communities in MFCs/BESs, it is critical to gain an understanding of the following:

1. The quantity and distribution of the microbes contributing to the reactions (community biomass quantification and morphology).
2. The types and relative abundance of microbes within the community that may be contributing to key reactions (community taxonomy).
3. How the reactions may be impacted or regulated by system operational changes (community function).

Biofilms in Bioelectrochemical Systems: From Laboratory Practice to Data Interpretation, First Edition.
Edited by Haluk Beyenal and Jerome Babauta.

These fundamental parameters are also critical for normalizing and comparing MFC/BES performance results. For example, temporal improvements in bioelectrochemical performance observed for a given engineered system may be a result of different biological parameters including increased total biomass per unit electrode area, increased per biomass electron transfer rates, changing biofilm morphology, biofilm age, and/or taxonomic composition. Therefore, a "systems-level" understanding of bioelectrocatalytic performance is necessary for accurately reporting results, identifying rate-limiting steps, and optimizing applied BESs (Fig. 4.1).

4.1.1 Characterizing Microbial Community Biomass

Biomass associated with BESs must be quantified in order to normalize performance results to *active* surface area and *active* reactor volume. Researchers often report electrochemical performance normalized to the total geometric surface area of the electrode, or total volume of the reactor. However, it is unknown what fraction of the total surface area (or reactor volume) is contributing to the observed electrochemical activity unless electrochemically active surface area, biofilm coverage, and planktonic biomass are quantified within the reactor. Furthermore, quantifying the total biomass coverage and contribution is critical when comparing the electrochemical performance of different microbial strains, microbial communities, or a microbial community versus a pure culture. For example, a reported comparison of the electrochemical performance of *Geobacter sulfurreducens* and a mixed microbial community indicated that the single strain of *G. sulfurreducens* produced higher power than the community [2]. However, these researchers did not quantify biomass in either reactor type. A second report comparing the performance of a mixed community and *G. sulfurreducens* in the same reactor style, but with the added data of total protein content, showed that when results were normalized by total biomass, the mixed community achieved higher electron transfer rates than *G. sulfurreducens* [3]. This example highlights the importance of normalizing electrochemical data by biomass.

Quantifying biomass is also important for planning subsequent experiments such as next-generation sequencing for characterizing community composition.

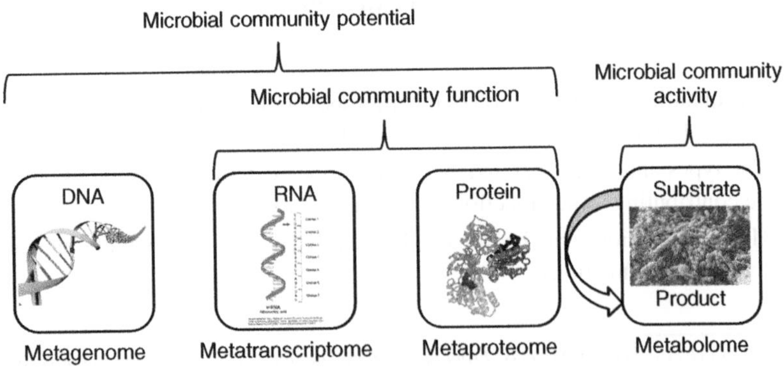

Figure 4.1 Systems-level approach for characterizing MFC/BES microbial communities. Adapted from Maron et al. [1].

Library construction for next-generation sequencing often requires a relatively high concentration (usually hundreds of nanograms) of high-quality DNA or RNA for postprocessing. Quantifying total biomass associated with BES samples being considered for sequencing will indicate if next-generation sequencing technologies can be used.

Several methods are available for estimating biomass in a BES system including quantifying biological components such as total protein, total organic carbon, or phospholipid content per unit area (or unit volume) or measuring cell numbers via direct cell counts per unit area using a microscope. Each method is utilized for estimating specific activity (e.g., current production) per a given biological unit; therefore, researchers must select the strategy that is most appropriate for answering their given scientific question.

Direct cell counting via a microscope can be used to estimate per cell activity (e.g., current per cell) and is a simple and accurate method for low culture densities and monolayers of cells attached to electrode surfaces; however, these studies require an optically accessible MFC system [4]. When complex biofilms are established at electrode surfaces (e.g., multiple cell layers and heterogeneous surface coverage), computational tools are required to estimate biovolumes and total cell densities of complex biofilms, which also have some associated errors.

Quantifying total protein per surface area is also an adequate indicator of biomass because protein is usually the most abundant component of biomass. In addition, protein extraction is simple, and there are several commercial kits to choose from.

It is important to note that the numbers yielded from quantifying total protein correlate to total biomass, but not necessarily to total cell numbers within a given biofilm. Cellular protein is correlated with cell number and is a relatively stable quantity; however, extracellular protein is associated with cell activity and can be variable. The two types of protein cannot be distinguished from total protein extraction kits; therefore, BES activities that are normalized by amount of protein may underestimate per cell activities. A protocol and recommendations for optimization of protein extraction from BES electrode samples are included in the subsequent sections.

4.1.2 Characterizing Microbial Community Taxonomy

Culture-independent methods for characterizing microbial community composition have rapidly advanced and enable a better understanding of microbial diversity. Several methods targeted to small subunit rRNA genes have been developed to characterize microbial diversity in different environmental and clinical samples. The small subunit (16S or 18S) rRNA gene is highly conserved, and all life forms that we have identified so far possess the gene in their genome. In addition, the occurrence of nucleotide substitutions in this gene is highly correlated with the microbial evolutionary rate. Therefore, the gene has been employed for environmental microbial community analyses for more than 15 years [5]. Several 16S rRNA-based methods have developed such as denaturing gradient gel electrophoresis (DGGE) [6], terminal-restriction fragment length polymorphism (T-RFLP) [7], automated ribosomal intergenic spacer analysis (ARISA) [8], clone library [9], and 16S rRNA-tagged pyrosequencing [10].

Molecular fingerprinting methods including DGGE, ARISA, and T-RFLP have proven useful especially to acquire community transitions with class- or genus-level resolutions. Clone library methods are able to determine nearly the full length of 16S rRNA sequences (~1500 bp) and generally provide more accurate taxonomy to each phylotype in the community with higher resolution (genus- or strain-level resolution). Meanwhile, 16S-tagged pyrosequencing can yield significantly higher numbers of sequence reads; thus, it can reveal rare populations in the community. However, as the length of sequence generated from next-generation sequencer (NGS) is short due to the instrument and enzyme limitations, the taxonomy of each phylotype cannot be assessed as accurately as that from the clone library approach (see the following section).

Although the 16S rRNA gene-based approaches are convenient and commonly used for characterizing microbial diversity in mixed communities, there are challenges associated with various biases. For example, the relative abundance of microbial taxa within a community cannot be accurately estimated based solely on the frequency of 16S rRNA genes because copy numbers of this gene vary between strains (1–15 copies). Therefore, the results from any 16S rRNA gene sequencing approach may potentially lead an overestimation or underestimation of the portion of a given species.

Furthermore, additional biases can be introduced by primer selection, polymerase chain reaction (PCR) amplification, and/or the number of 16S rRNA library amplicons or isolated gel bands (in the case of DGGE and T-RFLP) that are selected for sequencing. The issues caused from the biases must be carefully considered for characterizing the community using 16S rRNA gene-based approaches. Better estimations of community diversity and species abundance can be achieved by analyzing the coverage of single-copy genes recovered from metagenomic sequences of mixed microbial communities [11].

Metagenomic sequencing can be executed using conventional Sanger sequencing technology with BAC or shotgun libraries [12]. However, this approach has now been superseded by the advent of NGS technologies that are faster, more cost-effective, and can produce up to 600 GB of sequence reads in a single sequencing run. To date, the most-used NGS technologies include Roche 454, Life Technologies Ion Torrent and SOLiD, Illumina GAIIx, HiSeq and MiSeq, and PacBio RSII. There are several reviews available comparing the sequence quality of these technologies and their associated laboratory and analytical challenges [12b, 13]. However, the researcher should choose a sequencing technology, and associated library construction methods, on the basis of the type of analyses that are desired.

NGS-generated sequence data are bioinformatically assembled into long contigs with assembling software (e.g., CLC Genomics Workbench, Velvet [14], and Meta-Velvet [15]). Gene finding and/or open reading frame (ORF) calling software (e.g., MetaGene [16] and JCVI prokaryotic metagenomic pipeline [17]) are used to identify ORFs that encode putative peptides. The putative peptides are then analyzed against several different databases to determine taxonomy and then functional annotation [18]. Once single-copy genes are identified from this process, the coverage of single-copy housekeeping genes (e.g., *gyrB*, *recA*, and ribosomal proteins) from each metagenomic dataset is correlated with the portion of a given species [18, 19]. A disadvantage of this approach for taxonomic assignment is that databases for single-copy genes are not as extensive as those for the 16S rRNA gene.

The Roche 454 platform offers the capability to generate longer sequence reads (~450 bp), which are helpful when assembling data into long contigs, scaffolds, and assigning ORFs; however, these technologies do not generate as many reads so the sequence coverage of the community may be low, leading to different bioinformatic challenges. In addition, sequence errors in AT-rich and/or homopolymer regions are common to this technology, requiring bioinformatic corrections during the assembly process to avoid the overestimation of species diversity in a given community.

The Illumina technologies generate shorter sequence reads (100–300 bp) relative to 454, but many more reads can be acquired so genomic coverage can typically be higher for a given sample, and bioinformatic capabilities are rapidly improving to enable the assembly and analysis of shorter sequence reads. Additional information relative to sequencing technologies and their associated advantages or disadvantages can be found in the literature [10b, 12b, 13, 20].

4.1.3 Characterizing Microbial Community Function

Metagenomic analyses and 16S rRNA-based taxonomic profiles will yield useful information about the total genetic potential contained within a community (metagenomic analyses) and the estimated community function based on taxonomic assignments (16S and metagenomic analyses). However, these data sets only inform about the types of functions that are possible and do not give any information about what functions are actually active within a given microbial community. Therefore, additional methods relative to measuring community gene transcription (metatranscriptomics), what proteins the community ultimately transcribed (metaproteomics), and what the community metabolically produced (metabolomics) are necessary for a complete understanding of microbiological activity under a given condition.

The tools and technologies available for characterizing microbial community biomass, taxonomy, morphology, and function have rapidly advanced over the last few decades. These types of functional analyses are becoming more readily available and cost-effective; however, to date, there is no "standard" method for acquiring data sets that relate microbial taxonomy to microbial function within diverse communities. However, MFC/BES researchers can now take advantage of these powerful methods for the purpose of understanding and optimizing bioelectrochemical reactions.

Here, we offer protocols for BES-associated community analyses on the basis of 16S rRNA gene sequencing via Sanger Technology and some approaches for bioinformatic analyses.

4.2 PROTOCOLS

An exhaustive compilation of protocols and methods available for each type of microbial community characterization is beyond the scope of this chapter. However, subsequent sections provide protocols for quantifying protein from electrode-associated biofilms, suggestions for DNA extraction from electrode-associated biofilms, 16S rRNA library construction, and guidance relative to metagenomic and metatranscriptomic sequencing.

4.2.1 Biomass Quantification Using a Protein-Based Method

The protocol listed as follows is specific to recovering and quantifying protein from electrode-associated biofilms and is based on the bicinchoninic acid (BCA) assay, which has been found to be effective for use in quantifying protein concentrations from electrode-associated biofilm samples [3, 18, 21]. Figure 4.2 shows the basic scheme for determining protein concentration. Amendments to this protocol may be required for other types of microbial samples, and alternative assays may be selected on the basis of sample composition. Other colorimetric assays including the Bradford assay, Lowry assay, and Biuret assay may also be effective for measuring protein concentrations in a liquid sample. Additional guidance for selecting appropriate assays and determining protein concentrations by other colorimetric methods can be found in Olson and Markwell [22].

Note – Always run this protocol with sample replicates when possible, and regularly determine standard curves to check for accuracy.

Required materials:

- Disposable gloves.
- Eye guards.
- Laboratory coat.
- Water bath.
- −80 °C freezer.
- Spectrophotometer.
- Vortex.
- Dry bath incubator with heat block for microcentrifuge tubes.
- Sterile 2-ml microcentrifuge tubes.
- Sterile 15-ml conical tubes.
- Stopwatch or other timer.
- 1000-μl pipette and associated pipette tips (barrier tips recommended).
- 70% ethanol.
- BCA assay kit (e.g., Thermo Scientific Pierce BCA Protein Assay Kit).
- Sterile 0.2 N NaOH solution.
- Sterile scissors.

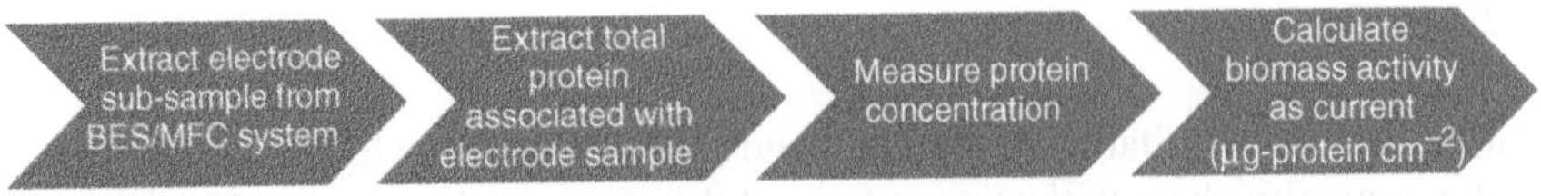

Figure 4.2 Scheme for determining protein concentration.

- Sterile deionized (DI) water (18 MΩ cm).
- Anaerobic chamber or portable anaerobic glove bag.

Collecting MFC/BES electrode samples:

Note – electrode subsamples should be collected under the same conditions as the MFC/BES is operated (e.g., anaerobic or microanaerobic with N_2, CO_2, H_2, or mixed gas headspace). Maintaining relatively the same conditions around the biofilm during sample extraction will minimize changes induced to the biofilm during sampling.

Porous electrodes including cloth, felt, or paper electrodes:

1. Sterilize scissors or razor blade using 70% ethanol. Take care to allow excess ethanol to evaporate prior to use.
2. If electrode is carbon paper, graphite cloth, or other woven fabric, cut electrode subsample as needed. Try to cut a 1 cm × 1 cm square to keep surface area estimations simple.
3. Freeze electrode subsample at −80 °C until ready to process.

Solid electrodes:

1. Use the entire electrode if possible, or remove a known area and follow the following protocols.
2. Alternatively, use a sterile razor blade to scrape a given area of the surface. Be sure to scrape the surface several times and carefully transfer the biomass on the razor's surface into a sterile 2-ml centrifuge tube filled with sterile buffer.
3. Rinse the blade several times with buffer solution to ensure that all biomass was transferred to the tube.
4. Pellet the biomass by centrifuging the 2-ml centrifuge tube for 2 min at 10,000 rpm, preferably at 4 °C.
5. Remove the supernatant and freeze pellet.

Note – The protein extraction protocol takes several hours; therefore, it is most time efficient to process several samples at once.

Protein extraction from a porous electrode sample:

1. Place sample into sterile 2-ml centrifuge tube and aliquot 1 ml of 0.2 N NaOH into tube. Vortex (maximum setting) for 10 s every 15 min for one full hour (four times in the hour).
2. Freeze/thaw the sample (in NaOH solution). Freeze/thaw should be done twice.
   a. −80 °C for 30 min.
   b. 90 °C for 10 min.
   c. Repeat.
3. Collect supernatant and transfer to a sterile 15-ml tube.
4. Repeat steps 1–3 using 1 ml of 0.2 N NaOH.
5. Repeat steps 1–3 using 1 ml of sterile DI water.

BCA assay using commercial kit:

Note – BCA kits are available from Sigma-Aldrich©, Pierce®, abcam®, and Novagen®, among others. This protocol is based on the Pierce BCA Protein Assay Kit available through Thermo Scientific©. We do not endorse this company. This protocol is solely meant to be a starting point for researchers.

1. Label microcentrifuge tubes.
2. Mix reagent parts A and B according to manufacturer's instructions.
3. Add 0.1 ml of sample to 2 ml of reagent mix and mix well (vortex for 5 s).
4. Incubate in a 60 °C water bath for 30 min.
5. Aliquot sample into cuvette and measure absorbance at 562 nm (use sterile DI as blank).
6. Calculate protein concentration on the basis of a standard curve using diluted bovine serum albumin (BSA) standards (usually provided in kit).

Note – The protein selected for determining standard curves may be different based on the type of sample being processed. Choose a protein standard that most closely matches your estimated sample composition.

Estimate current densities as a function of biomass:

1. Accurately measure electrode subsample surface areas and determine the amount of protein for that unit surface area (μg-protein cm^{-2}).
2. Calculate current production at the time immediately before electrode subsamples were extracted from the reactor, and report current densities as mA μg-protein^{-1} cm^{-2} or as μmol e^- μg-protein^{-1} cm^{-2}.

Determining protein concentration changes as a function of reactor operating time will yield more accurate information about system performance changes. See Figure 4.3 as an example of how to plot μg-protein^{-1} cm^{-2} and μmol e^- μg-protein^{-1} cm^{-2}.

4.2.2 Microbial Community Taxonomic Characterization

4.2.2.1 DNA Extraction The first step to any microbial community taxonomic analysis starts with DNA extraction. DNA quantity, quality, and fragment lengths are important factors for downstream analyses and sequencing. Several DNA extraction kits, reagents, and protocols are available to researchers. Each method features steps for cell lysis to release DNA, proteins, sugars, RNAs, and other small compounds from the DNA, and washing/cleaning the DNA (Fig. 4.4).

Cell lysis usually occurs through a series of different treatments including chemical, heat, and mechanical, and/or a combination of these treatments. Chemical lysis may include the use of detergents to breakdown the fatty acids comprising microbial cell walls; and salts and other ionic buffers, which increase the osmotic pressure outside of the cell and help to break apart the cell membrane. Heat treatment denatures proteins and helps to break apart cell membranes. Lastly, mechanical force through the use of specialized beads will fracture cells and membranes and help to release DNA.

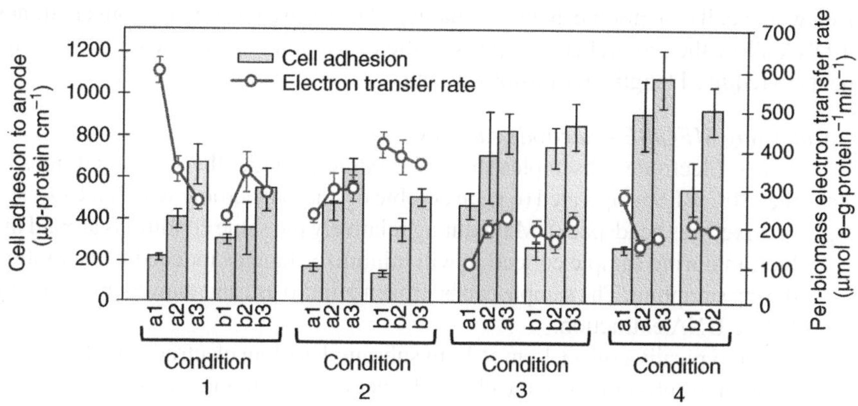

Figure 4.3 Example of how to plot biomass production and per biomass electron transfer rates. In this experiment, BES reactors were operated as MFCs. This study was focused on analyzing the effects of electron donors on community composition throughout biofilm enrichment. Condition 1 = acetate-fed; Condition 2 = butyrate/propionate-fed; Condition 3 = glucose-fed; and Condition 4 = Sucrose-fed reactors. Each condition was evaluated in duplicate reactors "a" and "b," and community analysis was performed monthly for each reactor over a 3-month enrichment period indicated as "a1," "a2," and so on. Experimental details can be found in Ishii et al. [21a].

Figure 4.4 DNA extraction scheme.

Because DNA is not the only material present in a given environmental or BES biomass sample, it is important to remove other contaminating compounds that may inhibit downstream processes. Inhibitors include proteins, humic acids, and other organic and inorganic materials.

Once the DNA has been isolated, it is important that any of the reagents used for cell lysis or inhibitor removal are completely washed away from the sample. DNA cleanup includes several washing steps to ensure that DNA is the only material present for subsequent experiments.

The type of sample (e.g., biofilm, soil, and water) and postprocessing requirements (e.g., 16S rRNA clone library or NGS) should be carefully considered when selecting a DNA extraction method. For example, DNA cleanup steps for purifying DNA and removing inhibitors such as humic acids and/or organics should be more rigorous for soil samples containing high concentrations of such contaminating compounds.

In addition, cell lysis methods for bacterial biofilm samples and/or other high-biomass samples (such as microbial mats) must be more rigorous than for certain low-cell-density planktonic communities (such as marine microbes). However, cell lysis must also be optimized for every sample so that the strands of DNA released from whole cells are not mechanically damaged or sheared too short. Longer strands of DNA allow the researcher to apply specific restriction enzymes to cut DNA fragments to required lengths for postprocessing.

Collecting MFC/BES electrode samples:

Note – Electrode subsamples should be extracted under the same conditions as the MFC/BES is operated (e.g., anaerobic or microanaerobic with N_2, CO_2, H_2, or mixed gas headspace). Maintaining relatively the same conditions around the biofilm during sample collection will minimize changes induced to the biofilm during sampling. The sample can be frozen prior to use and thawed immediately before DNA extraction.

We typically collect 1 cm × 1 cm subsamples of an electrode and freeze. At least three subsamples of the electrode, per time point, may be removed so that biofilm heterogeneities can be studied. For DNA extraction, we section each sample into 5 mm × 5 mm squares (under sterile conditions) and process multiple samples at one time. See Section 4.1 for more details relative to subsample extraction and processing.

Before you begin:

Good molecular technique is required to achieve desired results relative to DNA extraction and subsequent processing. Here, we note some factors that are important to consider throughout the execution of molecular protocols.

1. *Keep Work Areas Clean* – Before beginning any DNA-related protocol, you must ensure that your work area is clean. It is preferable to conduct molecular work in a UV-sterilized biosafety cabinet; however, work can be conducted at the lab bench if good sterile technique is utilized. For example, use 70% ethanol and DNA removal reagents (e.g., DNA Away) to wipe down the laboratory bench, and do this several times. You may want to place a clean sheet of aluminum foil on the laboratory bench where you will be working. Aluminum foil will prevent bench contamination in the case of spills and can be easily discarded and replaced at the end of an extraction. Use a Bunsen burner flame near the area where you are working. A flame will change air flow in the immediate area and help prevent contaminants from falling into your tubes while you are working. Exchanging your work gloves often, or at least when you think you've touched something that could be contaminated, is also an important habit for preventing contamination.
2. *Take Care of Pipettes* – Pipettes are a critical piece of laboratory equipment for a molecular biologist. Pipettes must be kept clean and calibrated so that cross-contamination does not occur and volume measurements are accurate. Calibration should occur once per year by an authorized representative familiar with your pipette brand. Pipette cleaning can be done by wiping the external surfaces with 70% ethanol and DNase eliminator. To keep the internal shaft of the pipette clean, it is preferable to only use barrier tips when transferring liquids.
3. *Always Use a New Pipette Tip* – Avoid contamination of chemicals and biological samples by always using a new pipette tip after each solution transfer.

This requires more pipette tips but is the best way to ensure that your chemical solutions stay sterile and your biological samples are not contaminated. The only exception would be if the same solution is being aliquoted repeatedly and the pipette tip does not contact any other chemical or surface during this process.

4. *Label Everything Clearly and Consistently* – Keeping track of your samples and associating your DNA sample to the correct experiment are critical for interpreting your results. It is important to use descriptive names and codes for labeling and tracking your biological samples through any community analysis protocol. Make sure that if a code is used, the definition of each code name is kept in multiple easy-to-find places for you to reference later. A DNA extraction may require up to five different tubes per sample, so prelabeling and organizing microcentrifuge/collection tubes ahead of time are helpful.
5. *Read the Instructions* – Most manufacturers providing kits for molecular biology research include detailed and descriptive protocols relative to how the kit should be used. Always read through a protocol before attempting to execute it. Some solutions may require preheating, you may need a special piece of equipment, or you may be out of a given reagent. Therefore, carefully read through the protocol and determine that you have all the necessary supplies, reagents, and equipment prior to getting started.
6. *Always Run Controls* – No matter what protocol you are executing, it is critical to always run at least one control to ensure that results are accurate and reagents are not contaminated.

Here, we discuss an optimized DNA extraction method specifically for electrode-adhered biofilm communities. We describe the use of a biofilm DNA extraction kit from MoBio Laboratories Inc. to extract DNA from electrode-adhered communities (PowerBiofilm™ DNA Isolation Kit, www.mobio.com). Here, we included a slightly modified alternative for cell lysis to include less mechanical bead-beating and multiple heat treatments. This method for cell lysis has been found to be successful for electrogenic community analyses [18, 21]. We do not endorse MoBio Laboratories. This protocol is solely meant to be a starting point for researchers.

Required materials:

- Disposable gloves.
- Eye guards.
- Laboratory coat.
- Fine tip permanent marker.
- Tip disposal containers with appropriate labels that follow institutional Health and Safety rules.
- Aluminum foil.

- Bunsen burner.
- Microcentrifuge tube rack.
- 70% ethanol in squeeze bottle.
- Water bath set to 55 °C.
- Water bath set to 65 °C.
- Analytical balance.
- Floating tube holder for 2-ml centrifuge tubes.
- Microcentrifuge.
- Pipettes (range, 100–1000 μl) and associated sterile barrier tips.
- Vortex and vortex adaptor for 2-ml centrifuge tubes.
- Ice bucket with shaved ice.
- Stopwatch or other timer.
- −20 °C freezer.
- PowerBiofilm DNA Isolation Kit, www.mobio.com, with the following components for a 50-sample kit:
  - PowerBiofilm™ Bead Tubes (50 tubes).
  - Solution BF1 (20 ml).
  - Solution BF2 (6 ml).
  - Solution BF3 (11 ml).
  - Solution BF4 (50 ml).
  - Solution BF5 (2 × 18 ml).
  - Solution BF6 (2 × 18 ml).
  - Solution BF7 (5.5 ml).
  - Spin Filters (50 tubes).
  - 2-ml collection tubes (250).

The MoBio PowerBiofilm DNA Isolation Kit includes detailed schematics and protocols for beginning and experienced users. The following is a summary of this protocol including some notes on the basis of our experience with DNA extraction from carbon cloth electrodes [23].

Cell lysis

1. Warm Solution BF1 in 55 °C water bath for 5–10 min to dissolve precipitates prior to use.
2. Place the electrode subsample (5 mm × 5 mm) in a 2-ml Collection Tube and centrifuge at 13,000×*g* for 1 min. Remove excess liquid from the collection tube using a pipette tip. If the sample is relatively dry, this step can be omitted.
3. Add electrode sample to the PowerBiofilm Bead Tube.
4. Add 350 μl of Solution BF1 directly to the Bead Tube containing biofilm material. Solution BF1 should be used while still warm. This step works toward breaking apart the fatty acids in cell membranes.

5. Add 100 μl of Solution BF2 to Bead Tube. Vortex on high for 1–2 s. This step works toward increasing the osmotic pressure outside of cells and helps to break apart the cell membrane.
6. Thermally lyse cells by incubating the Bead Tube in 65 °C water bath for 5 min.
7. Mechanically lyse cells using a bead-beating technique. Note that tubes can be secured horizontally onto a vortex adapter (MoBio catalog# 13000-V1) and vortexed at maximum speed for 10 min. A bead-beating step using a homogenizer is also an option.

Alternative to Steps 6 and 7: *To avoid excessive DNA sheering, we may run a repeated combination of temperature and mechanical lysis for shorter periods of time. For example, we will heat tubes at 70 °C for 5 min, vortex for 3–4 s, and repeat so that both heating and vortex steps occur twice. However, depending on biofilm density and electrode material, it may be preferential to follow the standard MoBio protocol.*

PCR inhibitor removal

1. Add 100 μl of Solution BF3 to Bead Tube and vortex for 3–4 s. Incubate tube on ice (or in a 4 °C refrigerator) for 5 min.
   Note – If the supernatant in the Bead Tube is very darkly colored, add 200 μl of Solution BF3.
2. Centrifuge Bead Tube at 13,000×*g* for 1 min at room temperature.
3. Remove from centrifuge. Carefully avoid the pellet at the bottom of the Bead Tube and transfer all of the supernatant (375–450 μl) to a clean, labeled 2-ml Collection Tube. Discard the Bead Tube and pellet.
4. Ensure that Solution BF4 does not have precipitate. Warm solution if necessary to dissolve precipitate. Add 900 μl of Solution BF4 to the 2-ml Collection Tube and vortex for 5 s.
5. Load 650 μl of supernatant onto a Spin Filter and centrifuge at 13,000×*g* for 1 min. Discard the flow through and repeat until all the supernatant has been loaded onto the Spin Filter. *DNA is now bound onto the Spin Filter.*

DNA cleanup

1. Place the Spin Filter into a clean 2-ml Collection Tube.
2. Shake Solution BF5 before use. Add 650 μl of Solution BF5 to Spin Filter and centrifuge at 13,000×*g* for 1 min at room temperature.
3. Discard the flow through and add 650 μl of Solution BF6 and centrifuge at 13,000×*g* for 1 min at room temperature.
4. Discard the flow through and centrifuge again at 13,000×*g* for 2 min at room temperature to remove residual wash.
5. Place Spin Filter into a clean 2-ml Collection Tube.
6. Add 100 μl of Solution BF7 to the center of the white filter membrane. Take care to add the entire volume only to the filter membrane and not lose any solution on the side-wall of the filter. This step releases the DNA from

the Spin Filter, and a lower volume of solution will result in lower yields of DNA.

7. Centrifuge the Spin Filter at 13,000×*g* for 1 min.
8. *Your DNA is now in the Collection Tube*. Discard the Spin Filter.
9. DNA should be stored frozen at −20 °C until further use.

To verify that DNA extraction was successful, the size of the DNA should be verified using gel electrophoresis. DNA concentrations and quality should be measured using a UV spectrophotometer (e.g., Nanodrop machine, Thermo Scientific, www.nanodrop.com). Double-strand DNAs can be measured using a PicoGreen Assay (www.lifetechnologies.com).

4.2.2.2 DNA Validation and Quantification It is important to know how successful the DNA extraction process was. The quality, quantity, and length(s) of DNA should be estimated prior to downstream processing. Here, we describe validation and quality measurement using a Nanodrop machine.

Determining DNA length is a straightforward process that requires gel electrophoresis to separate strand lengths of your DNA sample. By comparing the separated bands in your sample to a standard DNA ladder run on the same gel, you can determine the size(s) of DNA strands and roughly how much DNA is present in your sample.

Required materials:

- Disposable gloves.
- Eye guards.
- Laboratory coat.
- 250-ml glass serum bottle or flask.
- Tip disposal containers with appropriate labels that follow institutional Health and Safety rules.
- 1× concentration Tris–Acetate–EDTA (TAE) buffer (www.sigmaaldrich.com, www.neb.com, and www.lifetechnologies.com).
- Agarose powder (www.sigmaaldrich.com, www.neb.com, www.lifetechnologies.com, www.promega.com, or www.bio-rad.com).
- 10× concentration DNA loading buffer (www.neb.com, www.lifetechnologies.com, www.thermoscientificbio.com, or www.promega.com).
- 1-kbp DNA ladder (www.neb.com, www.lifetechnologies.com, www.thermoscientificbio.com, or www.promega.com).
- Nucleic acid gel stain (such as SYBR®Gold or SYBR®Green, www.lifetechnologies.com).
- Micropipettes (range, 0.1–20 μl) and associated barrier tips.
- Ice bucket with shaved ice.
- Plastic container with lid just larger than gel tray molds (e.g., laboratory-dedicated plastic food storage container).

- Microwave (laboratory only).
- Electrophoresis machine with power supply (able to run at 100 V).
- Gel tray and associated combs.
- UV gel documentation system (Gel Doc) or UV transilluminator.
- Nanodrop machine (Thermo Scientific, www.nanodrop.com).
- Timer.
- DNA sample(s).

Interpreting Nanodrop results:

A Nanodrop system is a UV–vis spectrophotometer that can measure absorption of compounds in microvolumes of a sample (0.5–2 μl). All Nanodrop or equivalent instruments should be operated according to the manufacturer's instructions relative to hardware and software. Informational videos are available on multiple websites for how to use a Nanodrop machine (e.g., http://www.nanodrop.com/Productnd2000coverview.aspx).

As with any spectrophotometer, it is important to have a high-quality "blanking" solution. For DNA samples, it is best to use either DNA-free water or the storage buffer solution contained in the DNA extraction kit. Water should be used to rinse the nanodrop surfaces between reading each sample.

The most important metric that is reported from a nanodrop measurement of a DNA sample is the 260/280 ratio, which is the ratio of the absorbance peaks at 260 and 280 nm. Absorbance at 260 nm measures nucleotides, and absorbance at 280 nm measures protein. A 260/280 ratio that is greater than 2 is indicative of a very clean DNA sample. A 260/280 ratio of 1.8 is acceptable for downstream PCR. Anything less than a 260/280 value of 1.8 may result in amplification (or other) issues. A Nanodrop system also estimates DNA concentrations.

Electrophoresis for validating DNA concentrations:

Even if a Nanodrop reading suggests a good yield of DNA and a 260/280 ratio greater than 1.8, it is important to still run a gel and quantify the length of DNA. The gel analysis will also give another estimate of DNA concentration.

Preparing an Agarose gel

(1) Weigh out 1.8 g of Agarose in a microwavable container and add 180 ml of 1× TAE buffer.

(2) Loosely cap the container or fill the mouth of the container with foam stopper.

(3) Microwave Agarose/buffer mixture for approximately 1 min, but watch the container the entire time so that the solution does not bubble over. After 1 min, tighten the cap and lightly shake to mix. Incrementally add 30 s of microwave time until the solution is completely clear.

(4) Let the solution cool to about 60 °C (roughly 10 min), and then pour it into gel tray with comb assembly.

(5) Wait for about 10–15 min for the gel to solidify.

(6) Pull out the comb from the gel, taking care not to rip the gel.

(7) Place gel (in gel tray) in the electrophoresis machine, oriented in the proper direction, and fill the well with 1× TAE buffer so that the gel is completely submerged. Usually, there is a fill line in the electrophoresis tray. *Note that the wells of the gel should be placed nearest to the negative terminal of the electrophoresis machine. DNA is negatively charged and will move toward the positive terminal when voltage is applied.*

Preparing and running the DNA sample

(1) Add a 1-μl aliquot of 10× DNA loading buffer to a 3-μl aliquot DNA sample and use the pipette to gently mix the loading buffer and DNA sample together.

(2) Load the 4 μl of sample and buffer into a well of the gel. Take care not to pierce the gel with the pipette tip or eject sample outside of the well.

(3) Load appropriate DNA ladder in a well adjacent to the DNA samples (e.g., 4 μl of high mass ladder, Life Technology).

(4) Hook up the electrophoresis machine to the power supply and run it for 20–40 min at 100 V. Take care not to let the DNA bands run off the gel.

Imaging the DNA sample

(1) Submerge gel in storage container containing 100 ml of 1× TAE buffer and 10 μl of DNA stain (e.g., SYBR Gold or SYBR Safe®, www.lifetechnologies.com). *Note that TAE and DNA stain mixture can be re-used for up to 2 or 3 days. Fresh solution is needed after 3 days.*

(2) Store gel in container, in the dark, for up to 20 min. A longer staining time is needed for lower concentrations of DNA.

(3) Carefully remove the gel from solution and excite stained DNAs using a UV lamp. For example, Gel Doc (www.bio-rad.com) can be used; however, be sure to use the Gel Doc according to manufacturer's instructions and take an image of your gel.

(4) Dispose off gel and staining solution in their appropriate containers after imaging.

If your DNA extraction and gel loading technique was successful, you should see several different bands in your DNA ladder and a single band for your DNA. If you see fragmented (smeared) DNAs, then it is indicative that some DNA degradation has occurred. You should now compare your gel results to the Nanodrop results to verify findings. If the DNA validation and quantification steps were successful, the DNA is ready for downstream processing.

4.2.2.3 Phylogenetic Analysis Based on 16S rRNA Sequencing The 16S rRNA gene has highly conserved primer binding sites and is commonly used to identify different phyla of bacteria and archaea [5]. A primer is a short sequence of DNA that can be thought of as either the beginning or end of the DNA segment that you want to copy. One primer binds to the left side of the selected sequence of DNA (forward primer), and the other primer binds to the right side (reverse primer). In other words, the DNA sequence to be copied is "fenced in" by the two primers.

16S rRNA gene sequences also contain hypervariable regions that can provide species-specific, or strain-specific, signature sequences that are useful for identifying uncultured microbes. Pioneering work by Woese and Fox led to the common use of the 16S rRNA gene in phylogeny studies [24].

Many different primer sets are available for 16S rRNA gene sequencing, and many universal primers (applicable to bacteria and archaea) cover hypervariable regions that provide information about species-level and strain-level taxonomy [23]. However, primer design and/or selection must be carefully considered so that conserved and variable regions are covered with enough sequence to accurately distinguish between genera, species, and strain [23b].

It should also be noted that many bacteria have multiple copies of the 16S rRNA gene, and therefore, biases are associated with using this type of sequence data to estimate the relative abundance of microbial taxa within a community. Researchers may use other single-copy marker genes, in addition to the 16S rRNA gene, to estimate diversity and relative microbial abundances [11, 18, 25]. Other methods such as metagenomics are also more robust in terms of accurately describing microbial diversity in a given community. Here, we offer a protocol for constructing 16S rRNA libraries for Sanger sequencing, which still provides the longest read lengths and most accurate sequence data.

It should be noted that there are several different reagents and many different protocols relative to constructing clone libraries. The protocol outlined here is offered as starting point for the researcher.

4.2.2.3.1 Clone Library Construction for Sanger Sequencing The basic scheme for clone library construction is shown in Figure 4.5. PCR is the first step in the process and is used to amplify a given nucleotide sequence from the DNA, producing millions of copies of that segment. The PCR product is then purified and ligated into a cloning vector. The next step is transformation, which is to force a nonpathogenic strain of *Escherichia coli* to take up the cloning vector and propagate the specific 16S rRNA gene fragment during colony formation. *E. coli* colonies are grown on selective media, and those colonies that successfully ligated the DNA fragment grow as white colonies. Those colonies that did not successfully ligate the fragment grow as blue colonies. A subset of white colonies is hand-picked for each sample, re-amplified using PCR, purified, and finally sequenced.

4.2.2.3.1.1 PCR PCR works by combining the DNA, a supply of nucleotides, two primers (forward and reverse), and DNA polymerase as the catalyst (an enzyme that

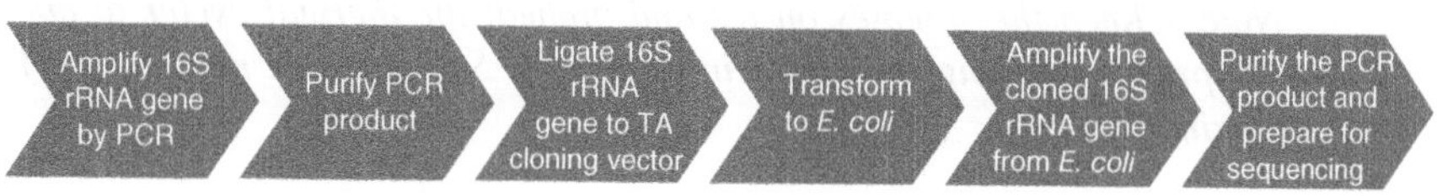

Figure 4.5 Scheme for 16S rRNA clone library construction.

adds the nucleotides). Many literature resources and educational tools are available to summarize how PCR works [26].

Required materials:

- Disposable gloves.
- Eye guards.
- Laboratory coat.
- 96-well plate or 0.2-ml tubes with illustra PuReTaq Ready-To-Go PCR Beads™ (www.gelifesciences.com).
  Note that other Taq DNA polymerases, Taq buffers, and Deoxynucleotide triphosphate (dNTP) mixes can be used instead of the PCR beads. The beads are suggested only for simplicity.
- Primer cocktail (mixture of the forward and reverse primers).
  Note – We use bacterial universal primer U27F (5′-AGAGTTTGATCCTG-GCTCAG-3′) and microbial universal primer U1492R (5′-GGTTACCTTGT TACGACTT-3′) for 16S rRNA gene amplification [27].
- Thermal cycler.
- Sterile 96-well plates with covers *or* 0.2-ml PCR tubes with caps.
- Micropipette (range, 10–100 μl) and associated barrier tips.
- Ice bucket with shaved ice.

PCR protocol (to be performed on ice):

1. Label PCR tubes and/or 96-well plates with DNA sample names.
2. Place 1 μl (~50 ng) of the DNA sample(s) in empty tray or empty 0.2-ml PCR tube(s).
3. Add 24 μl of the primer cocktail and 1 PCR bead to all samples.
4. Gently tap the tray or tubes so that all solution is in the bottom of the tube and put caps on with a cap-installing tool.
5. Place the samples into a thermal cycler (see the following note) and run the following cycling method:
   a. 94 °C for 3 min
   b. b.25 cycles of
      – 94 °C for 30 s.
      – 55 °C for 30 s.
      – 72 °C for 1 min 30 s.
   c. 72 °C for 10 min.
   d. 4 °C hold.

Note – Keep the samples on ice and preheat the thermal cycler to about 94 °C prior to putting the plates in the cycler. Samples will need to run for approximately 2 h.

4.2.2.3.1.2 GEL ELECTROPHORESIS Gel electrophoresis should be used to verify the results from PCR. See Section 4.2.2.2 for details. The protocol outlined as follows can be used instead of, or in addition to, Section 4.2.2.2.

Required materials:

- Disposable gloves.
- Eye guards.
- Laboratory coat.
- Agarose.
- 1× lithium borate (LB) buffer (http://www.fasterbettermedia.com/home.html).
- Amresco EZ-Vision Stain (http://www.amresco-inc.com) or SYBR® (www.lifetechnologies.com).
- EZ Load Precision Molecular Mass Ruler (DNA ladder). www.bio-rad.com.
- Gel electrophoresis box.
- Power supply.
- 250-ml Erlenmeyer Flask or glass serum bottle.
- Gel trays and associated combs.
- Microwave.
- Micropipette and tips (range, 1–20 μl).
- Scale/balance.

The 1× LB buffer is used both to make the agarose gel and as an electrophoresis buffer. It is made by diluting a more concentrated stock solution of LB buffer. For example, combine 100 ml of purchased 20× LB buffer with 900 ml of distilled water to make a liter of 1× LB. This should be prepared just before use:

1. Add 1 g of agarose to a 250-ml Erlenmeyer Flask containing 100 ml of 1× LB buffer.
2. Heat in a microwave with occasional swirling until the agarose melts completely. Handle the hot agarose carefully because it can easily become superheated and boil over.
3. Allow the agarose to cool slightly. If the agarose is poured while it is too hot, it will warp the plastic casting mold and eventually render it useless. Prepare and seal the ends of the gel mold according to the manufacturer's instructions. Also position the desired comb to cast the wells.
4. Pour the cooled, liquefied agarose into the gel mold and allow it to solidify. Gels should be 5- to 8-mm thick. Allow the gel to cool until solid at room temperature. *Note – Allow the agarose gel to cool until solidified. The gel will appear in a cloudy white color and will feel cool to the touch after*

approximately 20 min at room temperature. Gels can be stored, wrapped in plastic wrap, in the refrigerator for a few days.

5. After the gel has solidified, carefully remove the comb and place it in the electrophoresis chamber.
6. Pipet 10 μl of the Amresco DNA ladder into the first and last well of the gel.
7. Add 1 μl of DNA Amresco EZ-Vision (or SYBR) to 5 μl amplified PCR product.
8. Record which well you place your sample. Load 6 μl of your PCR product into designated well.
9. Add a sufficient volume of the 1× buffer used in step 1 to the electrophoresis chamber to cover the gel by 1–2 mm. Ensure that the wells of the gel are closest to the black electrode of the electrophoresis chamber.
10. Attach the electrodes, placing the negative (black) electrode at the end where the sample wells are located and the positive (red) electrode at the other end. Apply voltage from a direct-current power supply. Run the gel at 250 V until the loading dye has traveled a quarter to a half of the way down the gel (about 10 min). When finished running the gel, turn off the power supply and unplug the leads before opening the gel box. *Note – Very few electrophoresis buffers will work with such high voltage. When running at 250 V, be sure to use the correct buffer (e.g., LB buffer). If another buffer is used, the gel should be run at 100 V for 20 min.*
11. View gel using Gel Doc or UV light box. *Note – It is common to see an additional band lower on the gel representing an artifact of the PCR reaction that results from the primers overlapping one another and amplifying themselves.*

4.2.2.3.1.3 PCR PURIFICATION It is important to remove excess primers, nucleotides, enzymes, mineral oil, salts, and other impurities from DNA samples prior to executing other downstream processes. This DNA purification protocol is based on the QIAGEN QIAquick PCR Purification Kit (www.qiagen.com); however, several other DNA purification kits are available. We do not endorse this product but offer the outlined protocol as a starting place for the researcher.

Required materials:

- Disposable gloves.
- Eye guards.
- Laboratory coat.
- Micropipettes (range, 20–1000 μl) and associated barrier tips.
- Microcentrifuge.
- DNA-free water.
- Sterile 1.5-ml microcentrifuge tubes.

- QIAquick PCR Purification Kit (www.qiagen.com) containing the following components for a 50-sample kit:

Components	Quantity/Kit
QIAquick spin columns	50
Buffer PB	30 ml
Buffer PE (concentrate)	2 × 6 ml
Buffer EB	15 ml
pH Indicator I	800 μl
Collection Tubes (2 ml)	50
Loading dye	110 μl

Follow the DNA purification steps according to the manufacturer's instructions. We've included some modifications to the protocol as follows on the basis of our experience with BES samples.

1. In a clean 1.5-ml microcentrifuge tube, add 5 volumes of Buffer PB to 1 volume of the PCR sample and mix. For example, add 500 μl of Buffer PB to 100-μl PCR sample.
2. Place a QIAquick Spin Filter in a labeled 2-ml Collection Tube (provided in kit).
3. Apply the sample to the QIAquick Spin Filter and centrifuge for 1 min at 12,000 rpm at room temperature to bind DNA to the filter.
4. Discard flow-through liquid and place the QIAquick Spin Filter back in the same Collection Tube. *Note* – Collection Tubes are re-used to reduce plastic waste.
5. Add 700 μl of Buffer PE to the QIAquick Spin Filter and centrifuge for 1 min at 12,000 rpm at room temperature to wash the DNA sample.
6. Discard flow-through liquid and place the QIAquick Spin Filter back in the same tube. Centrifuge again for another 1 min at 12,000 rpm at room temperature.
7. Place QIAquick Spin Filter in a clean 1.5-ml microcentrifuge tube.
8. Add 30 μl of DNA-free water to the center of the QIAquick Spin Filter membrane and let the sample stand for 1 min, then centrifuge for 1 min at 12,000 rpm at room temperature. Discard the Spin Filter. *Your DNA is now in the centrifuge tube.*

4.2.2.3.1.4 LIGATION AND TRANSFORMATION Multiple purified rRNA amplicons exist in each sample tube. During the process, one amplicon can ligate to one plasmid in one *E. coli* cell. The transformed *E. coli* cell then makes a single colony. Single colonies with the single rRNA amplicons are then harvested, purified, and sequenced.

This protocol is based on the Invitrogen TOPO® TA Cloning® Kit from Life Technologies (www.lifetechnologies.com); however, there are many different kits and other protocols that can be used. We are not endorsing this product, but simply offering it as a starting point for the researcher.

Ligate amplicon to plasmid

Required materials:

- Disposable gloves.
- Eye guards.
- Laboratory coat.
- Incubator set to 37 °C.
- Micropipettes (range, 1–100 μl) and associated barrier tips.
- TOPO Cloning Reaction kit (www.lifetechnologies.com)

Component	Amount
pCR™2.1-TOPO vector	20 μl
10× PCR buffer	100 μl
Salt solution	50 μl
dNTP mix	10 μl
M13 forward primer	20 μl
M13 reverse primer	20 μl
Control template	10 μl
Control PCR primers	10 μl
Water	1 ml

- Rapid One Shot® Chemically Competent *E. coli* (www.lifetechnologies.com).
- SOC medium (www.lifetechnologies.com)

Component	Amount
SOC medium	6 ml
Mach1™-T1® chemically competent cells	21 × 50 μl
pUC19 Control DNA	50 μl

- Luria Bertani agar plates supplemented with 50 μg ml^{-1} ampicillin and 40 mg ml^{-1} X-gal in dimethylformamide.
- Bunsen burner.
- Aluminum foil.
- Stainless steel spreader/spatula.
- Clean 0.2-ml PCR tubes and caps.
- PCR amplicon.
- Ice bucket with shaved ice.

(1) Gently mix the following solutions together at room temperature in a clean, labeled, PCR tube. Do not vortex or use pipette tip to mix. Gently flick the bottom of the PCR tube after reagents have been added.

Fresh PCR amplicon	4 μl
Salt solution	1 μl
TOPO vector	1 μl
Final volume	6 μl

(2) Incubate for 5–30 min at room temperature (22–23 °C), depending on the number of colonies needed for analysis.

(3) Place the reaction on ice (or in −20 °C freezer) and prepare for the transformation protocol.

Transformation

1. Warm Luria Bertani agar plates (supplemented with 50 μg ml^{-1} ampicillin and 40 mg ml^{-1} X-gal in dimethylformamide) to 37 °C.
2. On ice, add 4 μl of the TOPOCloning reaction from Ligation step into one vial of the Competent *E. coli* and mix gently. *Do not mix by vortex or pipetting up and down.*
3. Incubate on ice for 5 min.
4. Sterilize spatula.
5. Aliquot 50 μl of cells on a prewarmed LB plate containing 50 μg ml^{-1} ampicillin and 40 mg ml^{-1} X-gal in dimethylformamide and spread evenly with a spatula until plate is dry.
6. Incubate overnight at 37 °C.

An efficient TOPO Cloning reaction should produce several hundred colonies. Subsample 24 colonies for PCR validation. Determine the number of white colonies required to accurately reflect community diversity from each sample for sequencing. (We use a 96-well plate format.) Do not pick dark-blue colonies.

4.2.2.3.1.5 COLONY PICKING AND PCR Confirm that there are sufficient white colonies for the desired analysis. The number of colonies picked per plate should correlate with the estimated diversity of the sample. For example, more colonies should be picked (>100) for analyzing high-diversity samples. Here, we describe the procedure for a 96-well plate format. In this case, prepare to pick at least 96 white colonies from each sample for sequencing. As a precaution, you can also re-plate these 96 colonies using a template that follows your 96-well labeling scheme. This will give you an extra plate of material to verify that only white colonies were selected, and material to re-sequence if any of the sequencing reactions should fail.

Required materials:

- Disposable gloves.

- Eye guards.
- Laboratory coat.
- Biosafety Cabinet.
- Incubator set to 37 °C.
- Thermal cycler.
- Multichannel pipettes (range, 1–50 μl) and associated barrier tips.
- Mixing wells for multichannel pipettes.
- 96-well plates.
- Autoclaved toothpicks.
- Luria Bertani agar plates supplemented with 50 μg ml^{-1} ampicillin and 40 mg ml^{-1} X-gal in dimethylformamide.
- Sterilized, DNA-free, water (Ambion or MoBio).
- TaKaRa Ex Taq (Clonetech).
- 10× Ex Taq Buffer.
- 25 mM $MgCl_2$ solution.
- dNTP mixture (2.5 mM).
- Primer 1 – M13F.
- Primer 2 – M13R.
- Template plate labels for each sample (see Fig. 4.6).

Setup all reactions in the biosafety cabinet and keep all reagents and reaction products on ice while conducting the following protocol:

1. Divide 5 μl of sterilized, DNA-free, water into a 96-well plate with a multichannel pipette.

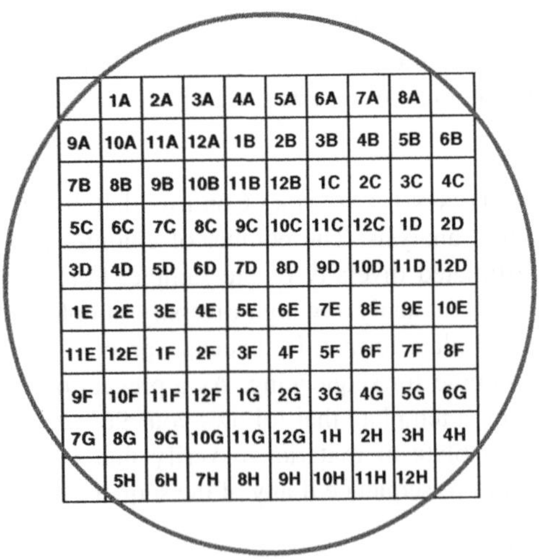

Figure 4.6 Template plate label for tracking colony PCR.

2. Tape a template label to the backside of LB plates with the numbers showing through the agar from the top of the plate.
3. Using a sterile toothpick, pick one white colony and suspend cells in the water in the 96-well plate. (A quick dip in the water will do.) Subsequently, streak the toothpick lightly within a square of the number-labeled plate. The numbered square must correspond with the numbered well of the 96-well plate.
4. Repeat step 3 until 96 colonies have been selected.
5. Place 96-well plate on ice.
6. Incubate the LB plates overnight at 37 °C, then remove and keep plates at 4 °C until sequences have been verified.
7. In a mixing well, on ice, prepare PCR mastermix for the 96-well plates using the following volumes. *Note that the total reaction mixture should total 25 μl in each well. Toothpicks will absorb some water, and we estimate a loss of 3 μl of water.*

Component	Concentration	Per 96-well plate
Ex Taq Buffer	10×	260 μl
$MgCl_2$ solution	25 mM	208 μl
dNTP Mix	2.5 mM each	208 μl
Primer 1 M13F	10 mM	104 μl
Primer 2 M13R	10 mM	104 μl
Sterilized, DNA-free water	–	1391 μl
TaKaRa Ex Taq	–	13

8. Using a multichannel pipette, divide 22 μl of the mastermix into the 96-well PCR plate. All transfers should take place on ice.
9. Amplify the cloned 16S rRNA gene from *E. coli* using the following thermal cycler method:
   a. 94 °C for 3 min.
   b. b.35 cycles of
      - 94 °C for 30 s.
      - 55 °C for 30 s.
      - 72 °C for 1 min 30 s.
   c. 72 °C for 10 min.
   d. 4 °C hold.
10. Confirm PCR amplification by electrophoresis. Select a subset out of the 96 samples for confirmation. Use the protocols outlined in Section 4.2.2.2.

The size of the PCR fragments must be about 1.5 kbp, and all samples should look the same on the agarose gel. Sequencing will be done to determine the actual nucleotide sequence of the fragments.

4.2.2.3.1.6 PCR PURIFICATION FOR SEQUENCING Excess nucleotides and reagents must be removed from the amplicons prior to sequencing. In addition, the last primer should be added for Sanger sequencing. To determine the nearly full length of the 16S rRNA sequences, three sets of primers are needed: (27F, 907R, and 1492R) or (M13F, 907R, and M13R).

Required materials:

- Disposable gloves.
- Eye guards.
- Laboratory coat.
- Thermal cycler.
- Micropipettes (range, 1–1000 μl) and associated barrier tips.
- Multichannel micropipettes (1–20 μl) and associated barrier tips.
- FastAP™ Alkaline Phosphatase (www.thermoscientific.com).
- Exonuclease 1 (Exo 1) (www.neb.com).
- Sterilized, DNA-free, water.
- Primer – 27F, 907R, and 1492R (1.6 mM).
- 96-well plate containing amplicons (25 μl volume in each well).
- Mixing wells for multichannel pipettes.
- Ice bucket with shaved ice.

All reagents and amplicons should be stored on ice while conducting the protocol.

1. In a multichannel mixing well, on ice, mix together the following reagents

Component	Per 96-well plate
FastAP Alkaline Phosphatase	208 μl
Exonuclease 1	104 μl
Sterilized, DNA-free water	728 μl

2. With a multichannel pipette, distribute 10 μl of the enzyme mixture to each amplicon for a total of 35 μl in each well.
3. Run the amplicon mixture through the following PCR method in a thermal cycler:
   a. 37 °C for 20 min.
   b. 80 °C for 15 min.
   c. 4 °C hold.
4. Place the plate on ice and aliquot 5 μl of each amplicon/enzyme mixture into a new 96-well plate. Always use fresh tips to avoid cross-contamination.
5. Add 2 μl of 27F, 907R, or 1492R primer into each well. Keep accurate notes about which well corresponds to which amplicon and sequencing read.
6. Keep plate on ice for immediate shipment to sequencing facility or at −20°C until ready for sequencing.

4.2.2.3.1.7 SANGER SEQUENCING Similarly to PCR, the sequencing primer is a small piece of DNA that will bind to the template and become a starting point for sequencing. Nucleotides are then randomly incorporated by the polymerase. There are normal nucleotides as well as dye-labeled, chain-terminating nucleotides. Both types of nucleotides will be incorporated by the polymerase. If a dye-labeled, chain-terminating nucleotide is incorporated, the chain stops growing and a color is now associated with that nucleotide. This process is repeated until there are thousands of fragments of DNA; all of them differ by one nucleotide in size.

These dye-labeled fragments are then loaded into a DNA sequencer, where the fragments are separated by capillary electrophoresis. Similarly to agarose gel electrophoresis, as DNA is negatively charged, in the presence of electricity, DNA will move toward the positive electrode. DNA will move according to size, with the smallest molecules moving the fastest. Unlike agarose gel electrophoresis, capillary electrophoresis separates the DNA based on one base pair size differences. The sequencer will detect the dye molecule attached to the DNA fragment as it is excited by a laser and then passes a camera that captures the fluorescence. Computer software then records that information with a colored peak and assigns nucleotides to each fragment.

Sanger sequence data is displayed in the form of an electropherogram, which shows base calls above the colored peaks. Each nucleotide is assigned a color. Normally A (Adenine) is green; C (Cytosine) is blue; G (Guanine) is yellow/black; and T (Thymine) is red. Figure 4.7 shows example electropherograms of "good" quality sequence data (a) and "bad" quality sequence data (b).

Sanger sequencing is an older sequencing method; however, it is still in use because it produces long sequence reads, which are necessary to more accurately estimate taxonomy.

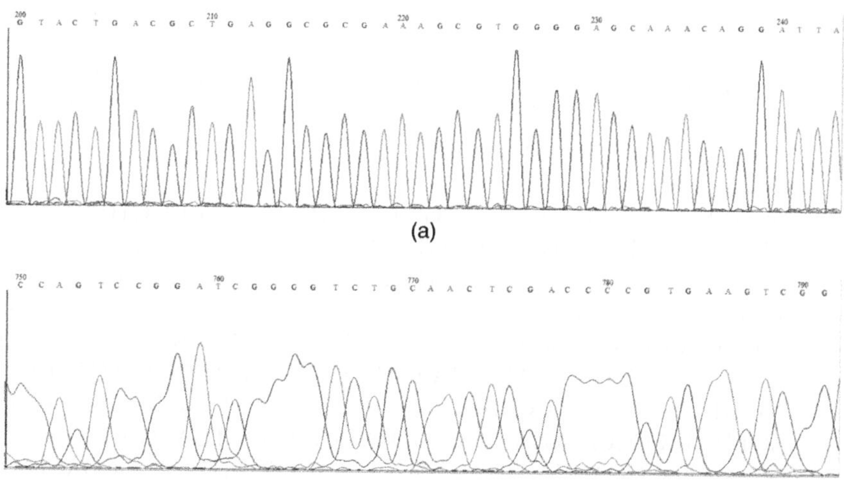

(a)

(b)

Figure 4.7 Examples of (a) "Good" quality and (b) "Bad" quality sequence data. *(See insert for color representation of this figure.)*

Pyrosequencing is now the common method for 16S rRNA gene sequencing [20g, 28]. Pyrosequencing requires primer design that will target the appropriate variable region of the 16S rRNA gene (e.g., V4). A basic overview for primer design, pyrosequencing using a 454 sequencer, and sequence postprocessing is available at http://rdp.cme.msu.edu/wiki/index.php/Pyrosequencing_Help.

4.2.2.3.1.8 SANGER SEQUENCE ANALYSIS AND INFORMATICS The resulting files from Sanger sequencing are generally AB1 files (sometimes SCF files) that contain trace information for each nucleotide A, T, G, and C (Fig. 4.7). However, FASTA format files are necessary for post-sequencing analysis and extracting DNA sequences of each 16S rRNA gene. FASTA file extensions vary and include .fasta, .fa, .fas, .fna, and so on; however, software will detect the FASTA format from the contents in the file. The following is an example of FASTA file format:

```
>2Bsub5_H24 (this is the clone ID)
  GGGCCTAAGCGTTCGTAGCCGGCTTGATAAGTCTCTGGTGAAATC
  CCATAGCTTAACTGTGGGAATTGCTGGAGATACTATTAGGCTTGAGG
  CCGGGAGAGGTTAGGGGTACTCCCAGGGTAGGGGTGAAATCCTATA
  ATCCTGGGAGGACCACCTGTGGCGAAGGCGCCTAACTGGAACGGA
  CCTGACGGTGAGTAACGAAAGCCAGGGGCGCGAACCGGATTAGATA
  CCCGGGTAGTCCTGGCCGTAAACGATGTGGACTTGGTGTTGGAATG
  GCTTCGAGCTGCTCCAGTGCCGAAGGGAAGCTGTTAAGTCCACCGC
  CTGGGAAGTACGGTCGCAAGACTGAAACTTAAAGGAATTGGCGGG
  GGAGCACCACAACGCGTGGAGCCTGCGGTTTAATTGGATTCAACGC
  CGGACATCTCACCAGGGGCGACAGTAGTATGATGGCCAGGTTGACG
  ACCTTGCTTGACAAGCTGAGAGGAGGTGCATGGCCGCCGTCAGCTC
  GTACCGTGAGGCGTCCTGTTAAGTCAGGCAACGAGCGAGACCCACG
  CCCTTAGTTACCAGCGGGTCCTCCGGGATGCCGGGCACACTAAAGG
  GACCGCCAGTGATAAACTGGAGGAAGGAGTGGACGACGGTAGGTC
  CGTATGCCCCGAATCCCCTGGGGCTACACGCGGGCTACAATGGGCT
  GGTGACAATTGGGTTCCGACACTTGAAAGGTGAAGGTAATCTCTTA
  AACATAGTCTTAATTTCGGATCGA
>2Bsub5_H08
  GGGTTTAAGGGTGCGTAGGCGGTTTGATAAGTCAGCGGTGAAAT
  ATTTCAGCTTAACTGGAAGGGTGCCGTTGATACTGTCAGGCTTGAAT
  TTAGTTGCTGTGGGTGGAATGTGTGGTGTAGCGGTGAAATGCA
  TAGATATCACACAGAATATCGATTGCGAAGGCAGCTCACAAAGCTAA
  AATTGACGCTGATGCACGAAAGTGTGGGGATCAAACAGGATTAGAT
  ACCCTGGTAGTCCACACTGTAAACGATGATAACTCGCTGTCGGCGA
  TACATAGTCGGTGGCCAAGCGAAAGCGATAAGTTATCCACCTGGGG
  AGTACGACCGCAAGGTTGAAACTCAAAGGAATTGACGGGGGCCC
  GCACAAGCGGAGGAACATGTGGTTTAATTCGATGATACGCGAGGAA
  CCTTACCCGGGCTCGAACGGTGCATGAATGATTAAGAGATTGATCAG
  TCCTTCGGGACATGTATCGAGGTGCTG ...
```

There are several free software packages that can be used to convert the Sanger trace files to FASTA files. One option for the process is BioEdit Sequence Alignment Editor [29], which can be downloaded at http://www.mbio.ncsu.edu/BioEdit/bioedit.html. Other free editors are also available and easily found. In addition, commercial software such as CLC Genomics Workbench (http://www.clcbio.com/products/clc-main-workbench/) can be used for this purpose.

Here, as an example, we show the process for converting trace files to FASTA files extraction by using BioEdit.

1. Save AB1 sequence files to an accessible directory.
2. Download and Open BioEdit.
3. Under *File*, select *Open*. Navigate to and select electropherogram files.
4. View the sequencing data to determine the quality. Please see Figure 4.7 for examples of good- and bad-quality sequencing data.
5. Select the appropriate region of good-quality sequence data. The maximum read-length from Sanger sequencing is approximately 1000 bp, but generally, sequence data beyond 800 bp shows bad quality. In addition, the primer region should be omitted from the selection.
6. Copy and paste to a text edit software (Notepad, MS word, or WordPad for Windows OS).
7. Add clone ID as shown previously (>clone ID) at the beginning of the sequencing and save the file as FASTA format.

The FASTA file is now ready for post-sequencing analyses of community composition and phylogenetic analysis of each clone (phylotype).

4.2.2.3.1.8.1 Community composition analysis via RDP The Ribosomal Database Project (RDP) provides data, tools, and services related to rRNA sequences to the research community [25]. As of October 2013, the RDP contained 2,809,406 aligned and annotated bacterial and archaeal small subunit rRNA gene sequences and 62,860 fungal large subunit rRNA gene sequences [30], and the database receives an average of 5000 new sequences each month [25]. To handle this volume of sequences, RDP developers constructed a naïve Bayesian classifier for classifying bacterial rRNA sequences into the new Bergey bacterial taxonomy [31].

The RDP Classifier is fast, does not require sequence alignment, and works well with partial sequences. This is important because the vast majority of rRNA sequences in the public databases is partial [25]. Here, we describe a simple method for community analysis using the RDP Classifier tool [25].

1. Save your FASTA files to an easily accessible directory.
2. Navigate to the RDP home page in your browser: http://rdp.cme.msu.edu/index.jsp. Figure 4.8 shows an example of the home page.

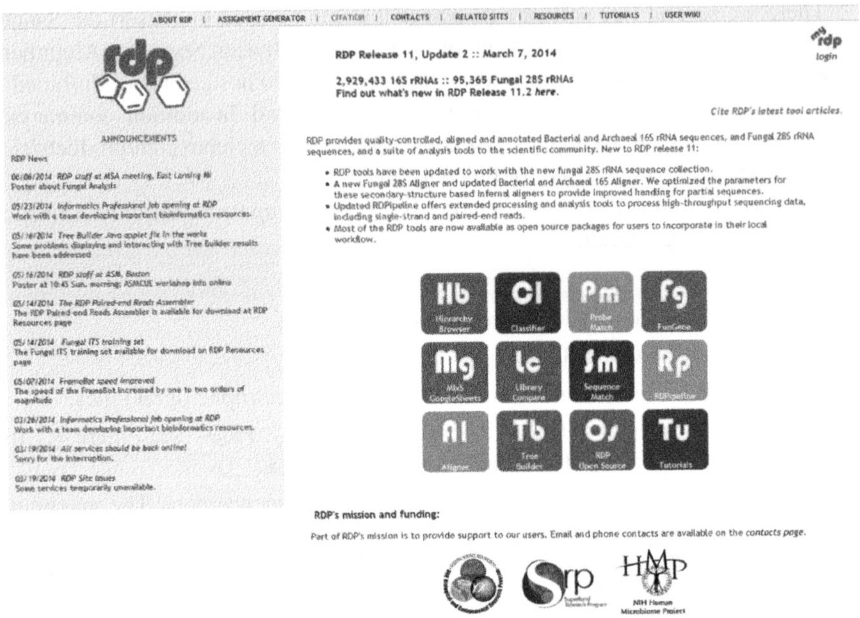

Figure 4.8 RDP home page, http://rdp.cme.msu.edu/index.jsp.

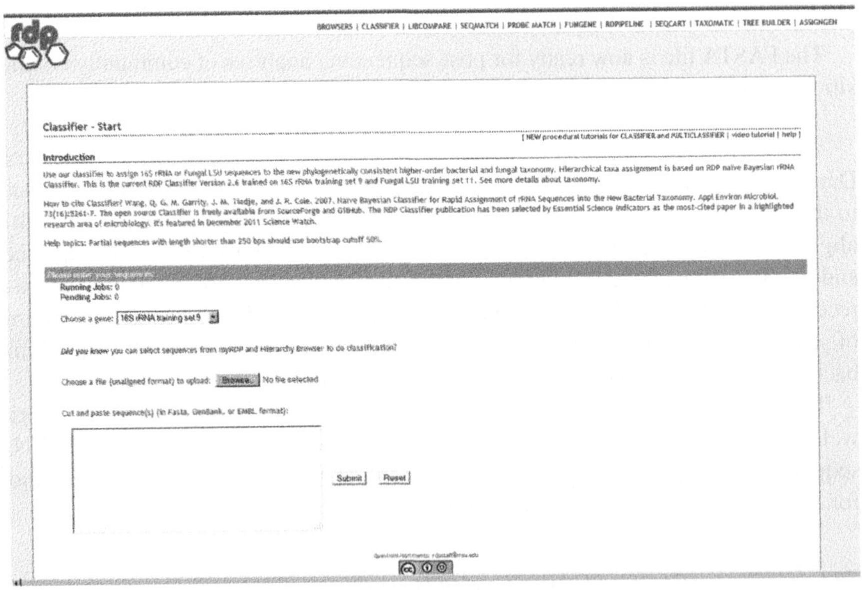

Figure 4.9 RDP Classifier tool.

3. Select the Classifier option. A new page will display (see Fig. 4.9).
4. For Bacterial and Archaeal sequences, select "16S rRNA training set 9" from the drop-down menu.
5. Use the "Browse" function to upload your FASTA file and choose "Submit." Figure 4.10 shows an example Hierarchy View of community composition from a FASTA file related to a methanogenic cathode-associated microbial community.
6. Download the text file and use the entire hierarchy for further graphical and statistical analyses toward characterizing the relative abundance of phylotypes in the microbial community. Researchers may also be interested in using the Tree Builder tool and other functions within RDP for further analyses.

4.2.2.3.1.8.2 BLAST search to identify phylotype close relatives The relatively abundant taxa within a given microbial community are identified from the RDP community composition analyses. To acquire a more detailed estimation of phylotype roles within the community, it is helpful to identify their closely related, cultured, microorganisms. The following is a method for using the Basic Local Alignment Search Tool (BLAST) tool to perform this type of analyses:

While new algorithms are constantly being developed, researchers still use the basic sequence comparison tool known as BLAST to gain a general understanding of the taxonomic composition within a given sample [32]. The BLAST algorithm will

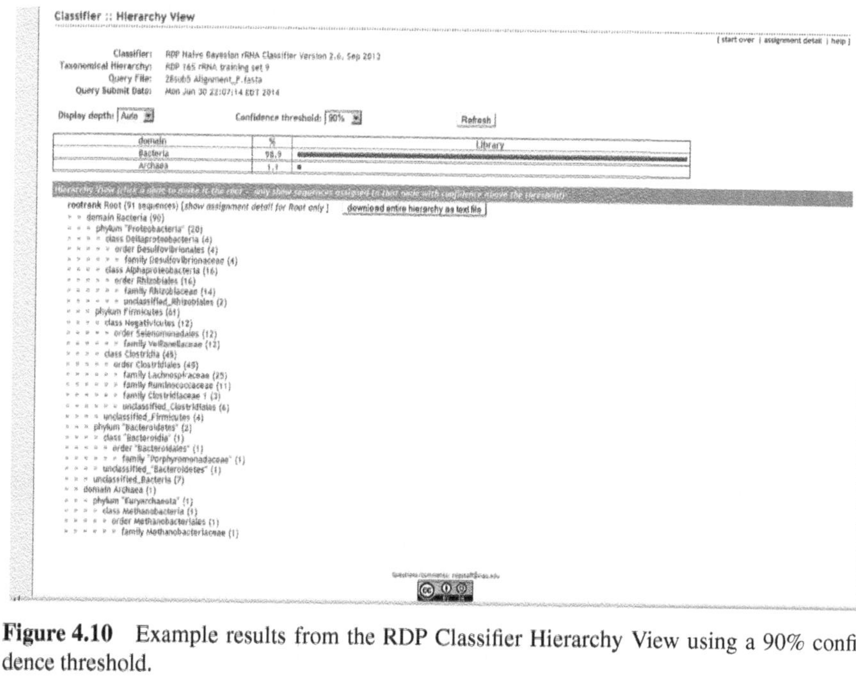

Figure 4.10 Example results from the RDP Classifier Hierarchy View using a 90% confidence threshold.

compare query sequences against those stored in GenBank, which is the NIH database maintained and distributed by The National Center for Biotechnology (NCBI) that stores all known public DNA sequences. Here, we provide a basic protocol for 16S rRNA sequence analysis using freeware.

1. Navigate with your browser to The National Center for Biotechnology Information (NCBI) BLAST website: http://blast.ncbi.nlm.nih.gov/Blast.cgi. Figure 4.11 shows the NCBI BLAST Home page.

There are a variety of specific searches that are available using the same basic underlying BLAST tool. As the 16S rRNA gene sequence data contain nucleotide information, first click on the nucleotide–nucleotide BLAST (blastn) link to open the appropriate search page. This search page compares the nucleotide sequence(s) you input to an entire database of previously identified nucleotide stored at NCBI (GenBank).

1. Select nucleotide blast.
2. In the web browser, upload your FASTA files.
3. Enter a job title.

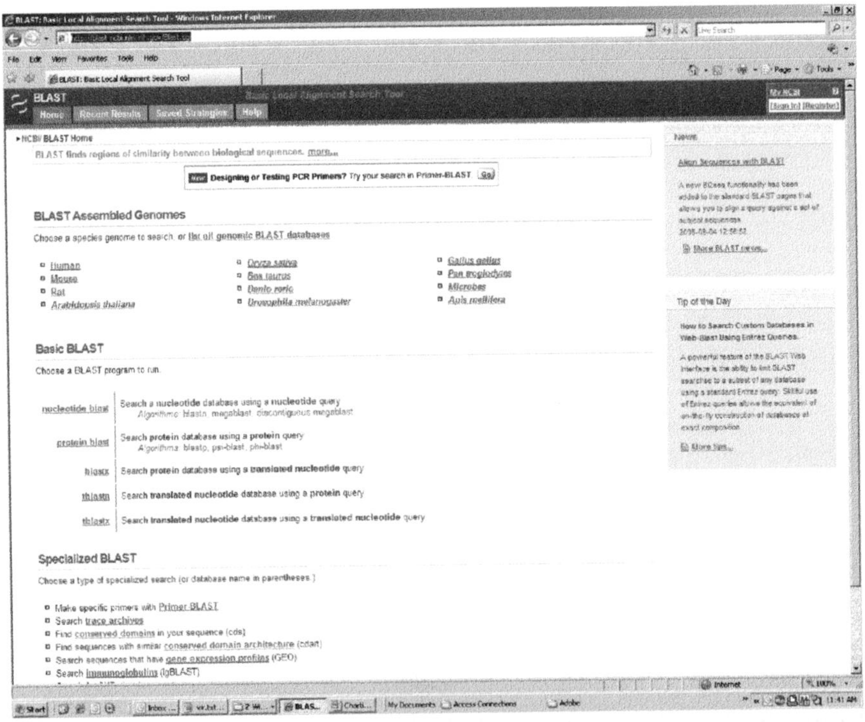

Figure 4.11 BLAST Home page.

4. Click on the drop-down box located next to the Database tab under the "Choose Search Set" section, and select "Nucleotide collection (nr/nt)."
5. Check the box for "Exclude Uncultured/environmental sample sequences."
6. Optimize for "Highly similar sequences" by choosing the megablast option under the "Program Selection" section.
7. Click BLAST at the bottom of the page to initiate the search. A new page will be displayed with the specified job title and the status of the search. This page will be updated every few seconds until the results are presented.

When the search is complete, a results page will appear. If no matches are found, this will be indicated on the page. An example is shown in Figure 4.12. If matches are found, a new page will display a graphic summary, sequence descriptions, and sequences alignments. Figure 4.13 shows an example.

The best match will be listed at the top of the "Sequence Descriptions" section, followed by matches that are of less quality. Click on the Accession numbers in the right-hand column to view a new page with the details of the matching sequence. Figure 4.14 shows an example.

Using the data generated from RDP and BLAST, the researcher is able to conduct statistical analyses of their communities. Such analyses may include estimations of

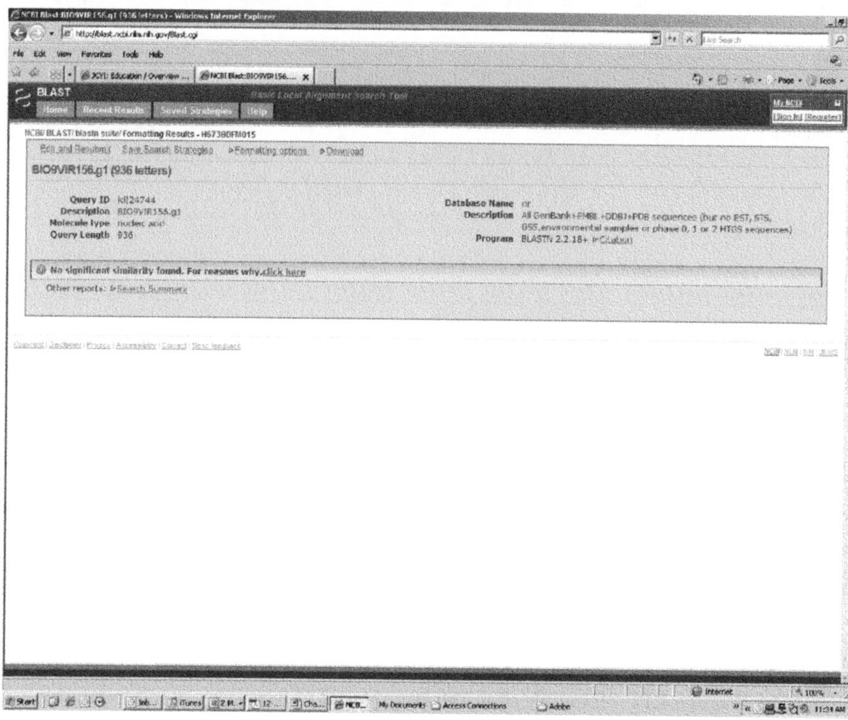

Figure 4.12 Example of results page in the case that no sequence matches were found.

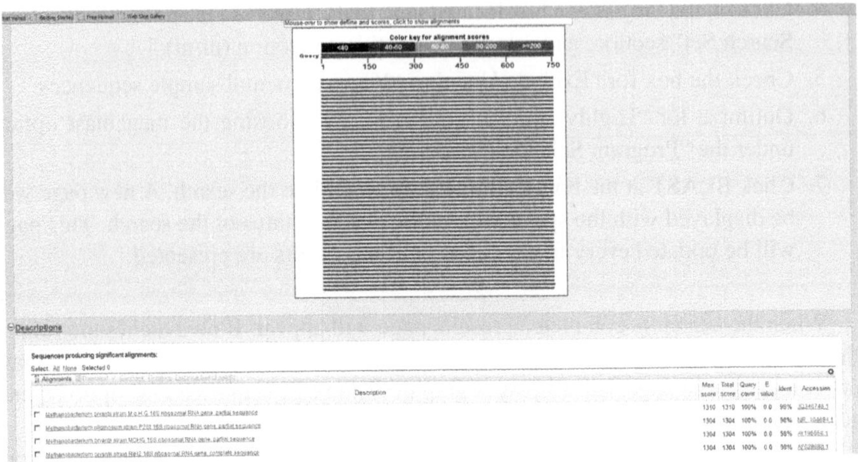

Figure 4.13 Example of successful BLAST results.

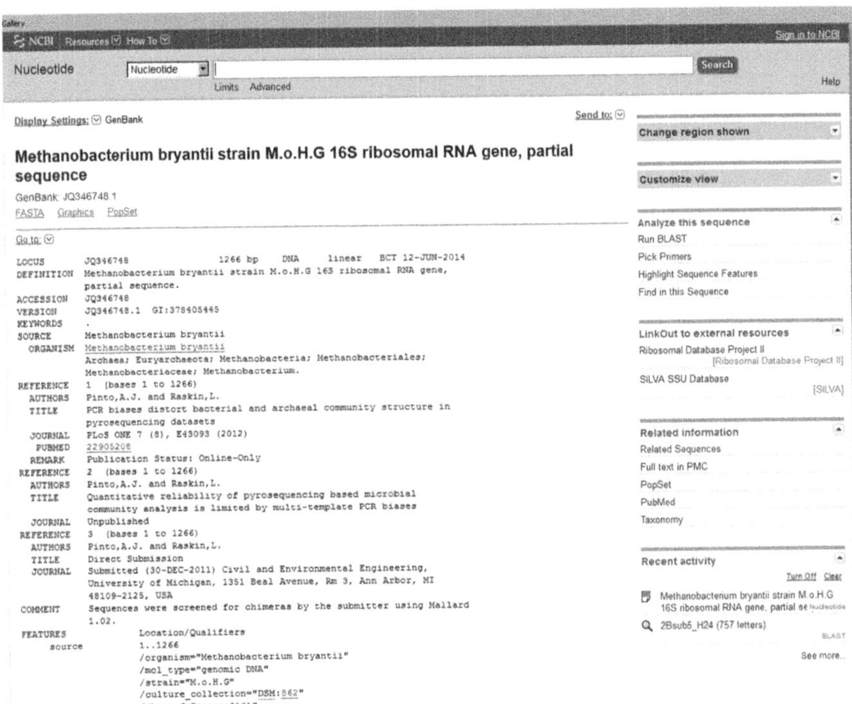

Figure 4.14 Example of results page linked to Accession numbers.

population diversity [33] and/or similarity [34] by determining the Shannon diversity index (http://www.changbioscience.com/genetics/shannon.html), Chao richness [35] (http://www2.biology.ualberta.ca/jbrzusto/rarefact.php), Rarefaction [36] curve (http://strata.uga.edu/software/index.html), and the Sorensen similarity index [37] (EstimateS, http://viceroy.eeb.uconn.edu/estimates/).

4.3 PERSPECTIVES

The culture-independent approach for taxonomic characterization via 16S rRNA gene sequencing has led to significant new discoveries and increasing knowledge relative to the microbial populations that exist and function in BES reactors. The advent of NGS technology has enabled an even greater understanding about the relative abundance of microbial taxa in BES reactors, their genetic potential, and actual function. It is now possible to describe species-level taxonomy, metabolic networks, and electron transport pathways within diverse BES-associated communities by combining metagenomic and metatranscriptomic sequencing information with BES operational data [18, 38]. The use of a full "systems-level" approach including metaproteomics and metabolomics for understanding BES communities is the next natural progression toward optimizing BESs.

REFERENCES

1. Maron P-A, Ranjard L, Mougel C, Lemanceau P. Metaproteomics: a new approach for studying functional microbial ecology. Microb Ecol 2007;53(3):486–493.
2. Nevin KP, Richter H, Covalla S, Johnson J, Woodard T, Orloff A, Jia H, Zhang M, Lovley D. Power output and columbic efficiencies from biofilms of *Geobacter sulfurreducens* comparable to mixed community microbial fuel cells. Environ Microbiol 2008;10(10):2505–2514.
3. Ishii Si, Watanabe K, Yabuki S, Yabuki BE, Sekiguchi Y. Comparison of electrode reduction activities of *Geobacter sulfurreducens* and an enriched consortium in an air-cathode microbial fuel cell. Appl Environ Microbiol 2008;74(23):7348–7355.
4. McLean JS, Wanger G, Gorby YA, Wainstein M, McQuaid J, Ishii S i, Bretschger O, Beyenal H, Nealson KH. Quantification of electron transfer rates to a solid phase electron acceptor through the stages of biofilm formation from single cells to multicellular communities. Environ Sci Technol 2010;44(7):2721–2727.
5. Lane DJ, Pace B, Olsen GJ, Stahl DA, Sogin ML, Pace NR. Rapid determination of 16S ribosomal RNA sequences for phylogenetic analyses. Proc Natl Acad Sci U S A 1985;82(20):6955–6959.
6. Muyzer G, De Waal EC, Uitterlinden AG. Profiling of complex microbial populations by denaturing gradient gel electrophoresis analysis of polymerase chain reaction-amplified genes coding for 16S rRNA. Appl Environ Microbiol 1993;59(3):695–700.
7. Liu W-T, Marsh TL, Cheng H, Forney LJ. Characterization of microbial diversity by determining terminal restriction fragment length polymorphisms of genes encoding 16S rRNA. Appl Environ Microbiol 1997;63(11):4516–4522.

8. Fisher MM, Triplett EW. Automated approach for ribosomal intergenic spacer analysis of microbial diversity and its application to freshwater bacterial communities. Appl Environ Microbiol 1999;65(10):4630–4636.

9. Hugenholtz P, Pace NR. Identifying microbial diversity in the natural environment: a molecular phylogenetic approach. Trends Biotechnol 1996;14(6):190–197.

10. (a) Huse SM, Huber JA, Morrison HG, Sogin ML, Welch DM. Accuracy and quality of massively parallel DNA pyrosequencing. Genome Biol 2007;8(7):R143; (b) Ronaghi M, Uhlén M, Nyrén P. A sequencing method based on real-time pyrophosphate. Science 1998;281(5375):363–365.

11. Wu M, Eisen JA. A simple, fast, and accurate method of phylogenomic inference. Genome Biol 2008;9(10):R151.

12. (a) Cole S, Brosch R, Parkhill J, Garnier T, Churcher C, Harris D, Gordon S, Eiglmeier K, Gas S, Barry CE. Deciphering the biology of *Mycobacterium tuberculosis* from the complete genome sequence. Nature 1998;393(6685):537–544; (b) Shendure J, Ji H. Next-generation DNA sequencing. Nat Biotechnol 2008;26(10):1135–1145; (c) DeLong EF. Microbial community genomics in the ocean. Nat Rev Microbiol 2005;3(6):459–469.

13. (a) Mardis ER. The impact of next-generation sequencing technology on genetics. Trends Genet 2008;24(3):133–141; (b) Metzker ML. Sequencing technologies – the next generation. Nat Rev Genet 2009;11(1):31–46; (c) Morozova O, Marra MA. Applications of next-generation sequencing technologies in functional genomics. Genomics 2008;92(5):255–264; (d) Quail MA, Smith M, Coupland P, Otto TD, Harris SR, Connor TR, Bertoni A, Swerdlow HP, Gu Y. A tale of three next generation sequencing platforms: comparison of Ion Torrent, Pacific Biosciences and Illumina MiSeq sequencers. BMC Genomics 2012;13(1):341; (e) Suzuki S, Ono N, Furusawa C, Ying B-W, Yomo T. Comparison of sequence reads obtained from three next-generation sequencing platforms. PLoS One 2011;6(5):e19534.

14. (a) Zerbino DR, Birney E. Velvet: algorithms for de novo short read assembly using de Bruijn graphs. Genome Res 2008;18(5):821–829; (b) Zerbino DR. Using the Velvet de novo assembler for short-read sequencing technologies. Curr Protoc Bioinformatics 2010:11.5. 1–11.5. 12.

15. Namiki T, Hachiya T, Tanaka H, Sakakibara Y. MetaVelvet: an extension of Velvet assembler to de novo metagenome assembly from short sequence reads. Nucleic Acids Res 2012;40(20):e155.

16. Noguchi H, Park J, Takagi T. MetaGene: prokaryotic gene finding from environmental genome shotgun sequences. Nucleic Acids Res 2006;34(19):5623–5630.

17. Tanenbaum DM, Goll J, Murphy S, Kumar P, Zafar N, Thiagarajan M, Madupu R, Davidsen T, Kagan L, Kravitz S. The JCVI standard operating procedure for annotating prokaryotic metagenomic shotgun sequencing data. Stand Genomic Sci 2010;2(2):229.

18. Ishii S i, Suzuki S, Norden-Krichmar TM, Tenney A, Chain PSG, Scholz MB, Nealson KH, Bretschger O. A novel metatranscriptomic approach to identify gene expression dynamics during extracellular electron transfer. Nat Commun 2013;4:1601.

19. Cuvelier ML, Allen AE, Monier A, McCrow JP, Messié M, Tringe SG, Woyke T, Welsh RM, Ishoey T, Lee J-H. Targeted metagenomics and ecology of globally important uncultured eukaryotic phytoplankton. Proc Natl Acad Sci U S A 2010;107(33):14679–14684.

20. (a) Scholz MB, Lo C-C, Chain PSG. Next generation sequencing and bioinformatic bottlenecks: the current state of metagenomic data analysis. Curr Opin Biotechnol 2012;23(1):9–15; (b) Koboldt DC, Steinberg KM, Larson DE, Wilson RK, Mardis

ER. The next-generation sequencing revolution and its impact on genomics. Cell 2013;155(1):27–38; (c) Mardis ER. Next-generation DNA sequencing methods. Annu Rev Genomics Hum Genet 2008;9:387–402; (d) Mardis ER. Next-generation sequencing platforms. Annu Rev Anal Chem 2013;6:287–303; (e) Metzker ML. Sequencing technologies – the next generation. Nat Rev Genet 2010;11(1):31–46; (f) Quail MA, Kozarewa I, Smith F, Scally A, Stephens PJ, Durbin R, Swerdlow H, Turner DJ. A large genome center's improvements to the Illumina sequencing system. Nat Methods 2008;5(12):1005–1010; (g) Ronaghi M. Pyrosequencing sheds light on DNA sequencing. Genome Res 2001;11(1):3–11.

21. (a) Ishii Si, Suzuki S, Norden-Krichmar TM, Phan T, Wanger G, Nealson KH, Sekiguchi Y, Gorby YA, Bretschger O. Microbial population and functional dynamics associated with surface potential and carbon metabolism. ISME J 2014;8:963–978; (b) Ishii S i, Suzuki S, Norden-Krichmar TM, Wu A, Yamanaka Y, Nealson KH, Bretschger O. Identifying the microbial communities and operational conditions for optimized wastewater treatment in microbial fuel cells. Water Res 2013;47:7120–7130; (c) Ishii S i, Suzuki S, Norden-Krichmar TM, Nealson KH, Sekiguchi Y, Gorby YA, Bretschger O. Functionally stable and phylogenetically diverse microbial enrichments from microbial fuel cells during wastewater treatment. PLoS One 2012;7(2):e30495.

22. Olson BJSC, Markwell J. Assays for Determination of Protein Concentration. In *Current Protocols in Protein Science*. John Wiley & Sons, Inc.; 2001.

23. (a) Baker GC, Smith JJ, Cowan DA. Review and re-analysis of domain-specific 16S primers. J Microbiol Methods 2003;55(3):541–555; (b) Watanabe K, Kodama Y, Harayama S. Design and evaluation of PCR primers to amplify bacterial 16S ribosomal DNA fragments used for community fingerprinting. J Microbiol Methods 2001;44(3):253–262.

24. Woese CR, Fox GE. Phylogenetic structure of the prokaryotic domain: the primary kingdoms. Proc Natl Acad Sci U S A 1977;74(11):5088–5090.

25. Case RJ, Boucher Y, Dahllöf I, Holmström C, Doolittle WF, Kjelleberg S. Use of 16S rRNA and rpoB genes as molecular markers for microbial ecology studies. Appl Environ Microbiol 2007;73(1):278–288.

26. (a) Mullis KB, Faloona FA. Specific synthesis of DNA in vitro via a polymerase-catalyzed chain reaction. Methods Enzymol 1987;155:335–350; (b) Saiki RK, Scharf S, Faloona FA, Mullis KB, Horn GT, Erlich HA, Arnheim N. Enzymatic amplification of beta-globin genomic sequences and restriction site analysis for diagnosis of sickle cell anemia. Science 1985;230(4732):1350–1354; (c)CSH Cold Spring Harbor Laboratory. DNA Learning Center. http://www.dnalc.org/resources/animations/pcr.html, XXX.

27. DeLong EF. Archaea in coastal marine environments. Proc Natl Acad Sci U S A 1992;89(12):5685–5689.

28. Liu Z, Lozupone C, Hamady M, Bushman FD, Knight R. Short pyrosequencing reads suffice for accurate microbial community analysis. Nucleic Acids Res 2007;35(18):e120.

29. Hall, T. A. BioEdit: a user-friendly biological sequence alignment editor and analysis program for Windows 95/98/NT, In: Nucleic acids symposium series, 1999; pp 95–98.

30. Cole JR, Wang Q, Fish JA, Chai B, McGarrell DM, Sun Y, Brown CT, Porras-Alfaro A, Kuske CR, Tiedje JM. Ribosomal Database Project: data and tools for high throughput rRNA analysis. Nucleic Acids Res 2014;42(D1):D633–D642.

31. Garrity GM, Bell JA, Lilburn TG. Taxonomic Outline of the Prokaryotes. Bergey's Manual of Systematic Bacteriology. New York, Berlin, Heidelberg,: Springer; 2004.

32. Johnson M, Zaretskaya I, Raytselis Y, Merezhuk Y, McGinnis S, Madden TL. NCBI BLAST: a better web interface. Nucleic Acids Res 2008;36(Suppl 2):W5–W9.
33. (a) Lieberson S. Measuring population diversity. Am Sociol Rev 1969;34:850–862; (b) Magurran AE, Magurran AE. Ecological Diversity and Its Measurement. Vol. 168. Springer; 1988; (c) Bianchi MG, Bianchi AM. Statistical sampling of bacterial strains and its use in bacterial diversity measurement. Microb Ecol 1982;8(1):61–69.
34. (a) Chao A, Chazdon RL, Colwell RK, Shen TJ. A new statistical approach for assessing similarity of species composition with incidence and abundance data. Ecol Lett 2005;8(2):148–159; (b) Gotelli NJ, Colwell RK. Quantifying biodiversity: procedures and pitfalls in the measurement and comparison of species richness. Ecol Lett 2001;4(4):379–391; (c) Colwell RK, Mao CX, Chang J. Interpolating, extrapolating, and comparing incidence-based species accumulation curves. Ecology 2004;85(10):2717–2727.
35. Simberloff D. Properties of the rarefaction diversity measurement. Am Nat 1972;106:414–418.
36. Heck KL Jr, van Belle G, Simberloff D. Explicit calculation of the rarefaction diversity measurement and the determination of sufficient sample size. Ecology 1975;56(6):1459–1461.
37. Wolda H. Similarity indices, sample size and diversity. Oecologia 1981;50(3):296–302.
38. Rittmann BE, Krajmalnik-Brown R, Halden RU. Pre-genomic, genomic and post-genomic study of microbial communities involved in bioenergy. Nat Rev Microbiol 2008;6(8):604–612.

5

BIOFILM ELECTROCHEMISTRY

Jerome T. Babauta and Haluk Beyenal
The Gene and Linda Voiland School of Chemical Engineering and Bioengineering, Washington State University, Pullman, WA, USA

5.1 INTRODUCTION

The electrochemist will tell you that cyclic voltammetry (CV) is absolutely necessary to diagnose an electrochemical reaction occurring in an electrochemical cell. By merely glancing at cyclic voltammograms, an electrochemist will arrive at a series of conclusions on the basis of previous experience and hypotheses that could be tested with further electrochemical techniques. Unfortunately, the inexperienced beginner will have no clue what the same set of cyclic voltammograms represents and will not have the experience or time to interpret those results. Biofilm researchers interested in experimenting with CV on what they believe is an electrochemically active biofilm (EAB) will be at an even greater disadvantage, as the complexities of metabolism in addition to biofilm heterogeneity make the simple case studies in classical electrochemical textbooks severely inadequate. An electrochemist meeting with a biofilm researcher to discuss the capabilities and limitations of electrochemistry in biofilms might say that it is impossible to study such a system without first reducing its complexity. The biofilm researcher would say that the complexity is of the utmost importance to understanding biofilm processes and that, far from being removed, it should be accounted for in any electrochemical experimentation. At this point, there is a mismatch in what should be coined "biofilm electrochemistry." The electrochemist will go through many steps to remove complications and generate "simple" cyclic voltammograms. The biofilm researcher will generate cyclic voltammograms with all the complications included and attempt to adapt electrochemical theory through various assumptions. At some point, the two approaches should reach common ground. In this chapter, we discuss and describe CV for the electrochemists and biofilm researchers wanting to study EABs on common ground.

Biofilms in Bioelectrochemical Systems: From Laboratory Practice to Data Interpretation, First Edition.
Edited by Haluk Beyenal and Jerome Babauta.

To begin, an EAB is simply a biofilm that can interact with a metal electrode by either transferring electrons to the electrode (anodic behavior) or removing electrons from it (cathodic behavior). Together, we refer to the biofilm/electrode composite as the biofilm electrode. The biofilm electrode can contain multiple species of bacteria and other microorganisms, each with its own unique metabolic activities. Although in principle, one could study electrochemistry at the biofilm electrode using a mixed culture biofilm, it is generally better to isolate strains of bacteria and grow monospecies biofilms. Working with monospecies biofilms simplifies the metabolic considerations of the biofilm electrode. Once the electrochemical behavior (i.e., anodic/cathodic behavior) of individual species has been sufficiently characterized, individual strains can be studied in co-culture. The electrochemical behavior of the biofilm electrode is governed by concentration gradients, or simply chemical and electrochemical gradients, formed within the biofilm. These include, but are not limited to: (1) pH, (2) solution redox potential, (3) biofilm redox potential, (4) oxygen, (5) light intensity, (6) carbon substrate, and (7) diffusing mediators. These are discussed in detail in Chapters 2 and 9. Although the initial electron-transfer studies of biofilm electrodes investigated macroscopic concentration changes in the bulk or "near" the biofilm electrode, recent studies have focused on studying microscale gradients within the biofilm. Currently, our knowledge of chemical and electrochemical gradients in living biofilms respiring on a solid surface is still limited.

In practice, the electrochemical behavior of a biofilm electrode is characterized by applying a potential with respect to a known reference potential. The polarization potential may or may not cause current to flow. Because of this, a wide range of polarization potentials are tested, giving rise to what is called voltammetry. At the practical level, voltammetry is an umbrella term that covers electrochemical techniques that utilize a potential sweep (potential waveform) to cause a current to flow in an electrochemical cell. The resulting current is measured and related back to the electrode potential. The current/potential relationship is observed and used to form hypotheses about the possible mechanisms controlling electron transfer in biofilms. Thus, voltammetry is a powerful tool for investigating electron-transfer mechanisms in biofilms using classical electrochemical theory to explain bioelectrochemical phenomena. Unfortunately, the complexity of biofilms limits the use of classical electrochemical theory, and several empirical correlations have been proposed [1, 2]. Voltammetry, along with other independent experimental data, is often used to support a hypothesized electron-transfer mechanism by noting changes to the voltammogram under specified experimental conditions (i.e., varying temperature, strain type, salinity, electrode material, electrode surface area, etc.). Therefore, *the crux of most arguments centered on voltammetric studies of biofilms lies in the "perceived" change in a voltammogram.* It is extremely important that proper steps were taken to achieve the most rigorously correct voltammogram for the given experimental system. Therefore, the practical considerations for CV of EABs are of paramount importance.

The goal of this chapter is to demonstrate the experimental steps necessary to conduct and optimize voltammetry of EABs successfully. A brief introduction to EABs and voltammetry is provided; however, the bulk of the chapter is concerned with describing the electrochemical cell and setup to be used. We focus on *Geobacter sulfurreducens* as a model organism to facilitate discussion; however, all of the topics covered here are applicable to other EABs. In particular, we provide a basic analysis of the effect of electrode materials on the limitations of voltammetry and show how the counter electrode (CE) reacts to the current generation in conjunction with the control of the working electrode (WE). The discussion is based on the use of a potentiostat and is therefore limited to potentiostatic electrochemical techniques. Nonaqueous electrochemistry is not considered here. For modeling of electrochemical systems, we refer the readers to Chapter 9 in this book. We describe how multiple voltammetric techniques can be complementary to each other in elucidating the electrochemical parameters to be determined.

In the following section, several commercial instruments that are useful for electrochemical experiments are shown. In our research laboratory, these instruments have worked well, but they can be substituted for by similar-functioning devices. Sometimes, we provide all the information about the manufacturer and instrument operation. These examples can be extended to the devices of other manufacturers, as the electrochemistry we discuss is independent of the instruments used. Firstly, we define relevant terminology we use in this chapter (Table 5.1). The reader should be familiar with this terminology prior to attempting voltammetry.

5.2 INSTRUMENTATION

Before studying voltammetry, it is critical to understand the practical considerations for the individual components required to run voltammetry on a biofilm electrode. There are a minimal number of components necessary to run voltammetry. In no particular order, the necessary components are (1) a potentiostat, (2) electrochemical cell/reactor, (3) ionic conductor, (4) WE, (5) CE*, (6) reference electrode* (RE), and (7) gas dispersion tube. The asterisks indicate that in a two-electrode configuration, the CE and RE are combined into one auxiliary electrode. Running voltammetry at a biofilm electrode requires only the basic components shown in Figure 5.1a, where a WE with biofilm (i.e., biofilm electrode) is placed in a cell with a CE and an RE. After the oxygen is purged using a gas dispersion tube, a potentiostat is hooked up and the user can run voltammetry. In this setup, the biofilm does not have to be grown in the cell: it can be transferred to it for short experiments. However, when the biofilm is to be grown and studied in the same reactor, the diagram takes on a more complex form, shown in Figure 5.1b. This particular reactor setup was used in Ref. [3]. The user must consider passing the purge gas through a 0.2-μm filter, flow breakers in the feed/waste streams, recycle rates, and the type of flow regime that the biofilm is grown in. In this case, the biofilm electrode and CE sit at the bottom of the shallow well, and a pseudo-reference electrode is threaded through a butyl rubber stopper. The flow across the biofilm is controlled by the recycle rate. In this setup, the biofilm is grown under well-controlled conditions, and voltammetry can be run at any time during its growth. Depending on the needs of the experiment, the user may opt for the simpler, short-term approach in Figure 5.1a or the complex, well-controlled approach

TABLE 5.1 Useful Terms and Their Working Definitions.

Voltammetry	Electrochemical techniques that utilize a potential sweep (potential waveform) to cause a current to flow in an electrochemical cell.
Electrochemically active biofilm	A biofilm that can interact with a metal electrode by either transferring electrons to the electrode (anodic behavior) or removing electrons from it (cathodic behavior).
Anodic biofilm	A biofilm that transfers electrons to an electrode. Anodic biofilms grow on the anodes of microbial fuel cells.
Cathodic biofilm	A biofilm that accepts electrons from an electrode. Cathodic biofilms grow on the cathodes of microbial fuel cells.
Biofilm electrode	The biofilm/electrode composite.
Working electrode (WE)	The electrode where the electrochemical half-reaction of interest occurs in a three-electrode electrochemical cell. When a biofilm is grown on the WE, it is referred to as the biofilm electrode.
Counter electrode (CE)	The current-collecting electrode where a known half-reaction (oxygen reduction, hydrogen evolution, etc.) completes the overall reaction occurring in an electrochemical cell.
Reference electrode (RE)	The non-current-collecting electrode with a stable, well-established reference potential that is used to measure the WE potential.
Microscale gradients	Concentration gradients occurring across a distance of microns. These are usually measured within the biofilm.
Polarization potential	The potential applied to an electrode, usually with a potentiostat, with respect to a known reference potential.
Potentiostat	An electronic device that controls the potential of the working electrode with respect to a reference electrode by controlling the counter electrode potential.

These should be understood before voltammetry is attempted.

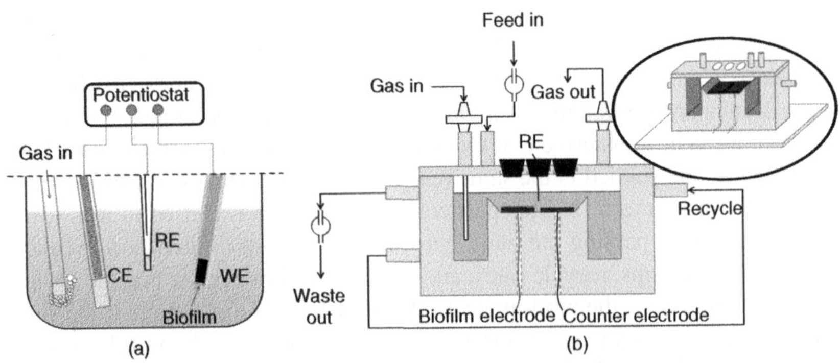

Figure 5.1 (a) Basic components needed to run voltammetry. (b) Extended components required to run a continuous biofilm reactor with recycle. Reproduced with permission from Ref. [3]. Copyright 2012 Wiley Periodicals, Inc.

in Figure 5.1b. In our experience, growing the biofilm and running voltammetry on the biofilm electrode in the same reactor yield more consistent results.

5.2.1 Potentiostat

A potentiostat controls the potential of the WE with respect to an RE by controlling the CE potential. The resulting current is measured. Although the function of a potentiostat is simple, the electronics behind modern potentiostats are so sophisticated that commercially available potentiostats are able to measure femtoamp (10^{-16}) current values, impedances up to 1 MΩ, and more. General discussion of potentiostat circuits, design, and operation can be found elsewhere [4, 5]. Users should also look to the potentiostat manufacturer's website for application notes tailored specifically to their device. Here, we only stress that users select a potentiostat suitable for their experimental needs. Generally, this requires prior knowledge of the electrochemical behavior of the cell/reactor. For example, you must know whether to expect femtoamps, microamps, or amps of current. Manufacturers typically identify their potentiostat products by the minimum and maximum currents that can be handled.

5.2.2 Electrochemical Cell/Reactor

The types of cells/reactors available are innumerous, but they all share the commonality that they were designed to grow biofilms on an electrode in a well-controlled aqueous environment. The geometry, size, and scope are determined by the corresponding application. For example, certain flow cells were developed with optical windows so that the biofilm growing on the electrode could be imaged using microscopy [6–8]. The open-channel flat plate reactor in Figure 5.1b was used for depth profiling with microelectrodes in biofilms on electrodes [3]. Packed columns, H-cells, plate-type, and other geometries exist as well [9]. For biofilm voltammetry, the cell/reactor must: (1) withstand sterilization, (2) be enclosed/sealed with all electrodes secured in place, (3) be chemically inert to (bio)corrosion, (4) resist chemical leaching, and (5) be electrically safe (insulated from any applied voltages). Glass, teflon, nylon, and various plastic resins such as polycarbonate are used as materials for cells/reactors; these are all essentially biocompatible materials. One may need to consider whether a particular material would promote the adsorption of certain chemicals more than another material and the possibility of using coated/composite materials. Finally, all cells/reactors *must be fully compliant with biosafety standards.* We emphasize the prevention or containment of aerosols contaminated with microorganisms. Although not all EABs are pathogenic, users should be aware of the dangers of sparging inert gas into solutions containing microbes.

5.2.3 Ionic Conductor

The ionic conductor for biofilm voltammetry will generally be growth medium with a biological buffering system. As growth medium contains various chloride salts, the ionic strength and therefore the conductivity are sufficient to run voltammetry without the problems associated with high-resistivity media such as nonaqeous solutions. At current values, approaching hundreds of milliamps or more, however, geometry, electrode positioning, and in general, the uncompensated resistance of the electrochemical cell are critical to control for studying electron transfer at the biofilm electrode. This is because the combination of a current passing through a non-negligible uncompensated resistance will cause a potential drop commonly referred to as the *iR* drop [10]. If both the current and uncompensated resistance are sufficiently large such that the potential drop is hundreds of millivolts, the applied potential at the biofilm electrode will be inaccurate. In general, we recommend that researchers measure the uncompensated resistance in electrochemical systems irrespective of the current densities observed. By doing so, researchers can determine if the *iR* drop in a particular cell configuration will obscure attempts at studying the biofilm electrode.

5.2.4 Working Electrode

As defined previously, the WE is the electrode we grow biofilm on and the primary electrode that we are concerned with. We refer to it as the biofilm electrode. The biofilm electrode material affects the measured current and open circuit potential (OCP), and the choice of electrode material is important for the standardization of reported values. Traditionally, cheaper graphite, carbon paper, carbon granule, carbon brush, or carbon felt electrodes are used in microbial fuel cell (MFC) practical applications [11, 12]. These carbon materials suffer from high background currents that can mask the electrochemical responses of redox species at low concentrations. This may be problematic when the catalytic current caused by the biofilm is low. In our laboratory, we often use glassy carbon electrodes to observe electrochemical activity. Glassy carbon has background currents that are practically zero in the potential ranges in which biofilm electrodes are studied, and it is nonporous. In addition, there is significant literature on electrochemistry utilizing glassy carbon electrodes, potentially opening up a vast amount of literature to biofilm studies. The use of glassy carbon electrodes would provide more universal current values from biofilm electrodes. For these reasons, we recommend users at least test their systems with glassy carbon electrodes. There are various glassy carbon types, and readers are referred to the literature for a more thorough review [13]. When glassy carbon electrodes are not compatible with an experiment, common substitutes include gold and indium tin oxide (ITO) electrodes [14]. Gold offers the advantage of a significant literature on self-assembled monolayers and the modification of surface functional groups [15–17]. Thin gold films on glass substrate have also been used in advanced spectroscopic techniques for direct electron-transfer studies [18]. ITO is used in spectroelectrochemical experiments where an optically transparent electrode is required [19, 20]. Users should be aware of the resistivity of ITO electrodes and their durability, as the conductive film is thin compared to glassy carbon [20]. Platinum and other catalytic electrode materials could have unanticipated effects on an experiment and are best avoided.

5.2.5 Counter Electrode

The CE completes the electrochemical cell by providing an electron sink/source in the form of a known abiotic half-reaction. For single-chambered anoxic biofilm experiments, this is usually hydrogen evolution (anodic biofilms) or oxygen evolution (cathodic biofilms). As detailed later in this chapter, in some cases, the reaction products generated at the CE cannot be neglected. Therefore, it is always best to sequester the CE and any associated reaction products. Usually, a cheaper carbon electrode with a larger surface area can be used. Graphite rods, fabric, and felt are typical examples.

5.3 BASICS OF CYCLIC VOLTAMMETRY

This section introduces basic theory behind the potential waveform applied during CV and the current response that should enable readers to follow the practical explanations given as follows for the case studies cited. A rigorous treatment and derivation of the diffusion equations relating transport phenomena with reaction rate kinetics at the electrode surface are not presented here; instead, these are only shown in their final form. There are many classical electrochemistry textbooks that cover these topics, and a trip to a nearby library or accessing an online collection is fully warranted here. However, where necessary, we cover important assumptions about the derivation of current–voltage/current–concentration relationships.

5.3.1 Applied Potential Waveform

CV uses a piecewise linear potential waveform that is described for one cycle (anodic or cathodic direction) as:

$$V(t) = vt + V_{\text{initial}} \quad \text{or} \quad V(t) = V_{\text{initial}} - vt(0 \leq t \leq t_{\text{rev}}) \tag{5.1}$$

$$V(t) = V_{\text{initial}} - vt + 2vt_{\text{rev}} \quad \text{or} \quad V(t) = vt + V_{\text{initial}} - 2vt_{\text{rev}}(t_{\text{rev}} \leq t \leq t_{\text{final}}) \tag{5.2}$$

where t is time, v is the scan rate, $V(t)$ is the electrode potential at time t, V_{initial} is the starting electrode potential at $t = 0$, and t_{rev} is the time at which the scan direction is reversed. $V(t_{\text{rev}})$ is the reversal potential or switching potential. $V(t_{\text{final}})$ is the final potential at the end of the cycle and is called V_{final}. Between $t = 0$ and $t = t_{\text{rev}}$ is the forward scan. Between $t = t_{\text{rev}}$ and $t = t_{\text{final}}$ is the reverse scan. One cycle is thus a combination of linear sweep voltammetry in both the forward and reverse directions. When plotted with multiple cycles, the applied potential waveform is often called a symmetrical sawtooth waveform or triangular waveform (see Fig. 5.2a). In Figure 5.2b, the current response is shown. Usually, there is an initial current transient that decays rapidly during cycle 1. By cycle 2, the current transient is undetectable, and we advise that at least two cycles be run per CV experiment (three is best). Figure 5.2c is the actual cyclic voltammogram showing the start of

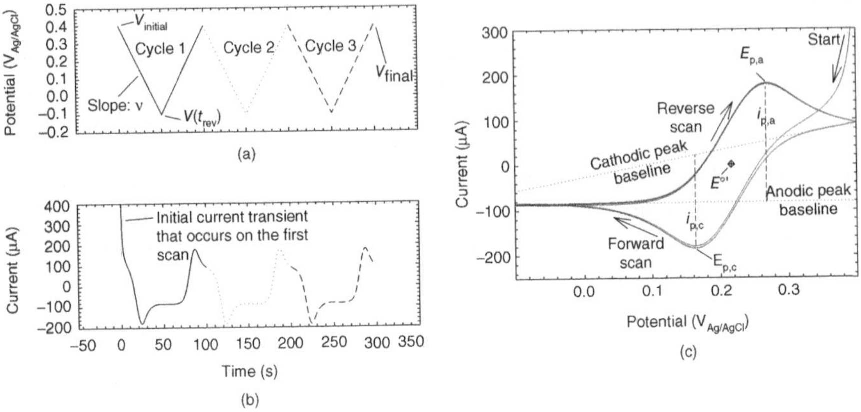

Figure 5.2 (a) Potential waveform applied during a CV. (b) The current response to the applied potential waveform during a CV. (c) Voltammogram showing selected parameters that can be calculated from CV. Note that this voltammogram deviates from the reversible case.

the CV and the effect of the initial current transient. Again, the initial current transient is undetectable after cycle 1. During the forward and reverse scans, cathodic and anodic peaks appear around the formal reduction potential ($E^{\circ\prime}$) of the redox couple. The characteristics of these cathodic and anodic peaks are discussed in the following section in *Case study* 5.2. One practical takeaway message from Figure 5.2a–c is that CVs tend to stabilize with increasing cycles. Usually, this behavior is related to a stabilizing concentration profile adjacent to the electrode surface. However, CVs can also destabilize with increasing cycles. New current peaks can develop after cycle 1, and peaks may also disappear in subsequent cycles. It is important that users observe and identify characteristic features of the CV as cycles 1, 2, 3, … , *n* are run.

When it comes to actually attempting to run a CV on a commercial potentiostat, the default script window for CV will typically look like Figure 5.3. Before operating a potentiostat, it is best to read through the help files and manual provided by the manufacturer. Here, the numbers represent parameters related to "setting" the applied potential waveform and are as follows: (1) the initial potential, at which the potentiostat begins the scan; (2) the first potential setpoint the potentiostat scans toward (determines scan direction); (3) the second potential setpoint the potentiostat scans toward (typically the reverse scan direction); (4) the final potential, at which the potentiostat ends the scan; (5) the scan rate, *v*; (6) the data sampling rate; (7) the number of cycles/repetitions the potentiostat is set to scan (typically at least three); (8) the time duration for which the potentiostat waits at the initial potential, allowing the system to reach a steady state or equilibrium; (9) an option to "condition" the electrode surface prior to a scan. Four potential setpoints are provided so that users can choose to begin and end a scan between the initial and final potentials, which determine the potential range. The advantages of doing so are covered in Figure 5.16c. We note that Figure 5.3 is a screenshot from Framework™ Data Acquisition Software

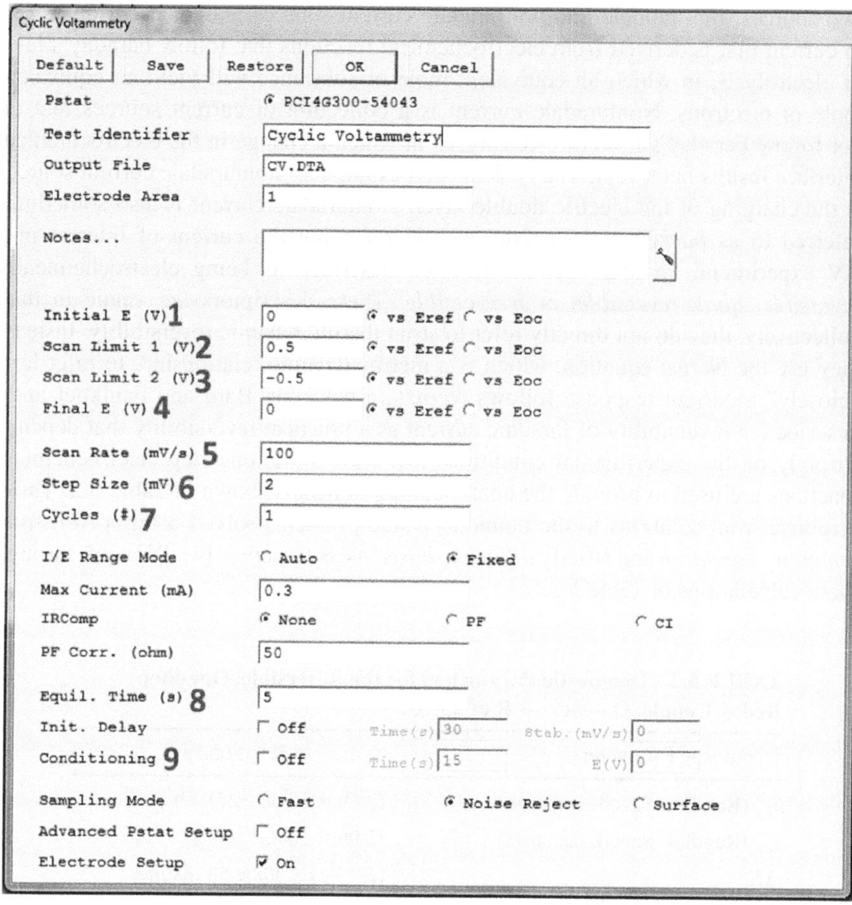

Figure 5.3 Screenshot of a typical cyclic voltammetry script showing the parameters that users need to configure. Here, we show a window from Framework™ Data Acquisition Software (v. 6.23), sold commercially by Gamry Instruments (Warminster, PA, USA).

(v. 6.23), sold commercially by Gamry Instruments (Warminster, PA, USA). Other software provided by potentiostat manufacturers will have similar, if not identical, parameters to configure prior to running CV. There is no fundamental advantage to using this or any other software. Gamry Instruments products are simply what we have available in our research laboratory.

5.3.2 Current Response to Applied Potential: Reversible, Quasi-Reversible, and Irreversible Reactions

As we discussed briefly previously, Figure 5.2b shows the current response during CV. The current response to the triangular potential waveform will be the sum of two sources, the faradaic and nonfaradaic current sources. Faradaic current refers to current that is derived from electrochemical reactions that follow Faraday's laws of electrolysis, in which an equivalent mole of substance will yield an equivalent mole of electrons. Nonfaradaic current is a collection of current sources that do not follow Faraday's laws of electrolysis, in which a change in the electrochemical interface results in current. The typical cited example of nonfaradaic current sources is the charging of the electric double layer. Nonfaradaic current is also sometimes referred to as *background current*, as it is often not the current of interest in a CV experiment. Faradaic current is often described as being electrochemically *reversible*, *quasi-reversible*, or *irreversible*. These descriptors are vague in that, collectively, they do not directly refer to strict thermodynamic reversibility. Instead, they use the Nernst equation, which is a thermodynamic relationship, to infer how "closely" a current response follows *Nernstian* behavior. Bard and Faulkner aptly describe the reversibility of faradaic current as a practical reversibility that depends strongly on the experimental conditions [4]. Reversible, one-step electrochemical reactions are used to provide the analytical relationships shown in Table 5.2. These originate from solutions to the boundary value problems solved using a *Nernstian boundary condition* and strictly *diffusion-based mass transport* [4]. We work through these calculations in Table 5.3.

TABLE 5.2 Diagnostic Parameters for the Reversible, One-Step Redox Couple, O + ne^- → R at 25 °C.

Diagnostic Parameters	Relationship
$i_{p,a}$ (Randles–Sevcik equation)	$(2.69 \times 10^5)n^{3/2}AD_o^{1/2}C_o v^{1/2}$
$i_{p,c}$ (Randles–Sevcik equation)	$(2.99 \times 10^5)\alpha^{1/2}AC_o D_o^{1/2} v^{1/2}$
ΔE	$(E_{p,a} - E_{p,c})/n \approx 58$ mV/n
$E^{\circ\prime}$	$\approx(E_{p,a} + E_{p,c})/2$
$E_{1/2}$	$E^{\circ\prime} - (0.059/n) \times \log[D_o^{1/2}/D_R^{1/2}]$
Peak/current ratio	$i_{p,a}/i_{p,c} \approx 1$

TABLE 5.3 Diagnostic Parameters Extracted from the Scan Rate Analysis in Figure 5.8.

Parameter	Scan Rate (mV s^{-1})						
	1	5	10	20	50	70	100
$E_{p,a}$	289	286	292	299	314	322	329
$E_{p,c}$	–	187	181	173	158	150	142
ΔE	–	99	111	126	156	172	187
$E^{\circ\prime}$	–	237	237	236	236	236	236
$i_{p,a}/i_{p,c}$	–	0.78	0.98	0.98	0.97	0.96	0.96

5.3.3 Caveat to the Use of Analytical Solutions of Cyclic Voltammetry

For new researchers wanting to use CV, it is important to realize that although elegant solutions to the boundary value problems exist for the electrode interface, most of these relationships will be diagnostic in practice. The reason for this is that significant effort is needed to ensure that a particular experimental setup will satisfy the assumptions described by the analytical solutions. The effort includes using high-purity (generally expensive!) electrode materials that have been polished to a mirror-like finish. Various chemical treatments and electrochemical pretreatments for differing electrode materials are then used by many researchers to obtain reproducible, ultrapure surfaces with homogenous surface chemistries. We encourage the readers to search for these methods to see for themselves the wide variety of treatments available, such as those used in gold electrode preparation [21]. Furthermore, the aqueous solutions/solvents used in the electrochemical cell must be equally of high purity! Thus, if a new researcher does not endeavor to make this level of effort, CV will generally yield diagnostic information rather than quantitative determination of electrochemical parameters. For example, experimental data such as those shown in Figure 5.2 will behave as predicted by the relationships shown in Table 5.2 when values such as bulk concentration, temperature, and scan rate are varied. However, deviations from reversibility will obscure quantitative determinations. In addition, factors such as the nonfaradaic current and background currents in general must be well behaved such that accurate baselines can be measured. Examples of these baselines are drawn in Figure 5.2c and are covered in more detail in the literature [4]. Although we are critical in our attitude toward the application of these analytical solutions, they are indispensable to researchers. They provide the basis for most electrochemical hypotheses. With enough statistical power (many experimental replicates) and variation in key experimental parameters such as scan rate, precise estimates of electrochemical parameters can be made even if the true value is obscured by nonideality. Scan rate analysis calculations that rely on the Randles–Sevcik equation shown in Table 5.2 are discussed in *Case study* 5.2.

5.4 CYCLIC VOLTAMMETRY CASE STUDIES

In the following section, three case studies are presented to demonstrate relevant CV experiments for studying biofilm electrodes. Firstly, the blank cell and the dummy cell are discussed in the context of the electric double layer. Secondly, the control case is presented using the ferri/ferrocyanide redox couple and the deviation from reversibility. Thirdly, the biofilm case is presented using a *G. sulfurreducens* biofilm respiring on an electrode. By examining the three cases together, readers can gain an understanding of the electrochemical concepts and limitations in translating traditional electrochemical analysis from simple redox systems to electrode-respiring biofilms. In addition, we purposefully combined the three case studies to demonstrate the importance of "calibrating" the electrochemical cell/reactor using relevant and known electrochemical redox systems before attempting to analyze electrode-respiring biofilms. The "blank" cell, or control "without biofilm," is not adequate for determining whether the electrochemical setup is capable of making key measurements. Thus, "calibrating" the cell/reactor is a critical first step. This is a routine process we follow in our research. We design a biofilm reactor system and test it using known electrochemical systems to verify its suitability for the particular electrochemical analysis to be used for the biofilm. We then grow the biofilm and are able to interpret the electrochemical results successfully.

Case study 5.1 The blank cell and the dummy cell

All electrochemical interfaces can be understood in terms of their current–voltage characteristics. One of the simplest current–voltage relationships that can be studied is that of a blank cell. "Blank" refers to the fact that no faradaic reactions are expected to occur. In a blank cell, an interface is formed between the electrode and an ionic conductor. At this interface, a balance of charge exists between the surface charge of the electrode and ions in solution with a separation distance of only a few angstroms [4]. The physical description and theory of this kind of electrochemical interface is known as the electric double-layer theory. We note that we purposefully do not go into greater detail about the fundamental theory behind the electric double layer and refer readers to classical electrochemical textbooks. The most important fact derived from the theory is that a separation of charge at the electrochemical interface causes capacitive behavior in actual electrochemical experiments. The separation of charge at the electrochemical interface, the transfer of electrons through that interface, and the energy barrier to transferring an electron dictate the shape of the current–voltage relationship. However, in a blank cell with no faradaic reaction, what is the expected current–voltage relationship and how can we describe it?

For a working knowledge of the electric double layer, it is easier to treat the separation of charge as a capacitor with a real capacitance that can be calculated using the ionic conductor as the dielectric. A rough approximation of the capacitance of the electric double layer is approximately 10 μF cm^{-2} [4]. As a result of the electric double layer, we expect the current–voltage relationship to have a capacitive characteristic. The ionic conductor also has an inherent resistance controlled by ionic strength (i.e., aqueous salt solutions). In addition, the route by which electrons pass through the interface can be thought of as a resistive pathway. Finally, we assume that faradaic and nonfaradaic currents occur simultaneously in practice. Using this overly simplified picture of the electrode interface, we can construct a model of electron pathways at the electrode interface. This model has often been described electrically as a parallel resistor–capacitor (RC) circuit. Figure 5.4 depicts this model, which will aid in our understanding of CV.

The advantage of this simple model is that it is easily constructed in the laboratory using resistors and capacitors. Some potentiostat manufacturers provide a high-quality cell that is similar to that shown in Figure 5.4 and is called a dummy cell. Through trial and error, one might learn to construct the circuit using a 2-MΩ resistor, 150-Ω resistor, and 10-μF capacitor. Figure 5.5 shows the

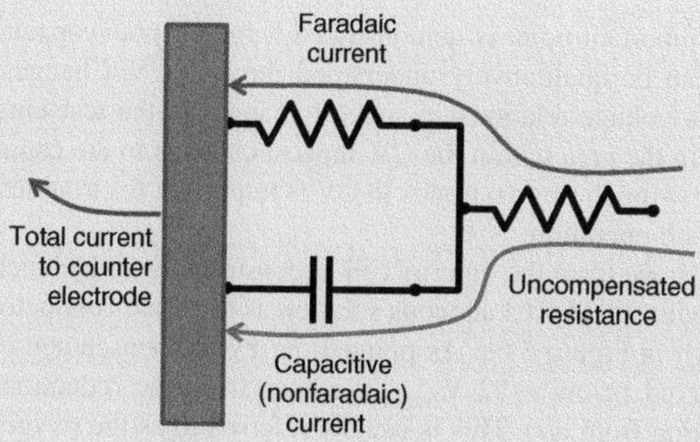

Figure 5.4 Simple parallel RC circuit model of the electrochemical interface.

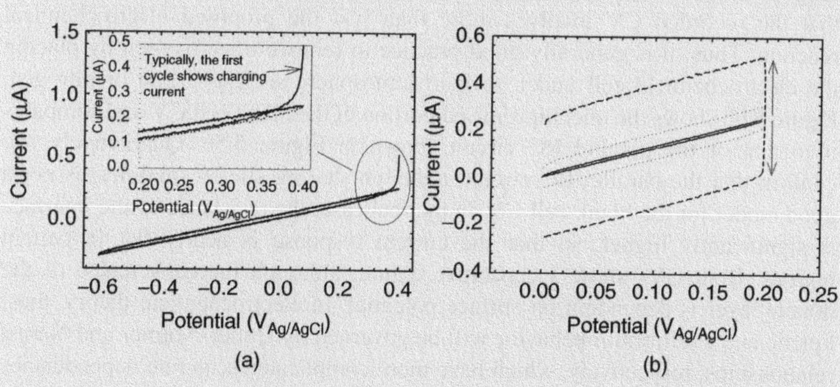

Figure 5.5 (a) CV of the parallel RC circuit showing two scans. Inset emphasizes the charging current in the initial scan. (b) Increase in capacitive current as scan rate is increased. Parallel RC circuit constructed with a 2-MΩ resistor, a 150-Ω resistor, and a 10-μF capacitor. The arrows show the increase in current with increasing scan rate.

result of running CV across the circuit. Figure 5.5a shows that, as in Figure 5.2c, the first scan has a transient that disappears with subsequent scans. The transient is attributed to charging of the capacitor. Figure 5.5b shows that increasing the scan rate increases the area within the CV. The increase in capacitive current is expected to follow the following relationship.

$$I = CdV/dt = Cv \tag{5.3}$$

Unlike capacitive current, resistive current does not change with scan rate, as there is no time dependence (i.e., $I = V/R$). At any given time, the total current passing through the circuit is

$$I_{\text{total}} = V/R + Cv \tag{5.4}$$

Thus, an important point is demonstrated here that both capacitive and resistive effects can be qualitatively understood through CV. Changes in the slope of the current–voltage relationship suggest changes in the resistance of the cell, and changes in the area within the CV suggest changes in the capacitance of the interface. Knowing these two trends in CV is important for troubleshooting electrochemical cell operation.

When we move from the dummy cell to constructing a blank cell as shown in Figure 5.1a with 0.1 M KCl aqueous solution as the ionic conductor, we observe the CV shown in Figure 5.6a. As pointed out by Vanbenschoten et al., [22] the current observed below −0.2 $V_{Ag/AgCl}$ comes from the reduction of dissolved oxygen (coming from air). This is usually referred to as the oxygen signal. Any trace oxygen will interfere with current measurements in the relevant potential range. Furthermore, oxygen may react chemically with redox mediators such that the recorded CV involves more than just the proposed electrochemical reaction. Thus, it is generally good practice to remove trace oxygen by placing the electrochemical cell under an inert atmosphere such as argon or nitrogen. Figure 5.6b shows the uncompromised portion of the blank cell CV and compares it to that of the parallel RC circuit shown in Figure 5.5b. Qualitatively, we confirm that the parallel RC circuit, although sharing similar qualities, is not a good model for the blank cell CV. In the real case, the resistance of the interface is significantly higher, so that the current response is nearly flat in certain regions. Both CVs show a capacitive nature, although the capacitance of the double layer is dependent on surface potential. In electrochemical theory, most kinetic and equilibrium behavior will be governed by Butler–Volmer and Nernst relationships, respectively, which have more complicated scan rate dependencies than the simple parallel RC circuit used to facilitate the electric double-layer discussion.

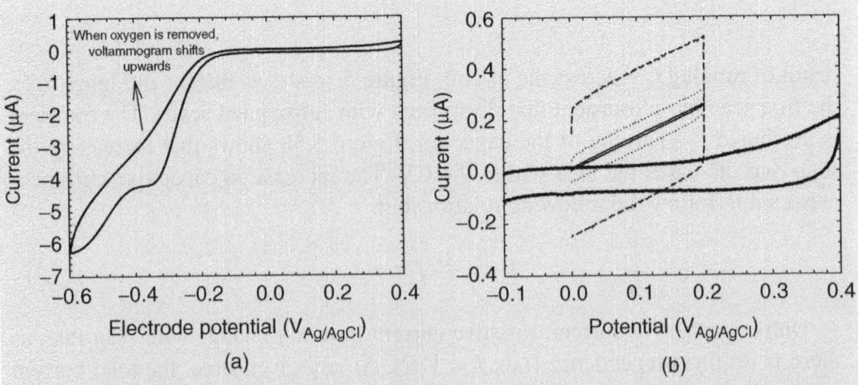

Figure 5.6 (a) CV in 0.1 M KCl solution showing the oxygen signal. (b) Comparison of the parallel RC circuit and 0.1 M KCl CVs.

Case study 5.2 Ferri/ferrocyanide redox system

In Case study 5.2, we add the complication of a known faradaic reaction to the CV of the blank cell. Ferricyanide is a well-known, relatively stable iron complex with experimentally observable, reversible electrochemical behavior. For simplicity, in this chapter, we use "ferricyanide" when we refer to potassium ferricyanide. Ferricyanide follows a single, one-electron reduction to ferrocyanide and has been used as an educational tool for electrochemistry. In particular, two articles cover the primary analyses for CV using ferricyanide under reversible conditions [22, 23]. Here, we follow the criteria outlined in the study by Kissinger and Heineman and use the data as a tool to understand biofilm CVs. We evaluate the scan rate dependence, electrode material and addition of rotation (to control mass transfer) and estimate some diagnostic parameters listed in Table 5.2. Figure 5.7 shows a picture of the fully assembled electrochemical cell with the yellow-colored solution containing ferricyanide. It was in this cell that all the ferricyanide results were obtained.

Figure 5.7 Picture of the fully assembled electrochemical cell with the yellow-colored solution containing ferricyanide. (*See insert for color representation of this figure.*)

Nearly all electrochemistry handbooks, textbooks, and reviews cover the CV current–voltage relationship of diffusing, reversible mediators at a planar electrode surface (i.e., ferricyanide reduction). Kissinger and Heineman provide a more than adequate description of CV, and we prefer readers look to this reference as well as electrochemistry textbooks for detailed theory. CV experiments are conducted such that the initial potential of the CV starts in a region where there is only nonfaradaic current. The potential is then scanned through the standard reduction potential (adjusted to the experimental conditions) to a reversal potential, upon which the reverse scan returns to the initial potential. Usually, the reversal potential is separated from $E^{\circ\prime}$ by several hundred millivolts. In the case of ferricyanide, a pair of peaks is observed as the potential is scanned across $E^{\circ\prime}$. The pair of peaks is referred to as the anodic and cathodic peak currents. Figure 5.2c shows the assignment of anodic and cathodic peaks in a ferricyanide CV. Although the standard reduction potential for ferricyanide is +0.162 $V_{Ag/AgCl}$, the apparent standard reduction potential (i.e., what is experimentally observed) is strongly dependent on the supporting electrolyte [24]. In potassium chloride solutions, the equilibrium potential value for equimolar ferri/ferrocyanide solutions is +0.231 $V_{Ag/AgCl}$. Thus, the apparent standard reduction potential in our data should center on the latter value of approximately +0.231 $V_{Ag/AgCl}$. For example, in Figure 5.2c, the cathodic and anodic peak potentials are +0.162 and +0.270 $V_{Ag/AgCl}$, respectively. According to Table 5.2, the formal potential, $E^{\circ\prime}$, can then be calculated and is +0.216 $V_{Ag/AgCl}$, which is reasonably close – for instructive purposes – to the anticipated value. Simple calculations such as this are what make CV a popular technique for qualitatively investigating an electrochemical reaction.

Peak currents require slightly more effort to obtain than peak potentials, because the peak baselines must be subtracted from the uncorrected peak currents. Figure 5.2c shows both the cathodic and anodic peak baselines for a ferricyanide CV. The baselines are simply a linear extrapolation of the baseline currents prior to current flow. From Figure 5.2c, the corrected cathodic and anodic peak currents are approximately 210 and 260 μA, respectively. According to Table 5.2, the peak current ratio, $i_{p,a}/i_{p,c}$, can now be calculated and is 1.2. Is 1.2 close enough to approximately 1? Or is a 20% deviation too large? A major factor in peak current analysis is first ensuring highly accurate prepared reagents and reproducible electrochemical cell conditions, which we have not attempted for our data meant for instructive purposes. Under such unknown conditions, how can the user determine whether a poor baseline subtraction [4, P. 239], nonfaradaic current, or coupled chemical reactions [23] were the cause of the 20% deviation? Through careful experiments, the user could methodically rule out each possibility. However, in biofilm systems, these experiments may be impractical and onerous.

A more useful application of peak currents is based on the realization that the Randles–Sevcik equation (see Table 5.2) predicts peak current will vary linearly with the square root of the scan rate [23]. With modern potentiostats, varying scan rates in CV is simple and practical ("5" in Fig. 5.3). By running consecutive CVs in a stable electrochemical system with varying scan rates, the user can determine the peak current dependence on scan rate. Figure 5.8a shows the result of consecutive CVs with increasing scan rates. In this particular experiment, both peak currents and peak potentials vary with scan rate. However, under reversible

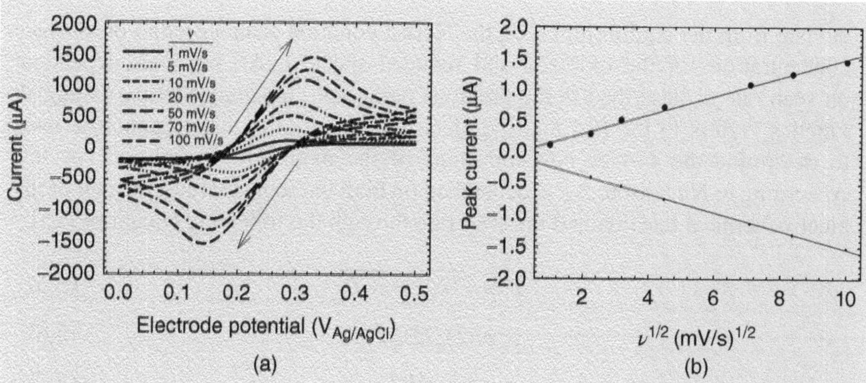

Figure 5.8 (a) CVs of ferricyanide at increasing scan rates. (b) Peak currents plotted against the square root of the scan rate. A linear trend indicates diffusion-based electrochemical reactions.

conditions, only the peak currents are dependent on scan rate (see Fig. 1 in Ref. [22]). In practice, the range of scan rates should be as wide as possible. Usually, the limitations are determined by unsteady behavior at slow scan rates and large capacitive currents at high scan rates. A good range to start with is 1–100 mV s^{-1} or, if possible, up to 1 V s^{-1}. Figure 5.8b plots the peak currents against the square root of the scan rate. The lines are regressions on the cathodic and anodic peaks. We see that even in the presence of quasi-reversible behavior, the Randles–Sevcik equation successfully predicts the linear dependence.

Table 5.3 lists the diagnostic parameters for each of the reported scan rates in Figure 5.8a. The variation in peak potentials with scan rate indicates sluggish kinetics at the electrode interface [23]. $E_{p,a}$ increases toward positive values, whereas $E_{p,c}$ decreases toward negative values. The formal potential varies minimally for the tested scan rates, and the peak current ratios were nearly identical at approximately 1 except for 0.78 at a scan rate of 5 mV s^{-1}. We use this poor demonstration of electrochemical kinetics to show and reiterate that even for well-known reversible systems such as ferricyanide, quasi-reversible behavior is readily observed. Together, Figure 5.8a, b and Table 5.3 provide simple calculations that can help diagnose diffusion-based electron transfer based on the relationships in Table 5.2.

Additional kinetic information and generalizations of an electrochemical reaction can be extracted from the peak potentials and peak separation, ΔE, reported in Table 5.3. Although not as ubiquitous as the use of peak current dependence on scan rate to estimate diffusion coefficients (Fig. 5.8b), the peak potential dependence on scan rate can be used to estimate the standard rate constant of an electrochemical reaction [25]. When an electrochemical reaction behaves reversibly, ΔE is independent of scan rate and holds the value shown in Table 5.2. However, when sluggish kinetics cause the electrochemical interface to deviate from the equilibrium (i.e., the Nernst equation cannot predict the surface concentrations of the oxidized and reduced species), ΔE becomes dependent on scan rate as described in the previous paragraph. In simpler terms, "sluggish kinetics" refers to the fact that the electrode potential by itself cannot be used to determine the surface concentrations of the oxidized and reduced species. According to Nicholson, ΔE is dependent on both the standard rate constant of the electrochemical reaction and the scan rate through the following equations [25]:

$$\psi = \gamma^{\alpha} k_s / \sqrt{\pi a D_o} \tag{5.5}$$

$$\gamma = (D_o / D_R)^{1/2} \tag{5.6}$$

$$a = nFv/RT \tag{5.7}$$

where α is the transfer coefficient, k_s is the standard rate constant, and D_o and D_R are the diffusion coefficients. The practical application of Equation 5.5 is shown in Figure 5.9a. When $\psi > 7$, ΔE is essentially independent of the scan rate and the electrochemical reaction behaves reversibly. When $\psi \leq 0.001$, the electrochemical reaction behaves irreversibly and the reverse reaction can be ignored. In practical terms, this means that slower scan rates yield reversible behavior and faster scan rates yield irreversible behavior. This explains why ΔE in Table 5.3 decreases with decreasing scan rates. As k_s is a multiplicative factor for ψ in Equation 5.5, larger k_s values shift the scan rate values needed to approach reversible behavior. For example, Nicholson [25] required scan rates of up to 120 V s^{-1} to achieve a ΔE of 115 mV, whereas here we only required a scan rate of 0.02 V s^{-1} to achieve a ΔE of 126 mV. The large gap in scan rates is largely attributed to the difference in k_s. Thus, scan rate analysis can be used to estimate k_s if ΔE is within the range where the kinetics are quasi-reversible ($0.001 \geq \psi > 7$). In this region, ΔE is approximately linear with $v^{1/2}$ as shown in Figure 5.9b.

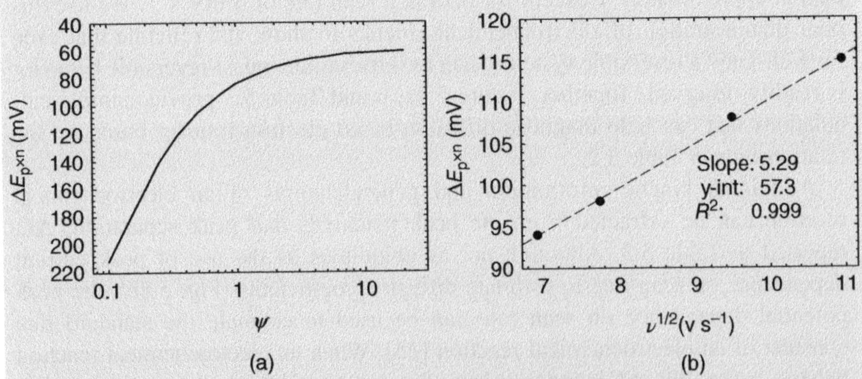

Figure 5.9 (a) Working curve showing the dependence of ΔE on ψ. (b) Dependence of ΔE on $v^{1/2}$ for cadmium reduction. Adapted with permission from Ref. [25]. Copyright 1965 American Chemical Society.

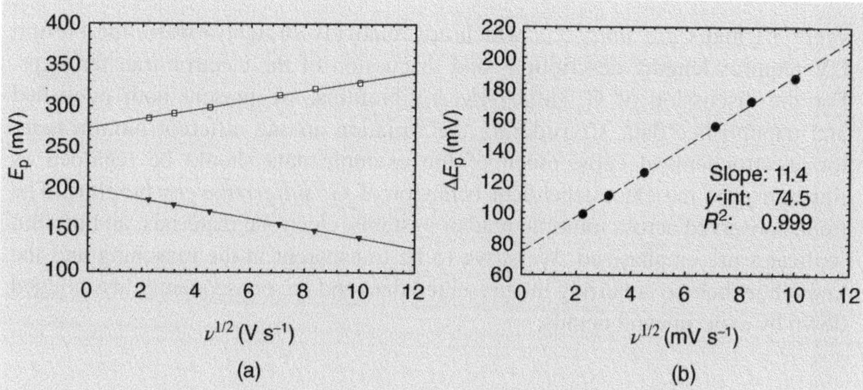

Figure 5.10 (a) Peak potential variation with $\nu^{1/2}$ for ferricyanide. (b) Dependence of ΔE on $\nu^{1/2}$ for ferricyanide. The data used in (a) and (b) are shown in Table 5.3.

There is something of a discrepancy in the scan rate analysis presented here. How is it possible to observe both reversible and quasi-reversible behavior for a single data set depending on whether peak currents or peak potentials are analyzed? If the peak potentials drift as shown in Figure 5.10a, peak currents are not predicted to be linear to $\nu^{1/2}$ for quasi-reversible behavior [4]. Conversely, if peak currents are observed to be linear to $\nu^{1/2}$, ΔE should not increase with scan rate, which Figure 5.10b clearly shows it does. We see both quasi-reversible and reversible behavior. The reason behind the discrepancy is that reversibility, quasi-reversibility, and irreversibility lie on a continuum that strongly depends on $\nu^{1/2}$. Therefore, although it may be difficult to determine confidently whether the electrochemical reaction of interest is behaving reversibly, quasi-reversibly, or irreversibly, it should be straightforward to observe the strong $\nu^{1/2}$ dependence. This is in contrast to the theory predicting electrochemical reactions involving strongly adsorbed, electroactive films in which a linear ν dependence is observed [4].

Practical application

Peak potentials and peak currents are useful diagnostic parameters to calculate in CV. In diffusing, reversible systems such as ferri/ferrocyanide, the straightforward relations in Table 5.2 can be used to extract estimates of formal potential, reversibility, and even diffusion coefficients. Understanding how to extract parameters from known electrochemical systems will help users identify trends in less well-known biofilm systems.

Case study 5.3 Biofilm electrochemical behavior

The following sections on anodic and cathodic biofilms provide a more detailed discussion of the electrochemical reactions occurring in EABs. Unlike *Case study* 5.1 and *Case study* 5.2, which are relatively straightforward, the biofilm CV requires lengthy descriptions and discussion of the electron-transfer steps. For the discussion of *G. sulfurreducens* biofilms, we present both published and unpublished data. Considering the variation among different biofilm reactors/electrochemical cells, much of the example data should be regarded as illustrative of the electrochemical behavior of *G. sulfurreducens* biofilms. The trends observed across multiple reactor systems, electrode materials, and biofilm replicates are emphasized. We strive to be transparent in the reasoning and the know-how behind acquiring the presented data and are consequently less bogged down by experimental details.

5.5 ANODIC BIOFILMS

5.5.1 Introduction to Electron Transfer in *G. sulfurreducens* Biofilms

Before discussing electron-transfer processes in biofilms, it is necessary to compare the easily identified one-step ferri/ferrocyanide electrochemical reaction based on diffusion to the unknown oxidation and reduction processes occurring in a biofilm with a finite thickness, Z (Fig. 5.11) [26]. In the former case, mass transfer is handled analytically through simple Fickian boundary conditions in the boundary value problem [4]. Electrochemical reactions are also confined to the electrode surface: any self-exchange reactions in the bulk are ignored. In the latter, biofilm, case, mass

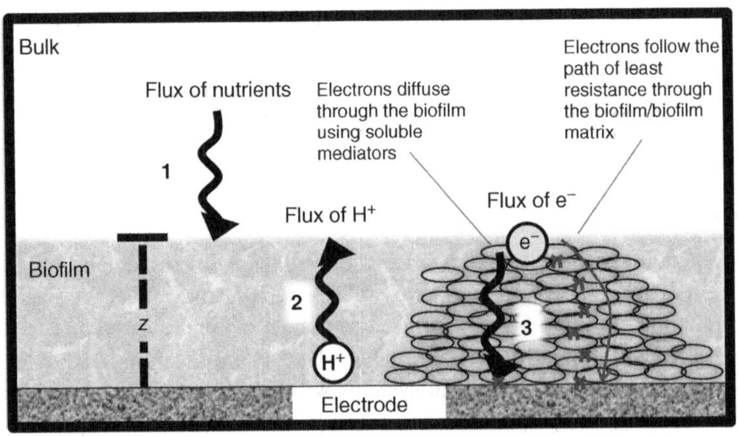

Figure 5.11 Biofilm cartoon showing: (1) the flux of nutrients, (2) flux of protons, and (3) flux of electrons inside a biofilm of finite thickness Z. This research was originally published in Ref. [26]. Copyright 2012 the Biochemical Society. (*See insert for color representation of this figure.*)

transfer through the biofilm is unpredictable because of varying diffusion coefficients related to biofilm heterogeneity [27]. Oxidation and reduction processes occur throughout the thickness of the biofilm, which can range from tens of microns to hundreds of microns. These two factors – completely different from the simple situation in Case study 5.2 – make biofilm electron-transfer processes difficult to study. The primary processes to consider in biofilm electron transfer are shown in Figure 5.11: (1) flux of nutrients into the biofilm, (2) flux of counter ions (i.e., protons) in the opposite direction of electron flow, and (3) flux of electrons through the biofilm toward the electrode caused by anodic reaction(s). Various multistep models that consider these steps have been developed and are covered in Chapters 2, 6, and 7 of this book. Many of these models have been successfully used to describe CV data from *G. sulfurreducens* biofilms [28–31]. What is not shown in Figure 5.11 are the (bio)kinetics of acetate turnover and the products of cell respiration such as metabolites. It is generally assumed that acetate is immediately consumed such that the flux of products (i.e., electrons, protons, and carbon dioxide) is the slower, rate-limiting step. Modeling of *G. sulfurreducens* CVs has shown that this is a reasonable assumption [29]. It is also generally accepted that the following overall half-reaction accounts for the total charge passed in batch experiments of *G. sulfurreducens* biofilms:

$$CH_3COO^- + 4H_2O \rightarrow 2HCO_3^- + 9H^+ + 8e^- \quad E^{\circ\prime} = -286\ \text{mV}_{\text{SHE}} \tag{5.8}$$

Therefore, *G. sulfurreducens* biofilm electron transfer is generally described as complete acetate oxidation to carbon dioxide, protons, and electrons in which electron transfer outside the cells is the rate-limiting step. The mass transfer of protons and carbon dioxide affects pH buffering and has been shown to affect the biofilm. Throughout the rest of this chapter, we consider only electron transfer and "pH effects," ignoring changes in metabolic activity of the biofilm. Only the metabolic respiration rate, approximated by the current measured at the biofilm electrode, is considered. This does not mean that changes in metabolic activity are less important. We refer the readers to Chapter 2 for more information on *G. sulfurreducens* metabolism.

It is important for readers to know that intrabiofilm gradients have already been measured for *G. sulfurreducens* biofilms [3, 7, 32, 33]. Several coupled techniques are capable of making such measurements [34], but in our laboratory, our choice of method (and area of expertise) is to couple microsensor depth profiling with electrode respiration [3, 33, 35]. Figure 5.12 shows example depth profiles of hydrogen, pH, and redox potential (for H_2) within and above a *G. sulfurreducens* biofilm respiring on an electrode. The reason for the choice of these particular measurements is as follows. As described for acetate oxidation, electron transfer results in proton transfer, which is expected to cause variations in pH adjacent to the electrode surface (see Fig. 5.11). Furthermore, if electrons are transferred by a redox mediator, the solution redox potential (E_h) should reflect the concentration profiles of the oxidized and reduced mediators developed adjacent to the electrode surface. Equations 5.9 and 5.10 illustrate the relationship among E_h, pH, and the concentrations of oxidized and reduced mediators:

$$[Ox] + 2e^- + 2[H^+] \leftrightarrow [H_2Red] \tag{5.9}$$

$$E_h = E^{\circ\prime} - (0.05916/2) \times \log([H_2Red]/[Ox]) - 0.05916 \times pH \qquad (5.10)$$

where E_h is the solution redox potential (V), $E^{\circ\prime}$ is the formal potential (V), [Ox] is the concentration of the oxidized species (M), and [H_2Red] is the concentration of the reduced species (M). By applying the Nernst equation to a general two-electron, two-proton redox reaction, we can see that any concentration gradient of protons or an oxidized/reduced mediator formed adjacent to the electrode surface will result in a corresponding solution redox potential gradient. Therefore, as Figure 5.11 diagrams, if the biofilm is respiring on an electrode while oxidizing acetate, both the pH and the E_h are expected to change. However, because the ratio, [H_2Red]/[Ox], is contained within a logarithm, E_h can only change significantly if a large amount of the mediator is oxidized/reduced near the surface of the biofilm electrode. This also implies that a significant current is passed by the redox mediator. Solution redox potential has been modeled in the literature for electrode-respiring biofilms [28]. As Equation 5.10 implies, E_h and pH are related and should be measured consecutively in electrode-respiring biofilms to determine whether mediated electron transfer critically controls overall electron transfer [35]. The hydrogen in this particular case originated at the CE that was placed in the same solution as the biofilm electrode. The decreasing trend in Figure 5.12 demonstrates hydrogen utilization by the *G. sulfurreducens* biofilms to generate current. The redox cycling of hydrogen from the CE to the biofilm electrode is generally neglected in the literature. In quantitative analysis, the electron flux resulting from hydrogen consumption must be accounted for when a charge balance is performed. In practical terms, the presence of hydrogen as an alternative electron donor to acetate could lead to erroneous redox potential

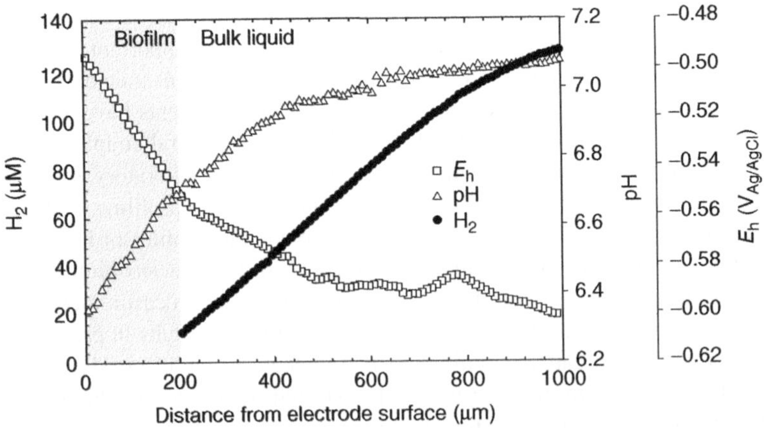

Figure 5.12 Hydrogen, pH, and redox potential depth profiles within (for H_2, above) a *G. sulfurreducens* biofilm respiring on an electrode. The current was 1.85 mA at a polarization potential of +300 $mV_{Ag/AgCl}$. This research was originally published in Ref. [26]. Copyright 2012 the Biochemical Society.

measurements as well [26]. In this particular case study, readers must recognize that in EABs, intrabiofilm gradients exist when current is measured at the electrode.

5.5.2 Operating the Biofilm Electrochemical Cell/Reactor

In order to demonstrate how CV is applicable to studying electron transfer in *G. sulfurreducens* biofilms or EABs in general, we first formulate the conditions prior to a CV being run: the reason(s) for acquiring CVs of the biofilm will become apparent to the reader. The following is a brief description of the conditions for a *G. sulfurreducens* biofilm:

Let us assume that an electrochemical cell was previously set up and a biofilm was to be grown long term on the WE as shown in Figure 5.1a. Acetate was continuously fed to the electrochemical cell such that the bulk volume properties (i.e., solution conductivity, acetate concentration, and pH) were constant. The electrochemical cell was purged with oxygen-free gas containing 20% CO_2; this was necessary to maintain a pH of 6.8 in the growth medium. The working, counter, and REs were connected to the potentiostat as per the manufacturer's instructions. We assume that we had prior knowledge of the polarization potential required to achieve *G. sulfurreducens* biofilm growth on the WE and that this value was +300 $mV_{Ag/AgCl}$. Thus, suspended cells of *G. sulfurreducens*, the inoculum, were introduced into the electrochemical cell at $t = 0$, at which time the WE was already polarized to +300 $mV_{Ag/AgCl}$. After inoculation, the current was observed to increase, and a pink-hued biofilm could be seen after some time had passed. After a long time (1–2 weeks in our experience, but this depends on electrode surface area), the measured current reached a maximum and proceeded to a steady value that did not change with time and was a significant, nonzero value (above background and noise levels). At this point in time, the *G. sulfurreducens* biofilm reached a steady state with respect to measured current.

The aforementioned description is important because the long-term electrode polarization experiment identifies sustainable current generation that can be systematically related to relevant experimental parameters. Steady-state current generation is usually the first confirmation of the electrochemical activity of a suspected EAB (in this case, *G. sulfurreducens*). The suspected EAB is expected to exchange electrons with the WE. However, to confirm that the current generated is related to the metabolism of the biofilm or its metabolic by-products, certain long-term electrode polarization experiments should be performed. In batch experiments, the total consumption of the electron donor can be correlated to the total charge transferred. If the current trends toward zero when the electron donor concentration goes to zero, then the oxidation of the electron donor is the source of electrons and the coulombic efficiency can be calculated [9]. For example, Bond and Lovley [36] showed that current production by *G. sulfurreducens* biofilms is directly related to the consumption of acetate in batch mode. In continuous experiments, the electron donor feed concentration can be altered and subsequent steady-state current values measured. Replacement of the bulk solution with fresh growth medium during electrode polarization experiments has been done to probe soluble extracellular electron-transfer mechanisms. For example, Marsili et al. [37] replaced the spent solution in a *Shewanella oneidensis* MR-1 biofilm reactor to show that soluble redox mediators were responsible for the steady-state current generated. Bond and Lovley [36] replaced the bulk solution of *G. sulfurreducens* biofilms with acetate-free growth medium to show acetate dependence. Other sophisticated experiments can be done to isolate controlled parameters that affect extracellular electron-transfer mechanisms. EABs can be genetically engineered to enhance/inhibit current generation. For example, Richter et al. [14] observed that no current could be produced by a ΔpilA *G. sulfurreducens* mutant on gold electrodes.

If the steady-state current were adequate for studying the electron-transfer rates of *G. sulfurreducens* biofilms, then it would be possible to choose a range of polarization potentials and determine the steady-state current on the basis of those polarization potentials. For example, a researcher could set up 10 electrochemical cells with 10 potentiostats in which 10 different polarization potentials were applied. The inoculum would then be added, and the current of each electrochemical cell would be monitored. At the end of several replicate experiments, the researcher would be able to plot a graph of steady-state current against polarization potential. Figure 5.13 shows an illustrative plot of such an experiment in which the researcher could make reasonable conclusions about the polarization potential required to generate sustainable current. No understanding of CV would be needed; instead, the biofilm researcher would only need to formulate a hypothesis around the complete oxidation of acetate to carbon dioxide, protons, and electrons; that is, a polarization potential more positive than the formal potential of acetate oxidation (−286 mV_{SHE}) would generate anodic current, and any additional potential would be considered some form of activation overpotential [38]. Here, the critical question is that "What advantage does CV have, then, for biofilm researchers interested only in sustainable current?"

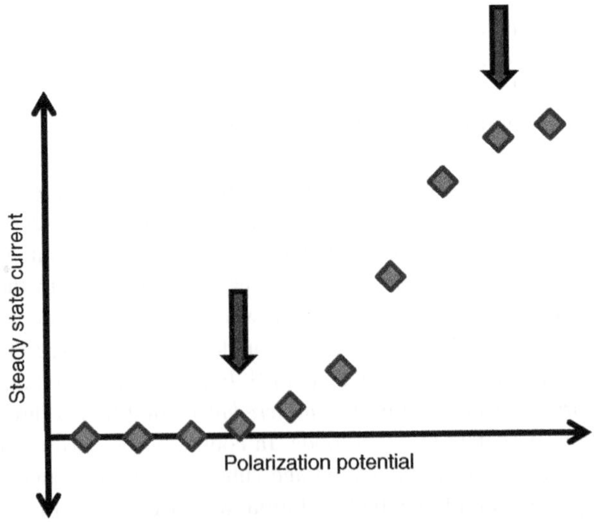

Figure 5.13 Illustrative plot of steady-state current versus polarization potential for multiple biofilms grown at different polarization potentials.

The obvious answer to this question is that CV provides the same information more quickly and with less experimental complexity. The voltammogram will also show kinetic information. More specifically, CV can be used to identify the biofilm electrode potential at which active redox couples secreted by the biofilm are oxidized or reduced. Under well-controlled conditions, CV can be used to determine whether freely diffusing species or surface-adsorbed species contribute to electron transfer and whether these electrode-respiring biofilms engage in catalytic activity toward specific substrates [29, 37, 39–47]. There are many studies that use CV in EABs, and they focus on changes in the voltammogram under controlled conditions. The following sections detail experimental CV data for *G. sulfurreducens* biofilms. We should note that the step-by-step procedures for growing *G. sulfurreducens* biofilms are given in our earlier publication [48].

5.5.3 Turnover CV in *G. sulfurreducens* Biofilms

When a CV is obtained for a *G. sulfurreducens* biofilm grown on an electrode, the WE potential is generally swept from approximately −0.7 to approximately +0.3 $V_{Ag/AgCl}$, or from a potential region without faradaic current to a region with faradaic current. Figure 5.14 shows a typical turnover, or catalytic, CV obtained from *G. sulfurreducens* respiring on an electrode while consuming acetate. Turnover conditions refer to microbial turnover of the electron donor (acetate) during respiration that results in the generation of current. Catalytic current refers to renewal of the reduced mediator by substrate turnover that leads to a sigmoidal voltammogram. Nonturnover conditions, or noncatalytic current, refer to biofilms without an electron donor, which is acetate for our case. Nonturnover is also referred to as "starving conditions." In the absence of microbial turnover of an electron donor or, equivalently, under nonturnover conditions, the voltammogram takes on a different characteristic, which is discussed in Figure 5.16. Turnover and nonturnover CVs were discussed in detail in as early as 2008 [40]. In Figure 5.14, we take a look at a turnover voltammogram (10 mV s^{-1}) of a *G. sulfurreducens* biofilm grown in a single-chambered electrochemical cell and also present the corresponding shift in CE potential during that voltammogram. In the top panel, we designate potential regions A–E, in which the current response varies as follows:

A. Region where no electrochemical/faradaic reaction takes place. Usually approximately 100 mV wide, this region is important in the voltammogram as an internal reference and system diagnostic. The slope of this region is controlled by the internal resistance of the electrochemical cell. Any nonfaradaic current measured here can be related to the double-layer capacitance of the WE. If oxygen is present in the electrochemical cell, a measurable cathodic current (negative current) will be observed and the oxygen should be stripped out immediately.
B. Region where current response is initially detectable. This region usually contains the OCP of the WE. The region appears to be flat and similar to A only because of the scaling of the *y*-axis.

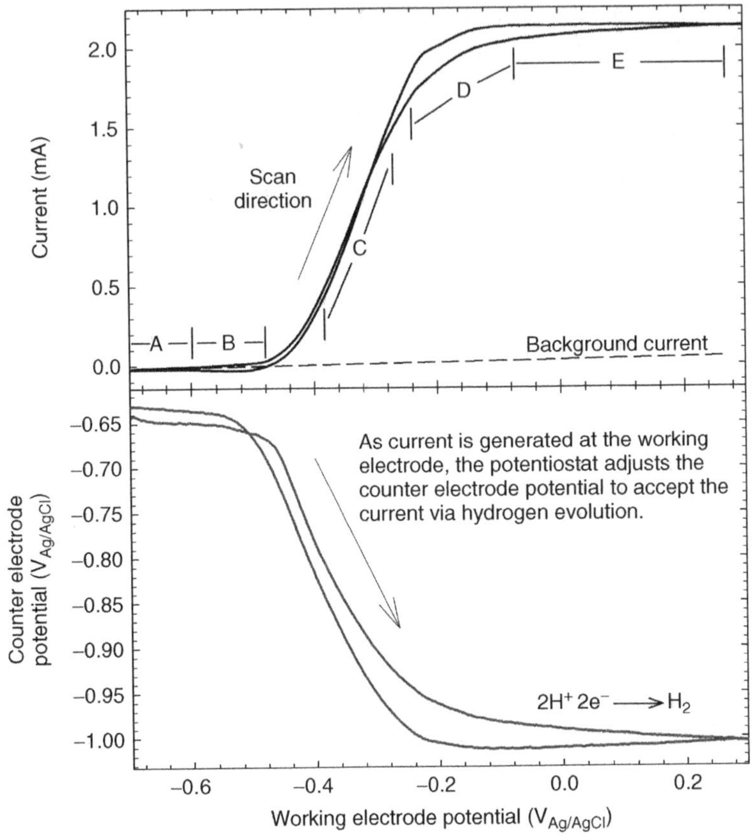

Figure 5.14 Typical turnover CV of a *G. sulfurreducens* biofilm respiring on an electrode and the corresponding counter electrode potential during the scan.

C. Region containing an important potential parameter called the potential at half-maximum current ($E_{1/2}$). The reaction rate is fastest at $E_{1/2}$, which can be understood by noting that the slope of the voltammogram is at its maximum [49]. $E_{1/2}$ is indicative of the formal potential of the dominating redox couple. As there may be a distribution of redox couples in the biofilm, it is appropriate to refer to the distribution as an apparent redox couple. We note that this $E_{1/2}$ is different from the definition presented in Table 5.2 because that $E_{1/2}$ was derived for diffusion-based electron transfer such as that of the ferricyanide in Case study 5.2.

D. Region beyond $E_{1/2}$, where the reaction rate is controlled by the electron flux to the apparent redox couple and not by the reaction rate of the apparent redox couple. In the case of *G. sulfurreducens* biofilms, this is not purely a mass transfer limitation [50].

E. Region of limiting current. Depending on the geometry of the WE, the geometry of the electrochemical cell, and the scan rate, a limiting current is observed between −200 and −100 $mV_{Ag/AgCl}$. In odd-shaped electrochemical cells and porous electrodes, this region may be shifted beyond 0 $mV_{Ag/AgCl}$. In the literature, this region can be as wide as 500 mV. In this region, the total current is the background current shown as the dashed line plus the turnover current.

For any EAB producing catalytic current, there are two potential parameters of interest. These are the OCP and $E_{1/2}$. The OCP can be measured accurately using a high-impedance ohmmeter. $E_{1/2}$ can be obtained from the turnover CV. Furthermore, the background current in region A and the limiting current in region E should be measured. The slopes in these two regions should be consistent for *G. sulfurreducens* biofilms. Researchers should strive to obtain estimates of these four parameters for all biofilm studies.

In this particular case, we paid special attention to the CE by monitoring its potential with respect to the RE simultaneously with the current produced. The bottom panel of Figure 5.14 shows the CE potential versus the RE potential. In regions A and B, the CE potential remained unchanged. Once current flowed in regions C, D, and E, the CE adjusted to a potential of −1 $V_{Ag/AgCl}$. The movement of the CE potential depends strongly on the surface area. In this case, the CE had approximately 3× more surface area than the WE. In most electrochemistry reference books, the CE is hardly ever mentioned beyond its role as the "current collector;" thus, new users rarely pay attention to what is occurring at the CE. This does not mean that the CE can be neglected or is less important. Experienced electrochemists by training know how to position the CE. Similar to how oxygen is generally unwanted and removed from electrochemical experiments, the CE must be sequestered from the WE. It must be able to support the current produced at the WE, and it must be placed behind a physical barrier such as: (1) a selective membrane, (2) a porous glass frit, or (3) a distal location as far from the WE as practically possible. The by-products (e.g., hydrogen) formed at the CE are fully realizable at the WE as we have shown in Figure 5.12. The purpose of this discussion was to bring transparency to the role of the CE in potentiostat-controlled biofilm electrochemical systems. The CE is an integral part of a three-electrode system, and biofilm researchers should take care when positioning it in their own experimental setups. In addition, the reactions occurring at the CE must be considered.

The turnover CV in Figure 5.14 was taken from an existing *G. sulfurreducens* biofilm on a graphite rod. The voltammogram of a biofilm can also be monitored during its growth from inoculation to a mature biofilm, as shown in Figure 5.15. Similar results have been published previously [46]. From background to the mature biofilm CV, the general shape of the voltammogram of *G. sulfurreducens* does not change appreciably. The approach outlined in Figure 5.14 can be applied at any time during biofilm growth. For example, the primary differences between curve I and curve II in Figure 5.15 are the limiting current and a shift in the onset potential of catalytic current. Curve III is slightly different from curves I and II, because the current continues

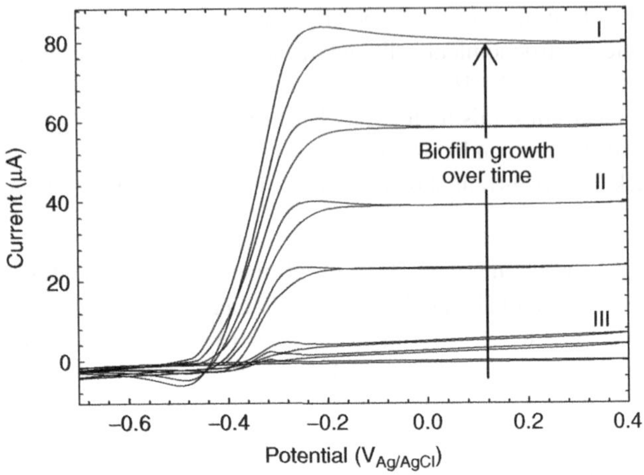

Figure 5.15 CVs taken over the growth of a *G. sulfurreducens* biofilm under turnover conditions.

to increase beyond $-0.2\ V_{Ag/AgCl}$ and a pair of peaks is more obvious at approximately $-0.3\ V_{Ag/AgCl}$. The mechanism behind these slight differences is currently unknown.

5.5.4 Nonturnover CV in *G. sulfurreducens* Biofilm

The results in Figure 5.16a are for the same biofilm that was used to obtain the CVs in Figure 5.15, but with the acetate slowly diluted out by replacing the initial growth medium containing 20 mM acetate with acetate-free growth medium. Several CVs (10 mV s^{-1}) were obtained while the biofilm headed toward nonturnover conditions. As the catalytic current decreased with decreasing acetate concentration, anodic peaks became visible. For example, in curve II in Figure 5.16a, the anodic peak at approximately $-280\ mV_{Ag/AgCl}$ is visible superimposed on the catalytic current of approximately 31 μA. However, cathodic peaks only became visible when the catalytic current reached approximately 4.5 μA in curve III. Curve IV was obtained after several days under starving conditions and is considered representative of the biofilm under nonturnover conditions. In our experience, curve IV can only be obtained several days after observing curve III, although the difference in catalytic current is only 2–3 μA. When acetate is completely depleted in the biofilm and the biofilm is under starving conditions, it is important to reduce the scan rate of the CVs to 1 mV s^{-1}. At a scan rate of 1 mV s^{-1} or less, individual anodic/cathodic peaks are observed, as shown in Figure 5.16b. We note that even with differing electrode materials such as glassy carbon and gold, the general character of the nonturnover CV for *G. sulfurreducens* biofilms does not change. Several larger peaks centered around $-0.3\ V_{Ag/AgCl}$ dominate the CV with smaller, broader, and overlapped peaks occupying the region between −0.4 and $-0.6\ V_{Ag/AgCl}$. The importance of the shape of the nonturnover CV for *G. sulfurreducens* is currently not well-established.

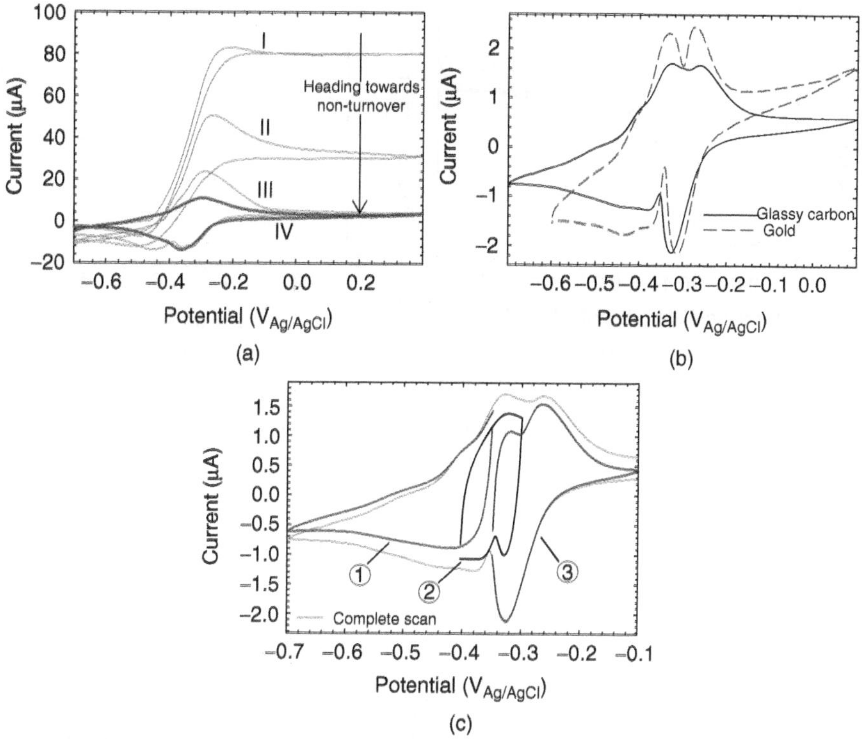

Figure 5.16 (a) CVs during acetate washout. (b) Nonturnover CVs taken at 1 mV s^{-1} for glassy carbon and gold working electrodes with biofilms. (c) Overlapping CVs that demonstrate peak coupling in the complex overlapping peaks of the complete scan. Reproduced with permission from Ref. [50]. Copyright 2013 Wiley Periodicals, Inc.

One underutilized approach to CV in EAB research that is particularly powerful under nonturnover conditions is to vary the initial potential and the potential window to study anodic and cathodic peak coupling. Figure 5.16c shows three curves, indicated by (1), (2), and (3), superimposed to form a complete curve. Curve 1 was taken for a potential range of −0.35 to -0.7 $V_{Ag/AgCl}$. Curve 2 was taken for a potential range of −0.4 to −0.3 $V_{Ag/AgCl}$. Curve 3 was taken for a potential range of −0.35 to −0.1 $V_{Ag/AgCl}$. By simply manipulating the initial potential of the CV, anodic peaks can be coupled to cathodic peaks.

5.5.5 Scan Rate Analysis for *G. sulfurreducens* Biofilms

Scan rate analysis in *G. sulfurreducens* biofilms is one of the most often used methods to determine a particular type of electron transfer (i.e., diffusion based). In Figure 5.17a, the CVs obtained at increasing scan rates of 1 up to 100 mV s^{-1} are shown. The corresponding peak currents are shown in Figure 5.17b. The line

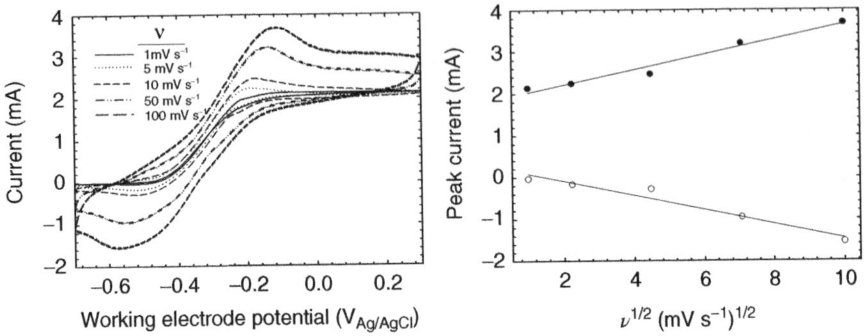

(a) (b)

Figure 5.17 Scan rate analysis of *G. sulfurreducens* biofilms under turnover conditions. (a) CVs at increasing scan rates. (b) Peak currents at each scan rate plotted against the square root of the scan rate.

is a regression on each anodic and cathodic peak. Compare the spread of the data points along the regression line in Figure 5.17b to that in Figure 5.8b. One of the difficulties in applying the Randles–Sevcik criterion to biofilms is the fact that it was derived for a single-step, reversible electron-transfer step for diffusing mediators. *G. sulfurreducens* biofilms do not fit the assumption of a single step. Purified c-type cytochromes of EABs such as *S. oneidensis* MR-1 have been studied in lipid bilayers [51], and it was shown that the multiple electron-transfer steps in the Mtr pathway result in electron transfer across the lipid bilayer. Electron transfer occurring across multiple biological complexes (i.e., proteins) within overlapping potential windows impinges on the user's ability to investigate any single electron-transfer step. Consequently, what appears as a single cathodic or anodic peak in Figure 5.17a is actually a superposition of peaks of an unknown number of mediators. Scan rate analysis and control experiments with isolated and purified biological complexes can show how a biofilm's characteristic CVs match or do not match those of controls. This critical information can be used to support or defeat proposed electron-transfer mechanisms.

Under nonturnover conditions, this complication is easily imaged as in Figure 5.18a. Visually, Figure 5.18a looks similar to Figure 5.17a except that the current range is reduced. However, in Figure 5.18b, the linear regression does not appear to fit the data well. The superposition of peaks causes the scan rate dependence to deviate from the expected behavior shown in Figure 5.17b. This is made clear in Figure 5.18c, where the CVs obtained at 1 and 10 mV s^{-1} are shown. The two anodic peaks denoted by the top arrow superpose into one. The three cathodic peaks denoted by the bottom arrow superpose into one larger cathodic peak. Under such tight formal potential distributions among the individual mediators, applying the Randles–Sevcik criterion is qualitative at best.

When we extended the scan rate analysis range from 320 to 1280 mV s^{-1} for the same *G. sulfurreducens* biofilm, we observed a peculiar scan rate dependence

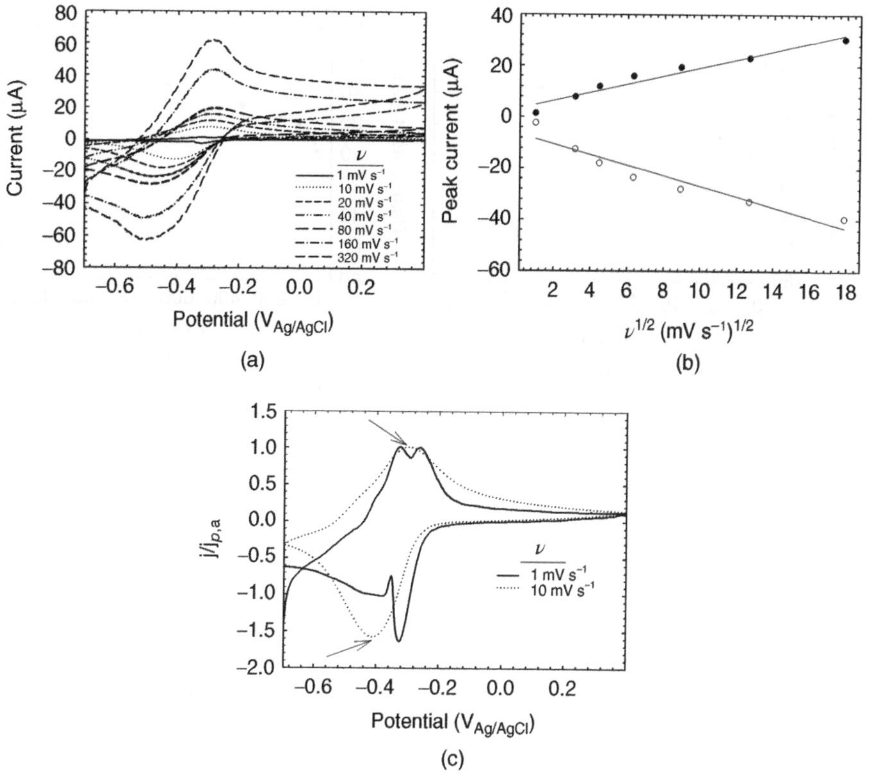

Figure 5.18 Scan rate analysis of a *G. sulfurreducens* biofilm under nonturnover conditions. (a) CVs at increasing scan rates. (b) Peak currents at each scan rate plotted against the square root of the scan rate. (c) Nonturnover CVs at 1 and 10 mV s^{-1} showing the superposition (arrows) of multiple anodic and cathodic peaks.

of peak currents, peak potentials, and ΔE. Unlike the results for ferricyanide shown in Figures 5.8–5.10, the scan rate dependence of peak current under nonturnover conditions for *G. sulfurreducens* biofilm (Fig. 5.19a) did not yield linear behavior across the entire scan rate range. Figure 5.19b plots peak current against ν, and no linear behavior is observed either. Figure 5.19c and d shows that the peak potentials and ΔE transition from a region of strong dependence on scan rate to a region of weak dependence on scan rate. We have already discussed the concept of a continuum of reversible, quasi-reversible, and irreversible behavior for electrochemical reactions such as the oxidation and reduction of ferricyanide. In this particular case, it is difficult to determine where we have operated our biofilm on that continuum. Because the biofilm has a combination of adsorption- and diffusion-based mass transfer, it may not be possible to achieve a strictly linear dependence on either $\nu^{1/2}$ or ν. The collective behavior seen in Figure 5.19 is likely what researchers will see when utilizing scan rate analysis in EABs over a wide range. Currently, there is no theory to treat this kind of scan rate dependence behavior.

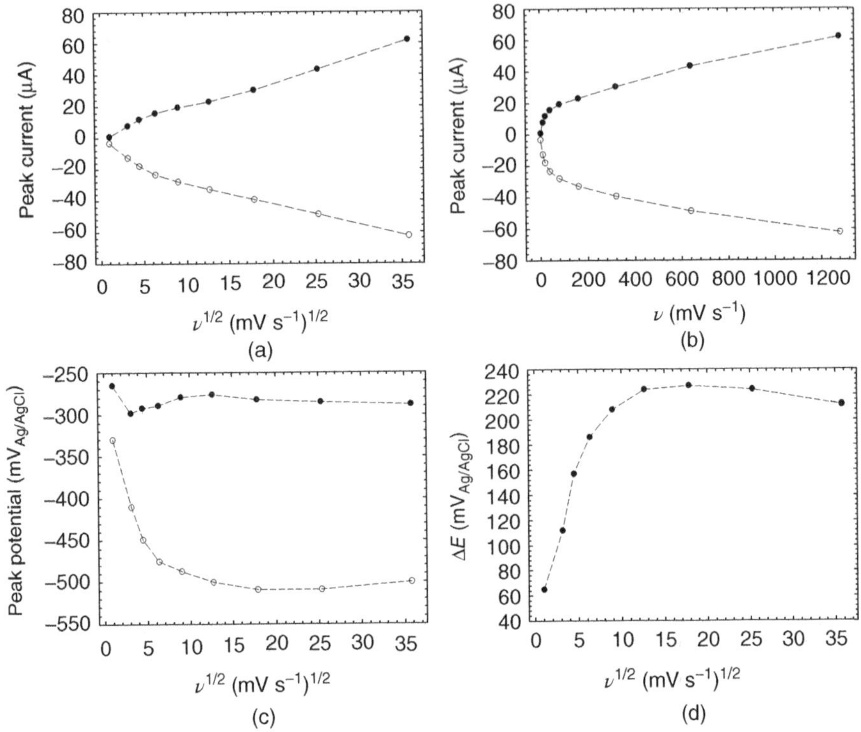

Figure 5.19 Extended scan rate analysis for the nonturnover CVs in Figure 5.18a. (a) Peak current versus $\nu^{1/2}$. (b) Peak current versus ν. (c) Peak potential versus $\nu^{1/2}$. (d) ΔE versus $\nu^{1/2}$. The scan rate was increased up to 1280 mV s^{-1}.

Practical application

Here, we have demonstrated the differences between turnover and nonturnover CVs for *G. sulfurreducens* biofilms. Turnover and nonturnover CVs can be used to correlate catalytic current under turnover conditions to redox peaks under non-turnover conditions. Scan rate analysis can provide a qualitative understanding of the biofilm electron-transfer mechanisms occurring within a biofilm. However, caution should be used when applying the Randles–Sevcik criterion to biofilms, as it was derived for a single-step, reversible electron-transfer step for diffusing mediators.

5.5.6 Linking Cyclic Voltammetry to Other Electrochemical Techniques

In certain cases, it may be advantageous for a researcher to use CV in conjunction with another electrochemical technique. One possibly useful technique is square wave voltammetry (SWV), which is a pulse voltammetric technique. It is used less

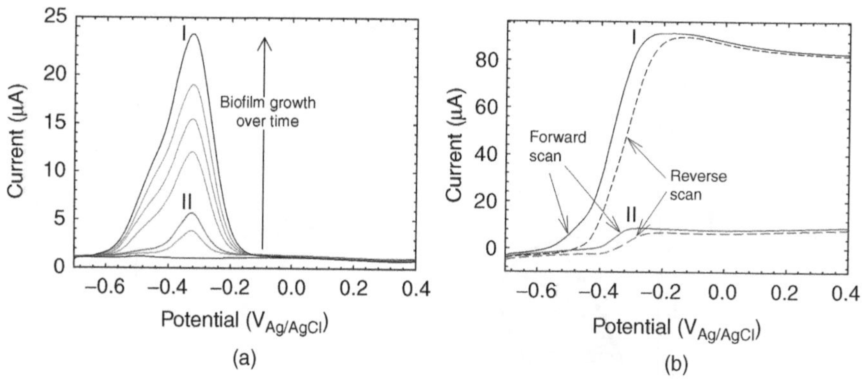

Figure 5.20 (a) SWVs of *G. sulfurreducens* biofilm during growth that correspond to Figure 5.15. (b) Forward and reverse scans for curve I and curve II.

frequently than CV to characterize electrode-respiring biofilms; however, it can detect low concentrations of electrochemically active species not easily detectable by CV. In addition, SWV provides quick diagnostic information that can help identify whether an electrochemically active reaction is reversible or not. Figure 5.20a shows the SWV response of the *G. sulfurreducens* biofilm described in Figure 5.15. The SWVs were obtained right after the CVs. In Figure 5.20a, only one large primary peak is observed in the early stages of biofilm growth (curve II). When the biofilm approaches the maximum current (curve I), a second peak, which appears to be a shoulder on the primary peak, develops. One of the significant advantages of SWV is apparent in Figure 5.20a. Outside of the range of −0.65 to −0.1 $V_{Ag/AgCl}$, there is no change in the SWV with biofilm growth. This can be better understood by looking at the forward and reverse scans of curves I and II in Figure 5.20b. Outside the region of electrochemical activity of the biofilm, the forward and reverse scans track each other exactly, so that the difference (Fig. 5.20a) is essentially constant. As these SWVs were obtained under turnover conditions, the forward and reverse scans behave similarly to the turnover CVs in Figure 5.15. However, the two techniques are fundamentally different in the potential waveform they apply, so the analytical/numerical treatments are very different. The capability of SWV to detect catalytic waves of EABs while resolving the midpoint potentials of electrochemically active species under turnover conditions is invaluable to EAB electron-transfer analysis.

5.5.6.1 Electrochemical Impedance Spectroscopy A more thorough review of electrochemical impedance spectroscopy (EIS) and its application to EABs has been given in Chapter 8, and the readers are referred there for detailed explanations. Briefly, in EIS, a sinusoidal potential waveform is applied to measure the real impedance (resistance) and imaginary impedance (capacitance) of an electrochemical system. The choice of wave amplitude and polarization potential is critical because the potential waveform should oscillate between relevant potentials, such

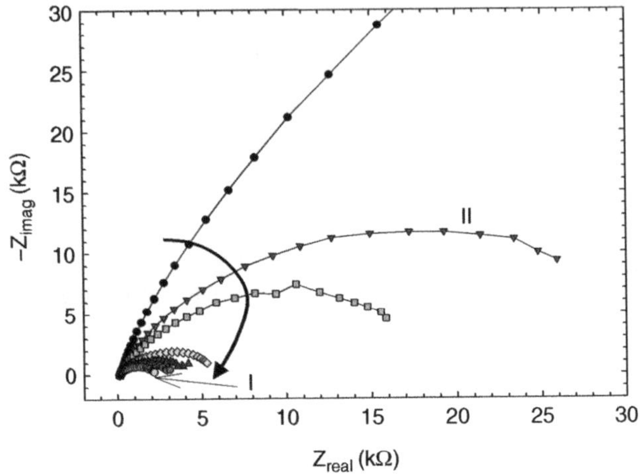

Figure 5.21 EIS of *G. sulfurreducens* biofilm corresponding to the CVs and SWVs shown in Figures 5.15 and 5.20a. The arrow indicates biofilm growth. Curve I and curve II were taken right after curve I and curve II in Figure 5.20a. Reproduced with permission from Ref. [50]. Copyright 2013 Wiley Periodicals, Inc.

as the midpoint potential of the EAB catalytic wave observed using CV and the OCP of the biofilm electrode after acclimatization [52]. In addition, potential and current should relate linearly. EIS has been used as a nondestructive electrochemical technique to study the development of charge transfer resistance in EABs [53–55]. Figure 5.21 shows a Nyquist plot corresponding to the CVs and SWVs shown in Figures 5.15 and 5.20a. Unlike turnover current, biofilm impedance decreases as the biofilm grows. Thus, the imaginary and real components of biofilm impedance decrease with time. The black arrow shows the direction of the trend with growth. These impedance spectra were taken using a DC bias of -0.34 $V_{Ag/AgCl}$. One of the strategies in our laboratories is to run a CV to find a linear region in the voltammogram (Fig. 5.15). Then, we run a SWV to find the approximate $E_{1/2}$ (Fig. 5.20a). Finally, we use the $E_{1/2}$ from SWV to set the DC bias of the EIS. This procedure has been tested many times in our laboratory and is the most foolproof way to obtain useful EIS data. In addition, when we follow these techniques that have been well established by electrochemists, the techniques we develop are reproducible and can be repeated in other research laboratories.

Practical application

Coordinating data from CVs, SWVs, and EIS provides strong electrochemical evidence that can be used to study EABs. CVs can be used to find linear regions in voltammograms and SWVs can be used to find $E_{1/2}$ of catalytic processes. These can then be used to choose the correct, more relevant parameters for EIS.

5.5.7 Mass Transfer Effect on the Biofilm Electrode Electron Transfer

In basic electrochemical textbooks, mass transfer coupled to electron transfer is handled through two methods: (1) using the semiempirical Nernst diffusion layer approximation, in which diffusion through a stagnant film is assumed to be linear and (2) using semi-infinite boundary conditions to obtain the analytical solution for diffusion in a stationary solution. In the first method, convection in the bulk solution is used to formulate the assumption of a stagnant film (i.e., mass transfer boundary layer). In our case, we use a rotating disk electrode (RDE) to form the hydrodynamic layer above the mass transfer boundary layer such that concentration gradients are restricted to within the mass transfer boundary layer. Essentially, the rotation rate controls the thickness of the boundary layer. The Levich equation (Equation 5.11) was derived for predicting the voltammetric response on RDEs [4]:

$$I = 0.62nFAD^{2/3}\omega^{1/2}\nu^{-1/6}C_{\mathrm{O,bulk}} \tag{5.11}$$

The Levich equation predicts the sigmoidal shape of the voltammograms concerning diffusing electroactive compounds, such as that for ferricyanide shown in Figure 5.22.

In the second method, diffusion is not restricted to within a thin film. Instead, it is considered as extending infinitely into the bulk solution from the electrode surface. As the electron-transfer rate reaches the maximum diffusion rate, a peak is observed in the voltammogram, as shown in Figure 5.22. The resulting decrease in current at/beyond the peak potential follows Cottrell behavior [4]. When the two methods are compared, we can simply state that the introduction of rotation causes the mass transfer boundary layer to be restricted to a finite thickness. In practical terms, this change in the mass transfer regime causes the change in shape of the voltammograms

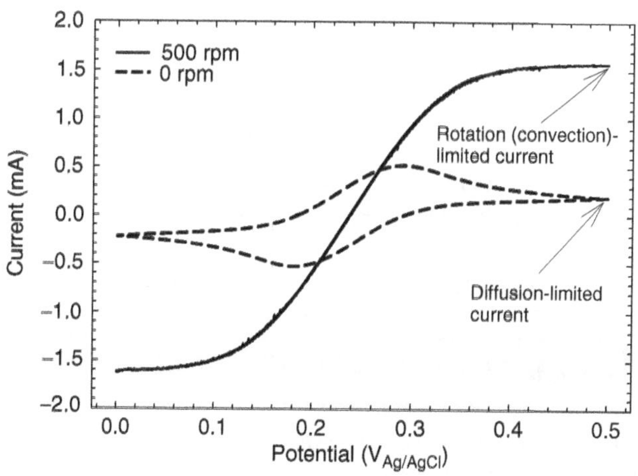

Figure 5.22 The effect of rotation from 0 to 530 rpm on the cyclic voltammograms of ferricyanide.

in Figure 5.22. Therefore, the addition of rotation and the consequent increase in current at the RDE tell us about the role of diffusion in a particular electron-transfer reaction.

In Figure 5.11, we described the interplay between electron transfer and mass transfer and how both are required to observe a current from a *G. sulfurreducens* biofilm. It is difficult to determine the role of mass transfer in biofilms simply from cyclic voltammograms; usually, certain electrochemical setups are required to investigate mass transfer via electrochemical methods. In our case, we used a combination of EIS and RDEs to study electron transfer and diffusional processes in *G. sulfurreducens* biofilms respiring on electrodes [50]. We tested the hypothesis that the RDE can be used as an electrochemical tool that controls diffusional processes when EABs are studied. We determined the film resistance, film capacitance, interfacial resistance, interfacial capacitance, and pseudocapacitance of *G. sulfurreducens* biofilms as shown in Figure 5.23. The details of the calculations and experimental procedures are given in the literature [50].

We expected that diffusion processes would affect the electron-transfer rates for mature *G. sulfurreducens* biofilms that had been allowed to reach a maximum steady-state current more than those for younger, thinner biofilms [56]. Therefore, after a steady-state current was obtained, we began rotating the biofilm. The RDE was spun from 0 to 530 rpm in discrete steps. Higher rotation rates did not add further useful information and were therefore not needed. Figure 5.24a shows that rotating the biofilm up to 530 rpm did not affect current generation, as two consecutive rotation sweeps yielded nearly identical current values. The largest increase in current occurred when the rotation was set to 10 rpm. Diminishing increases in current were observed upon increasing the rotation rate by a factor of 2, up to 160 rpm. A current of 102 μA at 530 rpm over a baseline current of

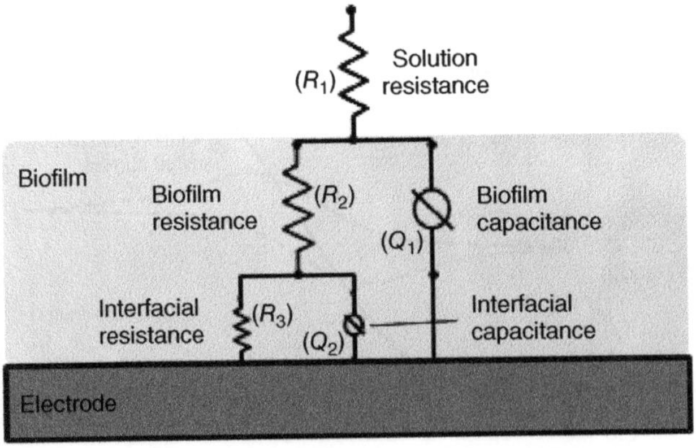

Figure 5.23 Equivalent circuit of *G. sulfurreducens* biofilm consisting of the film resistance, film capacitance, interfacial resistance, interfacial capacitance, and pseudocapacitance. Reproduced with permission from Ref. [50]. Copyright 2014 Wiley Periodicals, Inc. (*See insert for color representation of this figure.*)

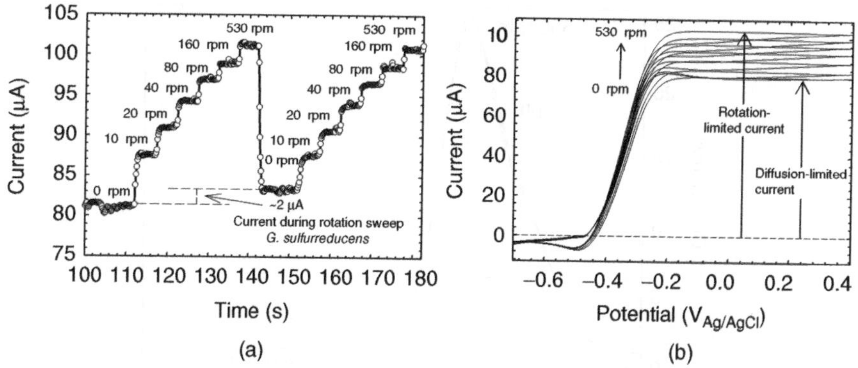

Figure 5.24 (a) Increasing current response to rotation of the *G. sulfurreducens* biofilm from 0 to 530 rpm. (b) Corresponding CV for each respective rotation rate from 0 to 530 rpm. Reproduced with permission from Ref. [50]. Copyright 2014 Wiley Periodicals, Inc.

82 μA at 0 rpm was measured. Figure 5.24b shows how the increased current is reflected in the cyclic voltammogram at each rotation rate. Unlike the change in shape of the voltammogram from 0 to 500 rpm in Figure 5.22, the biofilm voltammogram retained its characteristic shape. Only the limiting current was observed to increase with rotation rate. This observation is a strong indicator that diffusion-based mediators are not the primary source of current. It is also expected, given the numerous demonstrations of current from *G. sulfurreducens* biofilms not being dependent on the accumulation of mediators in the bulk solution [36]. From an electrochemical perspective, biofilms grown on RDEs can be simply tested for their electron-transfer mechanism by observing their voltammetric response before, during, and after rotation. It is also important to observe that the increase in catalytic current of the biofilm with rotation rate is relatively minor when compared to the order of magnitude change in the limiting current in Figure 5.22. When we plotted the SWV results in Figure 5.25 for the same rotation experiment, we observed that the voltammograms were relatively insensitive to the change in rotation rate. Because voltammograms obtained via SWV change appreciably during growth as shown in Figure 5.20a, we can state that the biofilm remained the same during the rotation experiment (i.e., it was not removed from the electrode surface because of shear stress). This is also another confirmation that the peak potential is insensitive to rotation rate.

Practical application

An RDE can be used to assess the biofilm electrochemical response to increased mass transfer rates of mediators, waste products, and nutrients. A typical sign of diffusion-based electron transfer is the shift from cyclic voltammograms containing peaks to the sigmoid-shaped cyclic voltammogram predicted by the Levich equation.

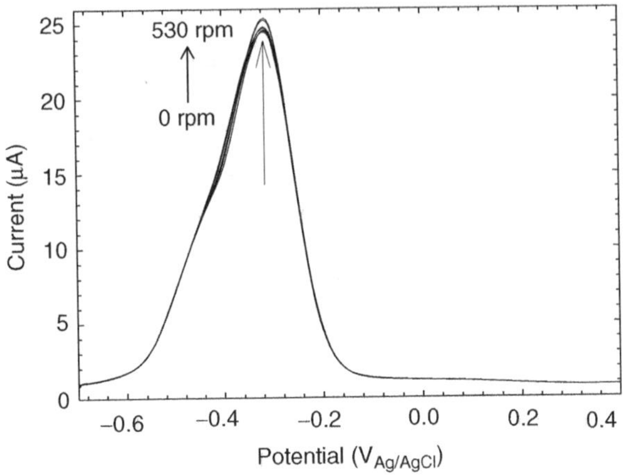

Figure 5.25 Square wave voltammograms of the rotation rate experiment shown in Figure 5.24b.

5.5.8 Changes to Biofilm Mass and Structure During Electron Transfer

Not all changes to a biofilm during electron transfer can be observed via CV with a coupled technique such as microscopy or rotation on an RDE. Some changes, such as changes in diffusion coefficients inside *G. sulfurreducens* biofilms, are only observable using nuclear magnetic resonance (NMR) methods [56]. Changes through growth in the diffusion coefficient and cell density inside live, electrode-respiring biofilms are indicators of changes in biofilm mass and structure. Biofilm mass and structure, although not depicted in Figure 5.11, are intimately tied to the overall electron-transfer processes occurring within the biofilm. In the case of *G. sulfurreducens* biofilms, the electron-transfer process is bound inside the biofilm, which is essentially a conductive film on an electrode surface. One unique tool for detecting and quantifying changes in the mass and structure of films attached to a surface is the quartz crystal microbalance (QCM). Combined with CV, QCM provides a way to quantify the "apparent" biofilm mass of each biofilm grown in our laboratory. "Apparent" refers to the fact that QCM measures both the mass of the biofilm attached to the exposed gold electrode and its energy-dissipating (viscoelastic) properties. Figure 5.26 shows a cartoon of QCM operation. QCM takes advantage of AT-cut quartz crystals operated near their fundamental frequency of vibration, which is predetermined by the construction of the quartz crystal electrode. Essentially, the quartz crystal oscillates at an initial resonant frequency (usually 5 or 10 MHz). When mass attaches or dampens the oscillation, the resonant frequency shifts from the initial value. Thus, QCM relates (small) changes in attached mass to (small) changes in resonant frequency. The change in resonant frequency is generally referred to as a frequency shift in QCM. Generally, the Sauerbrey relationship [57] is used to convert the changes in resonant frequency of an oscillating QCM electrode to mass:

$$\Delta f = -\frac{2f_0^2}{A\sqrt{\rho_q \mu_q}}\Delta m \qquad (5.12)$$

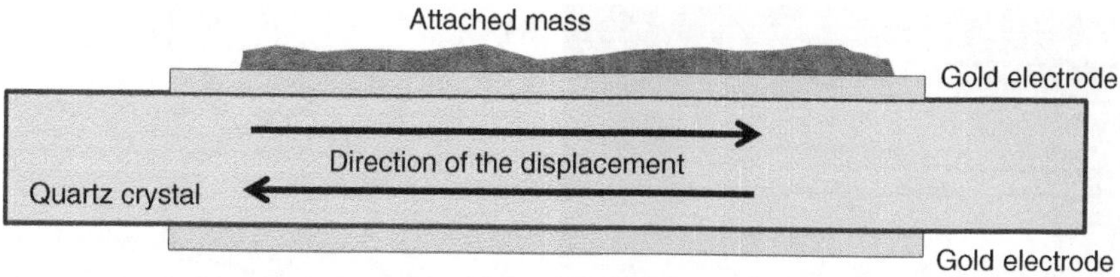

Figure 5.26 The QCM operates by applying a potential wave function across an AT-cut quartz crystal that produces a thickness-shear-type standing wave. The potential is applied through two gold layers, one deposited on each side of the flat quartz crystal surface. Thickness-shear-type deformation results in a measureable displacement parallel to the flat crystal surface and negligible displacement in the perpendicular direction.

where Δf is the frequency shift (Hz), f_o is the principal series resonant frequency (Hz), A is the area between electrodes (cm^2), ρ_q is the AT-cut quartz crystal density ($g\ cm^{-3}$), μ_q is the shear modulus of the AT-cut quartz crystal, and Δm is the mass change (g). For mass (where the small load approximation applies) on a QCM that can be considered theoretically an "extension" of the quartz crystal, Equation 5.12 applies. For viscoelastic films, a viscosity correction factor is used to modify Equation 5.12 [58]. In the absence of bandwidth measurements, it is difficult to employ such correction factors.

Growing *G. sulfurreducens* biofilms on a QCM electrode is similar to growing them on any common electrode. Figure 5.27a shows a biofilm that generally grew uniformly across a QCM electrode surface. Details of the biofilm surface are shown at magnifications of 800× and 6000× in Figures 5.27b and 5.28c, respectively. Because of the scanning electron microscope (SEM) sample preparation, cracks appeared and revealed the dense packing of cells in the biofilm: this is expected [20]. Figure 5.27d shows the QCM insert, which consisted of an AT-cut quartz disk, a 7-mm gold-coated quartz surface (the QCM electrode), and resin-coated quartz (inert). The pink hue on the QCM electrode is typical of attached *G. sulfurreducens* biofilm. Initially, upon the addition of cells to the QCM cell, there was an immediate response in both the current and the frequency shift of the oscillating crystal, indicating cell attachment and electron transfer. Figure 5.28 shows the current response and frequency shift upon the addition of cells. Initially, the current increased sharply from 100 to 130 nA with the addition of cells to the QCM cell. Simultaneously, the frequency shift dropped sharply by 30 Hz from a baseline frequency shift of 15 Hz. Within 6 h, the current and frequency shift reached values of 580 nA and −250 Hz, respectively. We expected that the frequency shift would decrease as the current increased. As cells attached and the biofilm grew, the mass loading on the oscillating crystal increased and caused the resonant frequency to decrease. As more biofilm attached, the frequency shift decreased further, indicating that even for small current responses (<1 μA), the QCM response could be detected. The immediate change in frequency shift and current corresponded to irreversible attachment of *G. sulfurreducens* cells to the QCM electrode. As *G. sulfurreducens* cells rely on direct electron transfer to solid surfaces, it is reasonable to

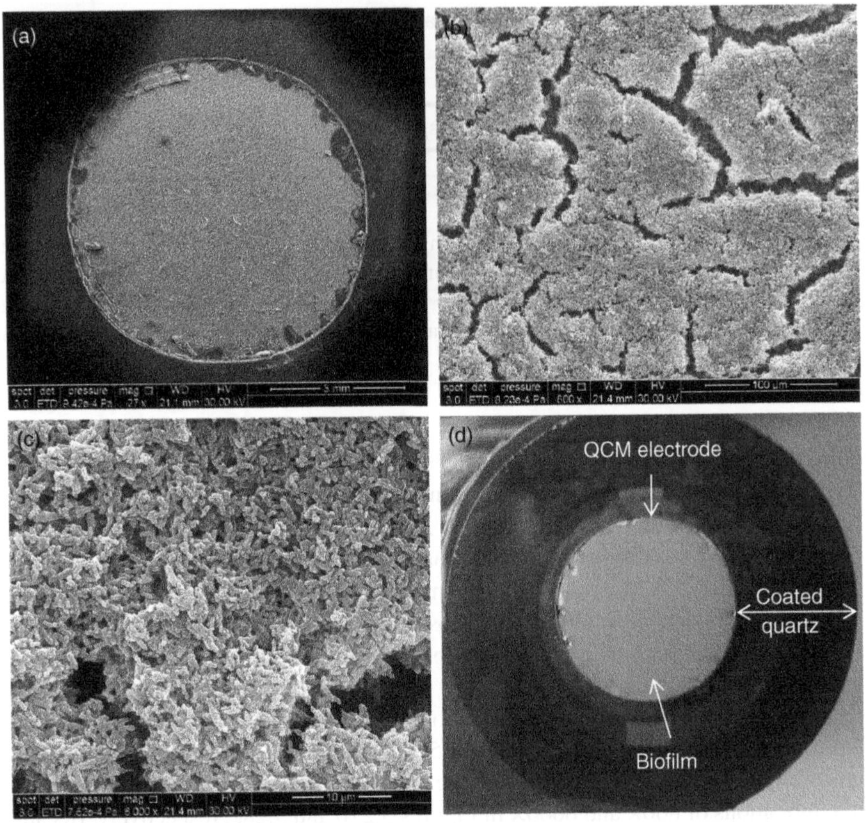

Figure 5.27 SEM images of a *G. sulfurreducens* biofilm grown on a QCM electrode at (a) 27×, (b) 800×, and (c) 6000× magnification. (d) Camera image of the biofilm before SEM sample preparation. *G. sulfurreducens* biofilms on QCM electrodes appeared to cover the electrode surface uniformly. Reproduced with permission from Ref. [59]. Copyright 2014 WILEY-VCH Verlag GmbH & Co. KGaA, Weinheim.

conclude that cells immediately and irreversibly attach to the QCM electrode surface in order to continue respiration and survive.

In addition to monitoring both frequency shift and current during biofilm growth, we coupled frequency shift to CV at discrete times during the growth of the *G. sulfurreducens* biofilm. When needed, polarization could be stopped and we could run CV. Interestingly, we observed a resonant frequency response as the biofilm was oxidized and reduced. In Figure 5.29a, three CVs are shown, corresponding to before inoculation (no biofilm), 11 h after inoculation, and 22 h after inoculation. The current/potential response was typical of early stages of *G. sulfurreducens* biofilm growth [46]. When there was no biofilm, the frequency shift/potential response was flat and unchanging. After 11 h, a slight decrease at oxidizing potentials to −19 Hz was observed. After 22 h, the frequency shift decreased further to −100 Hz.

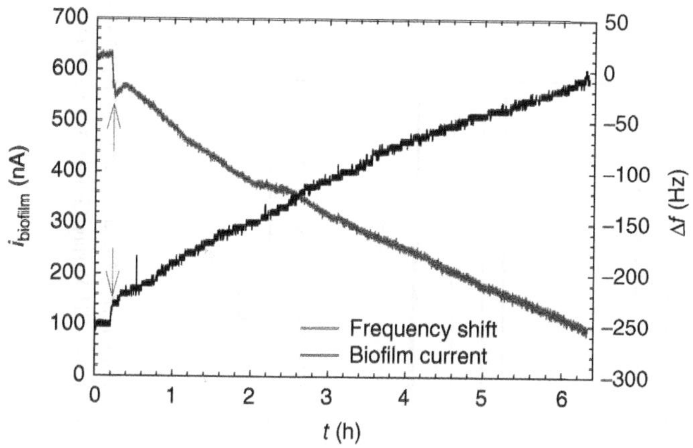

Figure 5.28 The current response and frequency shift during the initial biofilm growth on the QCM electrode. The arrows indicate the time of inoculation. Upon inoculation, both current and frequency shift responded sharply. Reproduced with permission from Ref. [59]. Copyright 2014 WILEY-VCH Verlag GmbH & Co. KGaA, Weinheim.

This response was characteristic of all *G. sulfurreducens* biofilms grown on the QCM electrode. To better understand the frequency shift/potential response, we removed a portion of the visible biofilm. Figure 5.29b shows the resulting response: the limiting current was reduced from 130 to 55 μA as approximately 50% of the visible biofilm was removed, and it was reduced to slightly above zero current when 100% of the visible biofilm was removed. At the same time, the frequency shift was reduced, from −92 to −32 Hz and finally to −4 Hz, when approximately 50% and 100% of the visible biofilm was removed, respectively. Therefore, from both Figure 5.29a and b, it is clear that the decrease in frequency shift toward oxidizing potentials (limiting current) was related to the state of the biofilm. Increasing biofilm attachment increased the frequency shift, and removal of biofilm decreased it. The decrease in frequency shift upon current flow suggests that the biofilm was increasing in apparent mass. Considering the timescale of the CV (˜6 min), it is unlikely that more biofilm attachment caused the change. In fact, when the visible biofilm was removed, the overall frequency shift changed correspondingly with the amount of visible biofilm removed.

To assess whether electron flux was the cause of the frequency shift during CV, we measured the frequency shift/potential response under both turnover and nonturnover conditions in which acetate was removed. Figure 5.30a shows a limiting current of 120 μA and a frequency shift of −440 Hz at oxidizing potentials under turnover conditions. When the biofilm was brought to nonturnover conditions, the usual *G. sulfurreducens* biofilm CV was observed, with multiple anodic and cathodic peaks (see Fig. 5.16b). Although catalytic current had decreased to background, a frequency shift of −85 Hz was still observed at oxidizing potentials. This means that although the frequency shift/potential response was predominately caused by electron

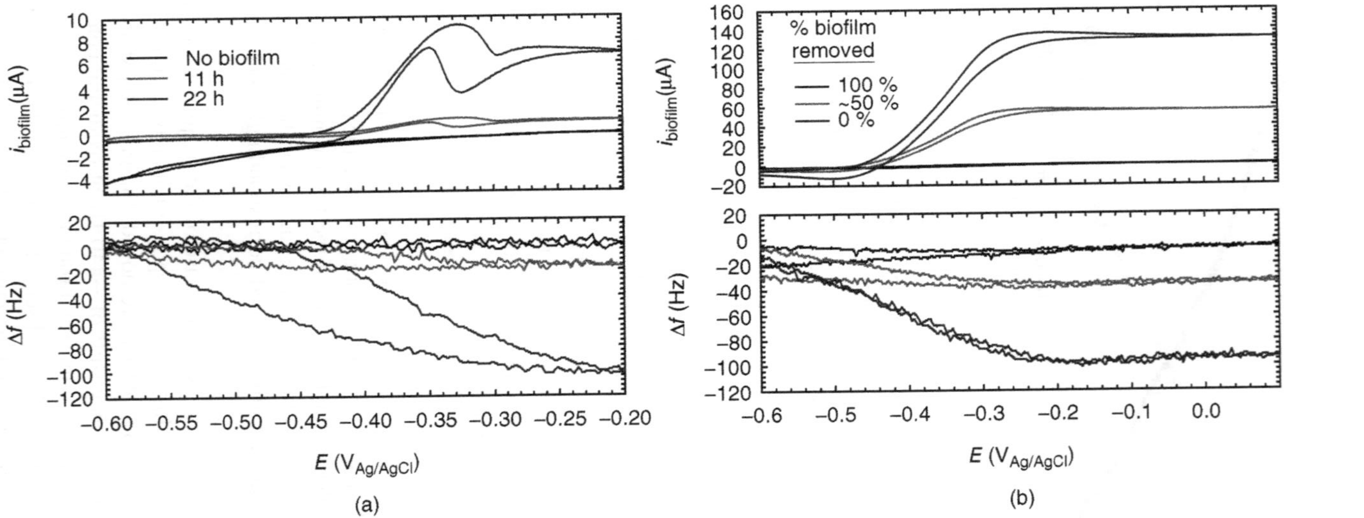

Figure 5.29 Cyclic voltammograms (top) and corresponding frequency shift responses (bottom) of *G. sulfurreducens* biofilm (a) during initial growth and (b) after visible biofilm was removed. Reproduced with permission from Ref. [59]. Copyright 2014 WILEY-VCH Verlag GmbH & Co. KGaA, Weinheim.

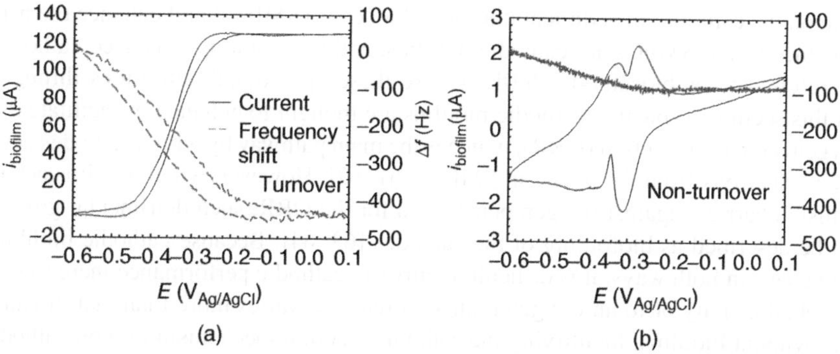

Figure 5.30 Cyclic voltammograms and corresponding frequency shift responses of *G. sulfurreducens* biofilm under (a) turnover and (b) nonturnover conditions. Reproduced with permission from Ref. [59]. Copyright 2014 WILEY-VCH Verlag GmbH & Co. KGaA, Weinheim.

flux through the biofilm (Δ of 355 Hz), there was still a residual amount of frequency shift caused by the change from a reduced to an oxidized state of the biofilm. For a more detailed interpretation of the frequency shift observed during CV, we refer the readers to the literature [59]. Within the context of this chapter, we emphasize that we used elements of what was discussed previously for CV of *G. sulfurreducens* to investigate the frequency shift response of the QCM. These elements included obtaining CV during the growth of the biofilm (Fig. 5.15), obtaining CV under nonturnover conditions (Fig. 5.16b), and monitoring the CV while biofilm was carefully removed. We used a similar approach in our use of the RDE in combination with CV. We stress that the outcome of any biofilm electrochemical experiment involving a coupled technique relies on the researcher's ability to incorporate the CV analysis (and any other electrochemical technique) into an overall research plan.

Practical application

A QCM can be used to assess changes in a biofilm that are not obvious in the electrochemical response in a CV. This reinforces the need for coupled techniques when electrode-respiring biofilms are studied.

5.6 CATHODIC BIOFILMS

Cathodic biofilms are biofilms that accept electrons from an electrode, as opposed to anodic biofilms, such as the *G. sulfurreducens* biofilms discussed previously, which deliver electrons to an electrode [34]. Cathodic biofilms are less well understood than their anodic counterparts despite being well documented in biocathode studies. Often, a more operational definition is given to cathodic biofilms. For example, biofilms that form on the cathodes of sediment microbial fuel cells (SMFCs) and enhance the performance of the SMFCs are cathodic biofilms, and the resulting enhanced cathodes are referred to as biocathodes [60]. We use these operational definitions throughout this section. Typically, cathodic biofilms are thought to enhance or facilitate the oxygen reduction reaction (ORR), which is the main pathway by which cathodes neutralize electrons from the anode in SMFCs [61, 62]. However, when biofilms act as diffusion barriers against oxygen penetration for the ORR, such detrimental growth can be described as biofouling of the cathodes [63, 64]. Because cathodic biofilms can behave in both ways, it is difficult to attribute cathodic performance increases to microbial activity or to an oxygen reduction rate that varies more than that of cathodes without biofilms. Identifying the cathodic reaction mechanism of a biocathode is not trivial, because the range of potential mechanisms is broad and may include the ORR [65], phototrophic activities [66, 67], biological H_2 production [68], denitrification [69, 70], manganese oxidation [71], and iron oxidation [72, 73]. It is likely that a complex web of microbial activities is present in cathodic biofilms and that not all activities contribute to generating current [74]. Regardless of all these mechanisms, the fate of oxygen inside cathodic biofilms is a key indicator in many of these microbe communities.

Because ORR can be quantified using noninvasive dissolved oxygen, hydrogen peroxide, and pH microelectrodes, we were able to couple microelectrode measurements to CV. For example, a hydrogen peroxide microelectrode (Fig. 5.31a) was placed approximately 100 μm above a glassy carbon electrode in river water. The glassy carbon electrode simulated a cathode in an SMFC in which oxygen reduction is the cathodic reaction. The hydrogen peroxide microelectrode was used to detect hydrogen peroxide generation during a CV scan. In this case, the microelectrode was held at a constant distance from the biofilm electrode, and the hydrogen peroxide concentration was monitored as a function of potential and current. Figure 5.31b shows that hydrogen peroxide was produced at potentials below $-200\ mV_{Ag/AgCl}$.

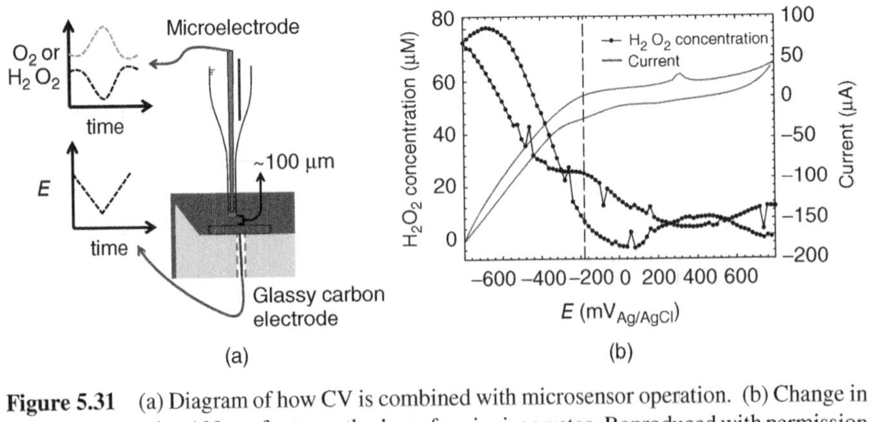

Figure 5.31 (a) Diagram of how CV is combined with microsensor operation. (b) Change in H_2O_2 concentration 100 μm from a cathode surface in river water. Reproduced with permission from Ref. [64]. Copyright 2013 WILEY-VCH Verlag GmbH & Co. KGaA, Weinheim.

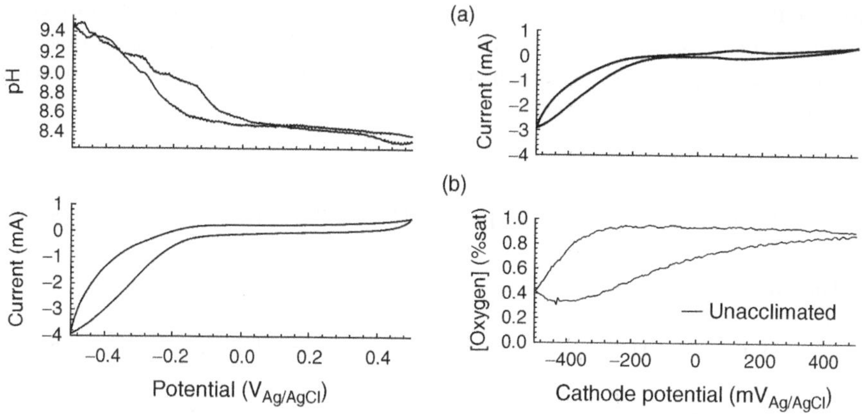

Figure 5.32 Changes in pH and dissolved oxygen at a cathode surface in seawater. Reproduced with permission from Ref. [75]. Copyright 2014 WILEY-VCH Verlag GmbH & Co. KGaA, Weinheim.

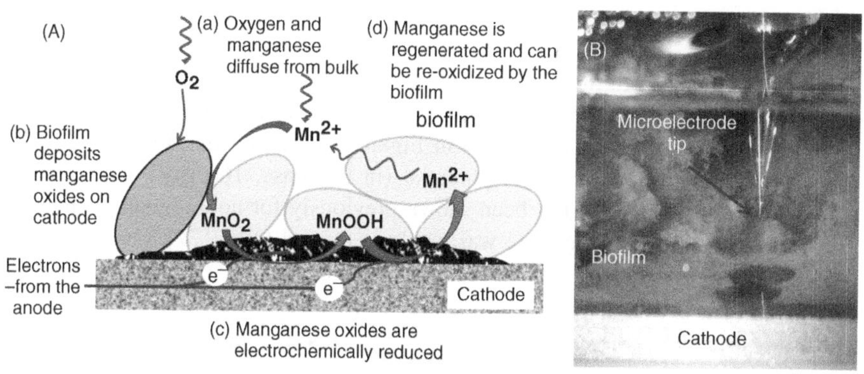

Figure 5.33 (a) Diagram of MOB mechanism. (b) Digital image of MOB on glassy carbon with a dissolved oxygen microelectrode in view. Reproduced with permission from Ref. [64]. Copyright 2013 WILEY-VCH Verlag GmbH & Co. KGaA, Weinheim.

We should note that we could not detect any hydrogen peroxide accumulating in the bulk solution.

In this particular case, we were measuring ORR in river water [64]. However, the same methodology can be used in seawater and is shown in Figure 5.32 [75]. Essentially, the behavior of ORR on unmodified carbon electrodes yields the same responses in dissolved oxygen, H_2O_2, and pH at the same potentials. This unique "fingerprint" allows us to study cathodic biofilms operating in SMFCs or other bioelectrochemical systems using oxygen.

As an example, we grew a *Leptothrix discophora* SP-6 biofilm on an initially nonpolarized cathode. The cathodic reaction mechanism is drawn schematically in Figure 5.33. Briefly, manganese-oxidizing bacteria (MOB) colonizing cathodes form biofilms that deposit biomineralized manganese oxides onto the cathode surfaces. Over time, the manganese oxides accumulate on the cathode surface and inside the biofilm. The manganese oxides can then be electrochemically reduced via a two-step, two-electron reduction as follows:

$$MnO_{2(s)} + H^+ + e^- \rightarrow MnOOH_{(s)} \quad E^0 = +0.81 \ V_{SCE} \tag{5.13}$$

$$MnOOH_{(s)} + 3H^+ + e^- \rightarrow Mn^{2+} + 2H_2O \quad E^0 = +1.26 \ V_{SCE} \tag{5.14}$$

The overall reaction is:

$$MnO_{2(s)} + 4\,H^+ + 2e^- \rightarrow Mn^{2+} + 2H_2O \quad E^0 = +1.28 \ V_{SCE} \tag{5.15}$$

The soluble manganese ions are then regenerated inside the biofilm, where they are re-oxidized to manganese oxides, which continue the cycle indefinitely. After a steady OCP value was reached, we measured stationary profiles similar to that in Figure 5.32b. As is shown in Figure 5.34, compared to the initial CV without biofilm, the CV with *L. discophora* SP-6 biofilm exhibited a large cathodic current at more positive potentials (−240 μA at −40 $mV_{Ag/AgCl}$). In addition, the onset of ORR began at approximately −40 $mV_{Ag/AgCl}$. The large cathodic current at more positive potentials can be attributed to the discharge of accumulated manganese oxides deposited on the electrode surface by the biofilm. Comparing the CV taken without biofilm with a biofilm CV (scan 1 or scan 2) in Figure 5.34 shows that ORR catalysis is associated with a decrease in overpotential (in this case, 160 mV). This form of manganese-catalyzed ORR has been shown previously for nonbiogenic manganese [76]. The presence of manganese was catalytic toward ORR [77], consistent with

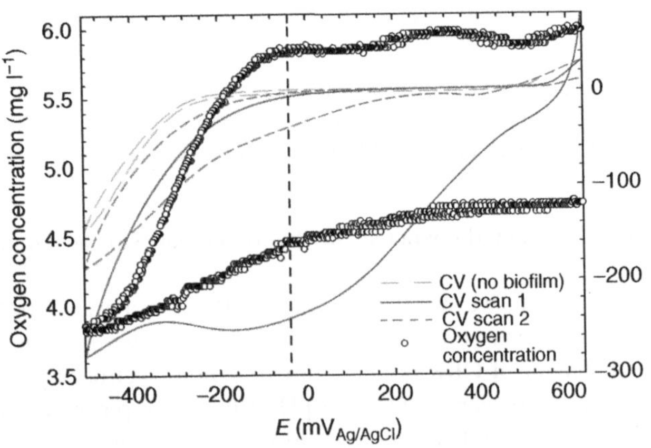

Figure 5.34 Stationary profile of oxygen at a glassy carbon electrode with *L. discophora* SP-6 grown on the surface. The dashed line is centered at −40 $mV_{Ag/AgCl}$. Reproduced with permission from Ref. [64]. Copyright 2013 WILEY-VCH Verlag GmbH & Co. KGaA, Weinheim.

previous observations of microbial catalysis of ORR [65]. However, the term "discharge" is purposefully used in this case, as the cathodic current was not sustainable. For CV scan 2 in Figure 5.34, the cathodic current attributed to manganese oxide reduction decreased to approximately 15% of its original value at 0 $V_{Ag/AgCl}$. While the biofilm oxidized Mn(II) in solution to Mn(IV), the electrode reduced the Mn(IV) back to Mn(III)/Mn(II) at a faster rate. The inability of the biofilm to regenerate the cathodic current highlights a potential weakness of manganese-catalyzed biocathodes: sustainability. The rate of manganese oxide deposition by *L. discophora* SP-6 cannot sustain the initial current and suggests that continuous energy harvesting may be less efficient than more adaptive energy-harvesting techniques, such as intermittent energy harvesting [78]. It could be argued, however, that the cathodic current was still 20 times larger than was the case without biofilm for CV scan 2 in Figure 5.34 and that the overpotential was reduced. This observation demonstrates an important facet of cathodic biofilm behavior: it can increase the ORR rate, decrease the overpotential, or generate a cathodic reactant such as manganese to enhance cathode performance.

5.6.1 Field-Relevant Biocathodes and Their Sustainability

Monospecies biofilms of MOB are relatively straightforward to investigate because the mechanism is manganese cycling. In field-ready SMFCs and biocathodes that acclimate in the open environment, the cathodic reaction mechanisms are less obvious. In this section, we discuss cathodes that were operated in SMFCs with sediment and seawater collected from San Diego Bay for approximately 165 days prior to microscale measurements, which were described earlier in this section. Local oxygen concentration and pH profiles were measured simultaneously along with the electrochemical responses of both acclimated and unacclimated cathodes following Figure 5.31a. Acclimation refers to time allowed for EABs to populate an electrode surface prior to its use in a BES and is often used for multispecies EABs.

Figure 5.35a shows the cyclic voltammograms of an acclimated cathode for both short (+500 to +100 $mV_{Ag/AgCl}$) and long (+500 to −500 $mV_{Ag/AgCl}$) scans. As a control, the unacclimated cyclic voltammogram is also included. Short scans are included because operational cathode potentials never exceed anode potentials when SMFCs generate power. Two distinct regions of limiting current (designated I and II), when current is independent of potential, are observed. Although two limiting currents are observed for the acclimated cathode, only limiting current II reduced oxygen concentration at the cathode surface to below our detection limit (Fig. 5.35b). Limiting current I reduced oxygen concentrations at the surface but did not cause severe oxygen depletion as seen with limiting current II. For a freely diffusing reactant, semiempirical treatment of the voltammetric equations under mass transfer limitation equates to a limiting current with a zero surface concentration of that reactant [4], that is, the electron flux reaches a maximum when the diffusive flux of the reactant toward the electrode surface reaches a maximum. Assuming that oxygen is freely diffusing in these systems, limiting current II is consistent with the direct, abiotic reduction of oxygen diffusing to the cathode surface. In contrast, limiting current I cannot be explained by the direct, abiotic reduction of oxygen diffusing to the cathode surface

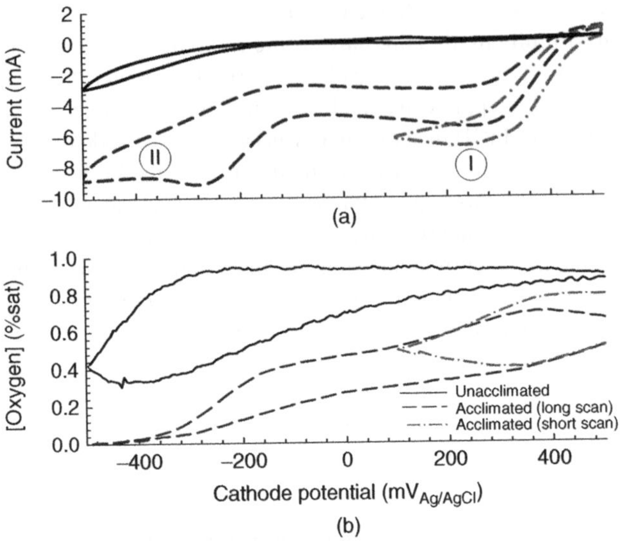

Figure 5.35 Dissolved oxygen measurements taken at an acclimated cathode surface during cyclic voltammetry. (a) Cyclic voltammogram. (b) Corresponding change in dissolved oxygen at the cathode surface. The short scan was taken under a realistic operating condition for cathodes, and the long scan includes the abiotic oxygen reduction reaction occurring at more negative potentials outside the realistic range. Unacclimated curves are shown for comparison. Reproduced with permission from Ref. [75]. Copyright 2014 WILEY-VCH Verlag GmbH & Co. KGaA, Weinheim.

because it is not associated with a zero surface concentration of oxygen. Limiting current I is likely another cathodic reaction mechanism that consumes oxygen. These example data sets show how multiple techniques can be used to identify ORR in cathodic biofilms. Without combining microelectrode measurements near the surface, it would be impossible to identify the ORR in cathodic biofilms.

Practical application

Microsensors and CV can be coupled to assess ORR in cathodic biofilms operating in aerobic environments and to investigate cathodic reaction mechanisms operating in biocathodes in SMFCs and other bioelectrochemical systems.

5.7 CONCLUDING REMARKS

In this chapter, we have presented electrochemical techniques for studying electron-transfer processes between biofilm and a solid electrode. To date, the questions of whether electrons are transferred through conductive nanowires, bound redox mediators, or diffusing redox mediators and of how electrons enter these pathways have yet to be unequivocally answered. It is believed that extracellular polymeric substances (EPS) play a critical role in electron transfer. For example, Cao et al. [79] found redox-active proteins in the EPS of *Shewanella* sp. HRCR-1, which were distinct from the redox-active proteins commonly found on the cell surface. EPS are known to facilitate oxidation/reduction reactions to minerals [80, 81]. How do EPS interact with the cells in EABs? There is a significant amount of knowledge missing on these electron-transfer mechanisms that have been proposed to account directly for the path of electrons from inside cells to the biofilm electrode. While many publications in the literature push various ideas and concepts, what is needed now is direct demonstration of key findings on the electron-transfer processes. For example, having conductive nanowires in a biofilm does not necessarily mean that they are involved in electron-transfer processes. Identifying the functions of key components of EABs in electron transfer could critically improve our understanding of these systems and allow us to improve the devices in which EABs are used. In addition, new methods of measuring the parameters reported in the past could add validity and create a functional base from which we can develop theory.

EAB processes are an interfacial phenomenon: EABs interact with the electrode inside the biofilm diffusive and reactive layers at the electrode surface. These microscale layers are directly related to extracellular electron transfer, whereas diffusion processes above these layers are linked indirectly. Thus, we expect that the surface concentrations of the redox-active compounds and local solution properties inside EABs are more relevant and critical than the corresponding values in the bulk. Correlating and fitting lines to bulk data may have little significance to the fundamental processes occurring inside EABs. Direct measurements inside EABs are needed, such as measurements of pH inside EABs and measurements of the spectroelectrochemical properties of EABs [7, 20, 35]. This is especially important because the cell density inside some EABs is not uniformly distributed and predictions based on simple diffusion may not apply [82].

The coupling of electrode polarization, CV, and SWV with methods that directly measure physical or chemical parameters varying inside EABs in situ addresses issues with the interpretation of electrochemical data. The goal of this coupling is to correlate the properties that vary within the EAB under various electrode-respiring conditions. Specifically, future EAB research and advanced techniques should focus on variation that occurs within the biofilm and not just in the bulk solution. In fact, the correlation of variation in biofilms and bulk solution will advance the field significantly. This correlation can be done by: (1) directly measuring the kinetics of redox mediators inside biofilms, (2) resolving local concentrations of chemical species inside the biofilm, (3) resolving the physical location of electrochemically active species in situ, and (4) correlating limiting current with local biofilm parameters. There are several available tools and techniques that can be used, such as microelectrodes, spectroelectrochemical methods, and microscopy.

In this chapter, we have presented how these coupled techniques should be used and have given several selected examples. Figure 5.36 shows how these multiple techniques can be integrated to study electron-transfer processes in EABs. In this chapter, we did not cover the use of NMR to measure acetate depth profiles in biofilms, a

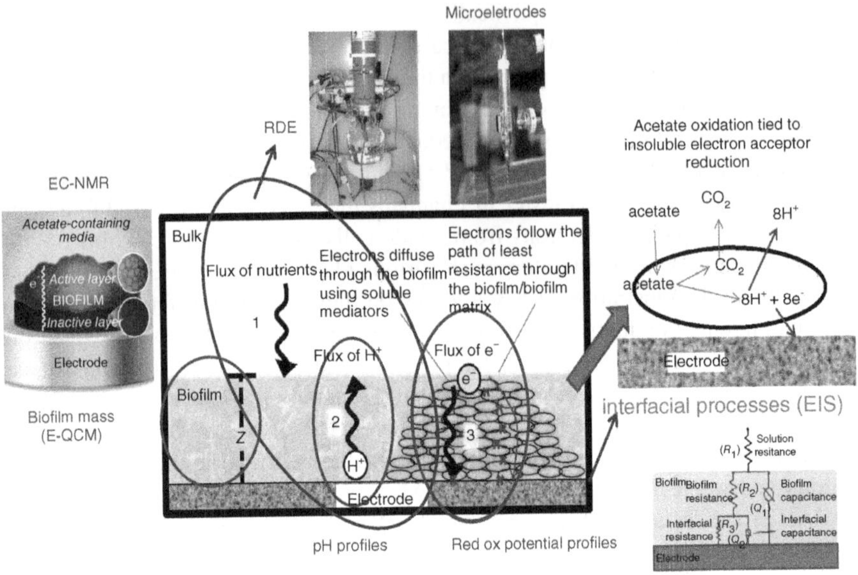

Figure 5.36 Combining techniques to understand electron transfer processes.

technique developed by our research group in collaboration with Pacific Northwest National Laboratory (PNNL) [33, 56]. The use of this technique is limited to one laboratory in the world (PNNL) because of the cost of the equipment. NMR can be an excellent research tool, and in the future, it may be made affordable for EAB research. However, if this is not possible, new microelectrodes measuring acetate or electron donor concentrations in biofilms should be developed. Having the ability to measure electron donor concentrations in biofilms and correlate them to underlying processes will potentially bring new insights to EAB research.

ACKNOWLEDGMENTS

This research was supported by NSF-CAREER award #0954186 and by US Office of Naval Research (ONR) grant #N00014-09-1-0090. Beyenal acknowledges additional support from Fundamental and Applied Chemical and Biological Catalysts to Minimize Climate Change, Create a Sustainable Energy Future, and Provide a Safer Food Supply (project #WNP00807).

REFERENCES

1. Torres CI, Marcus AK, Parameswaran P, Rittmann BE. Kinetic experiments for evaluating the Nernst–Monod model for anode-respiring bacteria (ARB) in a biofilm anode. Environ Sci Technol 2008;42(17):6593–6597.

2. Hamelers HVM, Ter Heijne A, Stein N, Rozendal RA, Buisman CJ. Butler–Volmer–Monod model for describing bio-anode polarization curves. Bioresour Technol 2011;102(1):381–387.
3. Babauta JT, Nguyen HD, Harrington TD, Renslow R, Beyenal H. pH, redox potential and local biofilm potential microenvironments within *Geobacter sulfurreducens* biofilms and their roles in electron transfer. Biotechnol Bioeng 2012;109(10):2651–2662.
4. Bard AJ, Faulkner LR, editors. Electrochemical Methods: Fundamentals and Applications. 2nd ed. New York (NY): John Wiley & Sons, Inc; 2001.
5. Kissinger PT. Introduction to analog instrumentation. In: Kissinger PT, Heineman WR, editors. Laboratory Techniques in Electroanalytical Chemistry. New York (NY): Marcel Dekker Inc.; 1996. p 165–194.
6. Istanbullu O, Babauta J, Duc Nguyen H, Beyenal H. Electrochemical biofilm control: mechanism of action. Biofouling 2012;28(8):769–778.
7. Franks AE, Nevin KP, Jia H, Izallalen M, Woodarda TL, Lovleya DR. Novel strategy for three-dimensional real-time imaging of microbial fuel cell communities: monitoring the inhibitory effects of proton accumulation within the anode biofilm. Energy Environ Sci 2009;2(1):113–119.
8. McLean JS, Wanger G, Gorby YA, Wainstein M, McQuaid J, Ishii SI, Bretschger O, Beyenal H, Nealson KH. Quantification of electron transfer rates to a solid phase electron acceptor through the stages of biofilm formation from single cells to multicellular communities. Environ Sci Technol 2010;44(7):2721–2727.
9. Logan BE, Hamelers B, Rozendal R, Schröder U, Keller J, Freguia S, Aelterman P, Verstraete W, Rabaey K. Microbial fuel cells: methodology and technology. Environ Sci Technol 2006;40(17):5181–5192.
10. Oelßner W, Berthold F, Guth U. The iR drop – well-known but often underestimated in electrochemical polarization measurements and corrosion testing. Mater Corros 2006;57(6):455–466.
11. Wei JC, Liang P, Huang X. Recent progress in electrodes for microbial fuel cells. Bioresour Technol 2011;102(20):9335–9344.
12. Zhou MH, Chia M, Luob J, Hea H, Jin T. An overview of electrode materials in microbial fuel cells. J Power Sources 2011;196(10):4427–4435.
13. Kissinger PT, Heineman WR, editors. Laboratory Techniques in Electroanalytical Chemistry. 2nd ed. New York (NY): Marcel Dekker Inc; 1996.
14. Richter H, McCarthy K, Nevin KP, Johnson JP, Rotello VM, Lovley DR. Electricity generation by *Geobacter sulfurreducens* attached to gold electrodes. Langmuir 2008;24(8):4376–4379.
15. Eckermann AL, Feld DJ, Shaw JA, Meade TJ. Electrochemistry of redox-active self-assembled monolayers. Coord Chem Rev 2010;254(15–16):1769–1802.
16. Gooding J, Yang WR. The self-assembled monolayer modification of electrodes – Some recent advances in biological application. Actual Chim 2008;320(321):85–89.
17. Mandler D, Kraus-Ophir S. Self-assembled monolayers (SAMs) for electrochemical sensing. J Solid State Electrochem 2011;15(7–8):1535–1558.
18. Busalmen JP, Esteve-Nuñez A, Berná A, Feliu JM. ATR-SEIRAs characterization of surface redox processes in *G. sulfurreducens*. Bioelectrochemistry 2010;78(1):25–29.
19. Jain A, Gazzola G, Panzera A, Zanoni M, Marsili E. Visible spectroelectrochemical characterization of *Geobacter sulfurreducens* biofilms on optically transparent indium tin oxide electrode. Electrochim Acta 2011;56(28):10776–10785.

20. Liu Y, Kim H, Franklin RR, Bond DR. Linking spectral and electrochemical analysis to monitor c-type cytochrome redox status in living *Geobacter sulfurreducens* biofilms. ChemPhysChem 2011;12(12):2235–2241.
21. Fischer LM, Tenje M, Heiskanen AR, Masuda N, Castillo J, Bentien A, Émneus J, Jakobsen MH, Boisen A. Gold cleaning methods for electrochemical detection applications. Microelectron Eng 2009;86(4–6):1282–1285.
22. Vanbenschoten JJ, Lewis JY, Heineman WR, Roston DA, Kissinger PT. Cyclic voltammetry experiment. J Chem Educ 1983;60(9):772–776.
23. Kissinger PT, Heineman WR. Cyclic voltammetry. J Chem Educ 1983;60(9):702–706.
24. Bard AJ, Parsons R, Jordan J, editors. Standard Potentials in Aqueous Solution. Monographs in Electroanalytical Chemistry and Electrochemistry. New York (NY): Marcel Dekker, Inc.; 1985. p 834.
25. Nicholson RS. Theory and application of cyclic voltammetry for measurement of electrode reaction kinetics. Anal Chem 1965;37(11):1351–1355.
26. Beyenal H, Babauta JT. Microscale gradients and their role in electron-transfer mechanisms in biofilms. Biochem Soc Trans 2012;40:1315–1318.
27. Stewart PS. Diffusion in biofilms. J Bacteriol 2003;185(5):1485–1491.
28. Renslow R, Babauta J, Kuprat A, Schenk J, Ivory C, Fredrickson J, Beyenal H. Modeling biofilms with dual extracellular electron transfer mechanisms. Phys Chem Chem Phys 2013;15(44):19262–19283.
29. Strycharz SM, Malanoski AP, Snider RM, Yi H, Lovley DR, Tender LM. Application of cyclic voltammetry to investigate enhanced catalytic current generation by biofilm-modified anodes of *Geobacter sulfurreducens* strain DL1 vs. variant strain KN400. Energy Environ Sci 2011;4(3):896–913.
30. Marcus AK, Torres CI, Rittmann BE. Conduction-based modeling of the biofilm anode of a microbial fuel cell. Biotechnol Bioeng 2007;98(6):1171–1182.
31. Picioreanu C, Head IM, Katuri KP, van Loosdrecht MC, Scott K. A computational model for biofilm-based microbial fuel cells. Water Res 2007;41(13):2921–2940.
32. Lebedev N, Strycharz-Glaven SM, Tender LM. Spatially resolved confocal resonant raman microscopic analysis of anode-grown G*eobacter sulfurreducens* biofilms. ChemPhysChem 2014;15(2):320–327.
33. Renslow RS, Babauta JT, Dohnalkova AC, Boyanov MI, Kemner KM, Majors PD, Fredricksond JK, Beyenal H. Metabolic spatial variability in electrode-respiring *Geobacter sulfurreducens* biofilms. Energy Environ Sci 2013;6(6):1827–1836.
34. Babauta J, Renslow R, Lewandowski Z, Beyenal H. Electrochemically active biofilms: facts and fiction: *a review*. Biofouling 2012;28(8):789–812.
35. Babauta JT, Nguyen HD, Beyenal H. Redox and pH microenvironments within *Shewanella oneidensis* MR-1 biofilms reveal an electron transfer mechanism. Environ Sci Technol 2011;45(15):6654–6660.
36. Bond DR, Lovley DR. Electricity production by *Geobacter sulfurreducens* attached to electrodes. Appl Environ Microbiol 2003;69(3):1548–1555.
37. Marsili E, Baron DB, Shikhare ID, Coursolle D, Gralnick JA, Bond DR. *Shewanella* secretes flavins that mediate extracellular electron transfer. Proc Natl Acad Sci U S A 2008;105(10):3968–3973.
38. Vetter KJ. Electrochemical Kinetics : Theoretical and Experimental Aspects. New York: Academic Press; 1967.

39. Rabaey K, Boon N, Siciliano SD, Verhaege M, Verstraete W. Biofuel cells select for microbial consortia that self-mediate electron transfer. Appl Environ Microbiol 2004;70(9):5373–5382.

40. Fricke K, Harnisch F, Schroder U. On the use of cyclic voltammetry for the study of anodic electron transfer in microbial fuel cells. Energy Environ Sci 2008;1(1):144–147.

41. Marsili E, Rollefson JB, Baron DB, Hozalski RM, Bond DR. Microbial biofilm voltammetry: direct electrochemical characterization of catalytic electrode-attached biofilms. Appl Environ Microbiol 2008;74(23):7329–7337.

42. Meitl LA, Eggleston CM, Colberg PJS, Khare N, Reardon CL, Shic L. Electrochemical interaction of *Shewanella oneidensis* MR-1 and its outer membrane cytochromes OmcA and MtrC with hematite electrodes. Geochim Cosmochim Acta 2009;73(18):5292–5307.

43. Richter H, Nevin KP, Jia H, Lowy DA, Lovleya DR, Tender LM. Cyclic voltammetry of biofilms of wild type and mutant *Geobacter sulfurreducens* on fuel cell anodes indicates possible roles of OmcB, OmcZ, type IV pili, and protons in extracellular electron transfer. Energy Environ Sci 2009;2(5):506–516.

44. Tender LM, Strycharz S, Yi H, Nevin KP, Kim BC, Franks A, Klimes A, Lovley DR. Cyclic voltammetry of electrode-bound biofilms of *Geobacter sulfurreducens*. Abstr Pap Am Chem Soc 2009;238.

45. Katuri KP, Kavanagh P, Rengaraj S, Leech D. *Geobacter sulfurreducens* biofilms developed under different growth conditions on glassy carbon electrodes: insights using cyclic voltammetry. Chem Commun 2010;46(26):4758–4760.

46. Marsili E, Sun J, Bond DR. Voltammetry and growth physiology of *Geobacter sulfurreducens* biofilms as a function of growth stage and imposed electrode potential. Electroanalysis 2010;22(7–8):865–874.

47. Carmona-Martinez AA, Harnisch F, Fitzgerald LA, Biffinger JC, Ringeisen BR, Schröder U. Cyclic voltammetric analysis of the electron transfer of *Shewanella oneidensis* MR-1 and nanofilament and cytochrome knock-out mutants. Bioelectrochemistry 2011;81(2):74–80.

48. Lewandowski Z, Beyenal H. Fundamentals of Biofilm Research. 2nd ed. Vol. 672. Boca Raton (FL): Taylor & Francis Group, LLC; 2014.

49. Snider RM, Strycharz-Glaven SM, Tsoi SD, Erickson JS, Tender LM. Long-range electron transport in *Geobacter sulfurreducens* biofilms is redox gradient-driven. Proc Natl Acad Sci U S A 2012;109(38):15467–15472.

50. Babauta JT, Beyenal H. Mass transfer studies of *Geobacter sulfurreducens* biofilms on rotating disk electrodes. Biotechnol Bioeng 2014;111(2):285–294.

51. White GF, Shi Z, Shi L, Wang Z, Dohnalkova AC, Marshall MJ, Fredrickson JK, Zachara JM, Butt JN, Richardson DJ, Clarke TA. Rapid electron exchange between surface-exposed bacterial cytochromes and Fe(III) minerals. Proc Natl Acad Sci U S A 2013;110(16):6346–6351.

52. Rabaey K, Angenent L, Schroder U, Keller J. Bioelectrochemical Systems: From Extracellular Electron Transfer to Biotechnological Application. London, England: IWA Publishing; 2010.

53. Borole AP, Aaron D, Hamilton CY, Tsouris C. Understanding long-term changes in microbial fuel cell performance using electrochemical impedance spectroscopy. Environ Sci Technol 2010;44(7):2740–2744.

54. Manohar AK, Bretschger O, Nealson KH, Mansfeld F. The use of electrochemical impedance spectroscopy (EIS) in the evaluation of the electrochemical properties of a microbial fuel cell. Bioelectrochemistry 2008;72(2):149–154.

55. Srikanth S, Marsili E, Flickinger MC, Bond DR. Electrochemical characterization of *Geobacter sulfurreducens* cells immobilized on graphite paper electrodes. Biotechnol Bioeng 2008;99(5):1065–1073.

56. Renslow RS, Babauta JT, Majors PD, Beyenal H. Diffusion in biofilms respiring on electrodes. Energy Environ Sci 2013;6(2):595–607.

57. Sauerbrey G. Verwendung von Schwingquarzen zur Wägung dünner Schichten und zur Mikrowägung. Z Phys 1959;155(2):206–222.

58. Johannsmann D. Viscoelastic, mechanical, and dielectric measurements on complex samples with the quartz crystal microbalance. Phys Chem Chem Phys 2008;10(31):4516–4534.

59. Babauta JT, Beasley C, Beyenal H. Investigation of electron transfer by *Geobacter sulfurreducens* biofilms using an electrochemical quartz crystal microbalance. ChemElectroChem 2014;1:2007–2016.

60. He Z, Angenent LT. Application of bacterial biocathodes in microbial fuel cells. Electroanalysis 2006;18(19–20):2009–2015.

61. Finkelstein DA, Tender LM, Zeikus JG. Effect of electrode potential on electrode-reducing microbiota. Environ Sci Technol 2006;40(22):6990–6995.

62. Erable B, Féron D, Bergel A. Microbial catalysis of the oxygen reduction reaction for microbial fuel cells: a review. ChemSusChem 2012;5(6):975–987.

63. De Schamphelaire L, Boeckx P, Verstraete W. Evaluation of biocathodes in freshwater and brackish sediment microbial fuel cells. Appl Microbiol Biotechnol 2010;87(5):1675–1687.

64. Babauta JT, Nguyen HD, Istanbullu O, Beyenal H. Microscale gradients of oxygen, hydrogen peroxide, and pH in freshwater cathodic biofilms. ChemSusChem 2013;6(7):1252–1261.

65. Bergel A, Feron D, Mollica A. Catalysis of oxygen reduction in PEM fuel cell by seawater biofilm. Electrochem Commun 2005;7(9):900–904.

66. Cao XX, Huang X, Liang P, Boon N, Fan M, Zhanga L, Zhang X. A completely anoxic microbial fuel cell using a photo-biocathode for cathodic carbon dioxide reduction. Energy Environ Sci 2009;2(5):498–501.

67. Liu XW, Sun XF, Huang YX, Li DB, Zeng RJ, Xiong L, Sheng GP, Li WW, Cheng YY, Wang SG, Yu HQ. Photoautotrophic cathodic oxygen reduction catalyzed by a green alga *Chlamydomonas reinhardtii*. Biotechnol Bioeng 2013;110(1):173–179.

68. Pisciotta JM, Zaybak Z, Call DF, Nam JY, Logan BE. Enrichment of microbial electrolysis cell biocathodes from sediment microbial fuel cell bioanodes. Appl Environ Microbiol 2012;78(15):5212–5219.

69. Clauwaert P, Rabaey K, Aelterman P, de Schamphelaire L, Pham TH, Boeckx P, Boon N, Verstraete W. Biological denitrification in microbial fuel cells. Environ Sci Technol 2007;41(9):3354–3360.

70. Ter Heijne A, Strik DP, Hamelers HV, Buisman CJ. Cathode potential and mass transfer determine performance of oxygen reducing biocathodes in microbial fuel cells. Environ Sci Technol 2010;44(18):7151–7156.

71. Rhoads A, Beyenal H, Lewandowski Z. Microbial fuel cell using anaerobic respiration as an anodic reaction and biomineralized manganese as a cathodic reactant. Environ Sci Technol 2005;39(12):4666–4671.
72. Summers ZM, Gralnick JA, Bond DR. Cultivation of an obligate Fe(II)-oxidizing lithoautotrophic bacterium using electrodes. MBio 2013;4(1):e00420-12.
73. Carbajosa S, Malki M, Caillard R, Lopez MF, Palomares FJ, Martín-Gago JA, Rodríguez N, Amils R, Fernández VM, De Lacey AL. Electrochemical growth of *Acidithiobacillus ferrooxidans* on a graphite electrode for obtaining a biocathode for direct electrocatalytic reduction of oxygen. Biosens Bioelectron 2010;26(2):877–880.
74. Strycharz-Glaven SM, Glaven RH, Wang Z, Zhou J, Vora GJ, Tender LM. Electrochemical investigation of a microbial solar cell reveals a nonphotosynthetic biocathode catalyst. Appl Environ Microbiol 2013;79(13):3933–3942.
75. Babauta JT, Hsu L, Atci E, Kagan J, Chadwick B, Beyenal H. Multiple cathodic reaction mechanisms in seawater cathodic biofilms operating in sediment microbial fuel cells. ChemSusChem 2014;7:2898–2906.
76. Valim RB, Santos MC, Lanza MRV, Machado SAS, Lima FHB, Calegaro ML. Oxygen reduction reaction catalyzed by ε-MnO_2: influence of the crystalline structure on the reaction mechanism. Electrochim Acta 2012;85:423–431.
77. Zhang L, Liu C, Zhuang L, Li W, Zhou S, Zhang J. Manganese dioxide as an alternative cathodic catalyst to platinum in microbial fuel cells. Biosens Bioelectron 2009;24(9):2825–2829.
78. Dewan A, Beyenal H, Lewandowski Z. Intermittent energy harvesting improves the performance of microbial fuel cells. Environ Sci Technol 2009;43(12):4600–4605.
79. Cao B, Shi L, Brown RN, Xiong Y, Fredrickson JK, Romine MF, Marshall MJ, Lipton MS, Beyenal H. Extracellular polymeric substances from *Shewanella* sp HRCR-1 biofilms: characterization by infrared spectroscopy and proteomics. Environ Microbiol 2011;13(4):1018–1031.
80. Sand W, Gehrke T. Extracellular polymeric substances mediate bioleaching/biocorrosion via interfacial processes involving iron(III) ions and acidophilic bacteria. Res Microbiol 2006;157(1):49–56.
81. Gralnick JA, Newman DK. Extracellular respiration. Mol Microbiol 2007;65(1):1–11.
82. Renslow RS, Majors PD, McLean JS, Fredrickson JK, Ahmed B, Beyenal H. In situ effective diffusion coefficient profiles in live biofilms using pulsed-field gradient nuclear magnetic resonance. Biotechnol Bioeng 2010;106(6):928–937.

6

THEORY OF REDOX CONDUCTION AND THE MEASUREMENT OF ELECTRON TRANSPORT RATES THROUGH ELECTROCHEMICALLY ACTIVE BIOFILMS

Darryl A. Boyd and Rachel M. Snider
National Research Council, Washington, DC, USA
Jeffrey S. Erickson
Center for Bio/Molecular Science and Engineering, Naval Research Laboratory, Washington, DC, USA
Jared N. Roy
George Mason University, Manassas, VA, USA
Sarah M. Strycharz-Glaven and Leonard M. Tender
Center for Bio/Molecular Science and Engineering, Naval Research Laboratory, Washington, DC, USA

6.1 THEORY

6.1.1 Electron Hopping

Geobacter sulfurreducens wild-type strain DL-1 can be grown as a multicell thick biofilm on the surface of a noncorrosive anode. In such a biofilm, electrons originating

Biofilms in Bioelectrochemical Systems: From Laboratory Practice to Data Interpretation, First Edition.
Edited by Haluk Beyenal and Jerome Babauta.

from intracellular metabolism of acetate are transported to the underlying anode surface that acts a metabolic electron acceptor [1]. Just how electrons are transported from cells distributed throughout the biofilm to the anode surface over distances that can exceed 100 μm is unresolved [2]. One proposed mechanism, referred to as redox conduction, also known as incoherent multistep electron hopping [3], puts forth that long-distance electron transport occurs by a sequence of short-distance electron transfer reactions through a network of immobile redox cofactors [4]. Redox conduction occurs in redox polymers, hydrogels that contain covalently attached redox cofactors such as osmium-based organometallic complexes [5]. The current versus potential dependencies derived as follows, which have been successfully applied to fit experimental electron transport rate measurements of *G. sulfurreducens* biofilms [6, 7], are based on those derived for redox polymers [5, 7–14].

6.1.2 Cytochromes: Redox Cofactors Involved in Biofilm Long-Distance Electron Transport

G. sulfurreducens possesses an abundance of multiheme c-type cytochromes on its outer membrane in extracellular polymeric substances and along secreted filaments (pili) [15–22]. Electrochemical, spectroscopic, and electron transport rate measurements indicate that redox conduction occurs in *G. sulfurreducens* biofilms in which hemes are likely redox cofactors involved in the electron transport process [2, 6, 9, 10, 23–34]. (*It is suggested that the reader first reads Bond et al.* [2] *when delving into this topic.*) Precedence exists for the organization of multiheme c-type cytochromes into immobilized electron conduits, such as the *Shewanella* CymA–MtrA–MtrC system where an assembly of three multiheme c-type cytochromes comprised of 24 hemes transport electrons from the inner membrane to the cell outer surface [32, 35]. There is no direct structural information, however, regarding the organization of cytochromes within *G. sulfurreducens* biofilms, and thus no direct evidence for domains of intercytochrome heme spacing of less than 10 Å required for redox conduction to occur. For this reason, the mechanism of biofilm electron transport is contested [36]. To date, the only results inconsistent with redox conduction occurring in *G. sulfurreducens* biofilms are those reported by Malvankar et al., which have been addressed elsewhere [36] and which have not been corroborated by others.

6.1.3 Conceptualizing Long-Distance Electron Transport in Electroactive Biofilms in a Two-Electrode Biofilm "Box" Geometry

Figure 6.1 is a schematic representation depicting redox conduction-based electron transport through a cross-section of a rectangular-box-shaped *G. sulfurreducens* biofilm spanning a gap separating two planar electrodes (source and drain). The biofilm contains redox cofactors (i.e., hemes of c-type cytochromes) that are assumed to be spaced closely enough such that electron transfer reactions can occur between neighboring cofactors. (*We use "transport" and "conduction" interchangeably in this chapter, although they can have distinct meanings depending on the context. We also draw no distinction between electron transfer among hemes residing in*

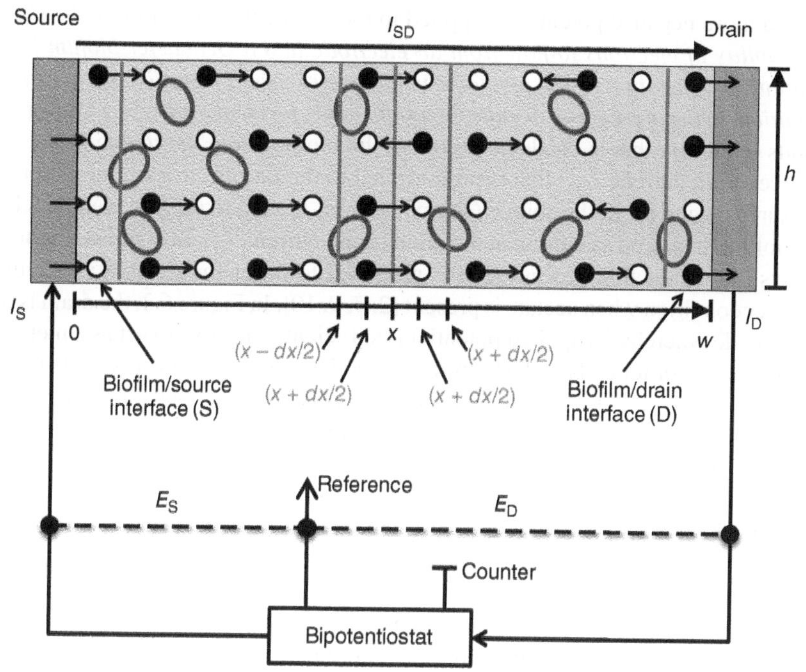

Figure 6.1 Schematic depiction of trans-biofilm electron transport by incoherent multistep electron hopping (also referred to as redox conduction) under nonturnover condition.

the same multiheme cytochrome and those residing in different cytochromes.) For simplification, the cofactors are depicted arranged in an ordered array [11]. This is most likely not representative of a real system such as redox polymer or a *G. sulfurreducens* biofilm but does not ultimately affect the current versus potential dependencies derived as follows.

The height of the biofilm is h, the width of the gap is w, and the length of the gap is l (perpendicular to the page). x indicates the position of an imaginary plane sliced through the biofilm between the electrodes that is at a distance x from the source, $dx/2$ is the thickness of imaginary thin biofilm layers (lamina [10, 37, 38]) on either side of plane x, and δ is the distance separating neighboring cofactors. In the limit that dx is made very small (as depicted in Fig. 6.1), the lamina on either side of distance x contains only cofactors that are $\delta/2$ on either side of x.

Cells are depicted as ovals distributed throughout the biofilm (not drawn to scale).

6.1.4 Two Electrodes under Nonturnover Condition Versus One Electrode under Turnover Condition

Figure 6.1 depicts the proposed conduction process under nonturnover condition when acetate is not present [9, 10]. Here, electrons enter the biofilm from the source electrode, are conducted across the biofilm, and exit the biofilm by the drain electrode when a more negative potential is applied to the source than the drain [6, 7, 11, 14]. (*The ability of electrons to flow from an electrode into a Geobacter biofilm is substantiated by the observation of anodic voltammetric peaks recorded under turnover condition, in the presence of acetate, at a sufficiently fast scan rate* [23].) This process results in current that is conducted through the biofilm from the source to the drain (source–drain current, I_{SD}) that is proportional to the rate electrons enter the biofilm measured as current at the source (source current, I_S) and to the rate electrons leave the biofilm measured as current at the drain (drain current, I_D). In contrast, Figure 6.2 depicts the conduction process as it is proposed to occur for a single anode under turnover condition when acetate is present [2, 7, 9, 10]. In Figure 6.2, the drain is used as an anode whereby its applied potential is sufficiently positive to act as a metabolic electron acceptor for cells in the biofilm metabolizing acetate, typically ≥−0.1 V versus SHE, −0.3 V versus Ag/AgCl. The source is disconnected (i.e., at open circuit so that no electrons enter or exit from it) [7] or nonexistent as in the case of a biofilm grown on an anode as typically reported in literature [9, 10, 23–25, 29, 39–46]. In Figure 6.2, the cells act as electron sources and some of the cofactors accept respired electrons from the cells while also engaging in long-distance transport of electrons through the biofilm to the anode in the exact same manner as in Figure 6.1, resulting in catalytic current measured at the anode [2, 7, 9, 10]. In Figure 6.2, catalytic current is proportional to the rate at which electrons enter the biofilm from all of the cells metabolizing acetate in the biofilm and to the rate at which electrons leave the biofilm by the anode.

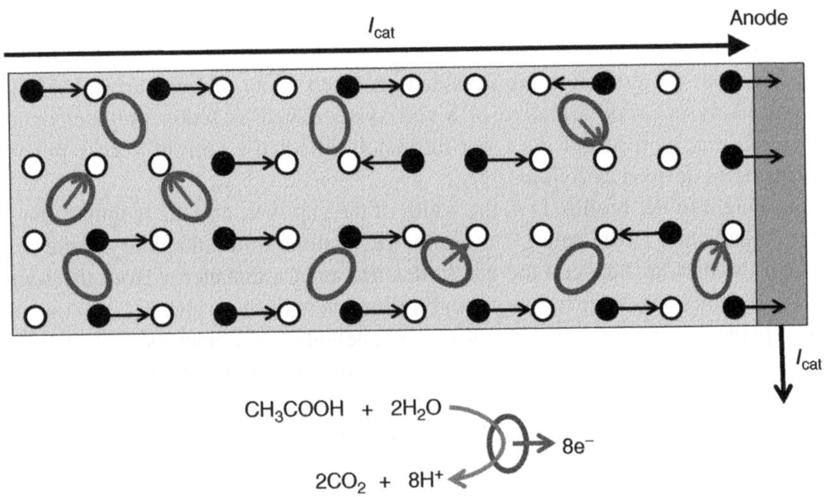

Figure 6.2 Schematic depiction of trans-biofilm electron transport by incoherent multistep electron hopping (also referred to as redox conduction) under turnover condition. Here, cellular oxidation of acetate generates eight electrons per molecule of acetate oxidized. These electrons are then subsequently transferred to the anode in the same manner as depicted in Figure 6.1 for transport to the drain.

6.1.5 Rationale for Measuring Electron Transport Rates Using Two Electrodes Under Nonturnover Condition

Performing electron transport rate measurements under nonturnover condition using two electrodes as depicted in Figure 6.1 allows external control of the potential energy of electrons both entering and exiting the biofilm, which are determined by the potentials applied to the source and drain electrodes [14, 47]. In this way, the dependency of the rate of electron transport through the biofilm on the source and drain potentials can be determined directly in the same manner as is done for other materials such as conducting polymers to investigate the mechanism of electron transport [14, 48–50]. (*The gold standard for performing electron transport rate measurements is the four-point probe measurement* [47]. *Four-point probe measurements, however, are extremely difficult to perform when the sample is immersed in an electrolyte, as in the case of in situ biofilm measurements described here, and the potentials of the voltage-sensing electrodes need to be determined vs. a reference electrode as would be the case with in situ electron transport measurements of biofilms. Moreover, in the case of a conductive film formed on an electrode such as a Geobacter biofilm, four-point probe electron transport rate measurements must be performed ex situ whereby the film is removed from the surface of the electrode it is formed on. This is "rather cumbersome, needs several working steps, and its reproducibility is not good."* [61] *For these reasons, we do not consider four-point probe measurements here, but rather more conventional two-point probe (source and drain electrodes) measurements are successfully used to perform in situ electron transport measurements of redox polymers.*) Assuming that the electron transfer reaction between cofactors and the electrodes is not the rate-limiting step in current flow across the biofilm (discussed in the following) [2, 10], current measured at either electrode is expected to be limited by the rate electrons are transported through the biofilm allowing direct measurement of the rate of this process. Moreover, as is described in the following, for the case of the two-electrode configuration under nonturnover condition depicted in Figure 6.1, the rate at which electrons cross the imaginary plane at distance x is the same for any distances x across the biofilm. This allows derivation of an explicit equation for the dependency of the rate of electron transport through the biofilm on the potentials applied to the electrodes (E_S and E_D), the concentration of cofactors in the biofilm (C_T), their spacing (δ), and the rate of electron transfer between neighboring cofactors (k^o) [6, 7, 11, 14, 51, 52]. In contrast, investigation of the electron transport process under turnover condition for a single anode as depicted in Figure 6.2 is complicated by the inability to externally control the potential energy of electrons entering the biofilm, which is determined by the cells metabolizing acetate. In addition, catalytic current measured at the anode may not be limited by the rate at which electrons are transported through the biofilm, but by the rate cells metabolize acetate and/or the rate electrons enter the biofilm from the cells, obscuring measurement of the electron transport process [10]. Moreover, in the case of the single-electrode configuration under turnover condition, the rate at which electrons cross the imaginary plane at distance x (not depicted in Fig. 6.2) increases as the distance between the plane and the anode decreases. This is due to the compounding effect of electrons being supplied to the biofilm by cells distributed throughout the biofilm, which are transported to the anode at one edge of the biofilm [10]. This can be understood by visualizing an imaginary plane sliced through the biofilm in Figure 6.2 whereby the rate that electrons flow across the plane will be depend on the number and metabolic activity of cells to the left of the plane that supply the electrons that cross the plane on their way to the anode. Moving closer to the anode, the number of cells to the left of the plane increases. For this reason, it is not possible to derive an explicit expression for dependency of the rate of electron transport on the potential applied to the anode under turnover condition [10]. In summary, while the more interesting question may be "just how are electrons transported to the anode under turnover condition (Fig. 6.2)?", the more tractable question to answer is "just how are electrons transported through the biofilm from one electrode to the other resulting in conducted current under nonturnover condition (Fig. 6.1)?". Regardless, the mechanisms for both processes are assumed to be the same, and answering the second question answers the first question.

6.1.6 Expressing Long-Distance Electron Transport by Chemical Equations

Figure 6.1 is a snapshot in time in which arrows represent individual electron transfer reactions (electron hops) that are about to occur involving reduced cofactors (black circles) and oxidized cofactors (white circles), after which reduced cofactors will be oxidized and oxidized cofactors will be reduced. These electron transfer reactions can be described by Reactions 1–3:

$$\text{Reaction 1: } \mathrm{Ox}_{(S)} + \mathrm{e}^- \underset{k_2}{\overset{k_1}{\rightleftarrows}} \mathrm{Red}_{(S)}$$

$$\text{Reaction 2: } \mathrm{Red}_{(x-x/2)} + \mathrm{Ox}_{(x+x/2)} \underset{k_4}{\overset{k_3}{\rightleftarrows}} \mathrm{Ox}_{(x-x/2)} + \mathrm{Red}_{(x+x/2)}$$

$$\text{Reaction 3: } \mathrm{Red}_{(D)} \underset{k_6}{\overset{k_5}{\rightleftarrows}} \mathrm{Ox}_{(D)} + \mathrm{e}^-$$

Reaction 1 indicates that cofactors at the source/biofilm interface (an imaginary thin layer containing cofactors close enough to exchange electrons with the source) are reversibly reduced with electrons entering from the source. Reaction 2 describes electron transfer between a reduced cofactor and an oxidized cofactor that is closer to the drain where $\mathrm{Red}_{(x-x/2)}$, $\mathrm{Ox}_{(x+x/2)}$, $\mathrm{Ox}_{(x-x/2)}$, and $\mathrm{Red}_{(x+x/2)}$ are reduced and oxidized cofactor residing immediately on either side of plane x. Reaction 3 indicates that cofactors at the drain/biofilm interface are reversibly oxidized with electrons exiting by the drain. k_1, k_2, k_3, and k_4 are the rate constants for the electron transfer reactions, and I_S and I_D (indicated in Fig. 6.1) are the source and drain currents, which are proportional to the rate electrons enter the biofilm from the source by Reaction 1 and exit by the drain by Reaction 3, and which are measured during electron transport rate measurements.

6.1.7 Expressing Long-Distance Electron Transport by Mathematical Equations

Following Reaction 2, the rate that electrons flow across plane x is:

$$J_{(x)} = k_3 C_{\mathrm{Red}_{(x-dx)}} C_{\mathrm{Ox}_{(x+dx)}} - k_4 C_{\mathrm{Ox}_{(x-dx)}} C_{\mathrm{Red}_{(x+dx)}} \tag{6.1}$$

where $C_{\mathrm{Red}_{(x-dx)}}$, $C_{\mathrm{Ox}_{(x+dx)}}$, $C_{\mathrm{Ox}_{(x-dx)}}$, and $C_{\mathrm{Red}_{(x+dx)}}$ are the concentrations of reduced and oxidized cofactor residing in thin biofilm layers of thickness dx on either side of plane x in the limit that dx is made very small, and $k_3 C_{\mathrm{Red}_{(x-x/2)}} C_{\mathrm{Ox}_{(x+x/2)}}$ and $k_4 C_{\mathrm{Ox}_{(x-x/2)}} C_{\mathrm{Red}_{(x+x/2)}}$ are the rates of the forward and backward electron transfer reactions occurring across plane x. If the biofilm is immersed in an electrolyte solution such as fresh water media typically used for *G. sulfurreducens* [6, 7], the application of different potentials to the source and drain electrodes (i.e., a voltage, V) will not create an appreciable electric field across the biofilm. This is because ion mobility as occurs within a *G. sulfurreducens* biofilm is expected to dissipate electric fields [11, 53–57]. As a result, k_3 and k_4 are assumed equal because an electron anywhere in the biofilm will have the same potential energy (i.e., migration effects are assumed negligible [6, 11, 55]). That is, because of the absence of an electric field, if there are oxidized cofactors on either side of a reduced cofactor, there is equal probability that the electron will transfer from the reduced cofactor to either oxidized cofactor, and hence some of the arrows are pointing backward to the source in Figure 6.1. (*This is a key feature of redox conduction – an individual electron transfer reaction between neighboring redox cofactors is random with respect to the direction that electron will go.*) If, in addition, the local concentration of cofactor regardless of oxidation state is the same throughout the biofilm such that:

$$C_{\mathrm{Red}_{(x-dx)}} + C_{\mathrm{Ox}_{(x-dx)}} = C_{\mathrm{Red}_{(x+dx)}} + C_{\mathrm{Ox}_{(x+dx)}} = C_{\mathrm{T}} \tag{6.2}$$

Equation 6.1 can be expressed as:

$$J_{(x)} = -k_{\mathrm{ex}} C_{\mathrm{T}} (C_{\mathrm{Red}_{(x-dx)}} - C_{\mathrm{Red}_{(x+dx)}}) \tag{6.3}$$

and further simplified to:

$$J_{(x)} = -k_{\mathrm{ex}} C_{\mathrm{T}} \left(\frac{dC_{\mathrm{Red}_{(x)}}}{dx} \right) dx \tag{6.4}$$

where $k_3 = k_4 = k_{\mathrm{ex}}$, and $dC_{\mathrm{Red}_{(x)}}/dx$ is the *redox gradient* that describes the difference in concentration of reduced redox cofactor residing in thin biofilm layers of thickness dx on either side of plane x. (*This is another key feature of redox conduction – the local flux of electrons through any imaginary plane in a redox conducting material is proportional to the local difference of reduced cofactors across the plane.*) C_{T} is assumed to be a constant whereby redox cofactors are assumed to be evenly distributed throughout the biofilm. While this assumption simplifies derivation of the dependency of current conducted through a redox conducting material on source and drain potentials (e.g., Equation 6.9), it is not necessary to do so. Mathematical treatment of redox conduction in redox polymers containing a nonuniform distribution of redox cofactors indicates that electron conduction is expected to concentrate in regions of relatively high cofactor concentration [52]. This process, referred to as percolation, does not qualitatively affect derivation of the current–potential dependency. We expect percolation to occur in *G. sulfurreducens* biofilms, owing to the volume of biofilm occupied by cells distributed throughout the biofilm through which long-distance electron transport is assumed not to occur, and to assumed heterogeneity in distribution of outer membrane and extracellular c-type cytochromes. In the limit that dx is made very small, $dx = \delta$, and multiplication of $dC_{\mathrm{Red}_{(x)}}/dx$ by δ^2 yields the difference in concentration of reduced cofactor immediately on either side of plane x in laminae of thickness δ:

$$J_{(x)} = -k_{\mathrm{ex}} C_{\mathrm{T}} \delta^2 \left(\frac{dC_{\mathrm{Red}_{(x)}}}{dx} \right) \tag{6.5}$$

6.1.8 Long-Distance Electron Transport by Sequential Short-Distance Electron Transfer Reactions Is a Diffusive Process

Equation 6.5 states that if the concentration of reduced cofactor immediately adjacent to plane x is different on one side versus the other, electrons will cross plane x at a rate that is proportional to this difference. Equation 6.5 is analogous to Fick's first law of diffusion whereby the local flux of a diffusive entity is linearly dependent on its local concentration gradient [58]. Here, the concentration gradient is a redox gradient, and the diffusive entity is an electron hopping among immobile redox cofactors [11, 52, 55, 56, 59]. The apparent diffusion coefficient for electrons, $D_{\mathrm{e}} = k_{\mathrm{ex}} C_{\mathrm{T}} \delta^2$, is analogous to that of physically diffusing molecules in response to a concentration gradient, $D_{\mathrm{P}} = dx^2/dt$, where dx is the mean distance increment that the molecules move and $1/dt$ is the frequency that the molecules move. Here, the mean distance increment that electrons hop is δ, and the frequency that electrons hop is $k_{\mathrm{ex}} C_{\mathrm{T}}$.

6.1.9 The Potentials Applied to the Two Electrodes Determines the Concentrations of Reduced Cofactor at the Two-Electrode/Biofilm Interfaces

If Reactions 1 and 3 are sufficiently fast (i.e., k_1, k_2, k_5, and k_6 are sufficiently large) [6, 7, 10], when potentials are first applied to the source and drain, Reactions 1 and 3 occur such that the concentration of reduced cofactor at the source/biofilm interface and drain/biofilm interface changes to achieve values described by the Nernst Equation:

$$C_{\mathrm{Red}_{(\mathrm{S})}} = \frac{C_{\mathrm{T}}}{\left(1 + e^{\left[\frac{nF}{RT}\left(E_{(\mathrm{S})} - E^{o\prime}\right)\right]}\right)} \tag{6.6}$$

$$C_{\text{Red}_{(D)}} = \frac{C_T}{\left(1 + e^{\left[\frac{nF}{RT}\left(E_{(D)} - E^{o'}\right)\right]}\right)} \tag{6.7}$$

where E_S and E_D are the potentials applied to the source and drain, $E^{o'}$ is the cofactor formal potential, $C_{\text{Red}_{(S)}}$ and $C_{\text{Red}_{(D)}}$ are the concentrations of reduced cofactor at the source/biofilm interface and the drain/biofilm interface. Equation 6.6 indicates that when $E_{(S)} \ll E^{o'}$, $C_{\text{Red}_{(S)}} = C_T$; when $E_{(S)} = E^{o'}$, $C_{\text{Red}_{(S)}} = C_T/2$; and when $E_{(S)} \gg E^{o'}$, $C_{\text{Red}_{(S)}} = 0$. Equation 6.7 indicates that when $E_{(D)} \ll E^{o'}$, $C_{\text{Red}_{(D)}} = C_T$; when $E_{(D)} = E^{o'}$, $C_{\text{Red}_{(D)}} = C_T/2$; and when $E_{(D)} \gg E^{o'}$, $C_{\text{Red}_{(D)}} = 0$. (*The assumption that Reactions 1 and 2 are fast (i.e., that they are not the rate-limiting steps in electron transport through the biofilm) is substantiated by the following two observations.* (1) *The theoretical current–potential dependency derived as follows* (*Equation* 6.9), *which has been successfully applied to fit experimentally measured current–potential dependencies for Geobacter biofilms* [6, 7], *results directly from the Nernst Equation.* (2) *The sigmoid-shaped dependency of catalytic current on anode potential, observed for Geobacter biofilms under turnover condition* [23], *also results directly from the Nernst Equation. This indicates that electron transfer between a Geobacter biofilm and an electrode is fast and not the limiting factor in catalytic current generation that is comparable in magnitude to that observed for conducted current under non-turnover condition for the same biofilm* [6, 7].)

Immediately after potentials are applied to the source and drain, electrons also begin to propagate across the biofilm due to Step 2. As a result, the concentration of reduced cofactor at each distance x within the biofilm changes resulting in transient source and drain currents. These transient currents will, however, subside, and a steady-state condition will be achieved in which the concentration of reduced cofactor at each electrode/biofilm interface and at each distance x in the biofilm remains constant as long as E_S and E_D remain constant.

6.1.10 The Local Rate of Electron Transport Across Any Cross-Sectional Plane Sliced Across the Biofilm Is the Same and Proportional to the Local Redox Gradient – The Redox Gradient Must Therefore be the Same Across the Biofilm

Under steady-state condition, the rate electrons enter the biofilm from the source equals the rate electrons exit the biofilm by the drain (i.e., source and drain currents have the same magnitude) because the total number of reduced cofactors in the biofilm is constant even though electrons are flowing through it. Moreover, the rate electrons cross plane x for any distance x is the same and equal to the rate electrons enter and exit the biofilm. (*By analogy, under steady-state condition, the biofilm is similar to a garden hose after it has been turned on for some time. The rates water enters the hose, exits the hose, and flows through any imaginary plane sliced through the hose are all equal, and the volume of water in the hose at any instant is a constant*

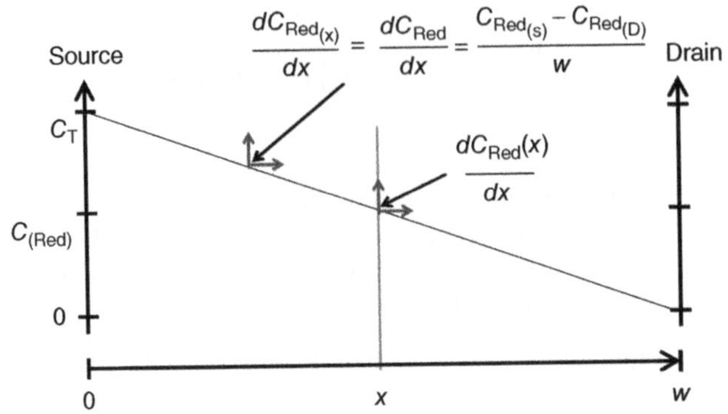

Figure 6.3 Schematic depiction of concentration profile of reduced cofactor across the biofilm under nonturnover condition for the source/biofilm/drain geometry depicted in Figure 6.1. The slope of the profile is the redox gradient. The rate of electron transport through any vertical plane sliced through the biofilm is the same and proportional to the redox gradient. The profile is, therefore, a straight line (i.e., has a constant slope) across the biofilm.

even though water is flowing through it.) Following Equation 6.5, a plot of concentration of reduced cofactor versus x must therefore be a straight line as depicted in Figure 6.3 such that for any distance x, the slope of this line, $dC_{\mathrm{Red}_{(x)}}/dx$, is the same. That is, the redox gradient at any distance x is the same because the rate of electron transport at any distance x is the same. This outcome occurs in the case of percolation when C_{T} is not the same throughout the biofilm, as we are considering the total rate of electron transport across cross-sectional planes sliced through the entire biofilm. *(Expanding upon the garden hose analogy, if there is an obstruction inside the hose, water will rush through it faster than through unobstructed sections of the hose, but the net flow of water through any cross-section remains the same.)*

As the concentrations of reduced cofactor at the source/biofilm interface and at the drain/biofilm interface are determined by the potentials applied to the source and drain described by Equations 6.6 and 6.7, $dC_{\mathrm{Red}_{(x)}}/dx$ can be expressed as:

$$\frac{dC_{\mathrm{Red}_{(x)}}}{dx} = \frac{dC_{\mathrm{Red}}}{dx} = \frac{C_{\mathrm{Red}_{(S)}} - C_{\mathrm{Red}_{(D)}}}{w} \tag{6.8}$$

where dC_{Red}/dx is the concentration gradient of reduced cofactor for any distance x, determined simply by the difference in concentration of reduced cofactor at the electrode/biofilm interfaces described by Equations 6.6 and 6.7 divided by the distance separating the electrodes (w). Although oxidized cofactor is apparently ignored in the aforementioned derivation, Equation 6.5 could just as easily been expressed in terms of oxidized cofactor. Owing to mass balance (Equation 6.4), there is also a linear concentration gradient of oxidized cofactor across the biofilm in which the concentration of oxidized cofactor is higher closer to the drain than to the source.

6.1.11 Expression for Rate of Electron Transport on Potentials Applied to the Electrodes

Combination of Equations 6.5–6.8 yields Equation 6.9, an expression for the rate of electron transport across the source/biofilm interface through the biofilm and across the drain/biofilm interface at steady-state condition as a function of the potentials applied to the source and drain under nonturnover condition. Here, the rate of electron transport is expressed as a current, commonly referred to as the source–drain current (I_{SD}):

$$I_{SD} = nFAk_0C_T\delta^2\left(\frac{C_{Red_{(S)}} - C_{Red_{(D)}}}{w}\right)$$

$$= nFAk_0C_T^2\delta^2\left(\frac{\left(\frac{1}{1+e^{\left[\frac{nF}{RT}\left(E_{(S)}-E^{o'}\right)\right]}}\right) - \left(\frac{1}{1+e^{\left[\frac{nF}{RT}\left(E_{(D)}-E^{o'}\right)\right]}}\right)}{w}\right) \quad (6.9)$$

where $I_{SD} = I_D = -I_S$; n, F, and R have their standard meanings [58]; T is temperature; and A is the cross-sectional area of the electrode/biofilm interfaces and biofilm box depicted in Figure 6.1 ($A = l \times h$). The convention is used in which source current is represented as negative current because cofactor is reduced by the source, and drain current is represented as positive current because cofactor is oxidized by the drain, where at any instant under steady condition, they have the same magnitude and either suffices to characterize the source to drain current. Equation 6.9 indicates that the maximum source–drain current occurs when $E_{(S)} \ll E^{o'}$ and $E_{(D)} \gg E^{o'}$, yielding $I_{SD} = nFAk_0C_T^2/w$. The product nFA, often seen in electrochemistry, converts the rate of an electrode reaction to current (Faraday's Law).

6.1.12 Source–Drain Measurements

Figure 6.4 is adapted from Figure 6.4a of Strycharz-Glaven et al. [6]. It depicts I_D and I_S plotted versus $E_{(D)}$ for values of $E_{(D)}$ from −0.3 to +0.3 V versus $E^{o'}$ while $E_{(S)}$ is fixed at −0.3 V versus $E^{o'}$ calculated using Equation 6.9. Here, the value of $E_{(S)}$ was chosen to ensure that all cofactors at the source/biofilm interface are reduced (i.e., $C_{Red_{(S)}} \approx C_T$ per Equation 6.6). Electron transport rate measurements performed in this manner, when the potential of one electrode is fixed while the other is changed, are often referred to as source–drain measurements [6]. Figure 6.5 depicts redox gradients (plots of C_{Red_x} vs. x where C_{Red_x} scales between 0 and C_T) resulting from the different values of $E_{(D)}$ indicated in Figure 6.4 that were calculated using Equation 6.8. For each value of $E_{(D)}$, the slope of the line, dC_{Red}/dx, is proportional to the corresponding source–drain current depicted in Figure 6.4. For example, when $E_{(D)}$ is less than or equal to −0.15 V versus $E^{o'}$ at room temperature (a), nearly all redox cofactors at the drain/biofilm interface are also reduced (i.e., $C_{Red_{(D)}} \approx C_T$ per Equation 6.7), resulting in little to no redox gradient and little to no current.

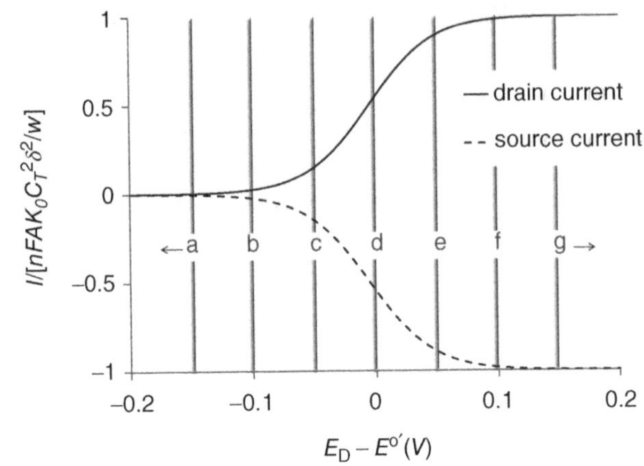

Figure 6.4 Schematic depiction of source and drain current versus drain potential when the source potential is fixed at a value much less than the formal potential of the redox cofactors calculated using Equation 6.9 (i.e., source–drain measurement) [6]. Lines labeled a–g indicate source and drain current for different drain potentials.

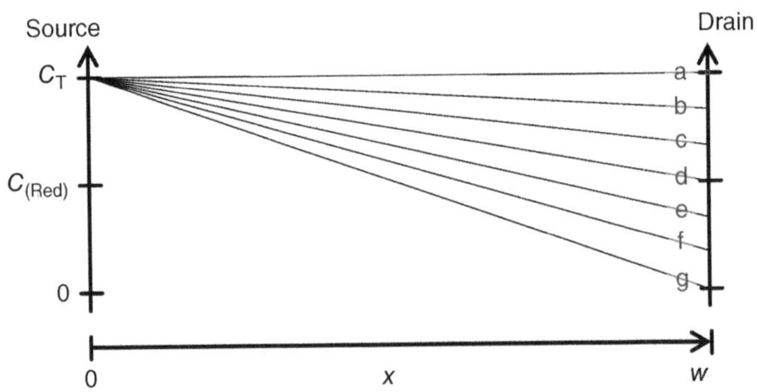

Figure 6.5 Schematic depiction of concentration profiles of reduced cofactor across the biofilm for different drain potentials corresponding to a–g in Figure 6.4 calculated using Equation 6.8.

Conversely, when $E_{(D)}$ is greater than or equal to +0.15˜V versus $E^{o'}$ (g), nearly all redox cofactors at the drain/biofilm interface are oxidized. This results in the maximum possible gradient across the biofilm (i.e., the maximum slope possible for C_{Red_x} vs. x as $C_{Red_{(S)}} \approx C_T$ and $C_{Red_{(D)}} \approx 0$), and thus the maximum possible current; further increasing $E_{(D)}$ does not increase the gradient nor current. When $E_{(D)} = E^{o'}$ (d), half of the cofactors at the drain/biofilm interface are reduced, resulting in half the maximum gradient and half the maximum current. The sigmoid-shape (Nernstian) dependency of current on potential applied to the drain directly reflects

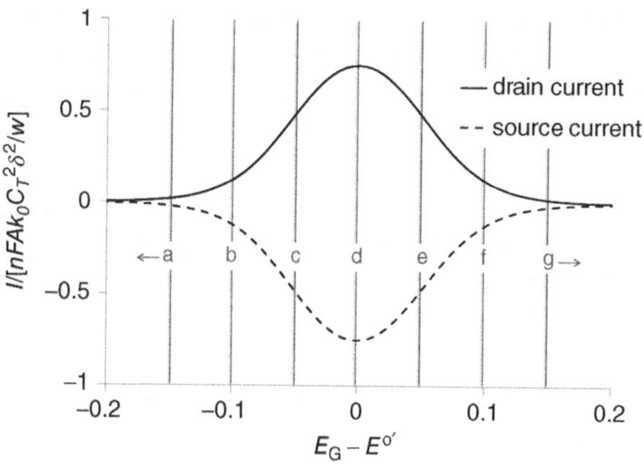

Figure 6.6 Schematic depiction of source and drain currents versus average of source and drain potentials (gate potential) when the source and drain potentials are changed simultaneously while maintaining a fixed potential offset between them (source–drain voltage) calculated using Equation 6.10 (i.e., gate measurement) [7]. Lines labeled a–g indicate source and drain current for different gate potentials. The source–drain voltage used here to calculate the source and drain current is 0.1 V.

Equation 6.7, whereby the change in $C_{\mathrm{Red}_{(\mathrm{D})}}$ caused by a small change in $E_{(\mathrm{D})}$ is the greatest when $E_{(\mathrm{D})} = E^{\mathrm{o}'}$. This dependency described by Equation 6.9 has been experimentally observed for generator-collector measurements of *G. sulfurreducens* biofilms [6].

6.1.13 Gate Measurements

Figure 6.6 is adapted from Figure 6.4a of Snider et al. [7] It depicts I_{D} and I_{S} plotted versus the average of $E_{(\mathrm{S})}$ and $E_{(\mathrm{D})}$, referred to as gate potential (E_{G}, Equation 6.10), for which $E_{(\mathrm{S})}$ and $E_{(\mathrm{D})}$ are changed in

$$E_{\mathrm{G}} = (E_{(\mathrm{S})} + E_{(\mathrm{D})})/2 \tag{6.10}$$

unison while maintaining a constant offset between them (V, commonly referred to as source–drain voltage). (Gate potential is used here to follow standard terminology when electron transport measurements are made across other materials.) Electron transport rate measurements performed in this manner are often referred to as gate measurements [7]. Expressing $E_{(\mathrm{S})}$ and $E_{(\mathrm{D})}$ in terms of E_{G} and V:

$$E_{(\mathrm{S})} = E_{\mathrm{G}} - \left(\frac{V}{2}\right) \tag{6.11}$$

$$E_{(\mathrm{D})} = E_{\mathrm{G}} + \left(\frac{V}{2}\right) \tag{6.12}$$

and substituting into Equation 6.9 yields Equation 6.13, which describes the dependency of source–drain current on E_G and V for gate measurements, which has also been experimentally observed for *G. sulfurreducens* biofilms [7]. Equation 6.13 was used to calculate the plots depicted in Figure 6.6:

$$I_{SD} = nFAk_{ex}C_T^2\delta^2\left(\frac{\left(\frac{1}{1+e^{\left[\frac{nF}{RT}\right]\left(E_G-\left(\frac{V}{2}\right)-E^{o'}\right)}}\right)-\left(\frac{1}{1+e^{\left[\frac{nF}{RT}\left(E_G+\left(\frac{V}{2}\right)-E^{o'}\right)\right]}}\right)}{w}\right) \quad (6.13)$$

In the case of Figure 6.6, $V = 0.100$ V. The peak-shape dependency of current on gate potential depicted in Figure 6.6 directly reflects Equations 6.6 and 6.7, whereby changes in $C_{Red_{(S)}}$ and $C_{Red_{(D)}}$ resulting from small change in $E_{(G)}$ are the greatest when $E_{(G)} = E^{o'}$.

Figure 6.7 depicts corresponding calculated redox gradients resulting from the different values of E_G indicated in Figure 6.6 that were calculated using Equation 6.13. For example, when E_G is less than or equal to −0.15 V versus $E^{o'}$ (a), nearly all redox cofactors at both electrode/biofilm interfaces are reduced, resulting in little to no gradient and little to no source current. Similarly, when E_G is greater than or equal to +0.15 V versus $E^{o'}$ (g), nearly all cofactors at both electrode/biofilm interfaces are oxidized, also resulting in little to no gradient and little to no current. Here, maximum current occurs when $E_G = E^{o'}$ for which $E_{(S)}$ is negative of $E^{o'}$ and $E_{(D)}$ is positive of $E^{o'}$.

Figure 6.8 depicts I_D and I_S plotted versus E_G for various values of V calculated on the basis of Equation 6.10, indicating that the maximum source–drain current for gate measurements occurs when $E_G = E^{o'}$ and V is sufficiently large (≥0.25 V at room temperature), yielding $I = nFAk_0C_T^2/w$.

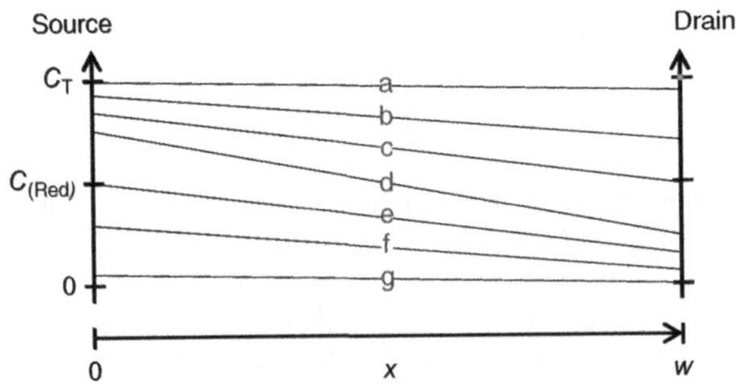

Figure 6.7 Schematic depiction of calculated concentration profiles of reduced cofactor across the biofilm corresponding to a–g in Figure 6.6 calculated using Equation 6.8.

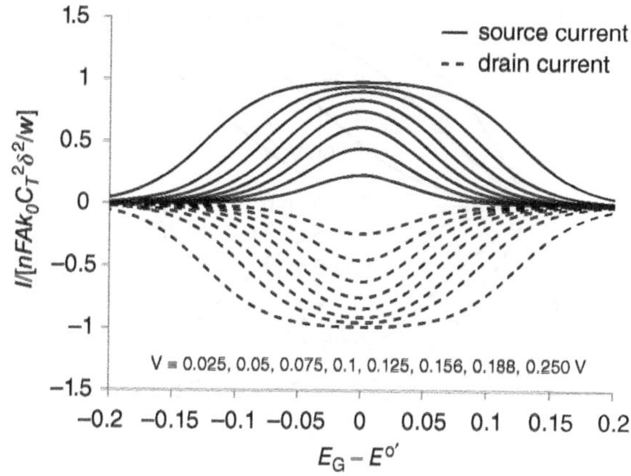

Figure 6.8 Schematic depiction of source and drain currents versus gate potential for different source–drain voltages calculated using Equation 6.10.

6.1.14 Case of Small Voltage and the Linear Approximation, Conductance, and Conductivity

When V is sufficiently small (≤0.05 V at room temperature), Equation 6.10 can be linearly approximated as (see Appendix):

$$I = GV \tag{6.14}$$

where

$$G = \sigma \frac{A}{w} \tag{6.15}$$

and

$$\sigma = n^2 F^2 k_{ex} C_T^2 \delta^2 \left[\frac{e^{\left[\frac{nF}{RT}\left(E_G - E^{o\prime}\right)\right]}}{RT\left(1 + 2e^{\left[\frac{nF}{RT}\left(E_G - E^{o\prime}\right)\right]} + e^{2\frac{nF}{RT}\left(E_G - E^{o\prime}\right)}\right)} \right] \tag{6.16}$$

Here, G, conductance, characterizes the dependency of source–drain current on source–drain voltage for gate measurements when V is sufficiently small to yield a linear (ohmic) dependency. G specifically applies to the box geometry depicted in Figure 6.1. Figure 6.9 depicts I_D and I_S plotted versus V for various values of E_G calculated using Equation 6.10, where G is the slope of the line tangent to the curve when $V \leq 0.05$ V per Equation 6.14. The purpose of G is that it enables a standardized comparison of different materials of their ability to conduct

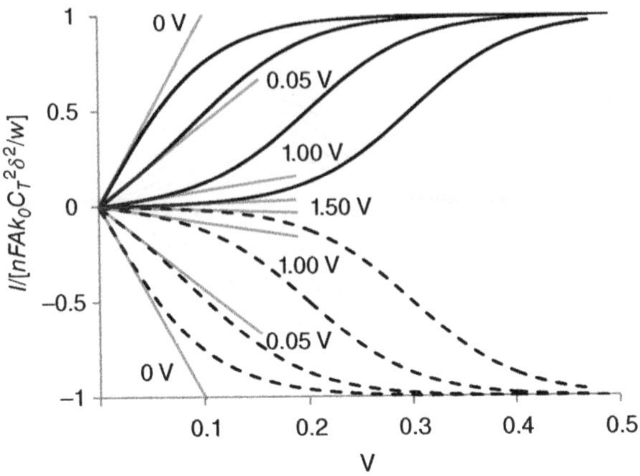

Figure 6.9 Schematic depiction of source and drain current versus source–drain voltage for indicated gate potentials calculated using Equation 6.10. For a given gate potential, the slope of the line tangent to current in the limit of small source–drain voltages is conductance for that gate potential based on Equation 6.15.

electrons when small voltage gate measurements are made using the box geometry depicted in Figure 6.1. In contrast, σ, conductivity, intrinsically characterizes the dependency of source–drain current on source–drain voltage independent of the geometry used. In the case of a redox conductor, such as a biofilm immersed in electrolyte, σ is dependent on the gate potential as described by Equation 6.16. Other mechanisms of conductivity will have different expressions for σ. Figure 6.10

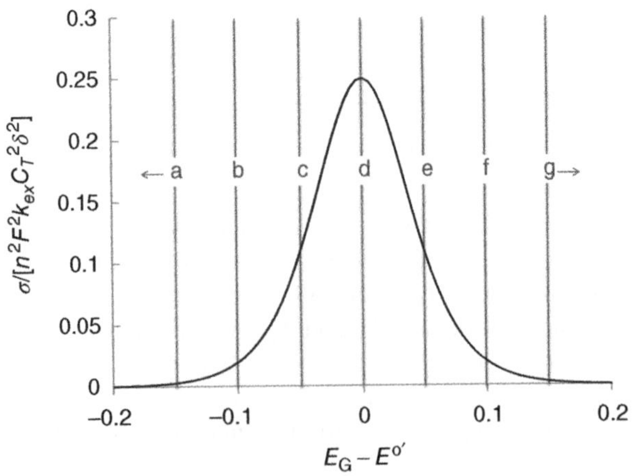

Figure 6.10 Plot of conductivity versus gate potential calculated using Equation 6.16.

depicts the dependency of conductivity on gate potential calculated on the basis of Equation 6.16, which is peak shaped, where maximum conductivity occurs when the gate potential is equal to the cofactor formal potential [11, 14]. Ideally, to determine conductivity of a material such as biofilm at a given gate potential, one would first determine conductance by measuring the dependency of source–drain current on gate potential over a range of small source–drain voltage such that I scales linearly with V (i.e., Equation 6.15) as depicted in Figure 6.9, then normalize conductance by A/w (Equation 6.16).

6.1.15 Box Geometry Is Good for Deriving Equations but Not Good for Performing Experiments

The box geometry depicted in Figure 6.1 is an idealized geometry often used for conceptualizing electron transport rate measurements. It is not practical, however, for actually making electron transport rate measurements for many materials. In the case of *Geobacter* biofilms, the gap width would need to be small enough to be able grow a biofilm between the source and drain, and the cross-sectional surface area would need to be sufficiently large so that source–drain current is large enough to be measureable. A more practical geometry for biofilms we contend is the one that has been used to study electron transport through redox polymers [11, 13, 14, 60] in which source and drain electrodes lay side by side on a common flat surface as depicted in Figure 6.11 (adapted from [61]) where g is the biofilm thickness, $2a$ is the gap width, $b - a$ is width of the electrodes, and l is the gap length (perpendicular to the page). This geometry could represent, for example, a *G. sulfurreducens* biofilm grown on an interdigitated microelectrode array (IDA) in which a pair of closely spaced electrode bands is first used as anodes to grow a biofilm that spans the gap separating them, then subsequently used as source and drain electrodes (described in the following) [6, 7]. While this geometry is more practical experimentally, derivation of the dependency of source–drain current on gate potential and source–drain voltage (Equation 6.13 in the case of the box geometry) is more complicated. Nonetheless, the

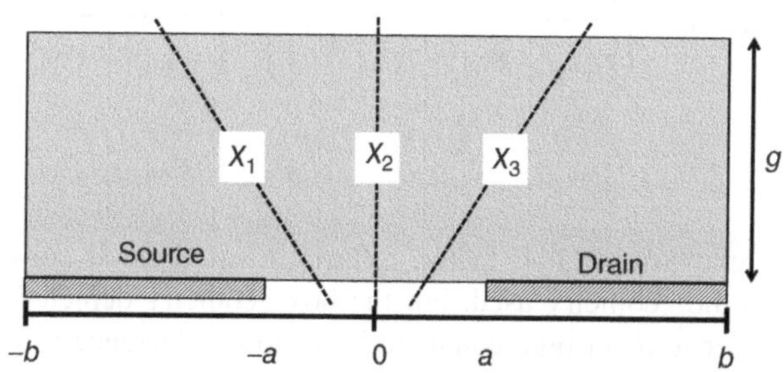

Figure 6.11 Schematic depiction of alternate source/biofilm/drain geometry based on that of Kankare and Kupila [61].

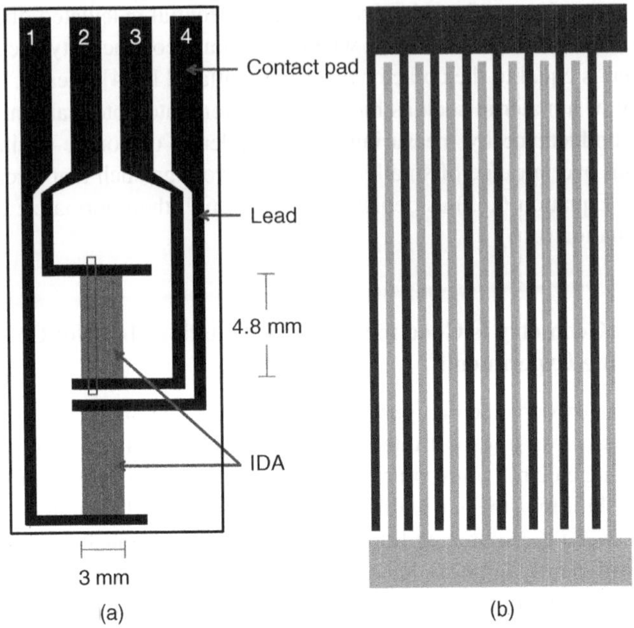

Figure 6.12 (a) Schematic depiction of interdigitated microelectrode array (IDA) chip fabricated on section of a glass microscope slide. Chip is composed of a gold pattern on glass slide consisting of two IDAs, electrical leads, and contact pads. (b) Magnified region of single IDA outlined in red in (a). Black bands represent portion of one interdigitated electrode of the IDA, and grey bands represent portion of the other interdigitated electrode of the IDA.

same approach applies to derive this relationship in which the rate of electron transport across imaginary planes sliced through the biofilm separating the source and the drain (depicted in Fig. 6.12) is derived, yielding equations similar to Equations 6.10 and 6.15:

$$I_{SD} = nFAk_{ex}C_T^2\delta^2 K\left(\left(\frac{1}{1+e^{\left[\frac{nF}{RT}\right]\left(E_G-\left(\frac{V}{2}\right)-E^{0\prime}\right)}}\right)-\left(\frac{1}{1+e^{\left[\frac{nF}{RT}\left(E_G+\left(\frac{V}{2}\right)-E^{0\prime}\right)\right]}}\right)\right) \tag{6.17}$$

$$G = \sigma K \tag{6.18}$$

where K is a geometric scaling factor, referred to as the cell constant [60], that is dependent on the geometry used. For the box geometry depicted in Figure 6.1, $K = A/w$. In order to determine conductivity from conductance as earlier, K must be calculated for the specific geometry used.

6.1.16 Mathematical Transformation from Box Geometry to Open IDA Geometry

In the case of the geometry depicted in Figure 6.11, K has been calculated for the following specific cases [61]:

$$K = \frac{l}{\pi} \ln \left(\frac{8g}{\pi a} \right) \tag{6.19}$$

when $g > a$ and $a < g < b$, and

$$K = \frac{l}{\pi} \ln \left(\frac{4b}{a} \right) \tag{6.20}$$

when $g > a$ and $g > b$.

For all cases in which the geometry depicted in Figure 6.11 is used, K will scale linearly with the gap length, and therefore, source–drain current will scale linearly with the gap length. K will also be larger for a smaller gap width ($2a$), and therefore, source–drain current will be larger for a smaller gap width. The form of the dependency of I on a depends on the thickness of the biofilm, width of the electrodes, gap width, and gap length (i.e., $I \propto \ln(1/a)$ for cases described by Equations 6.19 and 6.20). Regardless, a key experimental consideration in making electron transport rate measurements is generating a source–drain current that is sufficiently large to be readily measurable, favoring electrodes separated by a long, narrow gap. Finally, different geometries will affect K and thus conductance and source–drain current and I, but not conductivity.

6.2 EXPERIMENTAL

We have a found the IDA [13, 14, 62, 63] to be a valuable electrode geometry for measuring electron transport rates through *G. sulfurreducens* biofilms [6, 7]. Figure 6.12 is a schematic depiction of the specific IDA design we use where black and gray regions represent a thin layer of gold patterned onto one side of a section of a glass microscope slide. It is based on a commercially available design [6] that we fabricate in-house [7]. Each gray rectangular area represents an IDA (there are two per section of slide) that is composed of 100 parallel 15-μm-wide, 0.48-cm-long gold bands separated by 15-μm-wide gaps. Every other band of each IDA is electrically connected at opposite ends of the array, forming two sets of interdigitated electrodes, each interdigitated electrode comprised of 50 bands. Figure 6.12b is an expanded scale view of a portion of one of the IDAs depicting 16 of the 100 bands (not drawn to scale), in which red colored bands comprise one of the interdigitated electrodes and blue colored bands the other. The ends of each IDA connect to gold leads that terminate at gold contact pads at one end of the slide (numbered 1–4) for separate electrical connection each interdigitated electrode comprising each IDA.

The key feature of the IDA geometry is that the gaps separating adjacent bands effectively comprise a single long continuous gap that meanders back and forth across the array. In this way, the IDA is a variation of the geometry depicted in Figure 6.11 resulting in a very long gap length (here, $l = 99$ gaps $\times$ 0.48 cm per gap $\approx$ 48 cm total for each IDA) in a very manageable form (2.45 cm $\times$ 1.2 cm overall for two IDAs including leads and contact pads) enabling sufficiently large source–drain currents to be measured in a conventional electrochemical cell using a conventional bipotentiostat. The IDA dimensions can be varied to suit the specific experimental needs. For example, our use with *G. sulfurreducens* requires a gap width smaller than the typical thickness achieved by a *G. sulfurreducens* biofilm (~20 μm in our hands) so that a continuous biofilm can be grown over the entire IDA by using both electrodes first as anodes to grow the biofilm (see Figs. 1 and 2 in [7]). A gap width of less than approximately 5 μm, however, greatly complicates IDA fabrication.

6.2.1 Fabrication of Interdigitated Microelectrode Arrays (IDAs)

We utilize the following procedure to fabricate two separately addressable IDAs onto one side of a standard borosilicate glass microscope slide. The procedure involves thoroughly cleaning the glass slide, depositing and patterning a negative photoresist to mask areas of the slide that are not to be coated with gold, depositing gold over the entire slide, lifting off the photoresist and unwanted gold revealing the desired gold pattern, and depositing a photoresist blocking layer over the gold pattern leaving only the IDAs exposed.

To clean glass slides, piranha etch is used (2:1 concentrated sulfuric acid:30% hydrogen peroxide, 3 h; caution, piranha etch is a potent oxidant and great care should be taken when using it) followed by rinsing with deionized (DI), 18 MΩ, water and drying. The glass slides are further cleaned in an oxygen plasma before use (BENCHMARK 800-II® Plasma Processing System, Axic Inc., oxygen, 30 sccm, 30 mT, 350 W, 5 min).

In order to mask areas of the slide not to be coated with gold, a negative photoresist (NR71-1000PY, Futurex) designed to develop an undercut profile for lift-off processes is spin coated onto one side of the slide (1000 rpm, 40 s) and then baked on a hot plate (resist side up, 150 °C, 3.5 min). The photoresist coated slide is exposed at 365 nm (ABM Mask Aligner, 365 nm, 12 s) using a chrome mask fabricated in-house (low reflective chrome covered with AZ photoresist exposed in a Heidelberg DWL-66 laser pattern generator). The mask projects a negative image of the pattern depicted in Figure 6.12. (The exposure time is variable and must be tuned for the specific UV light source and photoresist used.) A post-exposure bake is performed (100 °C, 3.5 min). The slide is briefly (RD6, 10 s), immediately dunked in DI water to quench development and blown dry with N_2 gas. A very brief de-scum step is performed in an oxygen plasma (30 W, 10 s) to remove any residual photoresist from intended bare areas of the slide where gold is to be deposited. It is important to use low power and a short period of time during this step to avoid damaging the photoresist layer intended to mask areas of the slide where gold is not to be deposited.

Once the photoresist masking is completed, a 100-Å thick base layer of titanium is deposited followed by 1000-Å thick layer of gold over the entire photoresist patterned glass slide (Temescal Model FC-2000 E-Beam Evaporator). The slide is then sonicated in acetone to lift off the metal-coated photoresist mask revealing the desired gold pattern depicted in Figure 6.12. The slide is rinsed with DI water, blown dry using N_2 gas, and viewed under an optical microscope to visually inspect the quality of the gold pattern.

To electrically insulate portions of the gold pattern comprising the electrical leads and contact pads, a negative photoresist blocking layer (NR7-3000P, Futurrex) designed to protect a substrate during a wet etch procedure is spin coated overtop the gold pattern (3000 rpm, 40 s). The slide is then baked on a hot plate (150 °C for 3.5 min) and exposed (365 nm, 12 s) again using a second chrome mask that covers only the IDA regions of the pattern. The slide is placed in developer solution (RD6, 30 s) to lift off resist from the IDAs regions and immediately dunked in DI water to quench development. The slide is then dried and baked on hot plate (100 °C, 3.5 min), resulting in a robust photoresist coating on the leads and contact regions of the gold pattern. Finally, the slides are cleaned again in an oxygen plasma (350 W of power for 30 s) to clean the exposed gold IDAs and bare glass gaps. A multimeter is used to ensure that the two sets of electrodes comprising each IDA are not electrically shorted.

6.2.2 Packaging of Interdigitated Electrode Arrays

We utilize the following procedure to package the IDAs for use in aqueous media. Slides containing IDAs are mounted on a custom printed circuit board comprised of copper bands on FR-4 glass-reinforced epoxy laminate. Holes at either end of the circuit board (Fig. 6.13a) facilitate soldering to electrical leads at one end (Fig. 6.13b) and then to the gold contact pads of the IDAs at the other end (Fig. 6.14a). The circuit board is attached to the glass slide using super glue such that one of the sets of holes acts as reservoirs for soldering to the gold contact pads. Pure indium wire is used to solder to the contact pads because of its relatively low melting temperature, preventing thermal damage to the gold pattern. The entire circuit board, exposed wire leads, and soldered connections are then coated with two layers of liquid electrical tape. The exposed portions of the IDA are covered with a clean glass slide (noting that the IDAs are recessed with respect to the positive photoresist layer used to mask the leads and contact pads) – Fig. 6.14b. The electrical leads are threaded through a rigid plastic pen body (Fig. 6.15). The entire assembly with the exception of the bottom two-thirds of the gold patterned slide (as depicted in Fig. 6.12) is then encapsulated in water-insulating electrical splicing epoxy (3M™ Scotchcast™ 2130B Flame Retardant Compound) such that only the IDAs and portions of the photoresist-coated gold electrical leads are exposed (Fig. 6.16). A multimeter is used to ensure that interdigitated electrodes comprising each of IDAs did not become electrically shorted. Figure 6.17 depicts a pair of packaged IDAs pattered on glass slide in one of our water jacket electrochemical cells as used to measure electron transport rates through *G. sulfurreducens* biofilms [7].

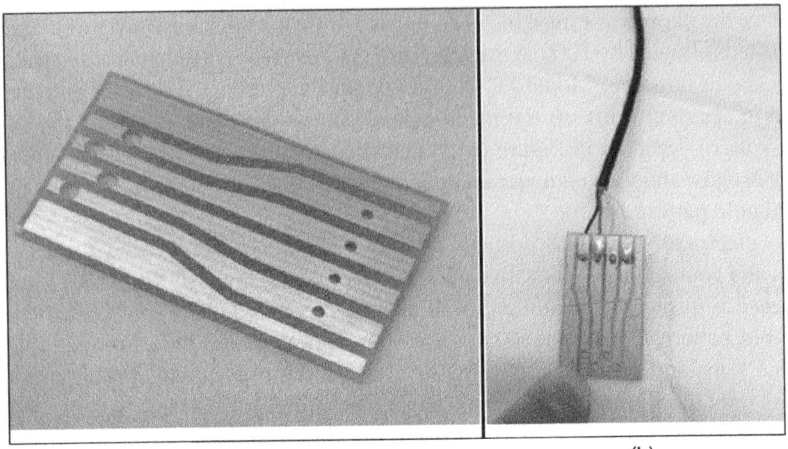

(a) (b)

Figure 6.13 (a) Circuit board used for making electrical contact to IDA chip. (b) Circuit board with electrical wires soldered in place. (*See insert for color representation of this figure.*)

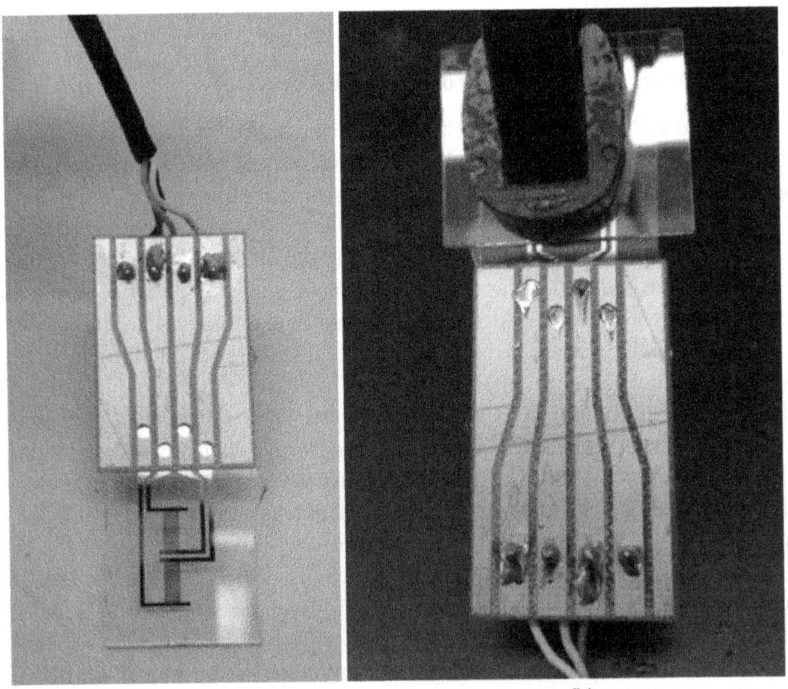

(a) (b)

Figure 6.14 (a) IDA chip glued to circuit board before being soldered together. (b) After being soldered together. (*See insert for color representation of this figure.*)

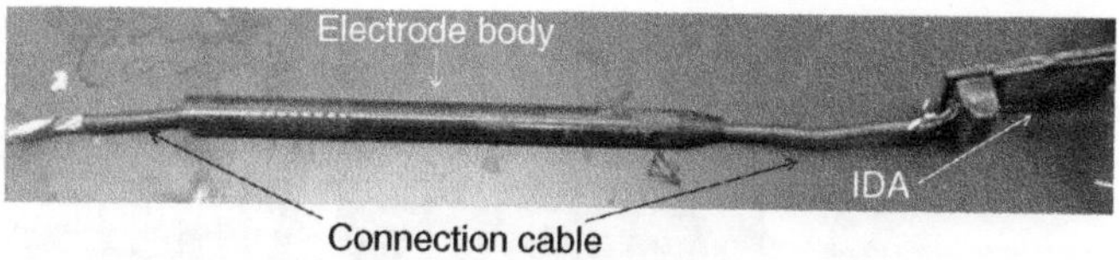

Figure 6.15 Packaging of IDA electrical wires in rigid plastic tube.

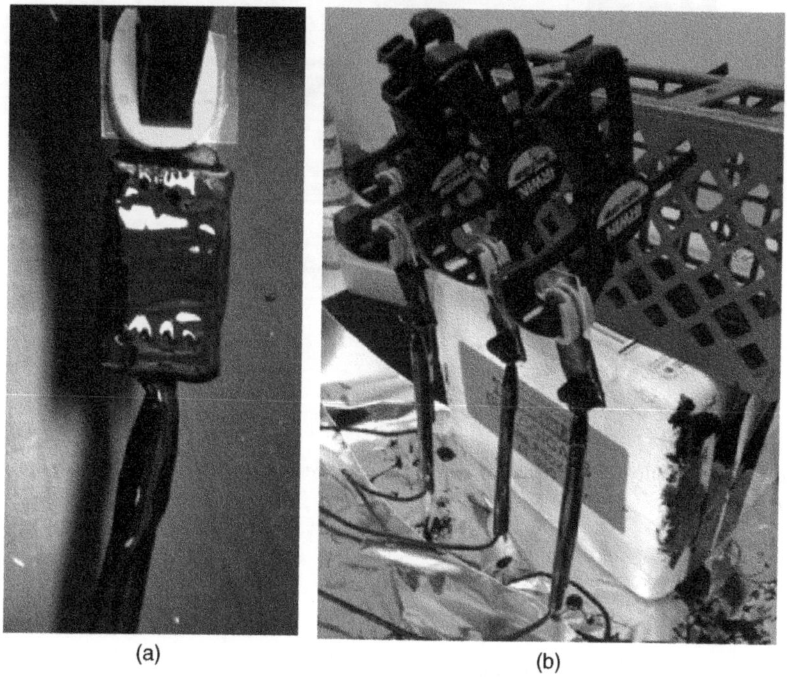

Figure 6.16 Sealing IDA packaging with water-insulating splicing epoxy. (*See insert for color representation of this figure.*)

6.2.3 Testing IDA Quality

Quality of the IDAs is tested by performing cyclic voltammetry for each set of interdigitated electrodes in 3-mM ferrocenemetbanol in 100-mM aqueous KCl (DI water) using a graphite counter electrode and an Ag/AgCl Ref. [6, 7]. Ferrocenemethanol exhibits a well-behaved reversible voltammetry with slightly larger than ideal peak splitting of 0.059 V at clean gold electrodes (Fig. 6.18) [64] depending on the voltammetric scan rate due to the microelectrode properties of IDA electrode bands [65]. If voltammetry that is not near perfectly reversible is observed, the IDAs are electrochemically cleaned by performing cyclic voltammetry in 0.1 M H_2SO_4 (DI water), cycling 200 times between 0 and 1.2 V versus Ag/AgCl at 0.1 V s^{-1} in order to remove any residual photoresist that might be contaminating the gold electrodes, then evaluated again by performing cyclic voltammetry in 3-mM ferrocenemethanol in

Figure 6.17 Sealed IDA in electrochemical cell. (*See insert for color representation of this figure.*)

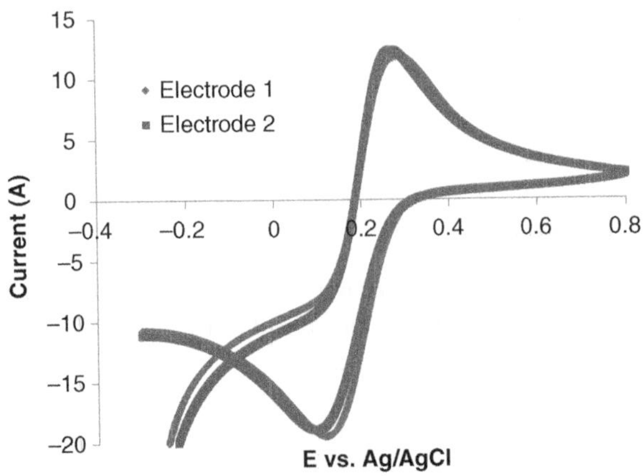

Figure 6.18 Representative cyclic voltammogram of one high-quality IDA electrode in 3-mM $K_3Fe(CN)_6$ in 100-mM aqueous KCl (DI water) using a graphite counter electrode and an Ag/AgCl reference at 10 mV s^{-1} performed to determine quality IDA electrodes before using to grow a biofilm.

100-mM aqueous KCl. Finally, IDA quality is validated by the ability to grow a biofilm using both electrodes as anodes that exhibits the expected voltammetric features [6, 7].

6.2.4 Utilizing IDAs to Measure Electron Transfer Rates Through Biofilms

Strycharz-Glaven et al. [6] and Snider et al. [7] provide ample information regarding the utilization of IDAs to make source–drain and gate current measurements of *G. sulfurreducens* biofilms, which should be directly applicable to other types of biofilms. There are two important considerations that we emphasize here.

1. *Instrumentation.* As indicated in Figure 6.1, we highly recommend using a bipotentiostat to make such measurements. We specifically use a Pine Instruments AFCBP1 bipotentiostat, which enables separate control of the source and drain potentials versus a common reference electrode with separate measurement of the source and drain currents. It does this without drawing current from the reference electrode, which can shift the potential of the reference electrode, causing uncertainty in the source and drain electrode potentials (see [36] for a more detailed explanation). Certain multichannel potentiostats can also be used, such as the Biologic VMP3, but we strongly suggest consulting with technical support to ensure proper configuration and operation.
2. *Approximating Steady-State Conditions.* The dependencies plotted previously assume steady-state conditions for each combination of source and drain potentials. Ideally, this would be accomplished by incrementally changing one or both electrode potentials depending on the type of measurement, waiting a sufficiently long time for current to become invariant with time, recording current, then repeating over and over again. In practice, this can be very tedious and error prone. Moreover, the microbes may adapt and change their properties during the time frame over which measurements are made. Often, electrochemical experiments are performed by approximating steady-state conditions. This is accomplished by changing the electrode potential at a slow enough rate such that the resulting current at each potential is independent of the rate of change of the electrode potential. In both Strycharz-Glaven et al. [6] and Snider et al. [7], we applied this approach and slowly changed the drain potential (for source–drain measurements) [6] and both source and drain potentials (for gate measurements) [7] at a sufficiently slow enough rate (0.002 $V\ s^{-1}$) such that source–drain current in each case was not dependent on the rate of change of the electrode potentials.

6.3 CONCLUSION

The long-distance extracellular electron transfer capabilities of *G. sulfurreducens* wild-type strain DL-1 have generated great interest in their use as electrode catalysts. Understanding extracellular electron transfer by *Geobacter* spp., particularly the role of the conductive, pilus-like filaments, will hopefully lead to rational strategies to improve current density of microbe-catalyzed electrode reactions because the rate of extracellular electron transfer is a contributing, if not dominant factor, limiting current density in the case of *Geobacter* spp. [2, 6, 7, 9, 10].

Glossary of Terms

A	cross-sectional area of biofilm-electrode configuration depicted in Figure 6.1
$C_{\mathrm{Red}_{(x-dx/2)}}$	concentration of reduced redox cofactor residing within distance x and $x - dx/2$
$C_{\mathrm{Red}_{(x+dx/2)}}$	concentration of reduced redox cofactor residing within distance x and $x + dx/2$
$C_{\mathrm{Ox}_{(x-dx/2)}}$	concentration of oxidized redox cofactor residing within distance x and $x - dx/2$
$C_{\mathrm{Red}_{(x+dx/2)}}$	concentration of oxidized redox cofactor residing within distance x and $x + dx/2$
C_{T}	local concentration of redox cofactor residing anywhere in biofilm
dC_{Red}/dx	redox gradient anywhere in biofilm
$dC_{\mathrm{Red}_{(x)}}/dx$	redox gradient at distance x
E_{D}	potential applied to drain
E_{G}	gate potential, average of E_{D} and E_{S}
E_{S},	potential applied to source
$E^{o'}$	redox cofactor formal potential
F	Faraday constant
I	current
I_{D}	drain current
I_{S}	source current
I_{SD}	source–drain current
G	conductance
h	height of biofilm-electrode configuration depicted in Figure 6.1
$J_{(x)}$	flux of electrons across plane x
K	cell constant
k_1	rate constant for electron transfer from source to oxidized redox cofactor residing at the source/biofilm interface
k_2	rate constant for electron transfer to source from reduced redox cofactor residing at the source/biofilm interface
k_3	rate constant for electron transfer from reduced redox cofactor residing at distance $x - dx/2$ to oxidized redox cofactor at distance $x + dx/2$
k_4	rate constant for electron transfer from reduced redox cofactor residing at distance $x + dx/2$ to oxidized redox cofactor at distance $x - dx/2$
k_5	rate constant for electron transfer from reduced redox cofactor residing at the drain/biofilm interface to the drain

k_6	rate constant for electron transfer to reduced redox cofactor residing at the drain/biofilm interface from the drain
k_{ex}	rate constant for electron transfer between neighboring redox cofactors
l	gap length
n	number of electrons transferred at time in Reactions 1–3
$Ox_{(D)}$	oxidized redox cofactor residing at the drain/biofilm interface
$Ox_{(S)}$	oxidized redox cofactor residing at the source/biofilm interface
$Ox_{(x+dx/2)}$	oxidized redox cofactor residing within distance x and $x + dx/2$
$Ox_{(x-dx/2)}$	oxidized redox cofactor residing within distance x and $x - dx/2$
$Ox_{(x+\delta/2)}$	oxidized redox cofactor residing within distance x and $x + \delta/2$
$Ox_{(x-\delta/2)}$	oxidized redox cofactor residing within distance x and $x - \delta/2$
x	distance of an imaginary plane sliced through the biofilm from the source
R	gas constant
$Red_{(D)}$	reduced redox cofactor at the drain/biofilm interface
$Red_{(S)}$	reduced redox cofactor at the source/biofilm interface
$Red_{(x+dx/2)}$	reduced redox cofactor residing within distance x and $x + dx/2$
$Red_{(x-dx/2)}$	reduced redox cofactor residing within distance x and $x - dx/2$
$Red_{(x+\delta/2)}$	reduced redox cofactor residing within distance x and $x + \delta/2$
$Red_{(x-\delta/2)}$	reduced redox cofactor residing within distance x and $x - \delta/2$
σ	conductance
T	temperature
V	potential difference between drain and source
w	gap width
δ	distance separating adjacent redox cofactors

APPENDIX

$$I = nFAk_0C_T^2\delta^2\left(\frac{\left(\frac{1}{1+e^{\left[\frac{nF}{RT}\left(E_{(S)}-E^{o\prime}\right)\right]}}\right)-\left(\frac{1}{1+e^{\left[\frac{nF}{RT}\left(E_{(D)}-E^{o\prime}\right)\right]}}\right)}{w}\right)$$

$$= nFAk_0C_T^2\delta^2\left(\frac{\left(\frac{1}{1+e^{\left[\frac{nF}{RT}\left(E_G-\frac{\Delta E}{2}-E^{o\prime}\right)\right]}}\right)-\left(\frac{1}{1+e^{\left[\frac{nF}{RT}\left(E_G+\frac{\Delta E}{2}-E^{o\prime}\right)\right]}}\right)}{w}\right)$$

$$= \frac{nFAk_0C_T^2\delta^2}{w}\left(\left(\frac{1}{1+e^{\left[a\left(b-\frac{c}{2}-d\right)\right]}}\right)-\left(\frac{1}{1+e^{\left[a\left(b+\frac{c}{2}-d\right)\right]}}\right)\right)$$

where $a = \frac{nF}{RT}$, $b = E_G$, $c = \Delta E$, $d = E^{o'}$

$$= \frac{nFAk_0C_T^2\delta^2}{w}\left(\frac{\left(1+e^{\left[a\left(b+\frac{c}{2}-d\right)\right]}\right)-\left(1+e^{\left[a\left(b-\frac{c}{2}-d\right)\right]}\right)}{\left(1+e^{\left[a\left(b-\frac{c}{2}-d\right)\right]}\right)\left(1+e^{\left[a\left(b+\frac{c}{2}-d\right)\right]}\right)}\right)$$

$$= \frac{nFAk_0{C_T}^2\delta^2}{w}\left(\frac{e^{\left[a\left(b+\frac{c}{2}-d\right)\right]}-e^{\left[a\left(b-\frac{c}{2}-d\right)\right]}}{1+e^{\left[a\left(b+\frac{c}{2}-d\right)\right]}+e^{\left[a\left(b-\frac{c}{2}-d\right)\right]}+e^{\left[a\left(b-\frac{c}{2}-d\right)\right]}e^{\left[a\left(b+\frac{c}{2}-d\right)\right]}}\right)$$

$$= \frac{nFAk_0C_T^2\delta^2}{w}\left(\frac{e^{ab}e^{\left(\frac{ac}{2}\right)}e^{-ad}-e^{ab}e^{\left(\frac{-ac}{2}\right)}e^{-ad}}{1+e^{ab}e^{\left(\frac{ac}{2}\right)}e^{-ad}+e^{ab}e^{\left(\frac{-ac}{2}\right)}e^{-ad}+e^{ab}e^{\left(\frac{-ac}{2}\right)}e^{-ad}e^{ab}e^{\left(\frac{ac}{2}\right)}e^{-ad}}\right)$$

where $e^x \approx 1 + x$ when $|x| \ll 1$

$$= \frac{nFAk_0C_T^2\delta^2}{w}\left(\frac{\left(1+\frac{ac}{2}\right)e^{ab}e^{-ad}-\left(1-\frac{ac}{2}\right)e^{ab}e^{-ad}}{1+\left(1+\frac{ac}{2}\right)e^{ab}e^{-ad}+\left(1-\frac{ac}{2}\right)e^{ab}e^{-ad}+\left(1-\frac{ac}{2}\right)\left(1+\frac{ac}{2}\right)e^{ab}e^{ab}e^{-ad}e^{-ad}}\right)$$

$$= \frac{nFAk_0C_T^2\delta^2}{w}\left(\frac{e^{ab}e^{-ad}+\left(\frac{ac}{2}\right)e^{ab}e^{-ad}-e^{ab}e^{-ad}+\left(\frac{ac}{2}\right)e^{ab}e^{-ad}}{1+e^{ab}e^{-ad}+\left(\frac{ac}{2}\right)e^{ab}e^{-ad}+e^{ab}e^{-ad}-\left(\frac{ac}{2}\right)e^{ab}e^{-ad}+\left(1+\frac{ac}{2}-\frac{ac}{2}-\left(\frac{ac}{2}\right)^2\right)e^{ab}e^{ab}e^{-ad}e^{-ad}}\right)$$

where $1 + x^2 \approx 1$ when $|x| \ll 1$

$$= \frac{nFAk_0C_T^2\delta^2}{w}\left(\frac{e^{ab}e^{-ad}+\left(\frac{ac}{2}\right)e^{ab}e^{-ad}-e^{ab}e^{-ad}+\left(\frac{ac}{2}\right)e^{ab}e^{-ad}}{1+e^{ab}e^{-ad}+\left(\frac{ac}{2}\right)e^{ab}e^{-ad}+e^{ab}e^{-ad}-\left(\frac{ac}{2}\right)e^{ab}e^{-ad}+e^{2[a(b-d)]}}\right)$$

$$= \frac{nFAk_0C_T^2\delta^2}{w}\left(\frac{(ac)\,e^{[a(b-d)]}}{1+2e^{[a(b-d)]}+e^{2[a(b-d)]}}\right)$$

$$= \frac{nFAk_0C_T^2\delta^2}{w}\left(\frac{\left(\frac{nF}{RT}\Delta E\right)e^{\left[\frac{nF}{RT}\left(E_G-E^{o\prime}\right)\right]}}{1+2e^{\left[\frac{nF}{RT}\left(E_G-E^{o\prime}\right)\right]}+e^{2\frac{nF}{RT}\left(E_G-E^{o\prime}\right)}}\right)$$

$$= \left[\frac{n^2F^2Ak_0C_T^2\delta^2e^{\left[\frac{nF}{RT}\left(E_G-E^{o\prime}\right)\right]}}{RTw\left(1+2e^{\left[\frac{nF}{RT}\left(E_G-E^{o\prime}\right)\right]}+e^{2\frac{nF}{RT}\left(E_G-E^{o\prime}\right)}\right)}\right]\Delta E$$

$$= G(\Delta E)$$

where

$$G = \left[\frac{n^2F^2Ak_0C_T^2\delta^2e^{\left[\frac{nF}{RT}\left(E_G-E^{o\prime}\right)\right]}}{RTw\left(1+2e^{\left[\frac{nF}{RT}\left(E_G-E^{o\prime}\right)\right]}+e^{2\frac{nF}{RT}\left(E_G-E^{o\prime}\right)}\right)}\right]$$

and

$$g = \left[\frac{n^2F^2k_0C_T^2\delta^2e^{\left[\frac{nF}{RT}\left(E_G-E^{o\prime}\right)\right]}}{RT\left(1+2e^{\left[\frac{nF}{RT}\left(E_G-E^{o\prime}\right)\right]}+e^{2\frac{nF}{RT}\left(E_G-E^{o\prime}\right)}\right)}\right]$$

ACKNOWLEDGMENTS

The authors are grateful for funding provided by the Naval Research Laboratory and the Office of Naval Research.

REFERENCES

1. Bond DR, Holmes DE, Tender LM, Lovley DR. Electrode-reducing microorganisms that harvest energy from marine sediments. Science 2002;295:483–485.
2. Bond DR, Strycharz-Glaven SM, Tender LM, Torres CI. On electron transport through *Geobacter* biofilms. ChemSusChem 2012;5:1099–1105.
3. Pirbadian S, El-Naggar MY. Multistep hopping and extracellular charge transfer in microbial redox chains. Phys Chem Chem Phys 2012;14:13802–13808.
4. Polizzi NF, Skourtis SS, Beratan DN. Physical constraints on charge transport through bacterial nanowires. Faraday Discuss 2012;155:43–62.
5. Forster RJ, Walsh DA, Mano N, Mao F, Heller A. Modulating the redox properties of an osmium-containing metallopolymer through the supporting electrolyte and cross-linking. Langmuir 2004;20:862–868.
6. Strycharz-Glaven SM, Snider RM, Guiseppi-Elie A, Tender LM. On the electrical conductivity of microbial nanowires and biofilms. Energy Environ Sci 2011;4:4366–4379.

7. Snider RM, Strycharz-Glaven SM, Tsoi SD, Erickson JS, Tender LM. Long-range electron transport in *Geobacter sulfurreducens* biofilms is redox gradient-driven. Proc Natl Acad Sci U S A 2012;109:15467–15472.

8. Smith DK, Tender LM, Lane GA, Licht S, Wrighton MS. Chemically induced release of charge from a rectifying polymer based on viologen and quinone subunits. J Am Chem Soc 1989;111:1099–1105.

9. Richter H, Nevin KP, Jia H, Lowy DA, Lovley DR, Tender LM. Cyclic voltammetry of biofilms of wild type and mutant *Geobacter sulfurreducens* on fuel cell anodes indicates possible roles of OmcB, OmcZ, type IV pili, and protons in extracellular electron transfer. Energy Environ Sci 2009;2:506–516.

10. Strycharz SM, Malanoski AP, Snider RM, Yi H, Lovley DR, Tender LM. Application of cyclic voltammetry to investigate enhanced catalytic current generation by biofilm-modified anodes of *Geobacter sulfurreducens* strain DL1 vs. variant strain KN400. Energy Environ Sci 2011;4:896–913.

11. Dalton EF, Surridge NA, Jernigan JC, Wilbourn KO, Facci JS, Murray RW. Charge transport in electroactive polymers consisting of fixed molecular redox sites. Chem Phys 1990;141:143–157.

12. Aoki A, Heller A. Electron diffusion coefficients in hydrogels formed of cross-linked redox polymers. J Phys Chem 1993;97:11014–11019.

13. Chidsey CE, Feldman BJ, Lundgren C, Murray RW. Micrometer-spaced platinum interdigitated array electrode: fabrication, theory, and initial use. Anal Chem 1986;58:601–607.

14. Natan MJ, Wrighton MS. Chemically modified microelectrode arrays. Prog Inorg Chem 1990;37:391–494.

15. Rollefson JB, Stephen CS, Tien M, Bond DR. Identification of an extracellular polysaccharide network essential for cytochrome anchoring and biofilm formation in *Geobacter sulfurreducens*. J Bacteriol 2011;193:1023–1033.

16. Mehta T, Coppi M, Childers SE, Lovley D. Outer membrane c-type cytochromes required for Fe(III) and Mn(IV) oxide reduction in *Geobacter sulfurreducens*. Appl Environ Microbiol 2005;71:8634–8641.

17. Leang C, Coppi M, Lovley D. OmcB, a c-type polyheme cytochrome, involved in Fe(III) reduction in *Geobacter sulfurreducens*. J Bacteriol 2003;185:2096–2103.

18. Leang C, Qian X, Mester T, Lovley DR. Alignment of the c-type cytochrome OmcS along pili of *Geobacter sulfurreducens*. Appl Environ Microbiol 2010;76:4080–4084.

19. Ding YHR, Hixson KK, Aklujkar MA, Lipton MS, Smith RD, Lovley DR, Mester T. Proteome of *Geobacter sulfurreducens* grown with Fe(III) oxide or Fe(III) citrate as the electron acceptor. Biochim Biophys Acta 2008;1784:1935–1941.

20. Inoue K, Qian X, Morgado L, Kim BC, Mester T, Izallalen M, Salgueiro CA, Lovley DR. Purification and characterization of OmcZ, an outer-surface, octaheme c-type cytochrome essential for optimal current production by *Geobacter sulfurreducens*. Appl Environ Microbiol 2010;76:3999–4007.

21. Qian XL, Mester T, Morgado L, Arakawa T, Sharma ML, Inoue K, Joseph C, Salgueiro CA, Maroney MJ, Lovley DR. Biochemical characterization of purified OmcS, a c-type cytochrome required for insoluble Fe(III) reduction in *Geobacter sulfurreducens*. Biochim Biophys Acta 2011;1807:404–412.

22. Ding YHR, Hixson KK, Giometti CS, Stanley A, Esteve-Nunez A, Khare T, Tollaksen SL, Zhu W, Adkins JN, Lipton MS, Smith RD, Mester T, Lovley DR. The proteome of dissimilatory metal-reducing microorganism *Geobacter sulfurreducens* under various growth conditions. Biochim Biophys Acta 2006;1764:1198–1206.
23. Strycharz-Glaven SM, Tender LM. Study of the mechanism of catalytic activity of *G. sulfurreducens* biofilm anodes during biofilm growth. ChemSusChem 2012;5:1106–1118.
24. Liu Y, Kim H, Franklin RR, Bond DR. Linking spectral and electrochemical analysis to monitor c-type cytochrome redox status in living *Geobacter sulfurreducens* biofilms. ChemPhysChem 2011;12:2235–2241.
25. Liu Y, Bond DR. Long-distance electron transfer by *G. sulfurreducens* biofilms results in accumulation of reduced c-type cytochromes. ChemSusChem 2012;5:1047–1053.
26. Busalmen JP, Esteve-Nunez A, Berna A, Feliu JM. c-Type cytochromes wire electricity-producing bacteria to electrodes. Angew Chem Int Ed Engl 2008;47: 4874–4877.
27. Busalmen JP, Esteve-Nunez A, Berna A, Feliu JM. ATR-SEIRAs characterization of surface redox processes in *G. sulfurreducens*. Bioelectrochemistry 2010;78:25–29.
28. Bonanni PS, Bradley DF, Schrott GD, Busalmen JP. Limitations for current production in *Geobacter sulfurreducens* biofilms. ChemSusChem 2013;6:711–720.
29. Robuschi L, Tomba JP, Schrott GD, Bonanni PS, Desimone PM, Busalmen JP. Spectroscopic slicing to reveal internal redox gradients in electricity-producing biofilms. Angew Chem Int Ed Engl 2013;52:925–928.
30. Millo D, Harnisch F, Patil SA, Ly HK, Schroder U, Hildebrandt P. In situ spectroelectrochemical investigation of electrocatalytic microbial biofilms by surface-enhanced resonance Raman spectroscopy. Angew Chem Int Ed Engl 2011;50:2625–2627.
31. Virdis B, Harnisch F, Batstone DJ, Rabaey K, Donose BC. Non-invasive characterization of electrochemically active microbial biofilms using confocal Raman microscopy. Energy Environ Sci 2012;5:7017–7024.
32. Hartshorne RS, Reardon CL, Ross D, Nuester J, Clarke TA, Gates AJ, Mills PC, Fredrickson JK, Zachara JM, Shi L, Beliaev AS, Marshall MJ, Tien M, Brantley S, Butt JN, Richardson DJ. Characterization of an electron conduit between bacteria and the extracellular environment. Proc Natl Acad Sci U S A 2009;106:22169–22174.
33. Leung KM, Wanger G, El-Naggar MY, Gorby Y, Southam G, Lau WM, Yang J. *Shewanella oneidensis* MR-1 bacterial nanowires exhibit p-type, tunable electronic behavior. Nano Lett 2013;13:2407–2411.
34. Lebedev N, Strycharz-Glaven SM, Tender LM. Spatially resolved confocal resonant Raman microscopic analysis of anode-grown *Geobacter sulfurreducens* biofilms. ChemPhysChem 2014;15:320–327.
35. Breuer M, Rosso KM, Blumberger J. Electron flow in multiheme bacterial cytochromes is a balancing act between heme electronic interaction and redox potentials. Proc Natl Acad Sci U S A 2014;111:611–616.
36. Strycharz-Glaven SM, Tender LM. Reply to the 'comment on "on electrical conductivity of microbial nanowires and biofilms"' by N. S. Malvankar, M. T. Tuominen and D. R. Lovley, Energy Environ. Sci., 2012, 5, DOI: 10.1039/c2ee02613a. Energy Environ Sci 2012;5:6250–6255.
37. Rittmann BE, McCarty PL. Model of steady-state biofilm kinetics. Biotechnol Bioeng 1980;22:2343–2357.

38. Rittmann BE, McCarty PL. Evaluation of steady-state biofilm kinetics. Biotechnol Bioeng 1980;22:2359–2373.
39. Strycharz-Glaven SM, Glaven RH, Wang Z, Zhou J, Vora GJ, Tender LM. Electrochemical investigation of a microbial solar cell reveals a nonphotosynthetic biocathode catalyst. Appl Environ Microbiol 2013;79:3933–3942.
40. Zhu X, Yates MD, Logan BE. Set potential regulation reveals additional oxidation peaks of *Geobacter sulfurreducens* anodic biofilms. Electrochem Commun 2012;22:116–119.
41. Katuri KP, Kavanagh P, Rengaraj S, Leech D. *Geobacter sulfurreducens* biofilms developed under different growth conditions on glassy carbon electrodes: insights using cyclic voltammetry. Chem Commun 2010;46:4758–4760.
42. Saravanan R, Kavanagh P, O'Flaherty V, Leech D, Katrui KP. Charge transport through *Geobacter sulfurreducens* biofilms grown on graphite rods. Langmuir 2012;28:7904–7913.
43. Torres CI, Marcus AK, Parameswaran P, Rittmann BE. Kinetic experiments for evaluating the Nernst–Monod model for anode-respiring bacteria (ARB) in a biofilm anode. Environ Sci Technol 2008;42:6593–6597.
44. Torres CI, Marcus AK, Rittmann BE. Proton transport inside the biofilm limits electrical current generation by anode-respiring bacteria. Biotechnol Bioeng 2008;100:872–881.
45. Torres CI, Marcus AK, Lee HS, Parameswaran P, Krajmalnik-Brown R, Rittmann BE. A kinetic perspective on extracellular electron transfer by anode-respiring bacteria. FEMS Rev 2010;34:3–17.
46. Lebedev N, Strycharz-Glaven SM, Tender LM. Spatially resolved confocal resonant Raman microscopic analysis of anode-grown *Geobacter sulfurreducens* biofilms. ChemSusChem 2014;15:320–327.
47. Logan MA. An AC bridge for semiconductor resistivity measurements using a 4-point probe. Bell Syst Tech J 1961;40:885–919.
48. Tans SJ, Verschueren ARM, Dekker C. Room-temperature transistor based on a single carbon nanotube. Nature 1998;393:49–52.
49. Dimitrakopoulos CD, Malenfant PRL. Organic thin film transistors for large area electronics. Adv Mater 2002;14:99–117.
50. Hoeben FJM, Jonkheijm P, Meijer EW, Schenning A. About supramolecular assemblies of pi-conjugated systems. Chem Rev 2005;105:1491–1546.
51. Anderson PW. Antiferromagnetism – theory of superexchange interaction. Phys Rev 1950;79:350–356.
52. Blauch DN, Saveant JM. Dynamics of electron hopping in assemblies of redox centers. Percolation and diffusion. J Am Chem Soc 1992;114:3323–3332.
53. Jernigan JC, Chidsey CED, Murray RW. Electrochemistry of polymer films not immersed in solution: electron transfer on an ion budget. JACS 1985;107:2824–2826.
54. Saveant JM. Electron hopping between fixed sites equivalent diffusion and migration laws. J Electroanal Chem 1986;201:211–213.
55. Buck RP. Coupled electron hopping anion displacement in plane sheet fixed-site polymer membranes. J Electroanal Chem 1989;258:1–12.
56. Andrieux C, Saveant J. Electron-transfer through redox polymer-films. J Electroanal Chem 1980;111:377–381.
57. Feldman B, Murray R. Electron diffusion in wet and dry Prussian blue films on interdigitated array electrodes. Inorg Chem 1987;26:1702–1708.

58. Bard AJ, Faulkner LR. Electrochemical Methods: Fundamentals and Applications. 2nd ed. New York: John Wiley and Sons, Inc.; 2001.
59. Saveant JM. Electron hopping between fixed sites – diffusion and migration in counterion conservative redox membranes at steady-state. J Electroanal Chem 1988;242:1–21.
60. Sheppard NF, Tucker RC, Wu C. Electrical conductivity measurements using microfabricated interdigitated electrodes. Anal Chem 1993;65:1199–1202.
61. Kankare J, Kupila EL. In-situ conductance measurement during electropolymerization. J Electroanal Chem 1992;322:167–181.
62. Zaretsky MC, Mouayad L, Melcher JR. Continuum properties from interdigital electrode dielectrometry. IEEE Trans Electr Insul 1988;23:897–917.
63. Sheppard NF, Day DR, Lee HL, Senturia SD. Microdielectrometry. Sens Actuators 1981;2:263–274.
64. Dowdy CE, Leopold MC. Enhanced electrochemistry of nanoparticle-embedded polyelectrolyte films: interfacial electronic coupling and distance dependence. Thin Solid Films 2010;519:790–796.
65. Wightman RM. Microvoltammetric electrodes. Anal Chem 1981;53:1125A–1134A.

7

ELECTRONIC CONDUCTIVITY IN LIVING BIOFILMS: PHYSICAL MEANING, MECHANISMS, AND MEASUREMENT METHODS

NIKHIL S. MALVANKAR

Department of Physics, University of Massachusetts, Amherst, MA, USA; Department of Microbiology, University of Massachusetts, Amherst, MA, USA

DEREK R. LOVLEY

Department of Microbiology, University of Massachusetts, Amherst, MA, USA

7.1 INTRODUCTION

Biofilms are commonly considered as electronic insulators: there are no free electrons that can flow under ordinary thermal conditions [1–3]. However, direct conductivity measurements revealed that the biofilms of pure culture [4] *Geobacter sulfurreducens* as well as mixed species derived from wastewater [5], when grown on the anodes of microbial fuel cell, are electronically conductive, enabling flow of free electrons within the biofilms. Surprisingly, measured biofilm conductivity was comparable to synthetic organic materials such as nanostructured polyaniline [4]. This conductivity enables microorganisms in the biofilms to utilize electron acceptors, such as anodes of microbial fuel cells, which are multiple cell lengths away from them [6] (Fig. 7.1). Modeling studies [7] as well as cyclic voltammetry measurements [8, 9] have also suggested that the anode biofilms are conductive. Direct conductivity measurements have further demonstrated that proteinaceous pili nanofilaments of *G. sulfurreducens*, referred to as microbial nanowires, confer conductivity to biofilms [4]. Surprisingly, the conductivity of pili is metallic-like rather than previously

Biofilms in Bioelectrochemical Systems: From Laboratory Practice to Data Interpretation, First Edition.
Edited by Haluk Beyenal and Jerome Babauta.

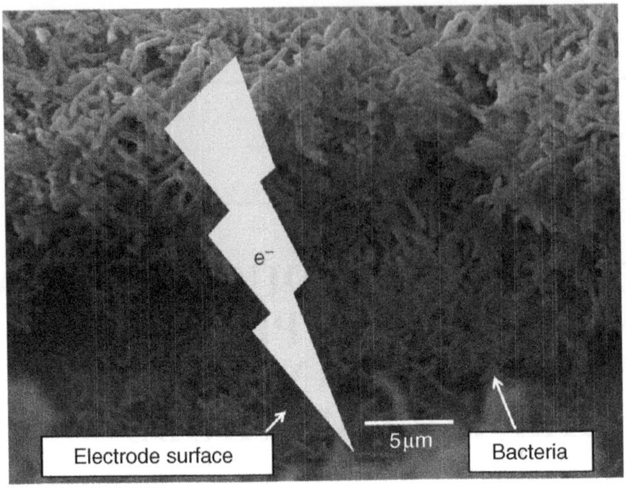

Figure 7.1 Electron conduction over multiple cell length in biofilm of *Geobacter sulfurreducens* strain KN400 grown on anodes of microbial fuel cell. Figure depicts SEM image of biofilm grown on split gold anode setup.

known methods of electron flow in proteins using redox-active co-factors such as c-type cytochromes [4, 10–12]. Electronic conductivity enables microorganisms to access electron acceptors that are many cell lengths away. In addition, biologically produced conductive films, which can be synthesized from inexpensive feedstocks and when alive, can self-repair and replicate, introduce new concepts and materials for bioelectronics [13]. This chapter summarizes the methods used to directly measure the newly discovered conductive properties of biofilms. Firstly, we describe the physical meaning of electronic conductivity and different mechanisms of conductivity. Then, we discuss in detail the experimental methods applied to directly measure conductivity in living biofilms and the results obtained using these methods.

7.2 PHYSICAL MEANING OF ELECTRONIC CONDUCTIVITY

7.2.1 Definition

For any material to be electronically conductive, it must possess free electrons that can move under ordinary thermal conditions [14]. At room temperature, voltage corresponding to the thermal energy $k_B T$ is approximately 25 mV. Therefore, in other words, a small bias voltage of about 25 mV applied at two different parts of the material should generate electronic current in a conductive material. The underlying atomic or molecular structure determines the presence of free electrons in a material.

In the context of biofilms, in order to utilize an electron acceptor present at a large distance (>10 μm), electrons released by the microorganism must be carried to the terminal electron acceptor. Electron flow occurs via two distinct mechanisms: electron *transfer* and *transport* [15]. Consider a donor–bridge (B_j)–acceptor system, where $B_j = B_1, B_2, \ldots, B_N$ are the N specific sites in the molecule bridging donor and acceptors. In a typical electron *transfer* process, donor donates electron in a single step to the acceptor. This exchange of electrons between donor and acceptor molecules is termed as electron transfer. While in a typical electron *transport* process, an electron injected by the donor in the bridge molecules is carried by the bridge to the acceptor in multiple steps. Injected electrons can enjoy long-range transport, that is, over tens of angstroms, on fast time scale of picoseconds or less [16]. This voltage-driven electron migration between two electrodes is referred as electron transport in semiconductor and device physics. The conductivity is a material property that enables electrons to flow within the material. Therefore, the conductivity is associated with the electron *transport* processes occurring in the material rather than the electron *transfer* processes. It is important to note that conductivity (electron transport) is proportional to the electron transfer rate, and calculating the precise value of conductance from electron transfer rate using the above-mentioned equation requires the coupling strengths of the molecule to the two electrodes and to the donor and acceptor, as well as the coupling of the donor (acceptor) to the nuclear coordinates [17]. Details of the relation between transfer and transport processes are described elsewhere [17].

7.2.2 Mechanisms of Conductivity

There are primarily three different modes of conductivity in materials [12, 18, 19]:

1. Tunneling.
2. Hopping.
3. Delocalization.

These three main models of electron transport are outlined in the following section. The discussion is presented in the context of organic solid materials because conductive biofilms are organic solids. These processes can occur simultaneously, and each may dominate at different values of the applied electric field or different temperature range [18, 19]. Figure 7.2 depicts the differences between these mechanism in the context of microbial nanowires and cytochromes aligned along the nanowires [12].

7.2.2.1 Quantum Mechanical Tunneling If the material of interest is an insulator, classical physics predicts that electrons incident on the metal/insulator interface will be reflected if the energy of an electron is less than the interfacial potential barrier of the interface. The electron cannot penetrate the barrier, and its passage from one electrode to the other is inhibited. However, quantum mechanics predicts that an electron will overcome this barrier due to the wave nature of the electrons. The electron wavefunction decays rapidly with depth of penetration from the electrode/insulator interface, and for barriers of microscopic thickness, the wavefunction is essentially zero at the opposite interface [18]. This means that the probability of finding the electron is essentially zero at the other electrode. However, if the barrier is very thin (<5 nm), the wavefunction has a nonzero value at the other electrode, allowing

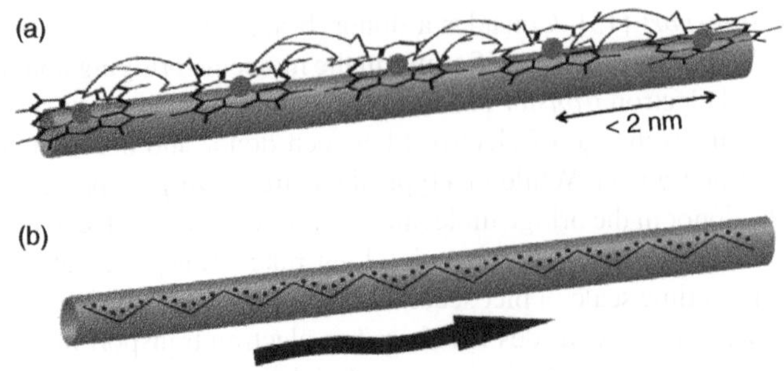

Figure 7.2 Pictorial representation of electron flow along pili filaments via (a) electron hopping and (b) metallic-like conduction. In the electron hopping mechanism, charges (shown as black dots) hop from cytochrome to cytochrome (shown as gray circles). In metallic-like conduction, charges are spread across the entire filaments, referred to as delocalized. Illustrated here is the charge delocalization in π-conjugated organic metal polyacetylene. Figure adapted from Ref. [7] with permission.

electron to pass from one electrode to the other by penetrating the barrier with finite probability [18, 20]. Further details on the tunneling are presented elsewhere [18].

The current–voltage relationship for tunneling depends on the magnitude of the applied voltage and whether the tunnel barrier is symmetric or asymmetric (i.e., whether the two electrodes are similar or different metals). When the Fermi level of the metal is aligned close to either highest occupied molecular orbital (HOMO) or lowest unoccupied molecular orbital (LUMO) states of the insulating film, the effect of other distant energy level on the tunneling transport is negligible and the Simmons model is widely used [21]. According to this model, the relationship between current density J and the voltage in the tunnel junction having similar electrodes at relatively low voltages (~25 mV) is given as:

$$J = \vartheta(V + \gamma V^3)$$

where

$$\vartheta = \left(\frac{e}{h}\right)^2 \frac{[m(\varphi_1 + \varphi_2)]^{1/2}}{s} \exp\left[-D\left(\frac{\varphi_1 + \varphi_2}{2}\right)^{1/2}\right]$$

$$\gamma = \{(De)^2/[48(\varphi_1 + \varphi_2)]\} - (De^2/32)[2/(\varphi_1 + \varphi_2)]^{3/2}$$

$$D = [4\varpi s(2m)^{1/2}]/h$$

φ_1 and φ_2 (in eV) are barrier height at the interface of the insulating film and electrode 1 and 2, respectively. At low voltages, barrier height $\varphi_B = (\varphi_1 + \varphi_2)/2$. s is the thickness of the film. Mass and charge of the electron are m and e. h is the Planck's constant. All the variables and constants are presented in their respective SI units.

The curves follow the linear + cubic barrier-bending form expected for tunneling of electrons across an insulating layer [21, 22].

When the applied bias is small compared to the barrier height (φ_B), the tunneling through an insulating layer is termed as direct tunneling ($V < \varphi_B/e$). The Simmons model expresses the tunneling current density through a barrier in this regime as:

$$J = \left(\frac{e}{4\pi\hbar d^2}\right)\left\{\begin{array}{l}\left(\Phi_B - \frac{eV}{2}\right) \\ \times\exp\left[-\frac{2(2m)^{1/2}}{\hbar}\alpha\left(\Phi_B - \frac{eV}{2}\right)^{1/2} d\right] - \left(\Phi_B + \frac{eV}{2}\right) \\ \times\exp\left[-\frac{2(2m)^{1/2}}{\hbar}\alpha\left(\Phi_B + \frac{eV}{2}\right)^{1/2} d\right]\end{array}\right\},$$

where d is the barrier width and α is the parameter used to modify the simple rectangular barrier model or to account for an effective mass. $\alpha = 1$ for a rectangular barrier and bare electron mass.

In this regime, the tunneling conductivity is given as

$$\sigma_t = A\exp(-\beta d)$$

where A is the constant, d is the thickness of the tunnel barrier, and β is the decay coefficient.

On the basis of the applied bias as compared to the barrier height (φ_B), the tunneling can be categorized into either direct ($V < \varphi_B/e$) or Fowler–Nordheim ($V > \varphi_B/e$) tunneling. These two tunneling mechanisms can be distinguished due to their distinct voltage dependences. For direct tunneling, current scales linearly with voltage at low bias, while for Fowler–Nordheim tunneling, $\ln(I^2/V)$ varies linearly with $1/V$. (We have previously demonstrated direct and Fowler–Nordheim tunneling in microcapsules formed by self-assembly and cross-linking of gold nanowires [23].)

At the nanoscale, tunneling current can be described by Landauer formula for the linear conductance, G, of the electrode–molecule–electrode junction [24]:

$$G = \frac{2e^2}{h}T$$

T is a function that reflects the efficiency of electron transmission from one contact to the other. This transmission function can be roughly divided into following components:

$$T = T_{lc}\, T_{rc}\, T_{mol}$$

where T_{lc} and T_{rc} give the efficiency of charge transport across left and right contacts and T_{mol} reflects the charge transport through the molecule. One can roughly approximate coherent, nonresonant tunneling through molecules as tunneling through a rectangular barrier, in which case

$$T_{mol} = \exp(-\beta l)$$

where l is the width of the barrier, that is, the length of the molecule and β is the tunneling decay parameter is the units of $(\text{length})^{-1}$ given by:

$$\beta = 2\sqrt{\frac{2m * \alpha\left(\phi - \frac{eV}{2}\right)}{\hbar^2}}$$

where V is the bias applied across the molecule and ϕ is the height of the barrier for tunneling through LUMO level ($\phi = E_F - E_{LUMO}$) or through the HOMO level ($\phi = E_{HOMO} - E_F$). E_F will be the Fermi level of the electrode or the electrode–molecule composite. The electron's charge and effective mass are given by e and m^*. α is the asymmetry in the potential profile across electrode–molecule–electrode junction.

7.2.2.2 ***Hopping*** The transport of electrons through macromolecular structures containing redox molecules occurs via physical displacement of the redox molecules and electron hopping from one reduced molecule to an adjacent oxidized molecule (Fig. 7.2a). The hopping process could be either unidirectional or involve one-dimensional diffusion. In solution, charge transport is dominated by physical diffusion of the redox molecules [25]. Enhancement of the charge transport rate by electron exchange is predicted to occur in solutions containing mixtures of the oxidized and reduced species as described by Dahms and Ruff equation [25]. The diffusion coefficient, D, resulting from a combination of physical displacement and electron hopping is given by

$$D = D_{phys} + \frac{k_{ex} C_E d^2}{6}$$

where D_{phys} is the diffusion coefficient for physical displacement of the redox molecules, k, is the bimolecular rate constant for electron self-exchange, C, is the total concentration of redox species, and d is the center-to-center distance between redox centers at the time of electron transfer.

Conductivity is related to diffusivity by linear expression [25]:

$$\sigma = \frac{ne^2 D}{k_B T}$$

Therefore, conductivity is proportional to the concentration of redox species.

Another kind of electron transport mechanism is thermally activated hopping [26]. It follows a classical Arrhenius expression:

$$k_{ET} = k_{ET}^0 \exp(-E_a / k_B T)$$

where k_{ET} is the electron transfer rate, E_a is the activation barrier, and k_B is the Boltzmann constant.

Similarly to multistep tunneling the electron traverses one or more sites to migrate over large distances. However, the main difference between multistep tunneling and hopping is the involvement of the nuclear motion. In hopping, electron transfer does not occur until the thermal motion of nuclei results in a favorable molecular geometry. Hopping involves electron motion over the barrier by rearrangement of the molecule, while tunneling involves electron motion through the barrier due to finite probability of finding the electron on the other side of barrier, without requiring nuclear motion [12, 26].

As hopping involves a series of transfers between relatively stable sites, it does not exhibit the exponential distance dependence of tunneling [26]. But instead, it varies as approximately $1/d$. For a homogeneous series with varying lengths, tunneling is the dominant mechanism for small d, but hopping for large d. For large d, the distance is too great for tunneling, and the electron can propagate more efficiently by a series of hops.

7.2.2.3 Molecular Band with Delocalized Energy States (Metallic-Like Conductivity) The simplest meaning of metallic conductivity is that electrons are free throughout the material [12, 27] (Fig. 7.2b). Metallic conductivity arises due to a process of electron delocalization or spreading of electron wavefunctions [18]. A liner arrangement of atoms similarly to a polymer chain can be described by Kronig–Penney model in which electrons are delocalized or spread over the entire lattice. The strength of interaction between overlapping orbitals determines the extent of delocalization. Firstly, atomic orbitals overlap to form discrete molecular orbitals. And then, for a very long chain, the overlapping molecular orbitals have energies so close together that the energy levels are indeed no longer discrete, but merge into a continuous band of allowed electron energies. In this way, covalent bonds between orbitals give rise to electronic bands. A partially populated band gives rise to conductivity.

For many organic materials, the bonding is sp^3 hybridized σ bonds, in which carbon atoms are strongly localized to inhibit the conduction. For unsaturated polymers (containing double and triple C–C bonds), each carbon atom along the chain has only one hydrogen atom attached to it. The spare electron in the p_z orbital of the carbon atom overlaps with those of carbon atoms on either side forming delocalized molecular orbitals of π symmetry.

Consider an example of a polymer composed of a linear chain of atoms, with N atoms, each separated by a distance d. Therefore, the total length of the chain is $(N-1)d \approx Nd$ for large number of atoms. We can treat this problem as electrons confined to a one-dimensional infinite potential well. The eigenenergies are given by

$$E_{\mathrm{n}} \approx \left(\frac{\hbar^2\pi^2}{2md^2}\right)\left(\frac{n^2}{N^2}\right)$$

Assuming that the π electrons from the N p orbitals are available, with two electrons per molecular orbital according to Pauli exclusion principle, the HOMO will be given by $n = N/2$, and the corresponding energy will be

$$E_{\text{HOMO}} \approx \left(\frac{N}{2}\right)^2 \left(\frac{\hbar^2\pi^2}{2md^2}\right)\left(\frac{1}{N^2}\right)$$

The lowest molecular orbital (LUMO) has the energy

$$E_{\text{LUMO}} \approx \left(\frac{N}{2}+1\right)^2 \left(\frac{\hbar^2\pi^2}{2md^2}\right)\left(\frac{1}{N^2}\right)$$

The energy required to excite an electron from the HOMO to the LUMO level is the band gap of the polymer given by [18]

$$E_g = E_{\text{LUMO}} - E_{\text{HOMO}} \approx \left(\frac{\hbar^2\pi^2}{2md^2}\right)\left(\frac{1}{N}\right)$$

Hence, longer the chain of atoms, the smaller will be the band gap, and it will practically vanish for macroscopic dimensions.

For example, if $d = 0.3$ nm and $N = 100$, then $E_g = 42$ meV.

The process of electron delocalization is previously observed in many natural and synthetic molecules. For example, in benzene, electrons are delocalized over the carbon ring. In synthetic molecular materials such as polyaniline, electrons become delocalized due to pi orbital overlap between phenyl rings [28]. However, it has long been considered, as proteins are electronic insulators, and metallic-like conductivity or electron delocalization is not possible in proteins [27]. In this chapter, we discuss methods and studies that have revealed metallic-like conductivity in anode biofilms of *G. sulfurreducens* due to proteinaceous pili nanofilaments present in these biofilms (Fig. 7.2).

7.3 METHODS TO MEASURE CONDUCTIVITY

The most direct way to measure electronic conductivity is to apply a small potential difference between two well-separated (>tens of micrometers) portions of biofilms with ion-blocking electrodes to inhibit ionic current and measure the resultant steady-state electronic current [14, 29, 30]. Electrodes composed of inert metals such as platinum or gold can serve as ion-blocking electrodes to inhibit the flow of ions and permit only the flow of electrons to determine electronic conductivity of the material. This approach enables measurement of long-range electron transport rather than short-range electron transfer process and facilitates the evaluation of conductivity [19]. One can use either AC or DC methods to measure conductivity, and the details are provided in the following sections. Although two-probe method using two electrodes is commonly used for conductivity measurements due to its simplicity, four-probe method using four independent electrodes is recommended to determine true conductivity of the sample by eliminating the contributions of the contacts to the measured conductivity.

Although cyclic voltammetry has been used to infer conductivity of biofilms [8, 9], catalytic current and electron transfer measurements cannot query conductivity in biofilms or pili because they involve a large potential range and the measured current results from capacitive charging as well as the faradic electron transfer reactions at the biofilm/electrode interface [10, 31, 32].

7.3.1 Strategy to Measure Biofilm Conductivity

In order to directly investigate the possible conductivity of *G. sulfurreducens* biofilms, biofilms of the DL-1 strain [33] were grown in a microbial fuel cell with 10 mM acetate as the electron donor [34], with the modification that the anode of the microbial fuel cell, which serves as the electron acceptor to support growth, was composed of two gold electrodes (as a split-anode) separated by a nonconductive gap of 50 μm, roughly 50 times the length of a cell (Fig. 7.3a). Experiments with the gap up to 100 μm also yielded similar results [6], whereas the gap was not bridged for electrodes with a wider gap, thus precluding the conductivity measurements.

Medium exiting the anode chamber, which still contained substantial acetate (~9 mM), was directed through another chamber, serving as a control, which had an identical arrangement of two gold electrodes, except that the electrodes were not connected to the cathode and thus could not serve as an electron acceptor for growth.

As expected [34, 35], the current between the anode and the cathode increased over time with cell growth (Fig. 7.4a). Confocal laser scanning microscopy (CLSM) revealed that the cells formed a confluent biofilm that spread across the nonconductive gap (Fig. 7.3b–e). In contrast, there was no growth on the control electrodes that were not connected to the cathode. If only one of the two anodes was connected to the cathode, a biofilm formed only on that side and did not bridge the gap.

Details of Experimental Procedure

Bacterial strains and culture conditions

G. sulfurreducens strains DL-1 (ATCC 51573) [33], strain KN400 [36], strain BEST [37], and strain CL-1 that forms highly cohesive biofilm [38] were obtained from our laboratory culture collection. These strains generated in our laboratory are available upon request. *Shewanella oneidensis* strain MR-1 (ATCC 7005500) was obtained from the American Type Culture Collection. The cultures were maintained in growth medium with fumarate (40 mM) as the electron acceptor with acetate (10 mM) as the electron donor for *G. sulfurreducens* or lactate (20 mM) for *S. oneidensis* [39]. *Pseudomonas aeruginosa* strain, obtained from the University of Massachusetts Amherst, Department of Microbiology Culture Collection, and *Escherichia coli* strain DH5α, obtained from our laboratory collection, were aerobically grown in M9 media [40] supplemented with 0.1% casamino acids (Difco Lab., Detroit, MI, USA). Anaerobic digester sludge from the Pittsfield, Massachusetts wastewater treatment plant served as the mixed-culture inoculum. The sludge from the digester was allowed to flow for a few seconds (>30 l min^{-1}) prior

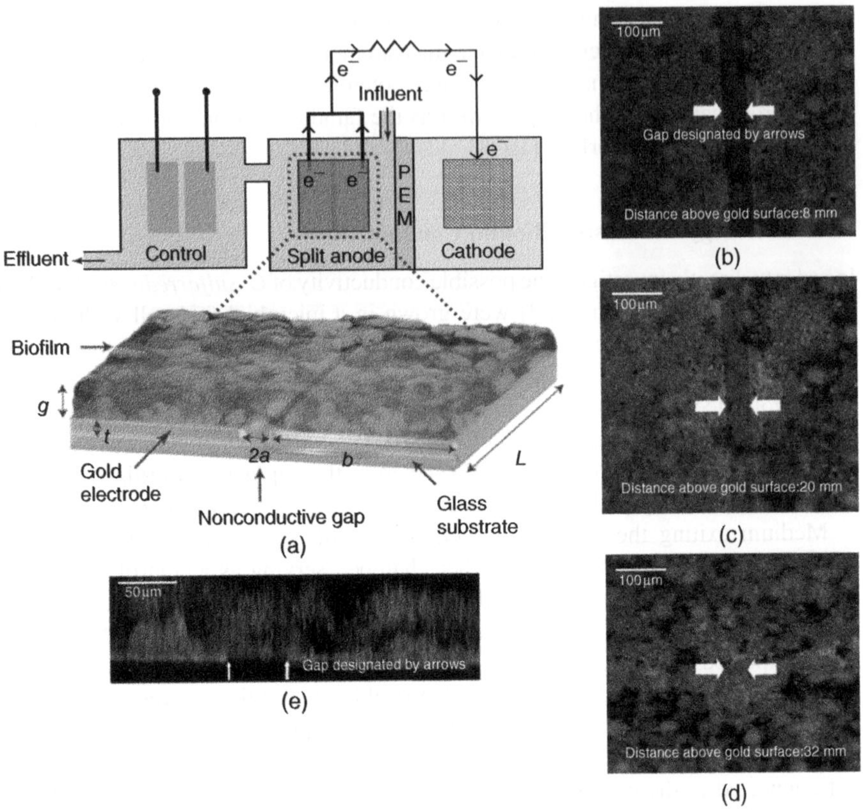

Figure 7.3 Strategy to measure in situ biofilm conductivity. (a) Schematic of microbial fuel cell with two gold electrodes serving as an anode, separated from cathode by a proton exchange membrane (PEM). The biofilm grows over the electrodes and the nonconducting gap between the two electrodes. No biofilm forms on the control electrode pair. Gap width $2a = 50$ μm, electrode width $b \approx 1.27$ cm, electrode length $L \approx 2.54$ cm, electrode thickness $t \approx 50$ nm, and biofilm height, g. (b–e) Representative fluorescent confocal scanning laser microscopy images of split-electrodes. Images were taken when microbial current was 0.25 mA and biofilm height was 36 ± 1.4 μm. (b–d) Top-down confocal image slices of biofilm spanning the nonconductive gap. X–Y image slices (parallel to the electrode surface). Scale bar 100 μm. (e) Cross-sectional image of biofilm spanning the nonconductive gap. X–Z image slice through the biofilm, in a direction perpendicular to the surface of the gold anode and across the 50-μm gap. Scale bar 50 μm. Figure adapted from Ref. [4] with permission. (*See insert for color representation of this figure.*)

to collection into a sterile, anaerobic bottle. The sample was transported on ice to the laboratory and used immediately to inoculate fuel cells [34]. Cells were anaerobically grown first and after reaching the mid-exponential phase (OD ~0.4), cells were inoculated in the anode chamber with the 10% final inoculum.

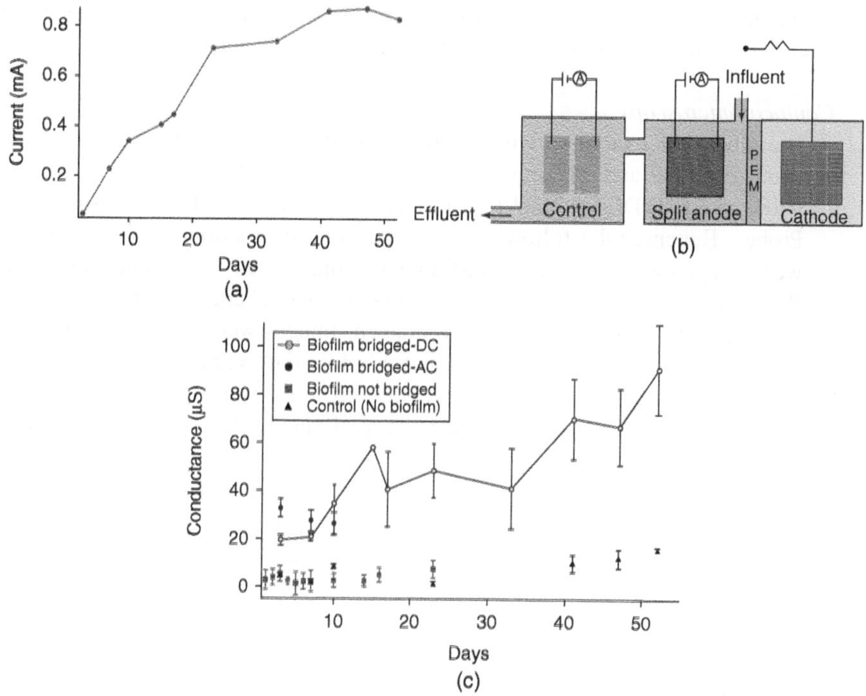

Figure 7.4 Measurement setup and conductance data. (a) Representative current produced by the DL-1 strain of *Geobacter sulfurreducens*. (b) Schematic of conductivity measurements. (c) Conductance measured across the gap-spanning biofilm as well as measurements from controls in which a biofilm did not span the gap. Error bars indicate SD of individual measurements for several biofilms. ($n = 5$ for biofilm bridging the gap and control electrodes, $n = 2$ for biofilm not bridging the gap.) Figure adapted from Ref. [4] with permission.

7.3.1.1 Two-Probe DC Current–Voltage Measurements The connection between the anodes and the cathode could be temporarily disrupted to connect the two anodes to electronics for in situ measurements of conductance across the nonconducting gap (Fig. 7.4b). There was substantial (>20 µS) DC conductance between the two electrodes when the biofilm bridged the gap between the electrodes (Fig. 7.4c). The conductance increased over time (~100 µS), consistent with the increase in biofilm thickness over time (Fig. 7.4c). The biofilm grew over the gap with thickness approximately 50 µm (Fig. 7.1b–e). When only one or neither of the electrodes had been connected to the cathode in the control chamber, and there was no biofilm bridging the gap, the conductance between the two electrodes was very low (<10 µS) and did not change over time (>50 days) (Fig. 7.4c). These results suggested that the conductance of the biofilm bridging the nonconductive gap was not due to the release of a soluble electron shuttle, which if present, would have also permitted the one-sided biofilm to transport electrons across the gap and would have provided higher conductivity in the medium entering the split-electrode control chamber. This is consistent with previous experimental [41] and computational work [42] suggesting that extracellular electron transfer in *Geobacter* is unlikely due to shuttles.

Confocal microscopy
Biofilms were examined with CLSM by removing the split-anode from the anode chamber and imaging the intact biofilm [34, 35]. Biofilms were stained with the LIVE/DEAD BacLight Bacterial Viability Kit (L7012) (Molecular Probes, Eugene, OR) following manufacturer's instructions. Anode biofilms were imaged with a Leica TCS SP5 microscope (Leica Microsystems GmbH, Wetzlar, Germany) with a HCX APO 63× (numerical aperture: 0.9) objective, and a minimum of five fields were imaged. Images were processed and analyzed with LAS AF software (Leica). A minimum of five random CLSM image stacks were used to determine average biofilm height using the biofilm analysis software Phobia Laser scanning microscopy Imaging Processor (PHLIP) [43].

Scanning electron microscopy (SEM)
Gold electrodes with the grown biofilms were removed from the anode chamber of the microbial fuel cell, and biofilms were fixed with 1% glutaraldehyde and 1% formaldehyde. Biofilms were dehydrated with graded ethanol series (30%, 50%, 70%, 80%, 95%, and 100%) for 10 min each step with a gentle periodic agitation. Then, the biofilms were CO_2-critical point dried from ethanol transitional solvent with a 3-h slow, continuous exchange. For SEM imaging, the biofilms were sputter coated in a Polaron E-5100 Sputter Coater (2 min at 2.2 kV) with argon at 13 Pa by using a gold target and observed in a JEOL JSM-5400 SEM with an accelerating voltage of 5 kV.

7.3.1.2 Four-Probe DC Current–Voltage Measurements A metal electrode in contact with an electrolyte develops a potential at its interface. When a current is passed through the metal–electrolyte interface, the potential at the interface gets modulated in proportion to the current induced. This effect is known as electrode polarization and can obscure the electrical measurements due to high polarization impedance [44]. A four-probe method was used to confirm that the measured conductance was an intrinsic property of the biofilm and could not be attributed to measurement artifacts due to electrode polarization [44] (Fig. 7.5). Firstly, the biofilm was grown on four gold electrodes, each separated by 50-μm nonconductive spacing (Fig. 7.5a). The four anodes were disconnected by cathode and connected to a sourcemeter to carry out four-probe measurements (Fig. 7.5b). This method allows measuring the potential between the inner two electrodes without drawing any current. The current is injected in outer two electrodes. Under such circumstances, the measured potential will not be influenced by electrode polarization, as the electrode polarization potential is proportional to the current through the electrode. Due to very high input impedance of the voltmeter, no current flows through the inner circuit (Fig. 7.5), thus eliminating artifacts of polarization. In a similar manner, the effects of contact resistances are eliminated with this method, as the resistance arises only when the current is flowing through the probe electrodes. The ratio of the current passing through the sample and the measured potential between two inner

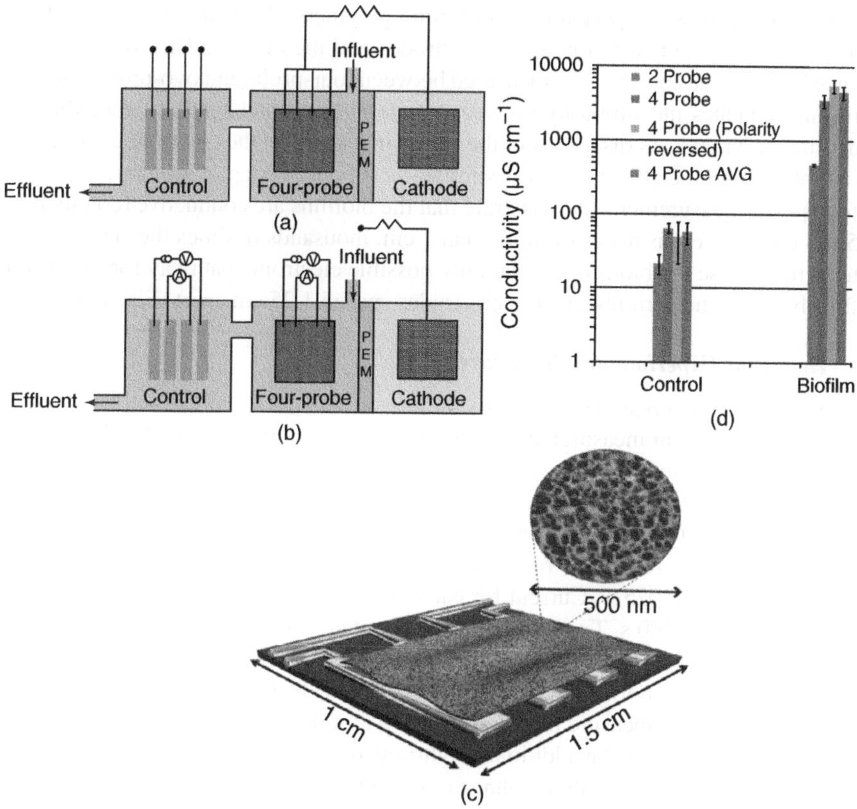

Figure 7.5 Four-probe measurements. (a) Schematic of microbial fuel cell with four gold electrodes, each separated by 50-μm nonconductive gap, serving as an anode (side-view). (b) Schematic of conductivity measurements using four electrodes (side-view). Current is injected through outer two electrodes, and voltage is measured across inner two electrodes. (c) Schematic of four-probe setup used for biofilm and pili measurements (top-view). Inset, AFM image of pili network placed on gold electrodes. (d) Comparison of conductivity measured using two-probe and four-probe methods. Error bars show SD of individual measurements for four biofilms of KN400. Figure adapted from Ref. [4] with permission. (*See insert for color representation of this figure.*)

electrodes is equal to the conductance of the sample. Note that the conductance measured using this approach is independent of the polarization impedance of the two outer current electrodes because the voltage of the two inner probes is used to determine the conductance [44]. Moreover, although the reference electrode is necessary to determine the anode potential at the electrode/electrolyte interface, it is not required for the four-probe approach mentioned here. The potential of the two inner electrodes is directly measured with a high-impedance voltmeter (Fig. 7.5b and c).

The conductivity values of the anode biofilms with the four-probe method were 10-fold higher than those with the two-probe approach (Fig. 7.5d). This is similar to results previously reported for conducting polymers [45] and can be attributed to the energy level mismatch between electrodes and the biofilm. This 10-fold difference between the conductivity measured between four-probe and two-probe approach further indicates the similarity between anode biofilms and synthetic metallic conducting polymers. As discussed in the following sections, the origin of conductivity in biofilms and metallic polymer is similar – π orbital overlap of aromatic rings. The four-probe measurements demonstrate that the biofilms are conductive (conductivity 5 mS cm^{-1}) over distances of more than 1 cm, thousands of times the size of a bacterium, because the biofilm was the only possible electronic pathway for the current flow between the current-carrying electrodes spaced 1.25 cm apart (Fig. 7.5c).

Details of Experimental Procedure

DC current–voltage measurements

Direct current measurements were made with a source meter (Keithley 2400). In case of biofilms, these experiments were performed both in the presence and in the absence of acetate, and there was no effect of acetate on the measured conductivity (Fig. 7.8), confirming that the conductivity is an intrinsic property of biofilms. A voltage ramp of 0–0.05 V was applied across split-electrodes in steps of 0.025 V. Current for each voltage was measured every second over a period of 100 s. Time-averaged steady-state current for each applied voltage was calculated to create the *I–V* characteristics. The Keithley 2400 source meter was used to apply a fixed current (~1–10 μA) between outer of the four electrodes and to measure the potential drop (~millivolts) between two inner electrodes (Fig. 7.5). An additional high-impedance voltmeter (Keithley 2000) was used to record the output voltage of the current source to calculate conductance [45]. For both two- and four- probe measurements, linearity of *I–V* characteristics was maintained by applying appropriate low voltage (~millivolts) and current (~1–10 μA) to determine the conductance in the ohmic regime. The dissipative power was kept under 10^{-6} W to eliminate self-heating effects. The validity of the four-probe measurements was checked by reversing the polarity of input current. Forward and reverse currents yielded similar conductance values verifying the ohmic contact of the junction. Each measurement was performed over the duration of few minutes, and no change in the biofilm structure was observed. The fuel cell current was returned to its value before the measurements (Fig. 7.4).

7.3.1.3 Two-Probe AC Impedance Spectroscopy Measurements In principle, the electrochemical processes of redox charge transfer and ion diffusion can contribute to the measured DC conductance [20, 31]. Therefore, to better define the intrinsic biofilm conductivity, it was desirable to use AC impedance measurements as an independent probe of conductance between the electrodes [46–48]. This method also serves to separate electronic and ionic conductivity in mixed-conductor system [49]. The applicability of this approach to anode biofilms was verified by evaluating the linearity and stability conditions described in the following section on the details of this experimental procedure.

Two types of processes occur at the electrode. Faradic processes involve charge transfer across the metal–solution interface, and nonfaradic processes involve transient current flow due to double layer charging [31]. Impedance due to faradic and nonfaradic processes is denoted as Z_F and Z_{NF}, respectively. Their frequency dependence is [20, 31]:

$$|Z_{NF}| = \frac{1}{\omega C}; |Z_F| = R_{ct} + \frac{\sqrt{2}}{\sqrt{\omega}}\sigma$$

$$\frac{1}{|Z|} = \frac{1}{|Z_{NF}|} + \frac{1}{|Z_F|} = \omega C + \left(R_{ct} + \frac{\sqrt{2}}{\sqrt{\omega}}\sigma\right)^{-1} \sim \omega C + \frac{\sqrt{\omega}}{\sqrt{2}\sigma}$$

where C is the capacitance, R_{ct} is the charge transfer resistance, and

$$\sigma = \frac{1}{nFA\sqrt{2}}\left(\frac{\beta_O}{{D_O}^{1/2}} - \frac{\beta_R}{{D_R}^{1/2}}\right); \qquad \beta = \left[\frac{\partial E}{\partial C}\right]_{i,C}$$

It can be seen from the aforementioned equations that faradic processes dominate over low frequencies due to inverse square root dependence. To estimate biofilm conductivity, we were interested in charge transport across the gap rather than charge transfer across the electrodes. (The latter describes a one-step tunneling process [19].) Hence, it was necessary to eliminate faradic reactions involving charge transfer. In the case of anode biofilms, a low-frequency cut-off of 100 mHz was used to eliminate any possible contributions from electrochemical reactions occurring at very low frequency [31, 46, 48]. Electrolytes with different concentrations were tested to verify this frequency range.

(*Note* – The aforementioned equations are valid only for high frequency. Complete expression is given by [20]

$$Z_w = R_{D0}\left[\frac{\tanh\left(\sqrt{js}\right)}{\sqrt{js}}\right]$$

where $s = l_e^2\left(\frac{\omega}{D}\right)$ and R_{D0} is the diffusion resistance, D is the diffusion coefficient of the diffusing particle (ions in the electrolyte in this case), and l_e the length of the diffusion region. For $s \gg 3$, the tanh term approaches unity and square root dependence on frequency appears.

The low-frequency impedance of the biofilm decreased by two orders of magnitude as the biofilm grew (Fig. 7.6a). In contrast, the impedance between the control electrodes, which did not have a biofilm, or in which only one electrode had a biofilm, did not change substantially over time (Fig. 7.4c). The AC impedance between the control electrodes without gap-spanning biofilms could be accurately modeled as

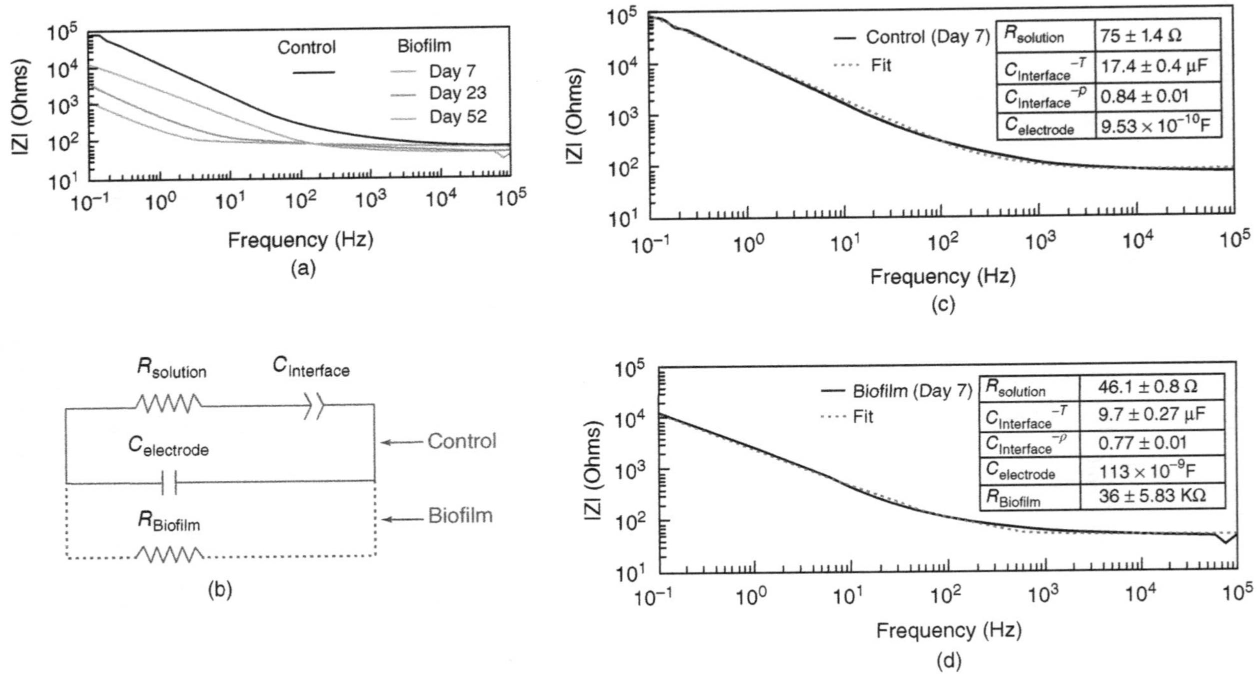

Figure 7.6 AC impedance spectroscopy – data, model, and fitting results. (a) Decrease in impedance between split-electrodes as biofilms grew over time. (b) Circuit models for control systems without gap-spanning biofilms and for systems with gap-spanning biofilms. (c) Modeling of representative impedance for control electrodes not bridged by a biofilm. (d) Modeling of impedance for corresponding electrodes on the same day that were bridged by a biofilm. Inset tables show model parameters. Error bars indicate SD ($n = 5$). Figure adapted from Ref. [4] with permission.

a series *RC* circuit (Fig. 7.6b), accounting for solution resistance (*R*) and interface capacitance (*C*), with an electrode capacitance in parallel [50]. Interface capacitance is due to the formation of electrical double layer at the interface of metal electrode and the solution. It is represented by constant phase element rather than capacitor to incorporate nonhomogeneity of the surface [20].

The interface capacitance was calculated as [20]:

$$C_{\text{Interface}} = (TR_s^{-(p-1)})^{\frac{1}{p}}$$

where T and p are parameters associated with the electrolytic double layer and R_s is the solution resistance, representing ionic conductivity. These parameters were obtained by fitting the circuit model (Fig. 7.6b) to the impedance data (Fig. 7.6a). Fitting results are presented in Figure 7.6.

The value of $C_{\text{electrode}}$ was extracted by fitting the model of the solution resistance and the electrode capacitance in parallel at high frequencies, and then the value of electrode capacitance was fixed for further fitting [50]. The double layer and electrode capacitance values obtained with this approach (Fig. 7.6c) agreed very well with results from the literature [20, 31, 49, 50], suggesting that this modeling approach was valid. Details of the fitting procedure are presented in the following section on the experimental procedure.

However, in order to fit the data accurately in the instances in which biofilms spanned the gap, it was necessary to add an extra resistor, representing an additional conductive pathway, in parallel to the *RC* model that described the controls without biofilms (Fig. 7.6b dotted line). This equivalent circuit method has been previously employed to measure conductivity of nanostructured films [51, 52] and conducting polymers [53] and serves to separate electronic and ionic conductivity in mixed conductors [49]. The need to include this extra conductance in the circuit model demonstrates that the biofilm is electronically conductive. Additional experiments in the absence of acetate further confirmed that the measured conductance is an intrinsic property of biofilm and does not arise from the charge transfer occurring at the biofilm/anode interface.

The capacitance of the biofilm measured by AC impedance measurements also increased as the thickness of the biofilm increased. The larger capacitance masked the biofilm conductance evaluated with comparatively high frequency (0.1 Hz–300 kHz) AC impedance measurements, but with DC conductivity measurements, which were performed over the timescale of 100 s, the biofilm conductance could still be measured at lower frequencies (10 mHz), suppressing the capacitive effect.

Details of experimental procedure for AC impedance spectroscopy measurements
Impedance was measured by sweeping the frequency between 100 mHz and 300 KHz with a 0.1-V amplitude AC voltage excitation using a Solartron 1252 frequency response analyzer (FRA) coupled to a 1287 electrochemical interface (Solartron Analytical, Hampshire, UK). To verify linear regime, AC signal amplitudes in the range 1–400 mV were applied (Fig. 7.7). No amplitude dependence was observed in the spectrums for the range 1–100 mV, confirming

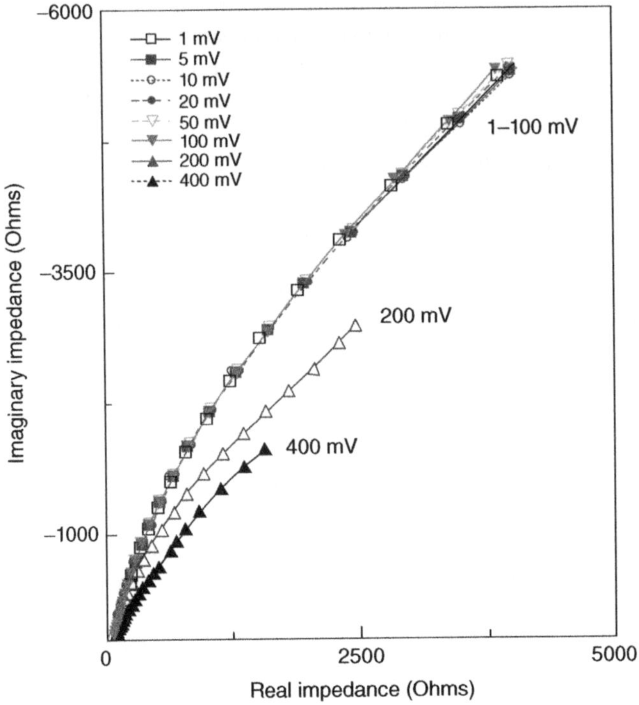

Figure 7.7 Amplitude dependence of *G. sulfurreducens* biofilm impedance showing the linearity in the impedance spectra for AC excitations in the range 1–100 mV, whereas amplitudes higher than 100 mV show the deviation from linearity. Figure adapted from Ref. [5] with permission.

the linear regime. Therefore, 100 mV was chosen as AC voltage excitation to maximize signal-to-noise ratio. To test the stability of the measurements, impedance tests were repeated one after another, and results were found consistent. There was no change in the impedance spectra measured just one after another. The Solartron 1252 was connected to an IEEE 488 GPIB interface on the personal computer. The fuel cells were electrically screened to keep out the interference using Steel Enclosure (Newark, IL, USA). The FRA was used to measure both the absolute value of impedance and phase shift. Short feature of autointegration mode was used with an integration time of 10 s to maintain the data stability. (*Note* – The FRA measures impedance by integrating the voltage and current signals over a specified amount of time and then dividing voltage by current to get impedance. Although longer integration time gives more accurate data, it results in slower measurements and can cause data stability problems. FRA must measure for at least one full cycle. A measurement of 0.1 Hz will take minimum of 10 s irrespective of specified integration time.

When autointegration is turned on, the FRA measures impedance for at most the specified time. If it determines that the data point is accurate only after a fraction of the integration time, it will automatically end the measurement and move on to the next frequency. Autointegration causes at least three full cycles to be measured. Short feature of autointegration specifies a maximum standard deviation SD of 10%, while long aims for 1%. A common compromise between noise level and measurement time is to set autointegration to short with an integration time of 10 s. This keeps the low frequencies from consuming too much time [54].) Z-Plot/Z-View software (Scribner Associates Inc., South Pines, NC, USA) was used to acquire and analyze the measurements. Equivalent circuits were designed using Z-view to match physical characteristics of the measured setup and to determine the values that best fit our data based on a designed model. Z-view equivalent circuits are based on the LEVM 6.0 program written by Dr. J. Ross Macdonald. 100 maximum iterations and an additional 100 optimization iterations were used to improve the quality of the fit. Complex type of fitting was used where both real and imaginary values of impedance were fit. In fitting mode, Z-view alters the values of the components and attempts to find the values that best fit the data. Instant fit feature was used to find initial estimates. Type of data weighting effects how much emphasis is applied to different portion of the data. Calc-modulus type of data weighting was used, which can give a better fit because each data point's weight is normalized by its magnitude in covering a wide range of frequencies. Weighted sum of squared (WSS) was used as a goodness of fit. The sum of squares is proportional to the average percentage error between the original data points and calculated values. This is particularly useful when comparing the goodness of fit of two different models to a single data set. Further details of equivalent circuit modeling are described elsewhere [20, 50].

Anode and cathode potentials were continuously monitored using Ag/AgCl reference electrode in the anode and cathode chambers, respectively. For all impedance measurements, open circuit potential was monitored until it reached a constant, stable value before over imposing AC signal. Typical open circuit potentials for the various strains were [55]: strain DL-1, −450 mV versus Ag/AgCl; strain KN400, −483 mV versus Ag/AgCl; strain ST, −441 mV versus Ag/AgCl; and strain BEST, −530 mV versus Ag/AgCl.

7.3.2 Distinguishing Between Electronic and Ionic Conductivity

Biofilms form a highly porous structure [3] with porosity close to 1 and water content 95–99%. This facilitates effective transport of nutrients to bacteria present in the biofilm structure and removal of toxic waste materials. This makes biofilms excellent ionic conductors [55]. If experiments were performed on dried biofilms, the biological components contributing the conductivity can be denatured and no longer in the physiologically relevant condition. Therefore, it is crucial to separate the ionic contribution to the measured conductivity from the electronic contribution in living biofilms.

The diffusion coefficient for ions in the biofilm is expected to be the same as the diffusion coefficient in the electrolyte [56] or reduced by 0.04–1.4 [57]. Ionic conductivity is related to diffusivity by linear expression [25]:

$$\sigma = \frac{ne^2D}{k_B T}$$

Therefore, the ionic conductivity across the gap will be either constant or decrease despite biofilm bridging the gap, and it is reasonable to assign the increase in conductance across the gap to an increase in electronic conductance due to the conductive biofilm bridging the gap.

Furthermore, ions traveling through the material under the influence of an applied electric field will accumulate at defects (e.g., grain boundaries) or at one of the solid electrodes. The resulting polarization will reduce the ionic current to zero over a period. Ionic conductivity is, therefore, expected to give rise to time-dependent currents [18]. The biofilm conductance was calculated by applying voltage over 100 s and by measuring the time-averaged steady-state current profile. Therefore, the measured conductance is predominantly electronic in nature.

AC impedance spectroscopy measurements further confirmed that the measured electron transport is electronic in nature. The necessity to include an additional current path to the ionic resistance shown in the equivalent circuit model of the control setup showed that the biofilm possesses significant electronic conductivity. This method has been employed successfully for mixed conductors [49].

Further evidence for electronic conduction was obtained by studying the dependence of current on applied voltage. For ionic conduction, a model based on the work of Mott and Gurney shows that ionic current is related to applied voltage by the following expression [18]:

$$I \propto \sinh\left(\frac{eaV}{2k_B T}\right)$$

where a is the distance between potential wells associated with the ionic movement. Current measured for biofilm experiments the above-mentioned response. Instead, current was linear at low voltages (<0.05 V) and decreased at high voltages, probably due to Joule heating.

Additional evidence that the measured transport is indeed electronic in nature came from measurements on biofilms performed under vacuum conditions (base pressure 10 Torr). These experiments are discussed in detail in later sections. Ionic conduction will require macroscopic ion reservoirs to generate a stable current as long as the voltage is applied [14]. There were no ion reservoirs present during the experiments on dried biofilm, further confirming the electronic nature of conductivity.

Although the measured conductivity must be electronic in nature, the possibility of water and counterions forming an ionic or covalent bond cannot be ruled out, even in a vacuum environment [14]. This can in principle affect the biofilm conductivity. In conjugated molecules, although inner π-electrons from the base stack carry the current, electric patch fields due to an outside ionic bond can influence the π-electron states [14]. Therefore, the influence of ions on the conductivity cannot be ignored.

7.3.3 Estimation of Conductivity Using Conformal Mapping

Conformal mapping (the Schwarz–Christoffel transformation) was employed to calculate conductivity from experimentally measured biofilm resistance. The details of the conformal mapping procedure are presented elsewhere [53, 58]. Variables are defined in Figure 7.3a. In the limiting cases $a < g \ll b$, conductivity formula reduces to

$$\sigma = \frac{\pi}{Rl}\left[\ln \frac{8g}{\pi a}\right]$$

where R is the experimentally measured resistance of the biofilm. Biofilm thickness g was measured by fluorescent microscopy. Using this technique, a conductivity of 5 mS cm^{-1} was determined for gap-spanning KN400 biofilms with a height of 40 μm. This is comparable to the conductivity of nanostructured conducting polymer films synthesized from polyacetylene and polyaniline [59].

7.3.4 Effect of Acetate Removal on Biofilm Conductivity

Deleting acetate from the media removed the potential for microbial current production, but did not change the biofilm conductance, confirming that the conductance measured was not associated with the current produced from the microorganisms (Fig. 7.8).

7.3.5 Control Experiments with Several Bacterial Strains

7.3.5.1 Low Conductance in Fumarate-Grown Biofilms When the gold electrodes were not connected to the cathode, but fumarate was provided in solution as an alternative electron acceptor, biofilms of wild-type and PilA mutant DL-1 strain were formed on the gold surfaces and bridged the nonconductive gap. However, these biofilms exhibited very low conductivity (Fig. 7.9). These results demonstrated that the conductivity of the biofilm depends on the physiological status of the cells.

7.3.5.2 Current-Producing PilA Mutant Biofilms Biofilms of PilA mutant of DL-1 strain of *G. sulfurreducens* and PilA mutant strain of *Geobacter* strain Speedy produced 10 times less current and did not bridge the nonconducting gap. This is consistent with previous studies that have demonstrated that because pili are electrically conductive [60], they are required for the development of the thick anode biofilms necessary for high-density current production by *G. sulfurreducens* [35, 61].

The pili of *G. sulfurreducens* may also be required for attachment of some surfaces [62], but a mutant lacking pili readily formed biofilms on split gold anode if fumarate was provided as an electron acceptor, further suggesting that the primary role of the pili was conduction of electrons through the biofilm. The cells not attached to electrode do not survive because they cannot transfer electrons to the electrodes in the absence of pili. Therefore, they do not form thick biofilms.

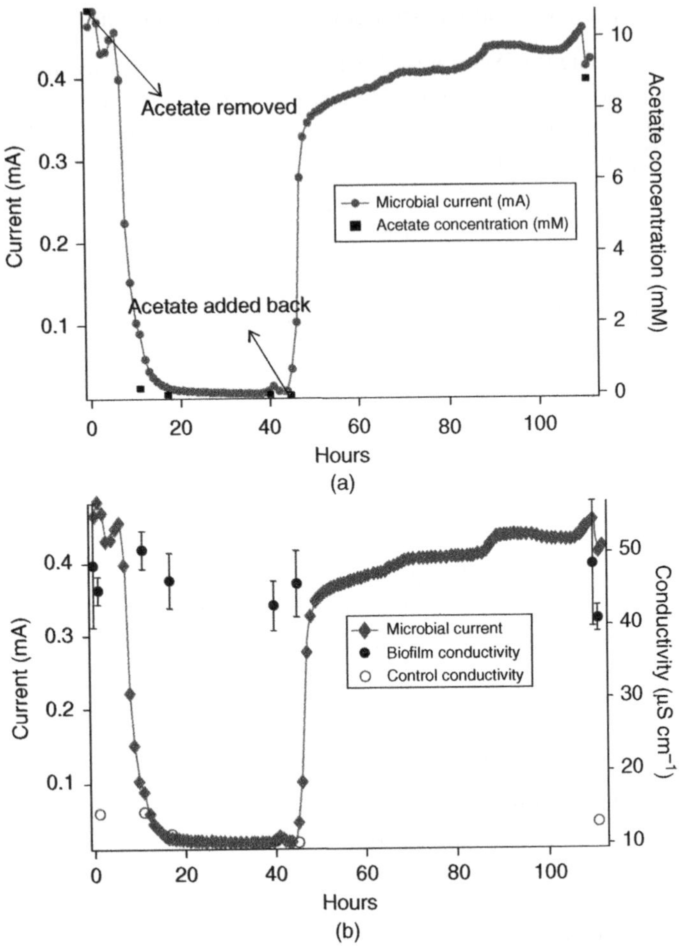

Figure 7.8 Acetate removal experiment. (a) Effect of acetate removal and addition on microbial current production and corresponding acetate concentration. (b) Conductivity measurements when acetate was removed and added back to the growth medium. Error bars indicate SD. Figure adapted from Ref. [4] with permission.

7.3.5.3* Shewanella oneidensis *Biofilm *S. oneidensis* is reported to possess conductive filaments [63], but *S. oneidensis* produced biofilms too thin (<10 μm) to bridge the nonconducting gap of our apparatus, consistent with previous electrode studies [39] and the finding that biofilms of the closely related *Shewanella loihica* are highly resistive [64].

7.3.5.4 Aerobic Biofilms Biofilms of well-known biofilm formers such as *P. aeruginosa* and *E. coli* grew across the nonconducting gap when their preferred electron acceptor, oxygen, was provided. However, the measured conductance

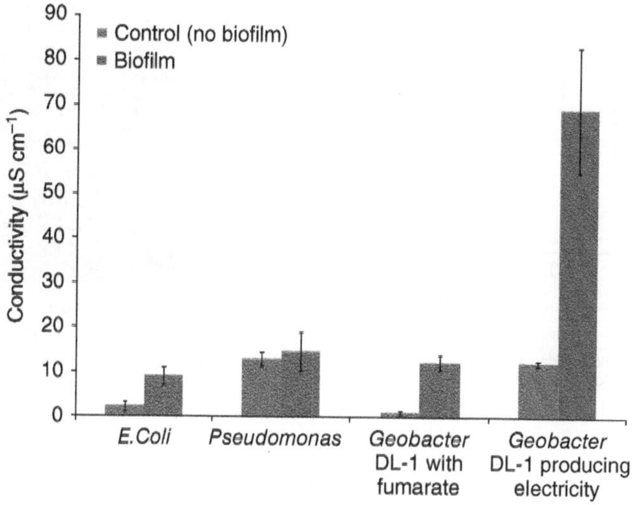

Figure 7.9 Comparison of conductivity of aerobically grown and *G. sulfurreducens* biofilm. Conductivity of biofilm and control for *Pseudomonas aeruginosa*, *Escherichia coli*, and *G. sulfurreducens* under various conditions. Error bars indicate SD for individual measurements for several different biofilms. Figure adapted from Ref. [4] with permission.

was not significantly greater than the control without a biofilm (Fig. 7.9). This is in accordance with previous reports of the biofilms functioning as insulators rather than as conductors [1–3]. The contrast between the electronically active biofilm of *G. sulfurreducens* and aerobically grown biofilm might be attributed to major biochemical differences. Under aerobic conditions, it is advantageous for microorganisms that are routinely exposed to oxygen to maintain electron transport proteins within the cell in order to reduce oxygen with energy-conserving electron transport chains and to minimize nonspecific interactions with oxygen, which can lead to free radical formation. In contrast, organisms such as *Geobacter* species that specialize in the utilization of poorly soluble minerals as electron acceptors must, by necessity, display electron transport components outside the cell [65, 66]. These observations suggested that biofilm conductivity is related to electron transport chain of the microorganisms.

7.3.6 Electrical Conductivity in Mixed-Species Biofilms

So far, all the measurements of biofilm conductivity were focused on pure culture strains. In order to evaluate whether mixed-culture biofilms are electronically conductive, wastewater inoculums was used as a mixed culture as described in experimental procedures in Section 7.3.1. Mixed-culture biofilms bridged the nonconductive gap in split-anode fuel cell. Interestingly, these biofilms were more conductive than biofilms of *G. sulfurreducens* wild-type (DL-1) strain (Fig. 7.10). These biofilms consisted of two distinct parts. Inner biofilm, which was pinkish in color, was firmly attached to

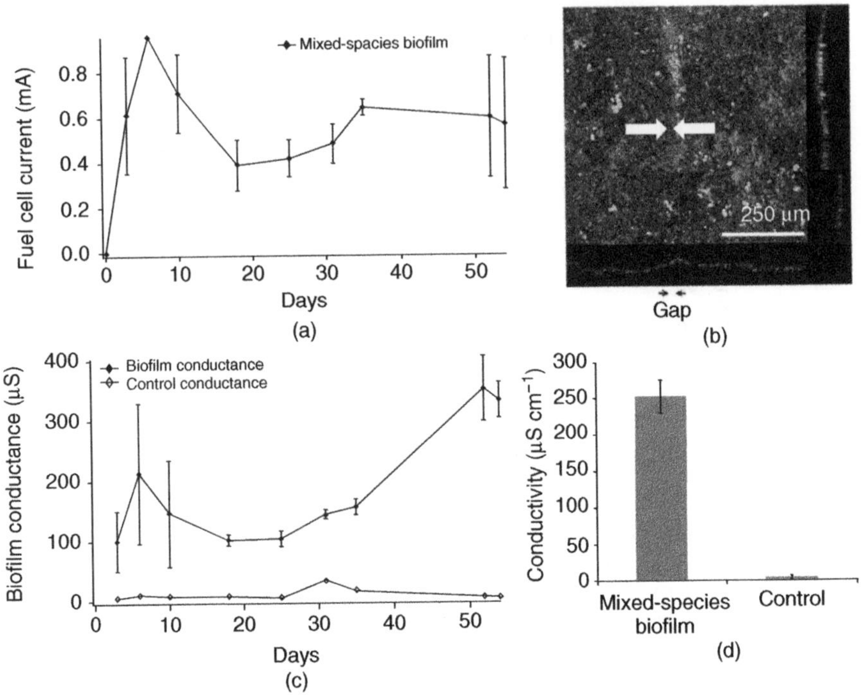

Figure 7.10 (a) Current production and (c) conductance data over days for mixed-species biofilm in split-anode microbial fuel cell and corresponding control. Error bars: standard deviation. The fuel cell was switched to flow through mode at day 10 that removes planktonic cells [21]. (b) Confocal laser scanning microscopy image (3D-reconstruction) of mixed-species biofilm showing that the biofilm bridged the nonconductive gap. Gap is designated by arrows. Scale bar, 250 μm. Biofilm thickness is 50.28 ± 8.13 μm. Biofilms were stained with the LIVE/DEAD BacLight Bacterial Viability Kit. (d) The conductivity of mixed-species biofilm and corresponding control, which lacked biofilm at day 54. Error bars: standard deviation. Figure adapted from Ref. [5] with permission.

the electrode, while outer biofilm, which was grayish in color, was loosely attached to the pink biofilm (Fig. 7.11). 16S rRNA analyses revealed the presence of many other microorganisms along with *Geobacteracea*. *Geobacteracea*, which is a part of Protebacteria, was only 8% in the inoculum (Fig. 7.12). Inner biofilm is the pinkish biofilm grown over the electrode, which contains 52% *Geobacteracea*. Outer biofilm is the greyish biofilm, which had only 10% *Geobacteracea*.

The finding that mixed-species biofilms can possess electrical conductivity comparable to pure culture biofilms of *G. sulfurreducens* with low charge transfer resistance provides an explanation for the capacity of mixed-species biofilms to produce the thick biofilms necessary for high current densities. Modeling studies have previously demonstrated that invoking a highly conductive biofilm could explain the effective function of high-current-density multispecies biofilms in which *Geobacter*

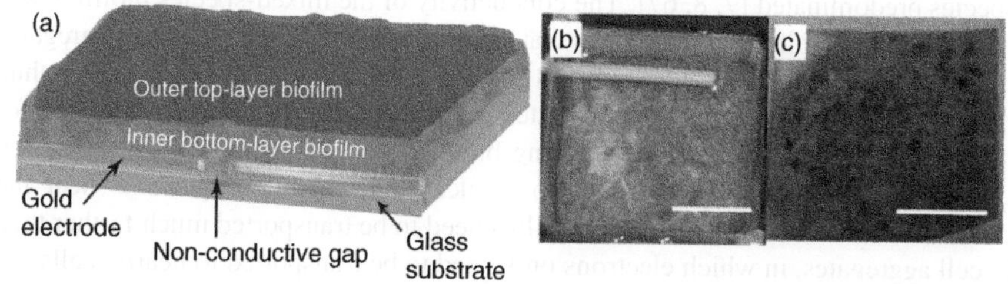

Figure 7.11 (a) Schematic of mixed-species biofilm formation. (b) Image of outer, top-layered brownish biofilm that is loosely attached to the anode. Scale bar 1 cm. (c) Image of inner, bottom-layered pinkish biofilm that is strongly attached to the anode. Scale bar 1 cm. Figure adapted from Ref. [5] with permission. (*See insert for color representation of this figure.*)

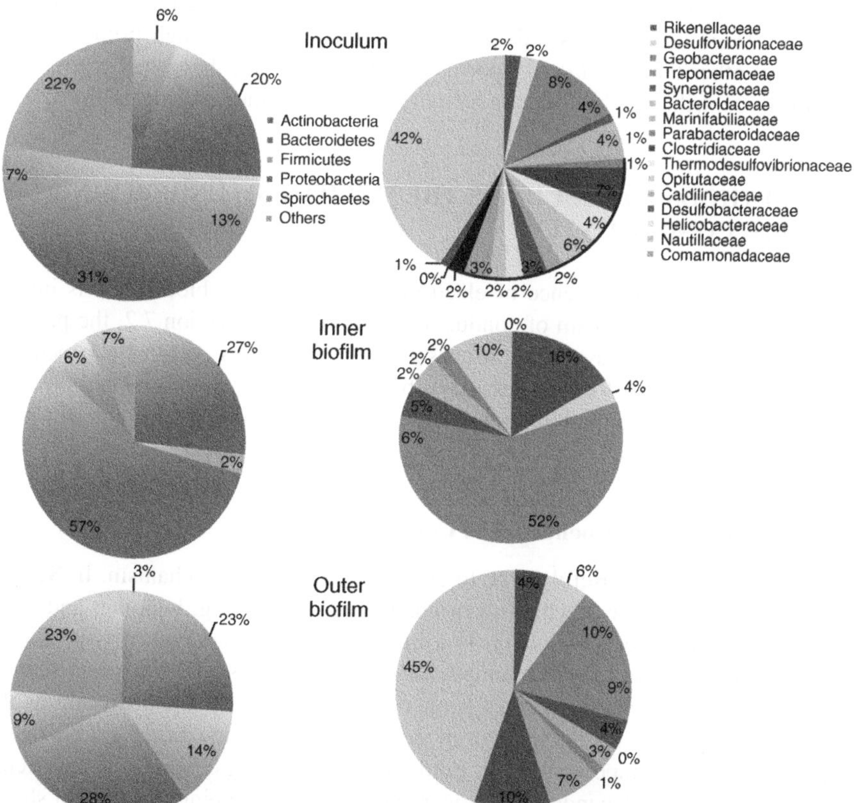

Figure 7.12 Community analysis of wastewater inoculum as well as in the inner and outer biofilms. Left-hand side charts show the division based on the phylums, and right-hand side charts show the division based on species. Figure adapted from Ref. [5] with permission. (*See insert for color representation of this figure.*)

species predominated [7, 8, 67]. The conductivity of the mixed-species biofilms was an order of magnitude higher than that of multispecies methanogenic aggregates derived from wastewater digesters [68] and two orders of magnitude higher than that of dual-species *Geobacter* aggregates [69]. There may be stronger selection for higher conductivity in current-producing biofilms than in the previously described conductive aggregates because electrons released from microorganisms near the outer surface of current-producing biofilms need to be transported much farther than in cell aggregates, in which electrons only need to be transported to nearby cells.

It was not possible to determine from the data available whether microorganisms other than *Geobacter* species contributed to the conductivity of the mixed-species biofilms. Biofilms of *E. coli* and *P. aeruginosa* grown on the two-electrode device described here were not conductive [4], and other microbial biofilms were also found to have poor conductivity [1–3]. Other current-producing microorganisms such as *S. oneidensis* [39] and *Thermincola* strain JR [70] do not form thick biofilms when producing current, suggesting that they are incapable of forming highly conductive biofilms. Thus, these results indicate that biofilms containing high proportions of organisms other than *Geobacter* species may be conductive, but whether the other organisms contribute to biofilm conductivity warrants further investigation.

7.4 METHODS TO ELUCIDATE THE MECHANISM UNDERLYING BIOFILM CONDUCTIVITY

After establishing the presence of electronic conductivity in biofilms, it is important to probe the mechanism of conduction in biofilms. In Section 7.2, the primary physical mechanisms have been discussed in detail. In this section, we focus on measurement methods for dependence of biofilm conductivity on two key physical probes – temperature and redox potential to reveal the conduction mechanism in *G. sulfurreducens* biofilms.

7.4.1 Temperature Dependence of Conductivity

Temperature is a very useful probe to elucidate conduction mechanism. In Section 7.2, we discussed three mechanism of conduction – tunneling, hopping, and delocalization. Delocalized conduction is a hallmark of metals. Conductivity of metals increases with lowering the temperature due to suppression of phonon scattering, while hopping conductivity of semiconductors decreases upon cooling as the thermal energy to excite electrons from valence band to conduction band decreases with lowering the temperature [18]. Furthermore, tunneling conduction that occurs in cytochromes is independent of temperature [71], and hopping conduction shows exponential decrease upon cooling with characteristic Arrhenius dependence [18].

Insight into the conduction mechanism in biofilm and pili of *G. sulfurreducens* was obtained by measuring conductivity as a function of temperature. Biofilm measurements were performed with the CL-1 strain of *G. sulfurreducens*, which forms highly cohesive biofilms that, once grown on graphite electrodes, could be peeled off the surface and placed on gold electrodes specifically designed for four-probe measurements. Intact wet biofilms of strain CL-1 exhibited conductance comparable to KN400; the conductance of peeled or dried biofilms was significantly lower. Upon cooling from room temperature, the biofilm conductivity increased by over an order of magnitude (Fig. 7.12). In the case of pili samples, the conductivity increased by three orders of magnitude. An increase in conductivity upon cooling is a hallmark of metallic conduction [28, 59, 72, 73]. Notably, the conductivity increased *exponentially* upon cooling as expected for high-performance, quasi-one-dimensional organic metals [28, 59, 72, 73], rather than linearly as in conventional metals, indicating that the conductivity is limited by back-scattering from thermal phonons [28, 59, 73]. In other words, the conductivity begins to decrease exponentially upon heating at temperatures where thermally excited back-scattering phonons begin to scatter the charge carriers. For temperatures such that the thermal energy ($k_B T$) is much less than the energy of the back-scattering phonons ($\hbar\omega_b$), exponential increase in conductivity upon cooling is expressed as [28, 73] $\sigma(T) = \sigma_m \exp(\hbar\omega_b/k_B T)$. The energy of back-scattering phonons calculated from these measurements (0.6 ± 0.2 eV for biofilm) is comparable to that observed experimentally for organic metals [72] ($\hbar\omega_b \approx 0.4$ eV).

At even lower temperatures, the conductivity decreased exponentially with cooling as expected for disordered metals [28, 72, 73], which is a signature of the thermally activated hopping mechanism with characteristic Arrhenius dependence [74] $\sigma(T) = \sigma_0 \exp(-T_0/T)$. The same conductivity was obtained for heating and cooling. A similar temperature dependence of conductivity has been observed previously for a number of nanostructured organic metals [28, 59, 72–75]. A maximum in conductivity or change in sign of temperature dependence at a crossover temperature T^* is a key characteristic of many organic metals [73]. The crossover temperature for pili and biofilm ($T^* \approx 260$–270 K) is close to that reported for organic metals ($T^* \approx 200$–250 K) [72, 73]. The observed conductivity maximum is likely a crossover between two competing mechanisms [28, 59, 73]: at higher temperatures, conductivity increases with cooling due to intrinsic metallic transport, and at lower temperatures, conductivity decreases with cooling due to the influence of disorder-induced charged traps. Chemical impurities, defects, structural imperfections, and inhomogeneities can act as traps to localize electrons [28, 76, 77]. The activation (detrapping) energy calculated from these measurements ($k_B T_0$) was 1.0 eV for biofilm, which compared favourably with organic metals [72, 74]. In the case of pili samples, the detrapping energy was reduced to 0.85 eV, the conductivity maximum shifted to lower temperatures, and a temperature dependence of conductivity was four orders of magnitude weaker compared to the biofilm (Fig. 7.13), all indicating reduced disorder and improved metallic nature [28, 59, 76]. These studies demonstrate that higher metallic conductivity can be achieved through better processing and further improvements can enhance the conductivity significantly [28, 73].

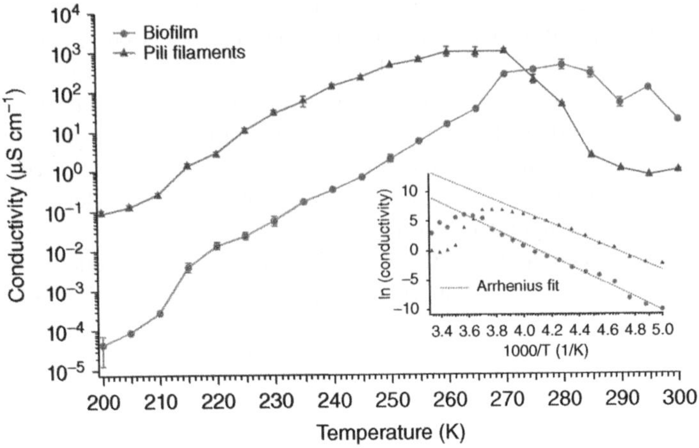

Figure 7.13 Temperature dependence of conductivity of free-standing biofilm formed by *G. sulfurreducens* strain CL-1 and pili filaments of strain KN400 measured with a four-probe method. Error bars represent SD. Data is a representative of several replicates ($n = 4$ for biofilm; $n = 3$ biological replicates for pili). The difference in conductivity between pili and biofilm at 300 K is due to lower pili concentration required on smaller electrodes for four-probe measurements. Control experiments containing media buffer or with biofilm, but not bridging the gap, showed very low conductivity ($<10^{-2}$ μS cm^{-1} at 300 K) that did not change with temperature. Inset: Arrhenius fit for exponentially decreasing conductivity. Figure adapted from Ref. [4] with permission.

Temperature dependence of conductivity of biofilms and nanowires is inconsistent with cytochrome-dependent conduction.

Redox conduction via cytochromes is a thermally activated process. Conductivity *decreases* with a decrease in temperature, as the diffusion coefficient that comprises the rate constant, k_{ex}, for the redox reactions is also temperature dependent and the rate decreases exponentially with a decrease in temperature [27] according to:

$$k_{ex} = A \exp\left(\frac{-\Delta G}{RT}\right)$$

where A is the equilibrium rate constant and ΔG is the change in free energy during the reaction.

However, biofilms and pili *increased* in conductivity when these materials were cooled from the room temperature, until a crossover temperature after which disorder dictates the response (Fig. 7.13) [4, 68]. It was suggested that potential contact effects negated the relevance of these measurements [78], but the studies were performed with a four-probe configuration that avoids such artifacts. Furthermore, the increase in conductivities was observed as the temperature was dropped through physiologically relevant temperatures of

27 to 0 °C, and the lack of chemical fixatives and the presence of bound water further ensured physiological relevance [11].

Details of experimental procedure for temperature variation experiments. A Physical Property Measurement System (PPMS-6000, Quantum Design) was used to vary the temperature of the sample stage with electrode size 1 cm × 1.25 cm. A four-probe approach was used to measure conductance. Biofilms grown in the microbial fuel cell were placed on the PPMS puck, and the electrical contacts were made using indium and gold wires. All of the temperature experiments were carried out under physiologically relevant conditions of low vacuum (≈10 Torr), which maintains sufficient water content for ion mobility [79]. Best fitting was obtained by testing linear regression (R^2) values for various temperature dependences.

7.4.2 Redox Potential Dependence of Conductivity Using Electrochemical Gating Approach

The large gating effect observed on biofilms is consistent with that previously observed in organic metals [59, 76, 80]. When electrochemical gating was performed on biofilms in an electrolyte-gated field-effect transistor configuration [59, 76, 80–82] (Fig. 7.14), the conductivity increased in a sigmoidal manner by more than two orders of magnitude with the increasing gate potential. This response was reversible, and no hysteresis was observed. This sigmoidal response is characteristic of the organic metals [59, 81, 82], and the conductivity plateau at higher gate voltages has been attributed to metallic conduction with a nearly constant density of states [59, 80]. At lower gate voltages, the charge carriers can be trapped by the Coulomb potential of anions. However, at higher gate voltages with sufficiently large carrier density, the carrier wavefunctions can overlap and screen the Coulomb attraction between the traps and the carriers, resulting in a transition to the metallic state [76].

Electrochemical gating analysis to measure the conductivity as a function of redox potential [83] offers another approach to evaluate the possibility of conduction through *G. sulfurreducens* biofilms via c-type cytochromes. When conduction is via cytochromes or other redox carriers, there should be a maximum when the surface potential is near the formal potential for the redox process [29, 30, 81, 82, 84]. This reflects the equal density of reduced and oxidized species at the formal potential, which allows maximum possibility for electron exchange. When the film is poised at a potential that is much more positive or negative than the formal potential, the conduction is poor due to an overwhelming majority of the reduced (electron-loaded) or oxidized (hole-loaded) redox carriers. For example, electrochemical gating on redox-active metalloproteins exhibited a conductance peak, whereas nonredox controls did not show any peak in conductance [85].

Electrochemical gating with an electrolyte-gated, field-effect transistor geometry revealed that the biofilm conductivity does not peak at the formal potential (Fig. 7.5), including at the formal potential of −0.350 V versus Ag/AgCl for the biofilms of DL-1 strain observed in this (Fig. 7.4a) and previous studies [55, 86]. There was no

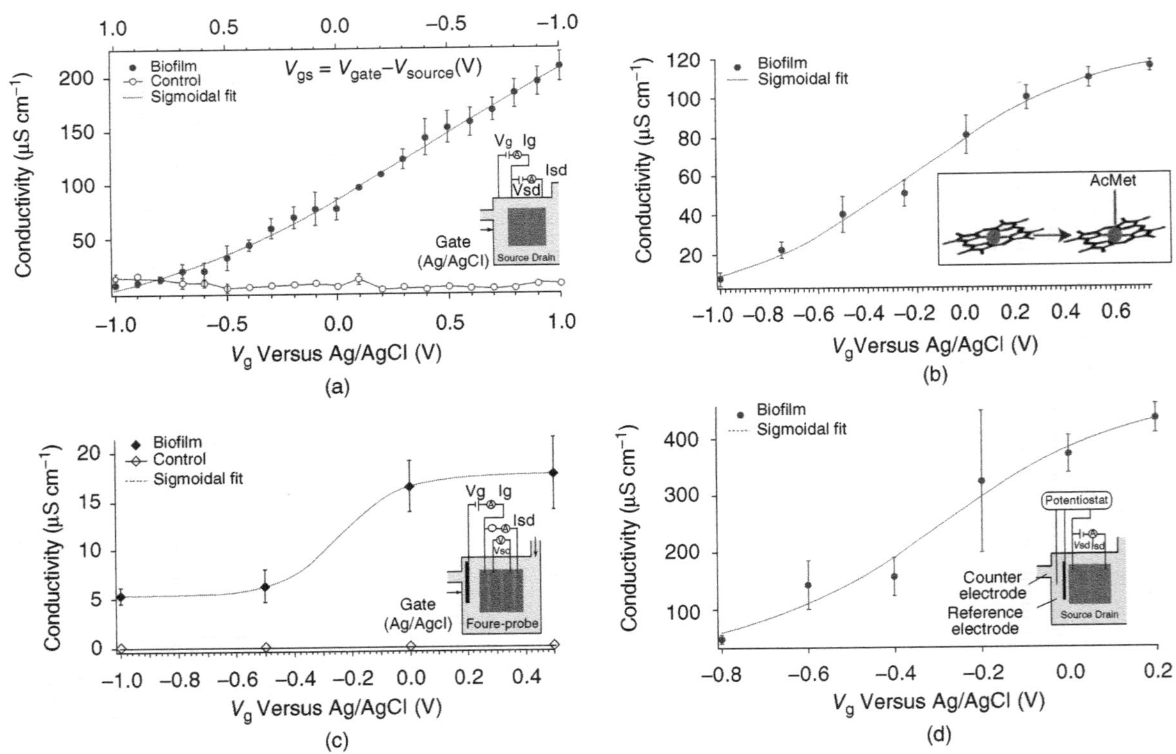

Figure 7.14 Electrochemical gating of *G. sulfurreducens* biofilms showing sigmoidal response with no peak at the formal potential. Gate voltage indicated is against Ag/AgCl, 3 M KCl reference electrode. (a) Gating response on the biofilm of strain DL-1 when gating and conductivity measurements were performed in two-electrode setup (inset). (b) Gating response on the biofilm of strain DL-1 in which cytochrome activity was inhibited with acetyl methionine. Inset: Ligand coordination reaction of AcMet to heme iron. (c) Four-probe conductivity of biofilm of KN400 strain (solid triangles) no biofilm control (open triangles) measured by electrochemical gating. Measurement setup is shown in the inset. (d) Conductivity of *G. sulfurreducens* biofilm strain ZMS123 measured by electrochemical gating using a three-electrode potentiostat configuration (inset). Error bars indicate standard deviation. Figure adapted from Ref. [8] with permission.

peak in conductivity for live biofilms of the DL-1 strain (Fig. 7.14a) as shown previously [4] or when acetyl methionine was added to denature cytochromes (Fig. 7.14b). This shows that the observed sigmoidal response in conductivity as a function of gate voltage, which is a characteristic of organic metals [1, references therein], cannot be attributed to cytochromes. The decrease in gating efficiency for the denatured biofilm is most likely due to the higher ionic strength of the electrolyte after addition of 120 mM acetyl methionine that screens the electric field [83] because the biofilm conductivity was unchanged after cytochrome denaturation.

The conductivity of biofilms of KN400 strain was also measured with four probes as a function of surface potential (Fig. 7.14c), which eliminated the possible contribution from the gate–voltage dependent contact resistance [50]. The formal potential for the biofilm of KN400 strain is −0.370 V versus Ag/AgCl [55, 87]. Similarly to studies with DL-1 biofilms, there was no peak in four-probe conductivity of KN400 biofilms at the formal potential. It has been argued that the current flowing through the reference electrode can cause artifacts in the response of electrochemical gating [78]. In order to evaluate any possible effect of current flowing through the reference electrode on the observed response, we used a potentiostat for electrochemical gating with independent counter and reference electrodes so that all the current was diverted to the counter electrode, with no current flowing through the reference electrode (Fig. 7.14d inset). Notably, the response of the biofilm was similar irrespective of the gating configuration with no peak at the formal potential (Fig. 7.14d). The differences in gating efficiency in four-probe and potentiostat gating experiments can be attributed to the differences in the electric field applied using differing geometries [45, 88]. At a highly oxidizing potential, the cytochrome-dependent conduction model predicts a large suppression in conductivity [14]. In contrast, there was an increase in conductivity, over two orders of magnitudes, at oxidizing potentials. Therefore, these studies further confirm that the observed conductivity is unlikely due to c-type cytochromes.

Details of experimental procedure for electrochemical gating.

For two-probe measurements, Keithley 2400 sourcemeter was used to apply voltage (V_g) between gate (Ag/AgCl, 3 M KCl reference electrode) and the source–drain to create the electrolyte gate field effect (Fig. 7.14a inset) [4]. As described earlier [4], gate current (I_g) was continuously monitored over the entire experiment. No monotonic dependence of the gate current on the measured biofilm current was observed, confirming that the largest part of this ion current does not flow through the biofilm across the gap but originates from the gold electrodes, which are partially covered by electrolyte [89, 90]. A second Keithley 2400 sourcemeter was used to apply voltage between source and drain to measure conductance as described in the method section about conductivity measurements.

For electrochemical gating using four-probe, gate voltage was applied between reference electrode and one of the outer electrode [45] and conductance was measured as described in the method section about conductivity measurements. For electrochemical gating using a potentiostat, Solartron 1287 was used. Platinum wire was used as a counter electrode (Fig. 7.14d inset) [91]. IGOR Pro software (WaveMetrics Inc., OR, USA) was used for all data analysis.

Note that all the electrochemical gating experiments were performed in the anode chamber, and the counter electrode in the cathode chamber was disconnected and unused. Reference electrode in the anode chamber was used as a gate electrode for all the experiments. For experiments with potentiostat, counter electrode was inserted in the anode chamber (Fig. 7.14d inset).

7.5 SUMMARY AND CONCLUSIONS

Discovery of electronic conductivity in living biofilms is relatively a recent finding, and it is important to employ a suite of complementary measurements such as DC and AC conductivity measurements to establish the electronic conductivity in these complex biological environments. In addition, it is necessary to carry out several control experiments to unambiguously confirm the presence of electronic conductivity in biofilms. Furthermore, to unravel the conduction mechanism, it is crucial to measure conductivity as a function of several key physical parameters such as temperature and redox potential. In addition, several genetic approaches such as construction of mutants and site-directed mutagenesis [92, 93] are necessary to elucidate the molecular basis underlying the conductivity of biofilms. The simple method described here to measure electronic conductivity of living biofilms is expected to be a useful tool for future studies of long-range electron conduction in a diversity of microbial systems such that those in sediments [94] and anaerobic digestors [68].

ACKNOWLEDGEMENTS

We thank Prof. Mark Tuominen, Prof. Madeline Vargas, and Prof. Kelly Nevin. The research is supported by the Office of Naval Research (grant no. N00014-12-1-0229 and N00014-13-1-0550), the Office of Science (BER), US Department of Energy (award no. DE-SC0006790), as well as the NSF Center for Hierarchical Manufacturing (grant no. CMMI-1025020). Nikhil S. Malvankar holds a Career Award at the Scientific Interface from the Burroughs Wellcome Fund.

REFERENCES

1. Dheilly A, Linossier I, Darchen A, Hadjiev D, Corbel C, Alonso V. Monitoring of microbial adhesion and biofilm growth using electrochemical impedancemetry. Appl Microbiol Biotechnol 2008;79(1):157–164.
2. Muñoz-Berbel X, Muñoz FJ, Vigués N, Mas J. On-chip impedance measurements to monitor biofilm formation in the drinking water distribution network. Sens Actuators B 2006;118:129–134.

3. Herbert-Guillou D, Tribollet B, Festy D, Kiéné L. In situ detection and characterization of biofilm in waters by electrochemical methods. Electrochim Acta 1999;45:1067–1075.

4. Malvankar NS, Vargas M, Nevin KP, Franks AE, Leang C, Kim BC, Inoue K, Mester T, Covalla SF, Johnson JP, Rotello VM, Tuominen MT, Lovley DR. Tunable metallic-like conductivity in microbial nanowire networks. Nat Nanotechnol 2011;6:573–579.

5. Malvankar NS, Lou J, Nevin K, Franks AE, Tuominen MT, Lovley DR. Electrical conductivity in a mixed-species biofilm. Appl Environ Microbiol 2012;78:5967–5971.

6. Malvankar NS, Tuominen MT, Lovley DR. Biofilm conductivity is a decisive variable for high-current-density microbial fuel cells. Energy Environ Sci 2012;5:5790–5797.

7. Marcus AK, Torres CI, Rittmann BE. Conduction-based modeling of the biofilm anode of a microbial fuel cell. Biotechnol Bioeng 2007;98:1171–1182.

8. Torres CI, Marcus AK, Parameswaran P, Rittmann BE. Kinetic experiments for evaluating the Nernst–Monod model for anode-respiring bacteria (ARB) in a biofilm anode. Environ Sci Technol 2008;42:6593–6597.

9. Snider RM, Strycharz-Glaven SM, Tsoi SD, Erickson JS, Tender LM. Long-range electron transport in *Geobacter sulfurreducens* biofilms is redox gradient-driven. Proc Natl Acad Sci U S A 2012;109:15467–15472.

10. Malvankar NS, Tuominen MT, Lovley DR. Lack of cytochrome involvement in long-range electron transport through conductive biofilms and nanowires of *Geobacter sulfurreducens*. Energy Environ Sci 2012;5:8651–8659.

11. Malvankar NS, Lovley DR. Microbial nanowires: a new paradigm for biological electron transfer and bioelectronics. ChemSusChem 2012;5:1039–1046.

12. Malvankar NS, Lovley DR. Microbial nanowires for bioenergy applications. Curr Opin Biotechnol 2014;27:88–95.

13. Qian F, Li Y. Biomaterials: a natural source of nanowires. Nat Nanotechnol 2011;6:538–539.

14. Fink HW, Schönenberger C. Electrical conduction through DNA molecules. Nature 1999;398:407–410.

15. Jortner J, Bixon M, Langenbacher T, Michel-Beyerle M. Charge transfer and transport in DNA. Proc Natl Acad Sci U S A 1998;95:12759–12765.

16. Endres R, Cox D, Singh R. Colloquium: the quest for high-conductance DNA. Rev Mod Phys 2004;76:195–214.

17. Chen F, Tao NJ. Electron transport in single molecules: from benzene to graphene. Acc Chem Res 2009;42:429–438.

18. Petty MC. Molecular Electronics: From Principles to Practice. Sussex, England: Wiley; 2002.

19. Porath D, Bezryadin A, De Vries S, Dekker C. Direct measurement of electrical transport through DNA molecules. Nature 2000;403:635–638.

20. Barsoukov E, Macdonald JR. Impedance Spectroscopy: Theory, Experiment and Applications. 2nd ed. New Jersey: John Wiley & Sons Inc.; 2005.

21. Simmons J. Electric tunnel effect between dissimilar electrodes separated by a thin insulating film. J Appl Phys 1963;34:2581.

22. Wang W, Lee T, Reed M. Mechanism of electron conduction in self-assembled alkanethiol monolayer devices. Phys Rev B 2003;68:35416.

23. Patra D, Malvankar N, Chin E, Tuominen M, Gu Z, Rotello VM. Fabrication of conductive microcapsules via self-assembly and crosslinking of gold nanowires at liquid–liquid interfaces. Small 2010;6:1402–1405.
24. Salomon A, Cahen D, Lindsay S, Tomfohr J, Engelkes VB, Frisbie CD. Comparison of electronic transport measurements on organic molecules. Adv Mater 2003;15:1881–1890.
25. Blauch DN, Saveant JM. Dynamics of electron hopping in assemblies of redox centers. Percolation and diffusion. J Am Chem Soc 1992;114:3323–3332.
26. McCreery R. Molecular electronic junctions. Chem Mater 2004;16:4477–4496.
27. Edwards PP, Gray HB, Lodge MTJ, Williams RJP. Electron transfer and electronic conduction through an intervening medium. Angew Chem Int Ed Engl 2008;47:6758–6765.
28. Heeger AJ, Sariciftci NS, Namdas EB. Semiconducting and Metallic Polymers. New York: Oxford University Press; 2010.
29. Chidsey CED, Murray RW. Electroactive polymers and macromolecular electronics. Science 1986;231:25–31.
30. Chidsey CED, Murray RW. Redox capacity and direct current electron conductivity in electroactive materials. J Phys Chem 1986;90:1479–1484.
31. Bard AJ, Faulkner LR. Electrochemical Methods: Fundamentals and Applications. 2nd ed. New York: John Wiley & Sons; 2001.
32. Dumas C, Basseguy R, Bergel A. Electrochemical activity of *Geobacter sulfurreducens* biofilms on stainless steel anodes. Electrochim Acta 2008;53:5235–5241.
33. Coppi MV, Leang C, Sandler SJ, Lovley DR. Development of a genetic system for *Geobacter sulfurreducens*. Appl Environ Microbiol 2001;67:3180–3187.
34. Nevin KP, Richter H, Covalla SF, Johnson JP, Woodard TL, Orloff AL, Jia H, Zhang M, Lovley DR. Power output and coulombic efficiencies from biofilms of *Geobacter sulfurreducens* comparable to mixed community microbial fuel cells. Environ Microbiol 2008;10:2505–2514.
35. Reguera G, Nevin KP, Nicoll JS, Covalla SF, Woodard TL, Lovley DR. Biofilm and nanowire production leads to increased current in *Geobacter sulfurreducens* fuel cells. Appl Environ Microbiol 2006;72:7345–7348.
36. Yi H, Nevin KP, Kim BC, Franks AE, Klimes A, Tender LM, Lovley DR. Selection of a variant of *Geobacter sulfurreducens* with enhanced capacity for current production in microbial fuel cells. Biosens Bioelectron 2009;24:3498–3503.
37. Voordeckers JW, Kim B-C, Izallalen M, Lovley DR. Role of *Geobacter sulfurreducens* outer surface c-type cytochromes in reduction of soil humic acid and anthraquinone-2,6-disulfonate. Appl Environ Microbiol 2010;76:2371–2375.
38. Leang C, Malvankar NS, Franks AE, Nevin KP, Lovley DR. Engineering *Geobacter sulfurreducens* to produce a highly cohesive conductive matrix with enhanced capacity for current production. Energy Environ Sci 2013;6:1901–1908.
39. Lanthier M, Gregory KB, Lovley DR. Growth with high planktonic biomass in *Shewanella oneidensis* fuel cells. FEMS Microbiol Lett 2008;278:29–35.
40. Webb JS, Thompson LS, James S, Charlton T, Tolker-Nielsen T, Koch B, Givskov M, Kjelleberg S. Cell death in *Pseudomonas aeruginosa* biofilm development. J Bacteriol 2003;185:4585–4592.
41. Marsili E, Rollefson JB, Baron DB, Hozalski RM, Bond DR. Microbial biofilm voltammetry: direct electrochemical characterization of catalytic electrode-attached biofilms. Appl Environ Microbiol 2008;74:7329–7337.

42. Picioreanu C, Head IM, Katuri KP, van Loosdrecht MCM, Scott K. A computational model for biofilm-based microbial fuel cells. Water Res 2007;41:2921–2940.
43. Mueller L, de Brouwer J, Almeida J, Stal L, Xavier J. Analysis of a marine phototrophic biofilm by confocal laser scanning microscopy using the new image quantification software PHLIP. BMC Ecol 2006;6:1.
44. Schwan HP. Electrode polarization impedance and measurements in biological materials. Ann NY Acad Sci 1968;148:191–209.
45. Lange U, Mirsky VM. Separated analysis of bulk and contact resistance of conducting polymers: comparison of simultaneous two-and four-point measurements with impedance measurements. J Electroanal Chem 2008;622:246–251.
46. Katz E, Willner I. Probing biomolecular interactions at conductive and semiconductive surfaces by impedance spectroscopy: routes to impedimetric immunosensors, DNA-sensors, and enzyme biosensors. Electroanalysis 2003;15:913–947.
47. K'Owino IO, Sadik OA. Impedance spectroscopy: a powerful tool for rapid biomolecular screening and cell culture monitoring. Electroanalysis 2005;17:2101–2113.
48. Wu J, Ben YX, Battigelli D, Chang HC. Long-range AC electroosmotic trapping and detection of bioparticles. Ind Eng Chem Res 2005;44:2815–2822.
49. Thangadurai V, Huggins RA, Weppner W. Use of simple ac technique to determine the ionic and electronic conductivities in pure and Fe-substituted SrSnO3 perovskites. J Power Sources 2002;108:64–69.
50. Hong J, Yoon DS, Kim SK, Kim TS, Kim S, Pak EY, No K. AC frequency characteristics of coplanar impedance sensors as design parameters. Lab Chip 2005;5:270–279.
51. Bisquert J. Physical electrochemistry of nanostructured devices. Phys Chem Chem Phys 2008;10:49–72.
52. Meulenkamp EA. Electron transport in nanoparticulate ZnO films. J Phys Chem B 1999;103:7831–7838.
53. Kankare J, Kupila EL. In situ conductance measurement during electropolymerization. J Electroanal Chem 1992;322:167–181.
54. Scribner. Zplot Operating Manual. Scribner Associates Inc.; 2001.
55. Malvankar NS, Mester T, Tuominen MT, Lovley DR. Supercapacitors based on c-type cytochromes using conductive nanostructured networks of living bacteria. ChemPhysChem 2012;13:463–468.
56. Cachet H, El Moustafid T, Herbert-Guillou D, Festy D, Touzain S, Tribollet B. Characterization of deposits by direct observation and by electrochemical methods on a conductive transparent electrode. Application to biofilm and scale deposit under cathodic protection. Electrochim Acta 2001;46:3851–3857.
57. Logan BE. Environmental Transport Processes. Wiley-Interscience; 1999.
58. Barsoukov E . (1996), Investigations of the redox kinetics of conducting polymers. Ph.D. Thesis.
59. Skotheim TA, Reynolds JR. Handbook of Conducting Polymers. Florida: CRC Press; 2007.
60. Reguera G, McCarthy KD, Mehta T, Nicoll JS, Tuominen MT, Lovley DR. Extracellular electron transfer via microbial nanowires. Nature 2005;435:1098–1101.
61. Nevin KP, Kim BC, Glaven RH, Johnson JP, Woodard TL, Methé BA, Didonato RJ, Covalla SF, Franks AE, Liu A, Lovley DR. Anode biofilm transcriptomics reveals outer surface components essential for high density current production in *Geobacter sulfurreducens* fuel cells. PLoS One 2009;4:e5628.

62. Reguera G, Pollina RB, Nicoll JS, Lovley DR. Possible nonconductive role of *Geobacter sulfurreducens* pilus nanowires in biofilm formation. J Bacteriol 2007;189:2125–2127.
63. Gorby YA, Yanina S, McLean JS, Rosso KM, Moyles D, Dohnalkova A, Beveridge TJ, Chang IS, Kim BH, Kim KS, Culley DE, Reed SB, Romine MF, Saffarini DA, Hill EA, Shi L, Elias DA, Kennedy DW, Pinchuk G, Watanabe K, Ishii S, Logan B, Nealson KH, Fredrickson JK. Electrically conductive bacterial nanowires produced by *Shewanella oneidensis* strain MR-1 and other microorganisms. Proc Natl Acad Sci U S A 2006;103:11358–11363.
64. Nakamura R, Kai F, Okamoto A, Newton GJ, Hashimoto K. Self-constructed electrically conductive bacterial networks. Angew Chem Int Ed Engl 2009;48:508–511.
65. Lovley DR. Microbial fuel cells: novel microbial physiologies and engineering approaches. Curr Opin Biotechnol 2006;17:327–332.
66. Lovley DR, Holmes DE, Nevin KP. Dissimilatory Fe(III) and Mn(IV) reduction. Adv Microb Physiol 2004;49:219–286.
67. Torres CI, Krajmalnik-Brown R, Parameswaran P, Marcus AK, Wanger G, Gorby YA, Rittmann BE. Selecting anode-respiring bacteria based on anode potential: phylogenetic, electrochemical, and microscopic characterization. Environ Sci Technol 2009;43:9519–9524.
68. Morita M, Malvankar NS, Franks AE, Summers ZM, Giloteaux L, Rotaru AE, Rotaru C, Lovley DR. Potential for direct interspecies electron transfer in methanogenic wastewater digester aggregates. MBio 2011;2:e00159–e00111.
69. Summers ZM, Fogarty HE, Leang C, Franks AE, Malvankar NS, Lovley DR. Direct exchange of electrons within aggregates of an evolved syntrophic coculture of anaerobic bacteria. Science 2010;330:1413–1415.
70. Wrighton KC, Agbo P, Warnecke F, Weber KA, Brodie EL, DeSantis TZ, Hugenholtz P, Andersen GL, Coates JD. A novel ecological role of the Firmicutes identified in thermophilic microbial fuel cells. ISME J 2008;2:1146–1156.
71. Wigginton NS, Rosso KM, Hochella MF Jr. Mechanisms of electron transfer in two decaheme cytochromes from a metal-reducing bacterium. J Phys Chem B 2007;111:12857–12864.
72. Dallas P, Stamopoulos D, Boukos N, Tzitzios V, Niarchos D, Petridis D. Characterization, magnetic and transport properties of polyaniline synthesized through interfacial polymerization. Polymer 2007;48:3162–3169.
73. Kaiser A. Systematic conductivity behavior in conducting polymers: effects of heterogeneous disorder. Adv Mater 2001;13:927–941.
74. Abthagir PS, Saraswathi R. Charge transport and thermal properties of polyindole, polycarbazole and their derivatives. Thermochim Acta 2004;424:25–35.
75. Lee K, Cho S, Park S, Heeger A, Lee C, Lee S. Metallic transport in polyaniline. Nature 2006;441:65–68.
76. Yuen J, Dhoot A, Namdas E, Coates N, Heeney M, McCulloch I, Moses D, Heeger A. Electrochemical doping in electrolyte-gated polymer transistors. J Am Chem Soc 2007;129:14367–14371.
77. Smart AG. Two experiments, two takes on electric bacteria. Phys Today 2010;63:18–20.
78. Strycharz-Glaven SM, Tender LM. Reply to the 'Comment on "On electrical conductivity of microbial nanowires and biofilms" by N. S. Malvankar, M. T. Tuominen and D. R. Lovley, Energy Environ. Sci., 2012, 5, DOI: 10.1039/c2ee02613a. Energy Environ Sci 2012;5:6250–6255.

79. Quantum_Design (2004), Physical Property Measurements System Hardware Manual, pp. 45.
80. Zotti G, Zecchin S, Schiavon G, Vercelli B, Berlin A, Dalcanale E. Potential-driven conductivity of polypyrroles, poly-*n*-alkylpyrroles, and polythiophenes: role of the pyrrole nh moiety in the doping-charge dependence of conductivity. Chem Mater 2003;15:4642–4650.
81. Zotti G, Zecchin S, Schiavon G, Berlin A, Pagani G, Canavesi A. Conductivity in redox modified conducting polymers. 2. Enhanced redox conductivity in ferrocene-substituted polypyrroles and polythiophenes. Chem Mater 1995;7:2309–2315.
82. Lindsay S. Molecular wires and devices: advances and issues. Faraday Discuss 2006;131:403–409.
83. Lindsay S. Introduction to Nanoscience. USA: Oxford University Press; 2009.
84. Heller A. Electron-conducting redox hydrogels: design, characteristics and synthesis. Curr Opin Chem Biol 2006;10:664–672.
85. Artes JM, Diez-Perez I, Gorostiza P. Transistor-like Behavior of Single Metalloprotein Junctions. Nano Lett 2012;12:2679–2684.
86. Richter H, Nevin KP, Jia H, Lowy DA, Lovley DR, Tender LM. Cyclic voltammetry of biofilms of wild type and mutant *Geobacter sulfurreducens* on fuel cell anodes indicates possible roles of OmcB, OmcZ, type IV pili, and protons in extracellular electron transfer. Energy Environ Sci 2009;2:506–516.
87. Strycharz SM, Malanoski AP, Snider RM, Yi H, Lovley DR, Tender LM. Application of cyclic voltammetry to investigate enhanced catalytic current generation by biofilm-modified anodes of *Geobacter sulfurreducens* strain DL1 vs. variant strain KN400. Energy Environ Sci 2011;4:896–913.
88. Tran E, Rampi MA, Whitesides GM. Electron transfer in a Hg-SAM//SAM-Hg junction mediated by redox centers. Angew Chem Int Ed Engl 2004;43:3835–3839.
89. Cronin SB, Barnett R, Tinkham M, Chou SG, Rabin O, Dresselhaus MS, Swan AK, Ünlü MS, Goldberg BB. Electrochemical gating of individual single-wall carbon nanotubes observed by electron transport measurements and resonant Raman spectroscopy. Appl Phys Lett 2004;84:2052.
90. Krüger M, Widmer I, Nussbaumer T, Buitelaar M, Schönenberger C. Sensitivity of single multiwalled carbon nanotubes to the environment. New J Phys 2003;5:138.
91. Tran E, Cohen AE, Murray RW, Rampi MA, Whitesides GM. Redox site-mediated charge transport in a Hg-SAM//Ru(NH(3))(6)(3+/2+)//SAM-Hg junction with a dynamic interelectrode separation: compatibility with redox cycling and electron hopping mechanisms. J Am Chem Soc 2009;131:2141–2150.
92. Vargas M, Malvankar NS, Tremblay PL, Leang C, Smith J, Patel P, Synoeyenbos-West O, Nevin KP, Lovley DR. Aromatic amino acids required for pili conductivity and long-range extracellular electron transport in *Geobacter sulfurreducens.* MBio 2013;4:e00105–e00113.
93. Liu X, Tremblay P-L, Malvankar NS, Nevin KP, Lovley DR, Vargas M. A *Geobacter sulfurreducens* strain expressing *Pseudomonas aeruginosa* type IV pili localizes OmcS on pili but is deficient in Fe(III) oxide reduction and current production. Appl Environ Microbiol 2014;80:1219–1224.
94. Malvankar NS, King GM, Lovley DR. Centimeter-long electron transport in marine sediments via conductive minerals. ISME J 2014;9:527–531.

8

ELECTROCHEMICAL IMPEDANCE SPECTROSCOPY AS A POWERFUL ANALYTICAL TOOL FOR THE STUDY OF MICROBIAL ELECTROCHEMICAL CELLS

Rachel A. Yoho and Sudeep C. Popat

Swette Center for Environmental Biotechnology, Biodesign Institute, Arizona State University, Tempe, AZ, USA

Francisco Fabregat-Santiago and Sixto Giménez

Grup de Dispositius Fotovoltaics i Optoelectrònics, Universitat Jaume I, Castellón de la Plana, Spain

Annemiek ter Heijne

Sub-department of Environmental Technology, Wageningen University, Wageningen Netherlands

César I. Torres

Swette Center for Environmental Biotechnology, Biodesign Institute, Arizona State University, Tempe, AZ, USA

8.1 INTRODUCTION

Microbial electrochemical cells (MXCs) represent a new class of technologies, within the realms of wastewater treatment, bioremediation, and bioenergy generation [1–3]. MXCs rely on the unique capability of certain bacteria to use electrodes in their metabolism, either as electron acceptors or as electron donors [4–6]. MXCs represent one of the most challenging types of bioreactors and electrochemical reactors to scale

Biofilms in Bioelectrochemical Systems: From Laboratory Practice to Data Interpretation, First Edition.
Edited by Haluk Beyenal and Jerome Babauta.

up. Unlike typical bioreactors, MXCs include electrochemical components that need to be considered and optimized. For example, scaling up MXCs in volume requires that very small distances be maintained between the anode and the cathode to minimize Ohmic losses from transport of ions between the two electrodes; this presents some restrictions on using traditionally scalable bioreactor designs. At the same time, MXCs differ from other electrochemical cells, such as the proton exchange membrane (PEM) fuel cell, because of the environmental conditions that are required to grow microorganisms (e.g., nutrient availability, water content, and neutral pH). In general, we want to achieve high efficiency in an MXC, which translates to large energy recovery in a microbial fuel cell (MFC) or small energy input in a microbial electrolysis cell (MEC). Unlike in typical bioreactors, where we would consider only the extents and kinetics of a given microbially catalyzed conversion in this regard, in MXCs, just like in electrochemical cells, the overpotentials at the electrodes, as well as those due to transport of ions and electrons, are also important in determining performance.

The characterization of overpotentials in MXCs is significantly different than in other electrochemical cells as well. In MFCs, for example, anode-respiring bacteria catalyze the oxidation of organic compounds, while oxygen is reduced at the cathode, usually on metal catalysts. Thus, classical activation and concentration overpotentials occur at the cathode, while complex metabolic and electron transport overpotentials occur at the anode. The fact that many microorganisms form biofilms on electrodes places the complication of a series of distinct intracellular cell reactions, which represents a more complex system than reactions on one or more catalyst sites on a planar or porous electrode in typical fuel cells. Recently, interest has increased in new types of MXCs in which bacteria accept electrons from the cathode (referred to as biocathodes) and reduce an electron acceptor of interest (e.g., CO_2, nitrate, perchlorate, and Cr(VI)) [7, 8]. The most attractive application of biocathodes is the synthesis of precursors to liquid fuel from CO_2, using electrical current derived from renewable sources [9]. The electrochemical characterization of biocathodes is even more complicated than that of bioanodes, as we know virtually nothing yet about the mechanisms of electron transport to microorganisms from a cathode [10].

The most common electrochemical techniques used to study overpotentials in MXCs are those that measure a change in current as the potential is changed. Cyclic voltammetry (CV) is the most popular technique, while others include linear sweep voltammetry, chronoamperometry, potential step-voltammetry, and so on [11]. CV involves changing the electrode potential (V) within a prespecified range of potentials at a defined linear rate and measuring the resulting electrical current (i). The result of a CV is an i–V relationship that often shows the total overpotential for the whole cell, or for a working electrode, when performed with respect to a reference electrode. However, CV does not specifically help identify the individual contribution of different phenomena that occur in an MXC to the overall i–V relationship. For example, the Ohmic loss that occurs between the reference electrode and the working electrode while performing CV measurements is inherently not distinguishable in the i–V response collected. When applied to studying electron transfer at the microbial anode, CV does not specifically help identify the contribution of different processes (e.g., microbial kinetics of substrate oxidation vs. extracellular electron transfer vs. reactant/product transport) to the total overpotential. These processes are lumped together in the response of the system studied as a whole. In addition, other techniques such as chronoamperometry and potential-step voltammetry have these same limitations.

8.1.1 Background on Electrochemical Impedance Spectroscopy

There are, however, electrochemical techniques available that are able to, theoretically, identify and quantify the individual contribution of various simultaneous processes to the overall overpotential in electrochemical cells. This chapter focuses on the use of electrochemical impedance spectroscopy (EIS), which has only recently found application in studies on MXCs [12]. Unlike direct current (DC) techniques such as CV, EIS is an alternating current (AC) method. Potentiostatic EIS involves applying small sinusoidal amplitude over a fixed voltage value at a range of frequencies and measuring the resulting magnitude and phase shift in current (Fig. 8.1). Because different processes govern the *i*–*V* response at different frequencies, it is thus theoretically possible to quantify individual contributions of different processes to the overall overpotential, provided that these processes have different characteristic time constants. While uses of EIS in research applications in many distinct fields are various, as reviewed elsewhere [13–16], it has also been used extensively in PEM fuel cell research to identify and quantify individual resistances, and thus overpotentials [17, 18].

Although the details of the theory behind EIS are available in literature [19], we review the concept of impedance in very simple terms, before discussing special requirements in experimental design, parameter selection, and protocol, in its application to MXCs, as well as highlight standard and alternative methods of data analysis.

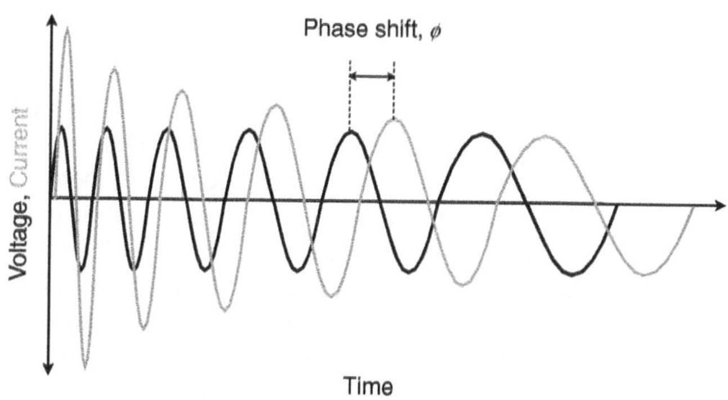

Figure 8.1 Sinusoidal current response to the sinusoidal voltage change made during potentiostatic EIS measurements. Each cycle represents a different frequency. A wide range of frequencies is usually used in EIS measurements.

Impedance is similar to resistance, except that it includes also a time domain linked to frequency of the sinusoidal amplitude applied to the system under study. Using Ohm's law, impedance (Z) can be represented as:

$$Z = \frac{E(t)}{i(t)} \tag{8.1}$$

Here, $E(t)$ and $i(t)$ are the potential and current as a function of time t. As we apply a sinusoidal amplitude on the potential as a function of frequency, $E(t)$ can be represented as:

$$E(t) = \Delta E \sin(\omega t) \tag{8.2}$$

Here, ω is the radial frequency (equivalent to $2\pi f$, where f is the frequency in Hertz) at time t, and ΔE is the amplitude in voltage.

An ideal resistor, following Ohm's law (Equation 8.1), would show constant impedance over time at the various frequencies, and thus the use of EIS would reduce to the measurement of a single point repeatedly over the entire frequency range. However, many processes that we want to study not only act as resistances to electron flow, but also act as capacitors, giving them the ability to store an electrical charge. A good example of a capacitor in a fuel cell is the electrical double layer at the interface of the electrode, where the charge transfer reaction occurs. As you can imagine, microbial cells, or even enzymes, have capacitance as well, due to their ability to store electrons in reaction centers or in redox compounds. Diffusion limitations also lead to a similar behavior. Because the electrons not only follow Ohm's law, but also are capable of charging (or discharging) the capacitor, there is a phase shift, ϕ between the voltage $E(t)$ and the current $i(t)$, which can be represented as:

$$i(t) = \Delta i \sin(\omega t + \phi) \tag{8.3}$$

Thus, the impedance can then be represented as:

$$Z = \frac{\Delta E \sin(\omega t)}{\Delta i \sin(\omega t + \phi)} = |Z| \frac{\sin(\omega t)}{\sin(\omega t + \phi)} \tag{8.4}$$

Here, $|Z|$ is the absolute magnitude of the impedance. What allows us to separate different electrochemical processes using EIS is the fact that capacitors have a frequency-dependent ability to be charged or discharged – a property called capacitive reactance. As a consequence, different capacitances within the system will respond at different frequencies, allowing us to separate temporally the electrochemical processes in EIS experiments on the basis of their capacitive reactance. Similarly, diffusion limitations usually also have a characteristic low frequency where they predominantly govern the system response and thus are easily distinguishable.

The information provided by EIS can be plotted in different graphical representations. Figure 8.2 shows one common representation of EIS data, called a Bode plot. In a Bode plot, the absolute magnitude of the impedance (Fig. 8.2a) and the phase shift (Fig. 8.2b), both of which are experimentally measured, are plotted against

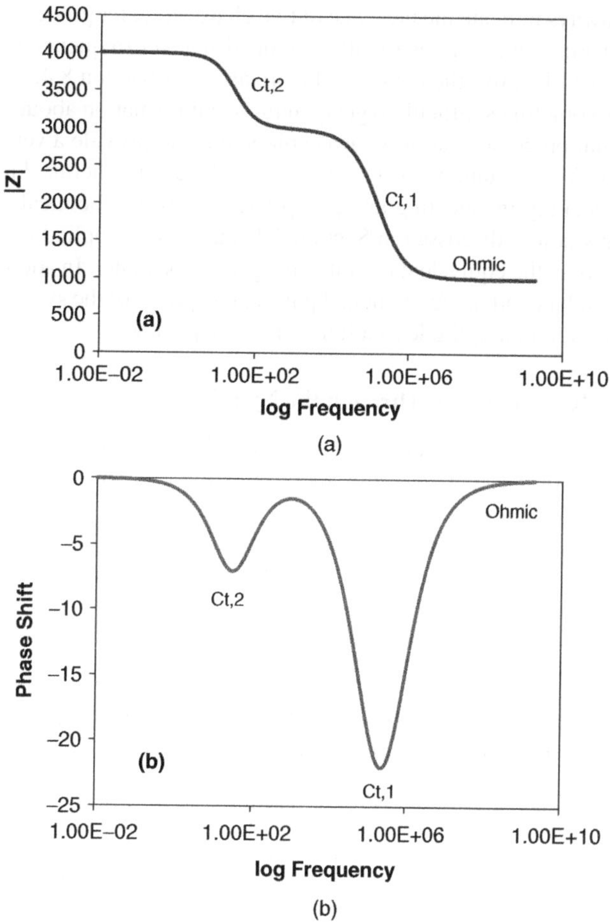

Figure 8.2 A representative Bode plot, showing: (a) absolute impedance (|Z|) versus frequency and (b) phase shift versus frequency. $C_{t,1}$, $C_{t,2}$, and Ohmic represent three major processes contributing to the overall impedance.

the frequency. In this example, two different processes can be identified, which are ascribed to different charge transfer events. More specifically, three clear regions are visible in Figure 8.2b. Firstly, the phase shift is zero at high frequencies; the high frequencies do not allow electrons to interact with any capacitor, and thus only Ohmic resistance is measured. Then two phase-shift peaks are apparent as the frequency is decreased, both corresponding to individual charge transfer processes (shown in the figure as $C_{t,1}$ and $C_{t,2}$). The phase shift ϕ occurs due to the existence of a capacitive process in the charge-transfer mechanism. Thus, from the Bode plot, we can conclude that there are in this example, at least two charge transfer processes in the system studied. It is important to note that the presence of two phase shifts does not necessarily always indicate the presence of two charge transfer processes but can also indicate diffusion limitations, especially if noticed at lower frequencies. We elaborate further on how to identify the nature of the processes in Section 8.4.

While it is sometimes difficult to get quantitative information about overpotentials from individual processes out of the Bode plots, they do provide a very good opportunity to identify important processes that govern the electrochemical response. This is helpful especially in selecting an appropriate biophysicochemical model for the system being studied (discussed in Section 8.4) that can be applied to such methods of data analysis as the equivalent circuits fitting described later. In addition, the Bode plot helps to understand the fundamental physical response of the system with respect to the perturbation in amplitude as a function of frequency.

8.1.2 Data Representation Through the Nyquist Plot

The impedance, as shown in Equation 8.4, can also be represented as a complex function as in Equation 8.6, derived using the correlation of a sine function to complex number theory (Equation 8.5):

$$\sin x = \frac{e^{jx} - e^{-jx}}{2j} \tag{8.5}$$

$$Z = |Z|\frac{e^{j\omega t} - e^{-j\omega t}}{e^{j(\omega t+\phi)} - e^{-j(\omega t+\phi)}} = |Z|e^{j\phi} \tag{8.6}$$

Here j equals the square root of -1. Impedance can be further represented as shown in Equation 8.8, derived using Euler's formula (Equation 8.7):

$$e^{jx} = \cos x + j\sin x \tag{8.7}$$

$$Z = |Z|(\cos\phi + j\sin\phi) \tag{8.8}$$

This relationship of the impedance as a complex function, shown in Equation 8.8, is often the reason that MXC researchers sometimes look at EIS as a complicated, mysterious technique. We believe that this may be a contributing factor to the sparse and incomplete application of EIS in MXC studies. The principles of EIS have foundations in basic mathematics of complex numbers. As we apply sinusoidal amplitude on voltage, both the voltage and the current with time have to be represented as a sine function, and it is through Euler's formula that these and the resulting impedance can be represented as complex functions.

In Cartesian coordinates, the impedance can also then be represented as,

$$Z = Z' - jZ'' \tag{8.9}$$

Here, Z' and Z'' (also often represented as Re(Z) and –Im(Z), respectively) represent the real and imaginary part of the impedance determined as follows:

$$Z' = |Z|\cos\phi \tag{8.10}$$

$$Z'' = -|Z|\sin\phi \tag{8.11}$$

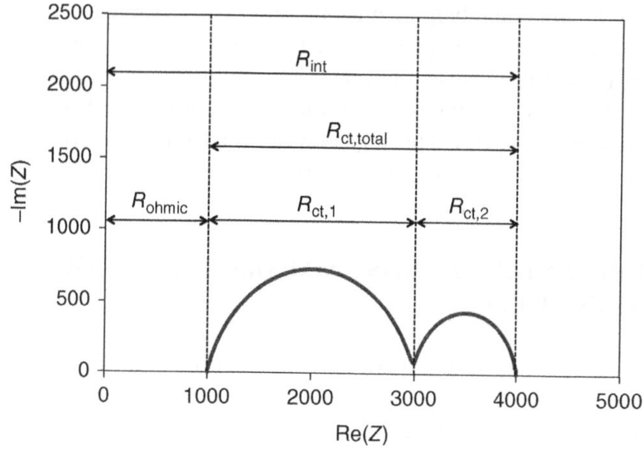

Figure 8.3 A representative Nyquist plot. R_{int} represents the total internal resistance of the cell, while R_{ohmic} represents the Ohmic resistance, and R_{ct} the charge transfer resistance.

The relationship of the impedance as a complex function is also what leads to the most common data representation in EIS, known as the Nyquist plot (Fig. 8.3). In the Nyquist plot, −Im(Z) is plotted versus Re(Z) over the entire frequency range of the EIS measurement. One of the main shortcomings of the Nyquist plot is that the frequency is not shown; there is only an implicit understanding that high frequencies are at the lower Re(Z) values and decrease in the positive X-direction. In the example provided in Figure 8.3, which is the same as the system represented in the Bode plot in Figure 8.2, three separate resistances are apparent as intercepts on the X-axis. Such a response is often observed in fuel cells where reactants and products are supplied in excess, and the only resistances governing potential losses are the Ohmic resistance and the activation losses at the anode and the cathode.

In the aforementioned example, Ohmic resistance is the first intercept on the X-axis (Re(Z) = 1000). Usually, Ohmic resistance is measured at a high frequency and results in a zero phase shift. The other two resistances can be interpreted as charge transfer resistances, and they show a characteristic semicircular shape on the Nyquist plot. This results from the capacitive nature of the interface at which the electrochemical reactions occur and is an indication of a characteristic phase shift distribution over a range of frequencies. As mentioned previously, low-frequency semicircles can sometimes be related to diffusion limitations. The Nyquist plot is used popularly because it is easy to obtain a quantitative value of the various resistances and capacitances by fitting the data to an equivalent circuit model (described in detail in Section 8.4). Despite its advantages in quantitation, the Nyquist plot also has several limitations. For example, oftentimes two charge transfer processes have characteristic frequencies so close to each other in the Nyquist plot that it is almost impossible to distinguish them, perform equivalent circuit fitting, and obtain accurate resistances. Sometimes, it even leads to the selection of incorrect or incomplete equivalent circuits. In this regard, the Bode plot could help identify these processes more effectively. But even so, if the two characteristic frequencies are too close to each other, usually less than a decade apart, the EIS results would measure only the combined effect of the two charge transfer processes. This is one of the main limitations of using EIS in complex systems, such as MXCs. The selection of an adequate experimental setup and EIS parameters can help minimize these limitations for fundamental EIS studies.

8.2 EXPERIMENTAL DESIGNS AND PARAMETERS FOR APPLICATION OF EIS

Before discussing the control experiments to perform to ensure that valid data is obtained through EIS measurements, we first discuss the various ways in which EIS has been and can be applied in the study of MXCs. The kind of information that can be obtained from EIS measurements depends largely on two important aspects of the experimental plan: the cell design and the measurement parameters. In the following section, we discuss both of these in detail, to help researchers in the MXC field use EIS to its maximum potential.

8.2.1 Two-Electrode Versus Three-Electrode Setups

The electrochemical investigation of all energy conversion devices in general must be carried out in two-electrode configuration, in order to account properly for the total losses in the system. However, when the subject of interest is centered on the response of either the anode or the cathode of the device, three-electrode measurements provide more accurate results, as the overpotential at the counter electrode can be neglected. In addition, reference electrodes provide an absolute reference for the electrochemical events, and therefore, energy diagrams can be constructed on the basis of this characterization.

The earliest application of EIS in MXCs was indeed with two-electrode setups [20]. In such a setup (Fig. 8.4a), one of the electrodes is used as the working electrode, with the other as the counter as well as the reference electrode. When performing EIS measurements in this configuration, the overall impedance of the cell is obtained. In fuel cells, this is usually referred to as the internal resistance. This resistance includes the overpotentials at both the anode and the cathode, resulting from both activation and concentration overpotentials at the individual electrodes, as well as the Ohmic resistance between the two electrodes. Ideally, EIS measurements should be done at a range of cell potentials to recreate the entire polarization curve. Using an applicable biophysicochemical model, translated into an equivalent circuit, and fitting the data to this model can help deduce the contribution of each individual process to the overall response of the system, as long as each of them have different characteristic time constants.

However, sometimes, it is difficult to identify each electrode process through EIS measurements with a two-electrode setup, and thus, it is important to use the three-electrode setup (Fig. 8.4b) to verify their identity and/or their magnitude

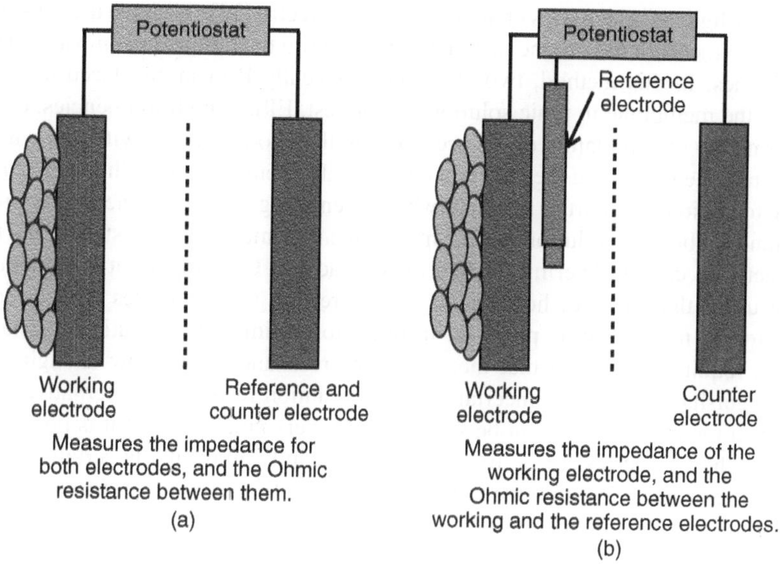

Figure 8.4 Cell configurations to use for EIS measurements: (a) two-electrode setup that measures impedance for both working and counter electrodes, and the Ohmic resistance between them and (b) three-electrode setup that measures the impedance of the working electrode, and the Ohmic resistance between the working and the reference electrodes.

[21]. In a three-electrode setup, a separate reference electrode is used, and the EIS measurements are performed for the working electrode with respect to the reference electrode. For example, to measure resistances from all anode processes, the anode would be selected as the working electrode, the reference electrode would be placed close to the anode, and the cathode would be used as the counter electrode. In this case, the impedance measured includes the resistance from all processes occurring at the anode, as well as the Ohmic loss that occurs between the anode and the reference electrode. The latter can also be used to perform *i–R* correction for CV measurements, which is an important step to interpret CV data accurately, especially when these resistances are large. Similarly, to obtain the cathode resistances, the cathode would be selected as the working electrode. While the three-electrode setup for EIS measurements has been used to support data primarily from two-electrode setups in MXC research, there have recently been a few studies published where the three-electrode setup is used to understand microbial electron transfer processes, both at the anode and at the cathode, at a fundamental level, in further detail [22, 23]. We believe that this represents one of the most promising applications of EIS.

We do, however, want to point out an application of the two-electrode setup that has not yet been used extensively in MXC research. We believe that EIS can be a simple tool to rapidly quantify the resistances from various membranes that are used in MXCs. As MXCs involve ionic solutions apart from the membrane, unlike in PEM fuel cells, traditional methods for measuring membrane conductivity with interdigitated electrodes cannot be applied directly [24]. We have used a simple method to measure membrane conductivity in relevant solutions to compare various membranes. In this method, two electrodes (typically Pt) can be placed on either side of the membrane in ionic solutions of interest. EIS, at high frequencies, can be performed to obtain Ohmic resistance between the two electrodes with and without the membrane separating them. When not using the membrane, the Ohmic resistance is due to the ion transport in solution, while when using the membrane, it is from the resistance of both the solution and the membrane. The membrane resistance, and thus conductivity, can be determined then by subtraction. One important consideration while using this method, however, is to ensure that the electrodes are placed as close to the membrane as possible, in order to minimize the solution resistance. This is required to ensure that the membrane resistance is not low enough to be indistinguishable from the uncertainty in the solution resistance. Alternatively, the projected membrane areas can be selected to be very small, so that it is possible to increase the Ohmic resistance from them significantly above the uncertainty in the solution resistance. We provide more details about this experiment in an example procedure described in Section 8.5.

8.2.2 Key Experimental Parameters

A typical parameter selection window in a potentiostat is shown in Figure 8.5. Most of the parameter selection options are similar across all well-known potentiostats used by MXC researchers. We discuss in detail in the following section the rationale behind selecting some of these parameters.

8.2.2.1 Working Electrode Potential Sometimes, EIS has been used only at open circuit conditions, in MXC research [25–27]. We believe that this form of measurement rarely gives data of any direct relevance to the system being studied in the context of characterizing individual resistances or overpotentials, other than the Ohmic resistance. This is because no reactions are occurring. The Ohmic resistance measurements, however, can be valuable for quickly evaluating the efficacy of the membrane being used or the effect of the distance between electrodes. It can be used also to distinguish between electronic and ionic conductivity [26]. Also, any measurements made of charge transfer resistances at open circuit can be a strong function of the amplitude applied, especially because most bacterial electrocatalysis kinetics characterized so far follow Nernstian behavior. In other electrochemical cells, such as photoelectrochemical cells, open circuit measurements often are used to measure the capacitance of the electrodes. However, it is well known that the largest fraction of capacitance in MXCs comes from biological redox cofactors that store charge [26, 28]. The capacitance of MXC electrodes thus changes significantly with electrode potential, also reducing the relevance of open circuit measurements. We believe that EIS measurements, when done in three-electrode setups, should be performed over a range of working electrode potentials that are known to result in an electrochemical response, as also predetermined through other electrochemical analyzes such as chronoamperometry and/or CV. We direct the readers to two excellent

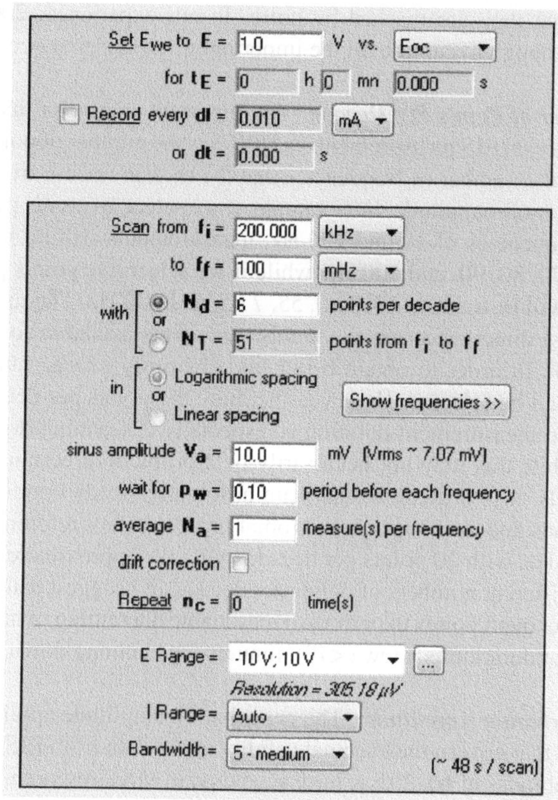

Figure 8.5 A dialog box for EIS parameter selection in EC-Lab software by Bio-Logic Instruments. Some of the main selecting parameters are working electrode potential (E_{we}), cell equilibration time (t_E), range of frequency from f_i to f_f, numbers of points per decade (N_D), and the sinus amplitude (V_a).

studies, by Jung et al. [23] and Ter Heijne et al. [22], where EIS measurements were performed over a range of potentials for an anode biofilm of *Geobacter sulfurreducens* and a biocathode, respectively. When performing EIS measurements with a two-electrode setup, a wide range of potentials that cover the entire polarization curve should be used.

8.2.2.2 ***Frequency Range*** Most modern potentiostats allow scanning the frequency between the ranges of tens of megahertz to tens of microhertz. Frequency ranges also have varied significantly. Generally, the EIS measurements should be carried out starting from high frequencies and finishing at low frequencies. It should be noted that at very high frequencies, the effects of inductance can cause the data to be not particularly useful, while at very low frequencies, measurements are often small enough that they are masked by noise. In our experience, 200–500 kHz to 10–50 mHz, is enough to capture all the important processes that occur in an MXC.

8.2.2.3 ***Number of Points Per Decade*** An important parameter that is rarely ever reported in studies on EIS measurements in MXCs is the number of points per decade. This represents the number of frequencies that the measurements are made between ranges of an order of magnitude. For example, if we select 10 points per decade, then between the frequencies of 10 and 100 Hz, measurements will be made at 10, 20, 30, 40, 50, 60, 70, 80, 90, and 100 Hz, while if we select five points per decade, the measurements will be made at 10, 32.5, 55, 77.5, and 100 Hz. The number of points per decade has a direct effect on the number of points available for data fitting in the Nyquist plot. In order to obtain better fits, obviously a large number of points per decade should be selected. However, the more the points per decade, the longer the experimental measurement duration is, especially if scanning down to very low frequencies. While this may not necessarily be a problem in certain kinds of electrochemical cells, it is difficult to maintain absolutely steady conditions in MXCs for long durations due to changes in microbial processes. For reference, a scan from 200 kHz to 1 mHz, with 20 points per decade, will take approximately 3 h. We have tested several different numbers of points per decade and suggest using 10 points per decade to have enough points to perform a reasonable data fitting, while ensuring that the experimental duration is a few (<10) minutes for scanning down to 10–20 mHz.

8.2.2.4 ***Perturbation Amplitude*** The perturbation amplitude applied is one of the key parameters that govern the eventual validity of the data from EIS measurements. In general, amplitudes of 10–20 mV have been used in MXC research. We believe that these amplitudes represent a good range to perform experiments in, but it is important to verify this through some of the control experiments we describe later.

In addition to the aforementioned parameters, another important parameter is the equilibration time at the selected working electrode potential prior to beginning the EIS measurements. This too, similar to the amplitude, has implications on the validity of the data, and thus we discuss in further detail in the following section. Many studies do not report on the selection of this parameter, but as we show later, it is important to select an appropriate equilibration time. Generally, the equilibration time can be chosen as the inverse of the lowest frequency selected for the analysis, in order to guarantee that the system is relaxed when the next perturbation is applied.

8.3 EXPERIMENTAL PROTOCOL TO ENSURE DATA VALIDITY

To be able to use the data generated during EIS measurements for any type of data analysis, it is important to consider that EIS measurements are only valid when the following four conditions are met:

1. *Linearity:* One of the first, fundamental assumptions of EIS measurements is that the system behaves linearly in change in current in response to change

in potential. This is illustrated in Equations 8.1 and 8.4. Of course, no electrochemical or biological system is perfectly linear, but the amplitude of the applied perturbation can be selected such that the system can be approximated to be linear. Usually, this amplitude is thus selected to be small, but it should be considered that the amplitude be large enough to distinguish real measurements over noise.

2. *Stability:* EIS measurements usually take anywhere from a minute to tens of minutes, depending on the range of frequencies selected. A key requirement of the experimental setup is that the system stays stable over the course of the experimental duration, that is, there are no inherent temporal changes in the system due to factors other than the applied perturbation in voltage. This can be particularly important with biological systems that can be affected by growth, transport of reactants/products, and cellular adaptations to the selected potential.
3. *Causality:* The system studied by EIS should not be affected by external perturbations other than the perturbation applied. Oftentimes, causality is a result of noise in measurements due to internal potentiostat controls, interfering instruments, or bad wire lead connections. When the noise in current is significant with respect to the signal obtained due to change in potential, the data quality can be considered poor.
4. *Finiteness:* One of the assumptions from Equation 8.4 is that the change in current as a function of change in potential not be zero. Thus, the value of impedance has to be finite at all frequencies applied. If there is no current response due to infinite impedance, the data analysis cannot be performed, as there is no signal response.

In this section, we discuss in detail how the selection of various experimental parameters affects each of these conditions. One of the first studies on EIS measurements in MXC applications by Strik et al. [29] covers some of these conditions very well, but we provide an expanded explanation here. While all these conditions are especially difficult to fulfill in a typical electrochemical cell, the conditions used in MXCs further exacerbate the problem. For example, it is known that polarization curves for microbial anodes exhibit nonlinear, Nernstian responses [30]. Thus, there are regions in the polarization curve where the system may not behave linearly even when small amplitudes are applied. The irreversibility of the enzymatic responses also leads to regions where finiteness is not met (Fig. 8.6). These cases would run also into difficulties in terms of the signal-to-noise ratio when small amplitudes are applied at potentials on the saturation region of the polarization curve. Similarly, as MXCs are biological reactors and can have changes in microbial responses due to small perturbations outside the control of researchers, conditions of both stability and causality are difficult to fulfill.

We have investigated various experimental procedures to ensure the accurate application of EIS to MXCs. In the following sections, we discuss EIS experiments performed with biofilms of *G. sulfurreducens* grown on gold or graphite anodes of various geometries in several three-electrode setups. These experiments were

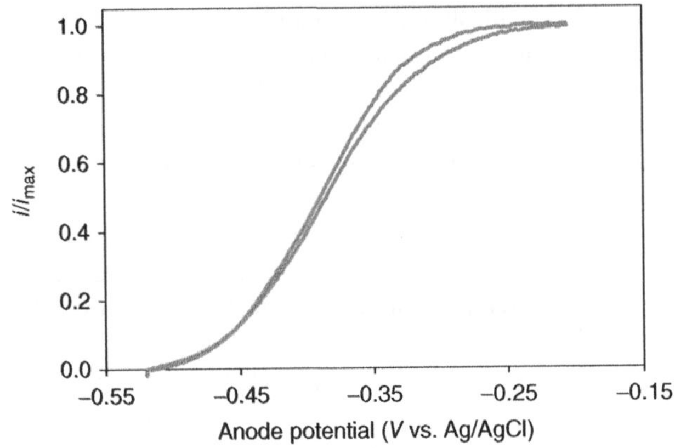

Figure 8.6 A typical cyclic voltammogram of a *Geobacter sulfurreducens* biofilm. The current (i) is shown normalized to the maximum current (i_{max}).

used to investigate several parameters, including perturbation amplitudes, working electrode potentials, equilibration times, and electrode materials and sizes to use to obtain accurate EIS data. Wherever needed, we compare experimentally collected data to the Kramers–Kronig relationship – a relationship that ensures that the data collected meet the four condition requirements of EIS [31, 32]. More details about the K–K relationship are available in literature cited previously. We used two different potentiostats, from Ivium (CompactStat) and Bio-Logic (VMP3), for the studies described here; the potentiostat used for the data collection is understood to be the Ivium CompactStat unless otherwise noted. While the data shown here were collected from three-electrode setups, that is, performing EIS specifically to understand anode phenomena, the same guidelines and principles outlined would apply for two- and four-electrode setups as well.

8.3.1 Linearity and Finiteness

There are two important parameters that affect the condition of linearity and finiteness required for accurate application of EIS. We discuss in the following sections how to ensure the appropriate selection of both parameters.

8.3.1.1 Perturbation Amplitude The perturbation amplitudes that can be used during EIS can be both theoretically and experimentally evaluated. In theory, the practical operating perturbation amplitude should be less than the product of the Boltzmann constant ($K_b = 8.617 \times 10^{-2}$ mV K^{-1}) and the temperature at which the experiment is performed. This requirement was shown clearly by Barbero et al. in their study on EIS measurements on electrolytic cells [33]. At 30 °C, a common temperature in MXC operation, the amplitude should be less than approximately 26 mV. While this theoretical guideline is generally valid, in case of MXCs, the amplitude selected also needs to take into consideration the potential of the working electrode. For example, amplitudes much smaller than 26 mV would be required if performing EIS on the anode in the regions that transition to or from the linear response around the midpoint potential (−0.4 vs. Ag/AgCl in Fig. 8.6).

In general, there are two considerations for the amplitude that can be used. Larger amplitudes can be advantageous because it provides a larger signal-to-noise ratio. However, smaller amplitudes may allow the system to fulfill the condition of linearity better, especially in the potential regions mentioned previously. This can be seen in Figure 8.7a, which shows the Nyquist plot at open circuit potential (−0.5 V vs. Ag/AgCl), which will be in the activation region of the Nernstian response. It appears that the 5- and 10-mV amplitudes provide similar results; however, as the amplitude is increased to 50 and 100 mV, progressively lower impedance responses as a result of nonlinear behavior are apparent. For most setups that we have tested in MXC anodes, we have observed that at open circuit, <20 mV is the optimum amplitude to use. We note here that such amplitude validation tests should be done with electrodes that already have the biofilms of interest developed on them, as bare electrodes may not necessarily follow the same kinetic behavior in relation to an applied overpotential.

Close to the midpoint potential, as determined from Figure 8.6, the system shows essentially linear behavior, and there is less dependence of the impedance response on the perturbation amplitude used (Fig. 8.7b). At this potential, it appears that it would be acceptable to perturb the potentials in the linear region with almost any of amplitudes we selected. However, the data collection should remain consistent with the amplitudes that meet the requirements toward the extremities of the appropriate potentials to investigate, as well as the theoretical limit of amplitude (using the $k_B T$ relationship).

8.3.1.2 Potentials Figure 8.8 shows a comparison of the Nyquist plots of the experimental data collected on the Bio-Logic (VMP3) potentiostat with amplitude

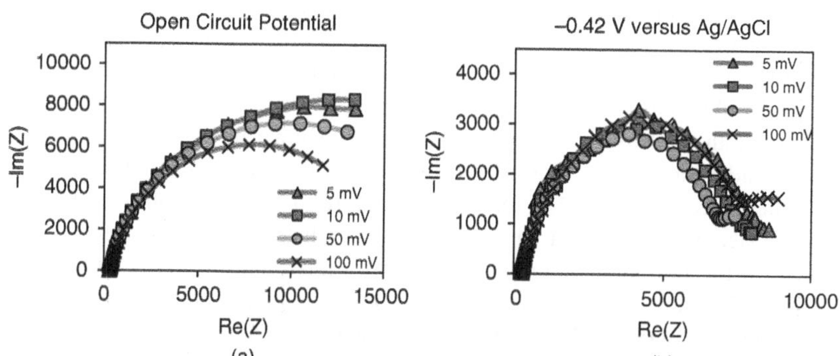

Figure 8.7 The effect of the voltage amplitude during EIS measurements on a *Geobacter sulfurreducens* biofilm on the Nyquist plots obtained under: (a) open-circuit conditions and (b) with the anode poised at −0.42 V versus Ag/AgCl, which is close the midpoint potential in the *i–V* relationship.

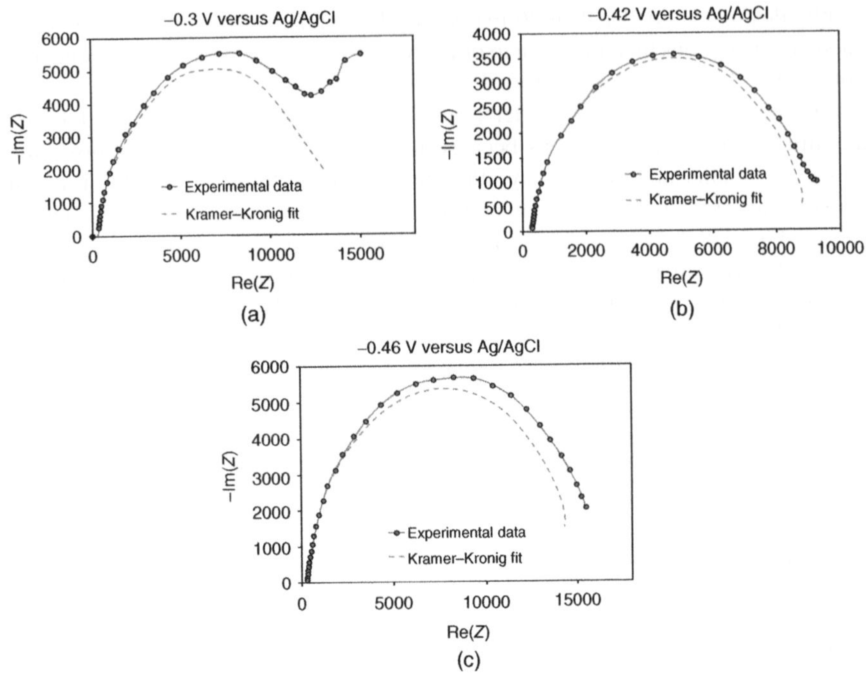

Figure 8.8 Nyquist plots from EIS measurements on a *Geobacter sulfurreducens* biofilm at three different potentials: (a) −0.3 V versus Ag/AgCl, which is close to saturation potential in the *i*–*V* relationship, (b) −0.42 V versus Ag/AgCl, which is close to the midpoint potential in the *i*–*V* relationship, and (c) −0.46 V versus Ag/AgCl, which is close to open-circuit anode potential. The dotted lines show Kramers–Kronig fits.

of 10 mV at three different working electrode potentials. The figure also includes the Kramers–Kronig relationship fits for each condition. The Kramers–Kronig relationship does not fit the actual data at the higher potential (−0.3 V vs. Ag/AgCl, Fig. 8.8a), suggesting that the system is not compliant to one of the conditions required for obtaining valid EIS data. In this case, however, we believe this condition is finiteness and not linearity. As seen in the CV shown in Figure 8.6, the system under consideration can be considered infinite with respect to the impedance, that is, in either direction of −0.3 V versus Ag/AgCl, for a 10-mV amplitude, there is no change in steady-state current, or it is smaller than the noise in the system.

Close to the midpoint potential (Fig. 8.8b), the Kramers–Kronig fit appears to validate the EIS data. At this potential, the system can be considered both linear and finite at the amplitude used. Toward the more negative end of the CV (−0.46 V vs. Ag/AgCl, Fig. 8.8c), the Kramers–Kronig fit again begins to deviate from the data collected. In this activation region, again the linearity is in question, and the data become less reliable, at least at the amplitude selected.

In summary, due to the Nernstian response of bacteria to the working electrode potential, it is important to perform several control studies, as described here, to select the optimum conditions under which both requirements of linearity and finiteness are met. Through the use of other electrochemical techniques such as CV, and also the use of data validation methods such as the Kramers–Kronig fitting, the appropriate perturbation amplitudes and working electrode potentials for EIS can be determined for the bacterium and reactor configuration under study.

8.3.2 Stability

A time delay for the equilibration of the system to the selected working electrode potential is required before performing an EIS measurement. This equilibration time allows the system to adjust to the specified condition and achieve a pseudo-steady state. It is of the utmost importance that the system has a stable current during the EIS measurement, as EIS measures small changes in current as a result of the AC voltage perturbation. Biological systems are difficult to get to a pseudo-steady state because of the dynamic nature of bacteria. Unlike electrochemical cells, MXCs can take several minutes to reach an equilibrium state while the cells adapt to the potential of the electrode by changing cellular processes such as intracellular conditions and protein expression. In addition, the presumably short time at which the system is at open circuit when changing between potentials for different scans allows charge to be built up in the system.

For example, while the method parameters are being changed, during which the system will be at open circuit, the heme groups of extracytoplasmic cytochromes that are known to provide a location for electron storage due to continued bacterial respiration might build up charge [34]. The re-polarization of the electrode creates a quickly decreasing higher current as compared to the steady state. In addition to these, there are a number of cellular processes, including changing electron donor concentrations, cell growth, and mixing patterns, among others that can affect the stability of the system during EIS. These changes in bioanode or biocathode performance in time (as a result of growth, for example) are an important aspect to take into account in EIS measurements, especially when the measurements take place over a longer period of time. In addition to the EIS measurements, it is important to provide information of the stability of the bioanode or biocathode performance. This can be done by recording the changes in current density as a function of time (at constant potential) and/or regular recording of polarization curves before and after the EIS measurement.

In Figure 8.9a, we show the equilibration time before an EIS experiment of a *G. sulfurreducens* biofilm as a result of an adjustment to the potential of the electrode (from −0.3 V vs. Ag/AgCl during chronoamperometry to −0.42 V vs. Ag/AgCl to perform EIS). The maximum in the transient current shown immediately after the polarization of the electrode is a result of the re-oxidation of cytochromes that had been reduced previously during the interruption due to built-up charge. Thereafter, the current reaches a minimum, presumably because the bacteria had changed their metabolic state during open circuit conditions. Approximately 10 min was required in this example to achieve a stable current density that is appropriate for EIS measurements. We stress that it is important to perform such equilibration time

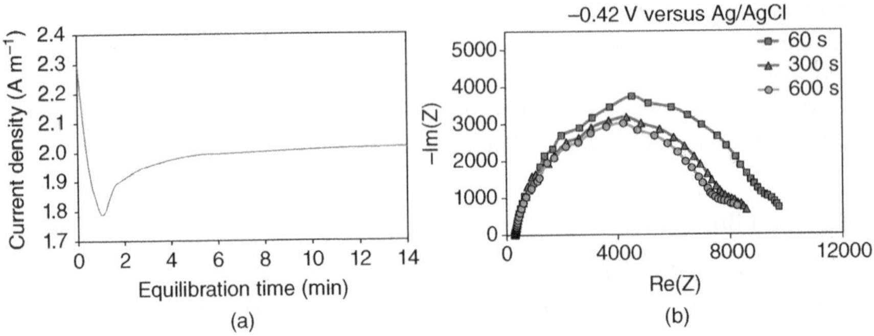

Figure 8.9 The effect of equilibration time on EIS measurements on a *Geobacter sulfurreducens* biofilm: (a) shows the time it takes to reach a steady current after a brief period at open circuit while the EIS method is input and (b) shows the effect of different equilibration times on the Nyquist plot obtained.

measurements before performing EIS measurements to ensure that the system is steady.

Figure 8.9b shows the Nyquist plot resulting from performing EIS measurements at different equilibration times. For each of these, the chronoamperometry was interrupted, allowing the cell to reach open circuit, before starting the equilibration time measurement at the selected working electrode potential and performing EIS measurements thereafter. The equilibration time selected affects the total impedance and capacitance of the system and can be seen in the changes in the Nyquist plot. Here, if we waited for only 1 min before beginning EIS measurements, we were essentially conducting EIS at the lower current density observed in Figure 8.9a, which is an artifact of the transient open circuit induced while changing the method on the potentiostat. Usually, the size of the electrode and the thickness of the biofilm affect the amount of charge collected when the system is not poised, thus affecting the equilibration time required. Thus, it is important to perform a stability analysis (as shown in Fig. 8.9a) for the MXC under study, before performing EIS measurements.

Overall, the stability principle applies to how stable the system appears not only before EIS collection, but also during a sweep. For more complicated systems, such as continuous flow cells, the flow rates must be kept constant during the scan to ensure a relatively constant concentration of electron donor and other media constituents. In general, every effort should be made to evaluate the experimental conditions and ensure that the system does not change prior to or during data collection.

8.3.3 Causality

The principle of causality for EIS measurements is related to many different sources that can cause interferences and noise in the data. For example, too many instruments running in a particular area can cause interferences. Orazem and Tribollet [35] recommends that particular attention be paid to avoiding frequencies including 60, 120, 50, and 100 Hz (all ±5 Hz), as at these frequencies, other instruments in the surrounding can affect the data collected.

8.3.3.1 ***Magnetic Stirring*** Although magnetic stirring is a common way to ensure proper mixing of media in a bioreactor and to minimize diffusion of species within the electrolyte, the magnetic field created may cause problems with data collection for certain reactor geometries. We have observed that there are certain cases where magnetic stirring does not interfere with the measurements, while some in which the noise in EIS measurement is large due to magnetic stirring. Whether magnetic stirring is acceptable during EIS data collection should be determined for each setup. One way to monitor noise of the system is the chronoamperometry method, which monitors the current output over time. In our experience, the noise in current should be <5% of the signal, to be able to perform EIS measurements at most working electrode potentials.

8.3.3.2 ***Faraday Cage*** In theory, a Faraday cage can be used to house the reactor and shield sensitive electrical connections from outside interferences. We have performed several measurements with and without a Faraday cage to determine its effect. In every situation we tested, the data suggested that the particular reactor setup we employed was not sensitive to the effects of external electromagnetic noise. However, other environments may have significant interferences and would require the use of a Faraday cage in order to obtain accurate EIS measurements. Outside interferences can increase the noise in the current measurements, making it more difficult to separate individual impedances from the results.

8.3.3.3 ***Electrode Size*** The size or type of electrode can alter the measurement performed with EIS. Larger and porous electrodes have higher capacitances at the electrode surface that can interfere with other measurements. In larger working electrodes, potential gradients also can lead to variable conditions depending on the location; thus, the recorded EIS measurement is a composite response of a range of potentials within the electrode. Larger electrodes are also more susceptible to variable fluid dynamics and substrate concentrations, which can affect the stability of the system.

Figure 8.10 shows the Bode phase angle plots for EIS measurements conducted for a working electrode potential of −0.36 V versus Ag/AgCl with an amplitude of 10 mV for three reactor configurations (an H-type MXC with 4.2 cm^2 cylindrical graphite working electrode, a single-chamber MXC with 4.2 cm^2 cylindrical graphite working electrode, and a single-chamber MXC with 3.14 mm^2 disc-shaped gold working electrode). It is important to compare the Bode plots in this case, because the Nyquist plots for the different set-ups show no significant differences (except obviously for the magnitude). Figure 8.10, however, suggests that processes can be manifested at different frequencies, depending on the size of electrode, or type of cell configuration used. In the high-frequency region of the figure, the Ohmic effects appear to be elongated for the larger sized electrodes. In addition, a final process is visible for the smaller electrode at the lowest frequencies (<1 Hz), suggesting that the two processes may be merged on the larger size electrodes and thus are inseparable. Although the

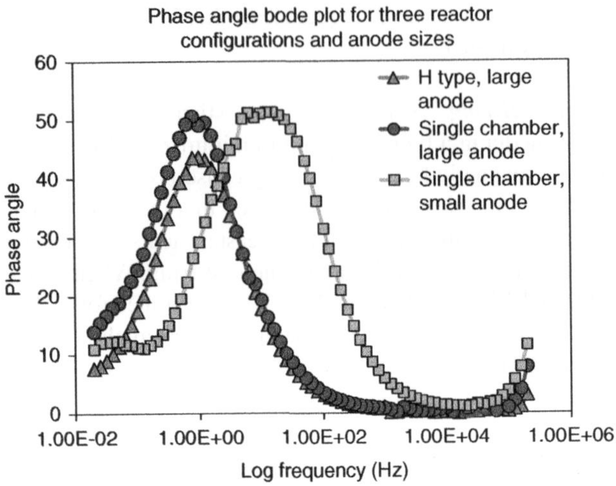

Figure 8.10 Bode plot for *Geobacter sulfurreducens* biofilms at the same anode potential, but with different cell configurations/anode sizes.

size of the electrodes to be used may depend on the intended application, our results suggest that smaller electrodes may be most effective in analyzing distinct bacterial processes using EIS.

8.4 DATA ANALYSIS

8.4.1 Equivalent Circuit Modeling

In order to correctly interpret the experimental information provided by EIS, conveyed in either Nyquist or Bode plots, the use of a sound physical model describing the relevant biophysicochemical processes taking place in the system is essential. A simple strategy to deal with the experimental information involves the implementation of the model into an equivalent circuit, which contains all the information of charge transport. In the equivalent circuit, the resistances and capacitances describe the charge loss and accumulation mechanisms that can take place in the system. In the following section, we first describe the most common circuit elements used in EIS data analysis, followed by the most common equivalent circuits used to describe typical electrochemical cells.

A resistor, along with a constant phase element (CPE, signifying a nonideal capacitor) in parallel, is usually used to describe a charge transfer process. Diffusion is usually accounted by a Warburg element, which contains information about the diffusion coefficient and diffusion length of the carriers. More complex situations take place when there is coupling between diffusion and resistive losses, and consequently distributed elements, such as the Gerischer element, can be introduced in the equivalent circuit to account for these complex phenomena. This is especially the case when porous electrodes or conducting polymers are studied. An extensive development

TABLE 8.1 List of Common Circuit Elements Used in Equivalent Circuit Modeling for Nyquist Plots Generated from Electrochemical Impedance Spectroscopy.

Denomination	Symbol	Scheme	Impedance
Resistance	R		R
Capacitance	C		$\frac{1}{i\omega C}$
Inductor	L		$i\omega L$
Constant phase element (CPE)	Q		$\frac{1}{(i\omega)^n Q_n}$
Warburg (conducting boundary)	W_S		$R_D \frac{\tanh[(i\omega/\omega_d)^{1/2}]}{(i\omega/\omega_d)^{1/2}}$
Warburg (blocking boundary)	W_O		$R_D \frac{\coth[(i\omega/\omega_d)^{1/2}]}{(i\omega/\omega_d)^{1/2}}$
Gerischer	G		$\frac{R_G}{(1 + i\omega/\omega_{rec})^{1/2}}$

of the theory of diffusion and recombination on thin layers has been carried out by Bisquert, to which we refer the readers [36]. Table 8.1 summarizes the basic elements that conventionally are employed to describe simple physicochemical processes and the impedance associated to these elements.

The most common approach to describe the physicochemical processes taking place within an MXC entails the use of equivalent circuits as those listed in Table 8.2. We also show in the table the corresponding Nyquist and Bode plots for each circuit. We describe some of these circuits in brief here. For example, circuit 1 is a simple circuit representing the Ohmic resistance in a system. This is a linear resistor, for which the Bode plot shows a straight line at a single impedance value throughout the range of frequencies.

A more relevant example is related to the charge transfer at an electrode, which is in contact with a liquid solution. This situation is best described by a parallel combination of a resistance and a capacitance (circuit 2), where the resistance conveys the charge transfer kinetics at the electrode/liquid interface and the capacitance reflects the accumulation of charge at the Helmholtz layer (Helmholtz capacitance). A series resistance (denoted here by R_2) associated to this RC combination reflects the Ohmic losses. This equivalent circuit is called a Randles' circuit. On the other hand, if the relevant capacitive processes identified in the electrochemical response of the system are not ideal, which is often the case for porous electrodes, CPEs (Q) can replace the capacitances (circuit 3). The parameter accounting for the nonideality of the system (n) must be comprised between 0.8 and 1 (to model a nonideal capacitance). Otherwise, for $n < 0.8$, the CPE can be related to a different process, which is not essentially capacitive. It is this circuit that is most commonly used to describe charge transfer

TABLE 8.2 List of Common Equivalent Circuits Used for Modeling Nyquist Plots Generated from Electrochemical Impedance Spectroscopy.

	Equivalent circuit	Nyquist	Bode
1	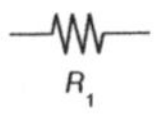 R_1 = 1 kΩ	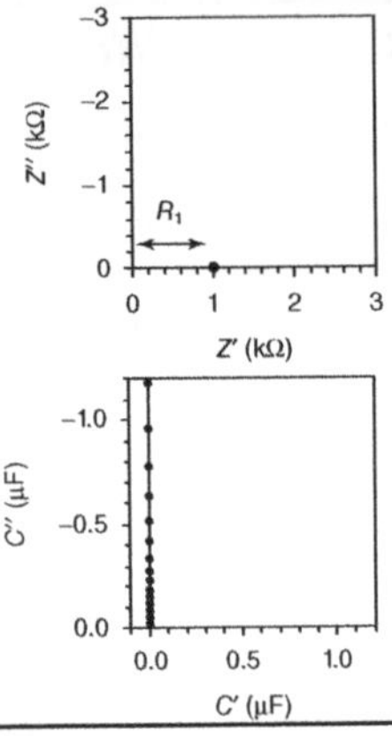	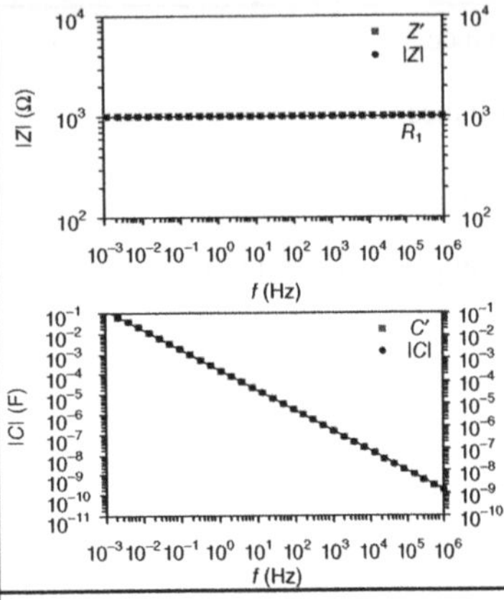
2	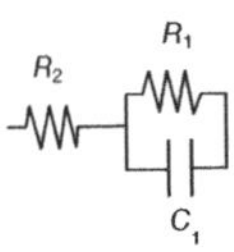 R_2 = 1 kΩ R_1 = 10 kΩ C_1 = 1 μF	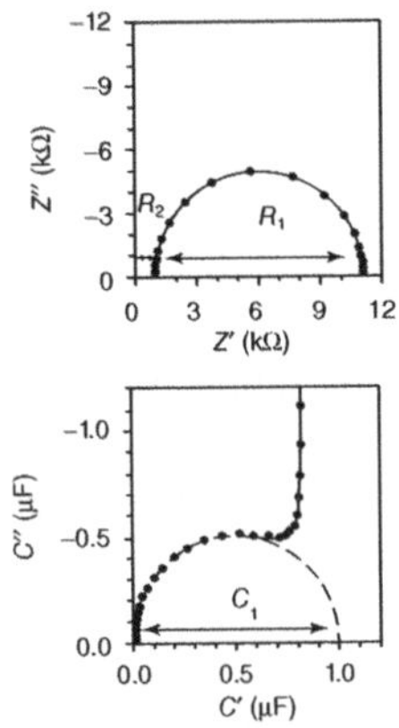	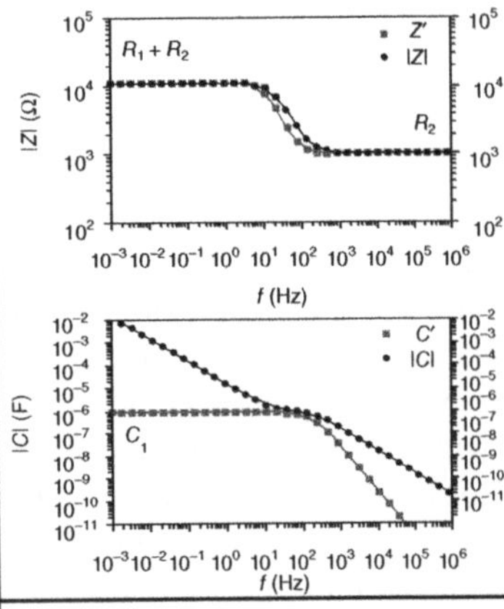
3	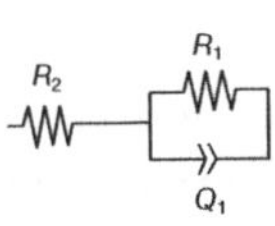 R_2 = 1 kΩ R_1 = 10 kΩ Q_1 = 1 μF s^{1-n} n = 0.8	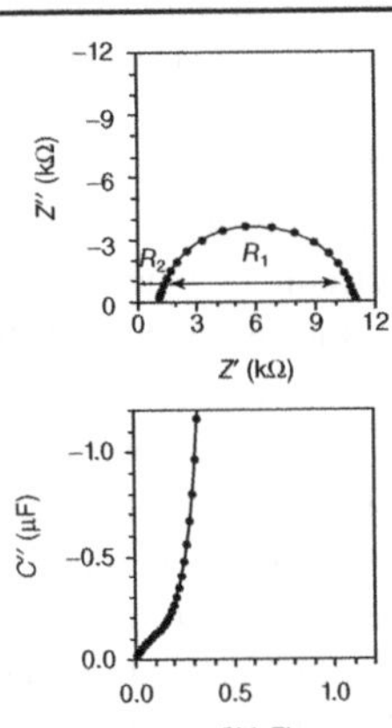	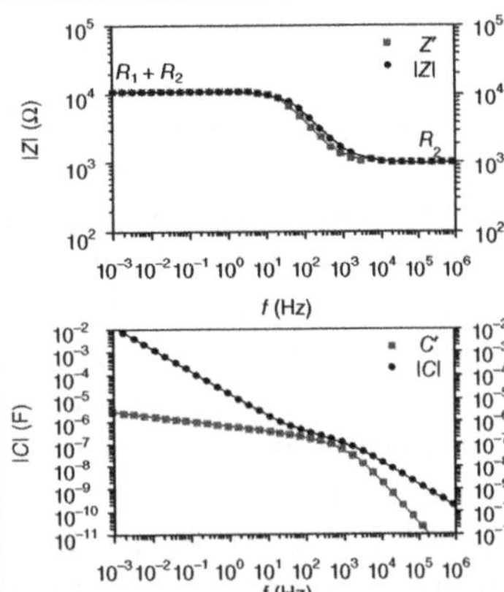

TABLE 8.2 (*Continued*)

4	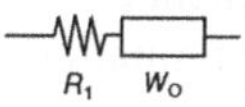 $R_1 = 1$ kΩ $R_D = 10$ kΩ $W_D = 100$ s^{-1} $C_D = 1/(R_D \cdot W_D)$	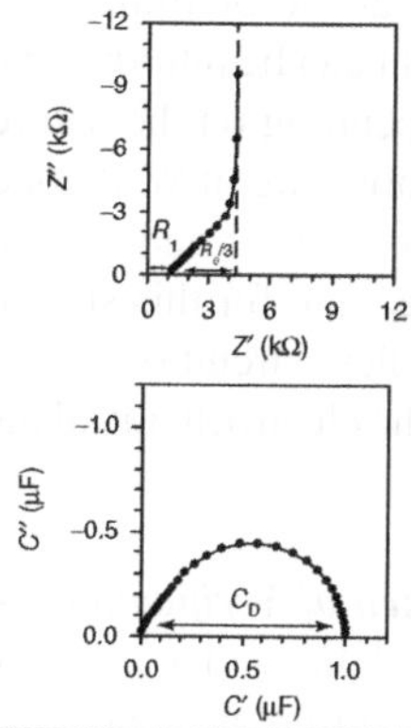	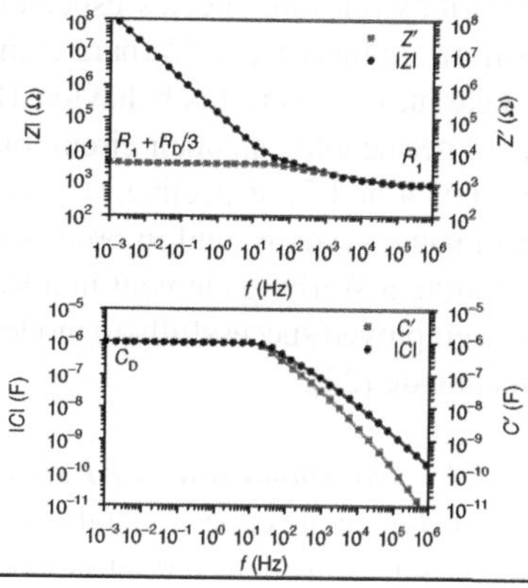
5	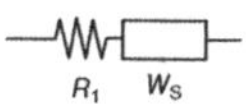 $R_1 = 1$ kΩ $R_D = 10$ kΩ $W_D = 100$ s^{-1} $C_D = a^2/(R_D \cdot W_D)$ $a = R_D/(R_1 + R_D)$	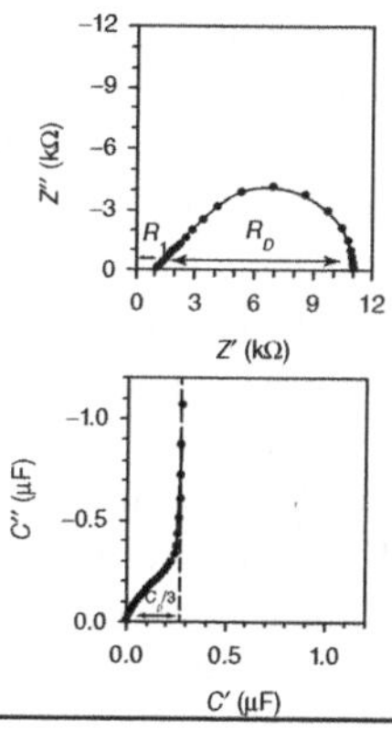	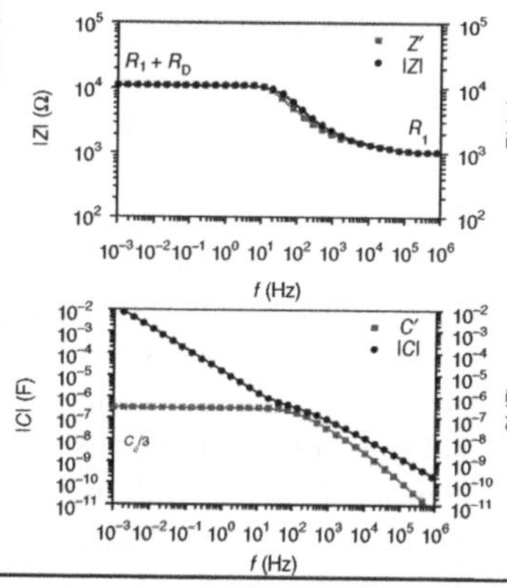
6	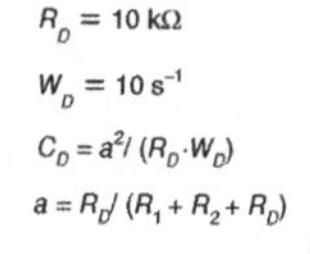 $R_1 = 1$ kΩ $R_2 = 2$ kΩ $C_1 = 0.1$ μF $R_D = 10$ kΩ $W_D = 10$ s^{-1} $C_D = a^2/(R_D \cdot W_D)$ $a = R_D/(R_1 + R_2 + R_D)$	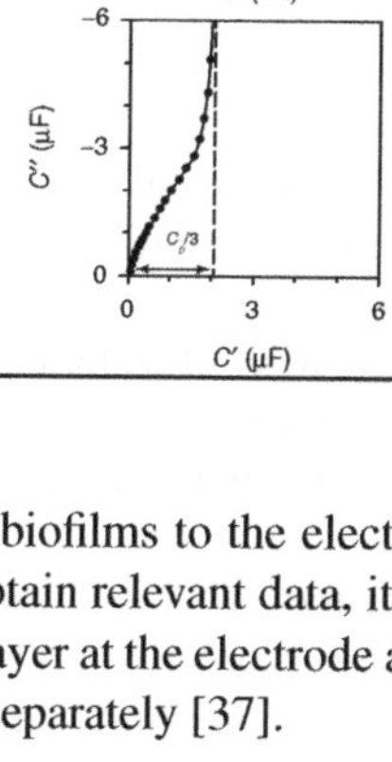	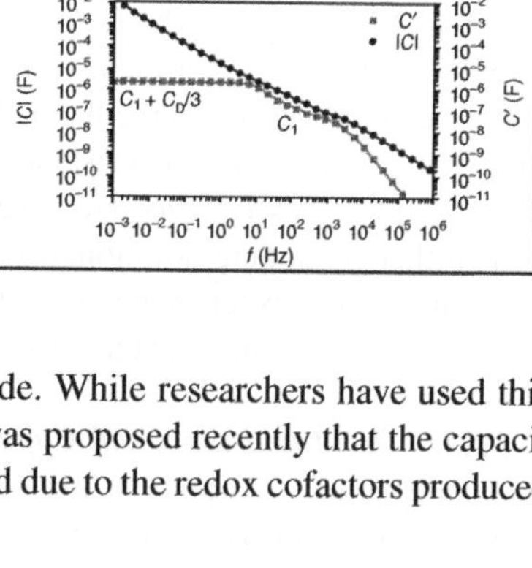

from electrode-attached biofilms to the electrode. While researchers have used this circuit successfully to obtain relevant data, it was proposed recently that the capacitance due to the double layer at the electrode and due to the redox cofactors produced by bacteria be modeled separately [37].

A more complex situation in MXCs involves the presence of diffusive limitations of various relevant species especially at the biofilm/liquid interface or within the biofilms. In this case, a Warburg element can be introduced into the equivalent circuit to take into account this behavior. Depending on the nature of the interface (blocking or permeable), an open-circuit or short-circuit Warburg element can be employed (circuits 4 and 5, respectively). Charge transfer and diffusion limitations can coexist in some systems, and in order to account for this situation, a simple approach is including a Warburg element in a Randles' circuit (see circuit 6). This approach has been employed successfully to model the electrochemical response of an O_2-reducing biocathode [22].

8.4.1.1 The Importance of Experimental Verification A good numerical agreement between the experimental data and the model fit does not necessarily mean that the chosen equivalent circuit is valid. Different models may give similar results in terms of numerical fit. The physical interpretation of the different parameters must be carefully taken into account in order to arrive at a valid interpretation of the EIS data. To verify the validity of the equivalent circuit to which the EIS results are compared, systematic changes in experimental conditions are of importance. For example, in the EIS work performed on biocathodes that catalyze oxygen reduction, the Nyquist plots showed a single arc [22]. In general, this single arc is interpreted as a single charge transfer resistance. Changes in the experimental conditions, such as the linear flow velocity, however, revealed that besides charge transfer, diffusion resistance also influenced the width of the arc. Analysis with the standard Randles circuit that includes both charge transfer and diffusion resistance (circuit 6 in Table 8.2) gave consistent results: with increasing linear flow velocity (improved mass transfer of oxygen), a decrease in diffusion resistance was observed. At the same time, as expected, charge transfer resistance remained unaffected at constant cathode potential. This showed that the standard interpretation of the single arc in the Nyquist plot, representing a charge transfer resistance, was not valid for the system under study. Similarly, measurements at different electrode potentials can be used to reveal the contribution of charge transfer resistances to the total impedance. We provide further details about this experiment in Section 8.5.

Another example of the use of experimental verification of a selected equivalent circuit includes the study by Ramasamy et al. [38] who performed EIS measurements on MFCs containing *Shewanella oneidensis* at the anode. As *Shewanella* spp. secretes flavins as soluble electron mediators between the cells and the anode [39], the authors included a charge transfer resistance in the equivalent circuit for flavins on the basis of their initial experiments. However, they also confirmed the selection of this circuit further by adding additional flavins, which resulted in a lower charge transfer resistance. Such experiments are vital in ensuring that equivalent circuits used are appropriate, rather than just an abstract mathematical formulation.

8.4.1.2 The Importance of Capacitance The identification of charge storage mechanisms is of paramount importance in order to understand charge transfer in MXCs. The information on charge storage is conveyed by the capacitance, and monitoring the evolution of this parameter with applied voltage can identify easily the origin of a capacitive response. Indeed, the different processes taking place simultaneously in the system can be resolved by the identification of their capacitive response. In addition, the capacitance provides a means to verify the validity of the model. For example, the double layer capacitance is not expected to change considerably during the experiment and should be independent of changes in the experimental conditions. So, if changes in double layer capacitance are found between different types of experiments, one might need to reconsider the chosen model that represents the system under study. On the other hand, electrodes with a high capacitance may complicate the interpretation of EIS data, particularly the distinction between the different processes.

8.4.2 Other Methods

Oftentimes, it is difficult to develop a first approximation of an equivalent circuit upon which to test various experimental conditions. This is because more than one process can have characteristic frequencies so close to each other that it is almost impossible to distinguish the two in Nyquist or Bode plots. To overcome this limitation, researchers in the field of solid oxide fuel cells have developed alternative techniques that allow for a better estimation of equivalent circuit to use to fit the data. Two examples of these are the distribution of relaxation times (DRT) method and the analysis of difference in impedance spectrum (ADIS). While these have never been applied to MXC studies, we believe that these could provide new information of processes that researchers may not have been able to identify previously. We direct the readers to references where these methods are discussed in more detail [40–42]. In addition to these two methods, Dominguez-Benetton and Sevda recently reviewed other alternative methods of data analysis that could be applied to MXCs that we strongly suggest the readers investigate also [37].

8.5 EXAMPLES OF EIS APPLICATIONS IN MXCs

We provide here a brief guide for performing EIS on MXCs to gain specific types of information, using three examples. It is difficult to review all the applications of EIS, but we provide here the three that we believe cover a significant breadth of information that can be obtained through EIS measurements. These examples focus on: (1) characterization of Ohmic resistance of membranes of interest, (2) revealing factors affecting the kinetics of electrode reactions catalyzed by bacteria, and (3) characterization of a complete MXC system.

8.5.1 Characterization of Ohmic Resistance

In this first example, we describe a simple method to easily and rapidly characterize the Ohmic resistance for a membrane used in an MXC. This method can be used to compare and select from various membranes, as we have reported recently [43]. A simple H-type MXC, with an electrode fixed in each chamber, is required for this measurement. The idea is to measure Ohmic resistance between the two electrodes in the presence and absence of the membrane. By subtraction, the resistance due to the membrane then can be calculated. We do want to caution here that the effective distance between the electrodes that is occupied by the solution should remain the same with or without the membrane. This is often difficult to ensure, because when a membrane is placed between two fixed electrodes, it occupies a small volume that the solution initially occupied. The effect this will have on the measurement depends on the relative magnitudes of the solution or membrane resistance.

It is important to optimize the surface area of the membrane and the distance between the two electrodes to get good measurements. We want the Ohmic resistance of the membrane to not be masked by noise and uncertainty and thus want it to be significantly higher than the Ohmic resistance of the solution. The best way to assure this is to use very small distances (<1 cm) between the two electrodes, but also a very small cross-sectional area of the membrane for ion flow. As resistance is inversely proportional to the cross-sectional area, ensuring a smaller area will result in a higher resistance. Note that for the electrodes, any conductive material is fine, as we are not looking specifically for any particular reactions on the electrodes. For the solution between the electrodes, it is recommended to perform the EIS experiment several times in solutions of different conductivities, as membrane resistance will change as the solution conductivity changes. In any case, the measurement at least should be done in the solution of interest to be used for further applications.

Once the electrochemical cell is ready, it needs to be connected to an EIS-capable potentiostat/channel in a two-electrode setup. This means one electrode is connected as the working electrode, while the other is connected as the counter and the reference electrode. Most potentiostats include a specific EIS method for measurement of Ohmic resistances. For reference, for the Bio-Logic potentiostats with EC-Lab software for channels with EIS capability, the method can be found under "Ohmic Drop Determination" as "IR compensation (PEIS) – ZIR." As Ohmic resistance is measured at high frequencies, the concept of the measurement is to apply an amplitude of 10–20 mV on the cell at open circuit, at high frequency (usually >100 kHz, but should be confirmed through measurements at multiple frequencies) and measure the resulting ratio of change in current to the change in voltage. We recommend doing at least 10 measurements and taking an average. The procedure must be repeated with and without the membrane of interest separating the two electrodes. It is absolutely essential not to disturb the location of the electrodes while doing so, and thus we recommend having the electrodes fixed in place in the chambers.

For potentiostats that do not include a specific method for Ohmic drop determination, a full EIS scan over a range of frequencies from high to low can be used. In the resulting Nyquist plot, the first intercept on the *X*-axis is the Ohmic resistance. In case this is not directly apparent, an equivalent circuit including an Ohmic resistor can be used to fit the data and obtain the Ohmic resistance. Once the resistance is determined, it should be reported in the form of area specific resistance (ASR), as Ohm square centimeter. Thus, the area of the membrane used should be multiplied with the subtracted value from the measurement with and without membrane.

8.5.2 Characterization of Electrode Processes

For the second example, we provide an overview of the study published by Ter Heijne et al. [22] on identifying the importance of charge and mass transfer resistances in an oxygen-reducing biocathode. In this study, EIS measurements were done on a three-electrode setup with the biocathode selected as the working electrode. This, as we discussed earlier, helps elucidate the resistances from all the processes occurring at the biocathode. We do want to point out in this case that this study was a follow-up to an earlier study [43] where the authors used chronoamperometry and CV to identify important phenomena that govern the performance of the biocathode. This highlights how EIS can be used in concert with, or to support, other techniques. This is a very important consideration when performing EIS measurements to understand electrode processes. Before performing the EIS measurements, it is important to have a good idea about what parameters affect performance and thus could affect the resistances that can be measured by EIS. For example, the initial study clearly laid out how cathode potential and oxygen availability affected biocathode performance, thus providing the conditions to change for each EIS measurement.

In this case, EIS measurements were performed at different cathode potentials and with different oxygen flow rates. The authors had hypothesized from the initial study that the cathode potential would affect the charge transfer resistance, and the oxygen flow rate the mass transfer resistance. For each EIS measurement done at a different cathode potential or oxygen flow rate from a frequency of 100 kHz down to 1 mHz, the perturbation amplitude (10 mV) was applied after 300 s at the condition selected. This should be done to ensure that the biofilm adapts to the new fixed potential or oxygen flow rate. We showed earlier how this equilibration time should be selected through a series of controlled studies. EIS measurements were done at five different cathode potentials that were selected from earlier *j–V* curves and CV scans obtained for the biofilms that showed the range within which catalytic current as observed for the biofilm. This stresses the importance of using voltammetric scans to determine exactly what range of potentials to obtain EIS measurements. Similarly, seven different oxygen flow rates were used.

The Nyquist plots obtained were fit to multiple equivalent circuits to determine what model works the best. As the authors had already determined earlier the importance of cathode potential and oxygen availability, it could easily be anticipated that the equivalent circuit should include at least one charge transfer and one mass transfer resistance, in addition to the Ohmic resistance between the working and the reference electrodes. Even so, changing the cathode potential and oxygen flow rates clearly confirmed which circuit should be used. We thus suggest that any time EIS measurements are done to understand electrode processes, such studies be performed to confirm the validity of the equivalent circuit used to model the data.

This study is also a good example of how EIS provides additional data to what can be obtained from chronoamperometry or CV. As the data can be fit to a circuit, parameters such as charge transfer resistance and diffusion lengths can be obtained from the fitting. In this case, charge transfer resistance was used to validate the Nernstian-based model used to fit the *j–V* relationship. Also, Dominguez-Benetton and Sevda [37] used the data later to extract important information related to the capacitance of the biofilm and how it could relate to biofilm thickness. Similarly, the experiments with different oxygen flow rates helped determine the diffusion lengths in each case, which can now help develop physicochemical models that not only include biofilm kinetics, but also include substrate transport.

A similar approach was also used by Jung et al. [23] where they looked at, through EIS measurements, the effect of anode potential and pH on anodic biofilms, which were both found to have an effect on the performance in chronoamperometric and voltammetric measurements. We encourage the readers to go through that study as well to learn more about performing EIS measurements to understand electrode processes.

8.5.3 Characterization of the Complete MXC System

For the final example, we select the study by Borole et al. [21], which is one of the first reports on the use of EIS to characterize completely an air-cathode MFC over time, using both two-electrode and three-electrode configurations. In this study, the MFC was characterized through EIS measurements at the cell voltage that resulted in the highest power density. A more expanded characterization would include also doing the measurements throughout the entire range of cell voltages in the polarization curve. However, the focus of this study was to do the characterization over time of operation. For the two-electrode configuration, the anode was used as the counter and the reference electrode, while the cathode was used as the working electrode. The same measurement could be done also with the anode as the working electrode and the cathode as the counter and the reference electrode. While not necessary, it is advisable for whole cell measurements to verify that the results of the measurements made through both of these configurations are similar. For the three-electrode configuration, a reference electrode was used, and either the anode or the cathode was used as working electrode and the other as the counter electrode. All EIS measurements here were done with an amplitude of 1 mV and between the frequency ranges of 100 kHz and 100 mHz. While most resistances appeared to be captured through the equivalent circuit modeling of the data obtained from scanning down to 100 mHz, it is important to note that scans to even lower frequencies could reveal resistances not detected.

The authors used an equivalent circuit model for the whole MXC that included an Ohmic resistance and two charge transfer resistances – one from the anode and one from the cathode. While this was an empirical selection during the time the study was published, Dominguez-Benetton and Sevda [37] showed later how certain graphical methods could be used to determine the number and types of circuit elements to use in the equivalent circuit. While this does not mean that there are only two governing resistances, it is possible that several of that several processes are merged on the Nyquist plot due to overlapping time constants. Performing three-electrode measurements could help reveal if this is really the case. Regardless, one of the important findings from this study was that over long-term operation, the cathode appears to limit the power densities that can be achieved in MFCs. This has since been confirmed through other studies using other techniques, such as that by Popat et al. [44], which identified a large concentration overpotential related to pH being the most dominant factor. Recent studies performed on air-cathodes for MFCs that include EIS measurements show a large diffusion resistance [45, 46], corroborating this limitation.

8.6 SUMMARY

In order to develop efficient MXC applications, it is imperative that we understand and predict overpotentials in these systems. The complexity of MXC processes requires that researchers use many techniques to achieve this goal. The use of EIS has been extremely helpful for other fuel cell applications to measure and characterize overpotentials; as such, EIS is also helpful in MXC research. Nevertheless, one must be aware that the interpretation of EIS data requires previous knowledge of the processes involved in order to build an adequate equivalent circuit. The validation of such equivalent circuits can in part occur through multiple EIS experiments in which external parameters are varied and impedances are measured as a function of these parameters. The use of alternate analytical techniques can be used also alongside EIS to build a better understanding of the electron transport processes. We are hopeful that this chapter will provide a background and guidelines suitable for MXC researchers to use EIS as part of their analytical toolbox.

Glossary of Terms

Activation overpotential	The overpotential associated with activation energy required to initiate an electrochemical reaction at an electrode.
Capacitance	The ability of a circuit element to store electrical charge.
Capacitive reactance	The opposition to change of voltage across a circuit element.
Concentration overpotential	The overpotential associated with the inability to provide reactants to or remove products from electrodes to maintain equilibrium thermodynamics.
Counter electrode	The electrode used in three-electrode electrochemical cells to either accept or donate electrons for the working electrode.
Impedance	The opposition to flow of electrons in a circuit when a voltage is applied. Unlike resistance, impedance also includes a time element, as it is related to the frequency at which the voltage is applied.
Ohmic losses/overpotential	The overpotential associated with resistance to flow of ions in an electrochemical cell.

Overpotential — The deviation of the electrode/cell potential from its equilibrium value when electrochemical reaction proceeds.

Potentiostatic EIS — EIS performed such that the amplitude applied is on the voltage, and the subsequent change in current is measured. There is also galvanostatic EIS, where the amplitude is applied on the current, and the resulting change in voltage is measured.

Resistance — The opposition to flow of electrons in a circuit. It can be related to overpotential as $\Delta V/\Delta I$, where ΔV is the overpotential.

Working electrode — The electrode of interest in an electrochemical cell where the reactions of interest are occurring. The potential of the working electrode is controlled against a reference electrode of choice.

REFERENCES

1. Logan BE, Hamelers B, Rozendal R, Schröder U, Keller J, Freguia S, Aelterman P, Verstraete W, Rabaey K. Microbial fuel cells: methodology and technology. Environ Sci Technol 2006;40(17):5181–5192.
2. Logan BE, Call D, Cheng S, Hamelers HV, Sleutels TH, Jeremiasse AW, Rozendal RA. Microbial electrolysis cells for high yield hydrogen gas production from organic matter. Environ Sci Technol 2008;42(23):8630–8640.
3. Rabaey K, Verstraete W. Microbial fuel cells: novel biotechnology for energy generation. Trends Biotechnol 2005;23(6):291–298.
4. Lovley DR. Bug juice: harvesting electricity with microorganisms. Nat Rev Microbiol 2006;4(7):497–508.
5. Logan BE. Exoelectrogenic bacteria that power microbial fuel cells. Nat Rev Microbiol 2009;7(5):375–381.
6. Lovley DR. Powering microbes with electricity: direct electron transfer from electrodes to microbes. Environ Microbiol Rep 2010;3(1):27–35.
7. Rabaey K, Rozendal RA. Microbial electrosynthesis – revisiting the electrical route for microbial production. Nat Rev Microbiol 2010;8(10):706–716.
8. He Z, Angenent LT. Application of bacterial biocathodes in microbial fuel cells. Electroanalysis 2006;18(19–20):2009–2015.
9. Lovley DR, Nevin KP. Electrobiocommodities: powering microbial production of fuels and commodity chemicals from carbon dioxide with electricity. Curr Opin Biotechnol 2013;24(3):385–390.
10. Rosenbaum M, Aulenta F, Villano M, Angenent LT. Cathodes as electron donors for microbial metabolism: Which extracellular electron transfer mechanisms are involved? Bioresour Technol 2011;102(1):324–333.
11. Fricke K, Harnisch F, Schröder U. On the use of cyclic voltammetry for the study of anodic electron transfer in microbial fuel cells. Energy Environ Sci 2008;1(1):144.

12. He Z, Mansfeld F. Exploring the use of electrochemical impedance spectroscopy (EIS) in microbial fuel cell studies. Energy Environ Sci 2009;2(2):215–219.
13. Katz E, Willner I. Probing biomolecular interactions at conductive and semiconductive surfaces by impedance spectroscopy: routes to impedimetric immunosensors, DNA-sensors, and enzyme biosensors. Electroanalysis 2003;15(11):913–947.
14. Pejcic B, De Marco R. Impedance spectroscopy: over 35 years of electrochemical sensor optimization. Electrochim Acta 2006;51(28):6217–6229.
15. Lisdat F, Schäfer D. The use of electrochemical impedance spectroscopy for biosensing. Anal Bioanal Chem 2008;391(5):1555–1567.
16. Amirudin A, Thieny D. Application of electrochemical impedance spectroscopy to study the degradation of polymer-coated metals. Prog Org Coat 1995;26(1):1–28.
17. Yuan X, Wang H, Colinsun J, Zhang J. AC impedance technique in PEM fuel cell diagnosis – a review. Int J Hydrogen Energy 2007;32(17):4365–4380.
18. Wu J, Yuan X, Wang H, Blanco M, Martin J, Zhang J. Diagnostic tools in PEM fuel cell research: Part I Electrochemical techniques. Int J Hydrogen Energy 2008;33(6):1735–1746.
19. Macdonald JR, Johnson WB. Fundamentals of impedance spectroscopy. In: Barsoukov E, Macdonald JR, editors. Impedance Spectroscopy: Theory, Experiment, and Applications. Hoboken, NJ: John Wiley & Sons, Inc.; 2005. p 1–26.
20. He Z, Wagner N, Minteer SD, Angenent LT. An upflow microbial fuel cell with an interior cathode: assessment of the internal resistance by impedance spectroscopy. Environ Sci Technol 2006;40(17):5212–5217.
21. Borole AP, Aaron D, Hamilton CY, Tsouris C. Understanding long-term changes in microbial fuel cell performance using electrochemical impedance spectroscopy. Environ Sci Technol 2010;44(7):2740–2745.
22. Ter Heijne A, Schaetzle O, Gimenez S. Identifying charge and mass transfer resistances of an oxygen reducing biocathode. Energy Environ Sci 2011;4(12):5035–5043.
23. Jung S, Mench MM, Regan JM. Impedance characteristics and polarization behavior of a microbial fuel cell in response to short-term changes in medium pH. Environ Sci Technol 2011;45(20):9069–9074.
24. Lefebvre MC, Martin RB, Pickup PG. Characterization of ionic conductivity profiles within proton exchange membrane fuel cell gas diffusion electrodes by impedance spectroscopy. Electrochemistry 1999;2(6):259–261.
25. Malvankar NS, Tuominen MT, Lovley DR. Biofilm conductivity is a decisive variable for high-current-density *Geobacter sulfurreducens* microbial fuel cells. Energy Environ Sci 2012;5(2):5790–5797.
26. Malvankar NS, Mester T, Tuominen MT, Lovley DR. Supercapacitors based on c-type cytochromes using conductive nanostructured networks of living bacteria. Chemphyschem 2012;13(2):463–468.
27. Manohar AK, Bretschger O, Nealson KH, Mansfeld F. The polarization behavior of the anode in a microbial fuel cell. Electrochim Acta 2008;53(9):3508–3513.
28. Busalmen JP, Esteve-Núñez A, Berná A, Feliu JM. c-Type cytochromes wire electricity-producing bacteria to electrodes. Angew Chem Int Ed Engl 2008;47(26):4874–4877.
29. Strik DP, Ter Heijne A, Hamelers HVM, Saakes M, Buisman C. Feasibility study on electrochemical impedance spectroscopy for microbial fuel cells: measurement modes & data validation. ECS Trans 2008;13(21):27–41.

30. Torres CI, Marcus AK, Parameswaran P, Rittmann BE. Kinetic experiments for evaluating the Nernst–Monod model for anode-respiring bacteria (ARB) in a biofilm anode. Environ Sci Technol 2008;42(17):6593–6597.
31. Macdonald DD. Application of Kramers–Kronig transforms in the analysis of electrochemical systems. J Electrochem Soc 1985;132(10):2316.
32. Boukamp BA. A linear Kronig–Kramers transform test for immittance data validation. J Electrochem Soc 1995;142(6):1885–1894.
33. Barbero G, Alexe-Ionescu AL, Lelidis I. Significance of small voltage in impedance spectroscopy measurements on electrolytic cells. J Appl Phys 2005;98(11):113703-1–113703-5.
34. Bonanni PS, Schrott GD, Robuschi L, Busalmen JP. Charge accumulation and electron transfer kinetics in *Geobacter sulfurreducens* biofilms. Energy Environ Sci 2012;5(3):6188.
35. Orazem ME, Tribollet B. Electrochemical Impedance Spectroscopy. John Wiley & Sons; 2011.
36. Bisquert J. Theory of the impedance of electron diffusion and recombination in a thin layer. J Phys Chem B 2002;106(2):325–333.
37. Dominguez-Benetton X, Sevda S. The accurate use of impedance analysis for the study of microbial electrochemical systems. Chem Soc Rev 2012;41(21):7228–7246.
38. Ramasamy RP, Gadhamshetty V, Nadeau LJ, Johnson GR. Impedance spectroscopy as a tool for non-intrusive detection of extracellular mediators in microbial fuel cells. Biotechnol Bioeng 2009;104(5):882–891.
39. Marsili E, Baron DB, Shikhare ID, Coursolle D, Gralnick JA, Bond DR. *Shewanella secretes* flavins that mediate extracellular electron transfer. Proc Natl Acad Sci U S A 2008;105(10):3968–3973.
40. Sonn V, Leonide A, Ivers-Tiffée E. Combined deconvolution and CNLS fitting approach applied on the impedance response of technical Ni/8YSZ cermet electrodes. J Electrochem Soc 2008;155(7):B675.
41. Jensen SH, Hauch A, Hendriksen PV, Mogensen M, Bonanos N, Jacobsen T. A method to separate process contributions in impedance spectra by variation of test conditions. J Electrochem Soc 2007;154(12):B1325.
42. Schichlein H, Müller AC, Voigts M, Krügel A, Ivers-Tiffée E. Deconvolution of electrochemical impedance spectra for the identification of electrode reaction mechanisms in solid oxide fuel cells. J Appl Electrochem 2002;32(8):875–882.
43. Ter Heijne A, Strik DP, Hamelers HV, Buisman CJ. Cathode potential and mass transfer determine performance of oxygen reducing biocathodes in microbial fuel cells. Environ Sci Technol 2010;44(18):7151–7156.
44. Popat SC, Ki D, Rittmann BE, Torres CI. Importance of OH-transport from cathodes in microbial fuel cells. ChemSusChem 2012;5(6):1071–1079.
45. Hutchinson AJ, Tokash JC, Logan BE. Analysis of carbon fiber brush loading in anodes on startup and performance of microbial fuel cells. J Power Sources 2011;196(22):9213–9219.
46. Lanas V, Ahn Y, Logan BE. Effects of carbon brush anode size and loading on microbial fuel cell performance in batch and continuous mode. J Power Sources 2011;247:228–234.

9

MATHEMATICAL MODELING OF EXTRACELLULAR ELECTRON TRANSFER IN BIOFILMS

RYAN RENSLOW

Environmental Molecular Sciences Laboratory, Pacific Northwest National Laboratory, Richland, WA, USA

JEROME BABAUTA, CORNELIUS IVORY AND HALUK BEYENAL

The Gene and Linda Voiland School of Chemical Engineering and Bioengineering, Washington State University, Pullman, WA, USA

JIM SCHENK[#]

The Department of Chemistry, Washington State University, Pullman, WA, USA

ANDREW KUPRAT AND JIM FREDRICKSON

Biological Sciences Division, Pacific Northwest National Laboratory, Richland, WA, USA

9.1 INTRODUCTION

Earlier chapters in this book describe electrochemically active biofilms (EABs) and provide tools and techniques to study them. Although there are differences in proposed mechanisms and experimentally measured parameter values, the mechanisms and measured data can be explored using mathematical models to help us understand

[#] In memory of our collaborator and friend James "Jim" O. Schenk (August 30, 1952 – January 31, 2013). We are grateful for his significant contribution to this chapter.

Biofilms in Bioelectrochemical Systems: From Laboratory Practice to Data Interpretation, First Edition.
Edited by Haluk Beyenal and Jerome Babauta.

electron transfer. EABs use a unique mode of respiration, in which terminal electrons derived from their metabolism can be transferred to extracellular, insoluble electron acceptors, such as minerals or electrodes [1–4]. This phenomenon has been explored for use in several applications, including the production of electrical power [5, 6], the production of hydrogen in microbial electrolysis cells [7, 8], and the improvement of wastewater treatment [9, 10] as discussed in detail in this book. Extracellular electron transfer (EET) has also been researched for its use in bioremediation [11–13], for understanding microbially influenced corrosion [14, 15], and for its role in subsurface sediment biogeochemistry [16, 17].

As discussed in this book, to date, there have been two main mechanisms proposed for long-range EET, a diffusion-based and a conduction-based mechanism [18]. Diffusion-based EET relies on the migration, diffusion, and/or advection of soluble electrochemically active molecules (mediators) to carry electrons from cells to the electron-accepting surface [19–21]. Conduction-based EET relies on the transmission of electrons through a conductive biofilm matrix composed of extracellular polymeric substances (EPS) [22–24]. These mechanisms are summarized in Figure 9.1. Theoretical calculations have shown that diffusion-based EET alone cannot account for the current produced by EABs because of mass transport limitations of soluble mediators [18]. Furthermore, there is evidence that some species are able to utilize both mechanisms. *Shewanella oneidensis*, a dissimilatory metal-reducing bacterium, was chosen as the focus of this chapter because of supporting evidence for both diffusion- and conduction-based EET in *S. oneidensis* biofilms [21, 22, 26]. Currently, it is not known to what extent *S. oneidensis* uses each EET mechanism or what

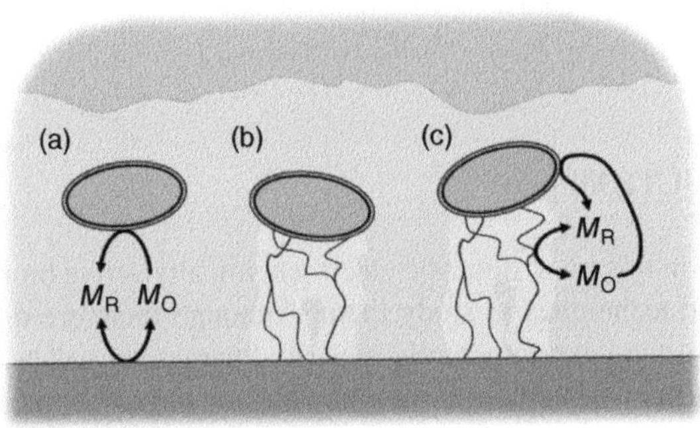

Figure 9.1 Schematic of the cases of extracellular electron transfer explored in this chapter. (a) Diffusion-based EET: redox-active mediators, which move through the biofilm by diffusion, are reduced by the cell and can exchange electrons at the electrode surface. (b) Conduction-based EET: electrons are transported through the conductive biofilm matrix directly to the electrode surface. A noninteracting combination of diffusion- and conduction-based EET (a and b) is called isolated-dual EET. (c) Interacting-dual EET: mediator interactions in a conductive biofilm. The conductive biofilm matrix acts as an extension of the electrode, allowing the matrix and mediators to exchange electrons. This figure and several subsequent figures were originally published in Renslow et al. [25] – Reproduced by permission of the PCCP Owner Societies.

effect different environmental and phenotypic conditions have on EET mechanism expression [18].

To date, all EAB modeling efforts have focused on individual EET mechanisms. In the last 8 years, several conduction-based models have been proposed. In one of the first biofilm-based microbial fuel cell (MFC) models, Marcus et al. developed the Nernst–Monod equation to relate substrate utilization and anode electrical potential to current production for conduction-based electron transfer [27]. This model assumed that the solid anode and the conducting biofilm could take the place of a soluble electron acceptor in the Nernst equation. It was further assumed that biofilm resistivity, ohmic loss, and proton concentration changes were negligible. Marcus et al. found that electron transfer was a primary limiting factor for high power production, principally because of the purportedly high resistance of the biofilm matrix. Pinto et al. developed a conduction-based model [28] for multispecies MFCs that generated power from wastewater via anodophilic and methanogenic microbes. The defining feature of this model is its potential application of controlling the external resistance of an MFC in real time, facilitating maximum power point tracking. This procedure is expected to find the optimum condition for a given MFC to generate maximum power. Also, Zeng et al. developed the first full dual-chamber MFC model [29]. Prior to this model, reduction reactions at the cathode were assumed to be nonlimiting. In this study, Zeng et al. demonstrated that the performance of the anode and that of the cathode are intimately connected. Concurrent with the development of these conduction-based models, several diffusion-based models were also published. Picioreanu et al. modeled an MFC with both biofilm and planktonic organisms with electron transfer via diffusion-based EET [30]. This multidimensional individual-based model considered the diffusion of redox mediators from the cells to a planar anode. Picioreanu et al. were able to use their model to describe experimental results obtained from Bond and Lovley [31]. Incidentally, the experiment by Bond and Lovley, which utilized *Geobacter sulfurreducens*, most likely did not have electron mediators such as flavins. This initial study clearly indicates that modeling should not be used blindly for fitting parameters unless multiple relevant experimental data sets under various conditions are used to verify the model. We should also note that Picioreanu and his collaborators were pioneers of theory and simulation of electron transfer processes. Their work led to the development of more advanced models based on very innovative experimental data, which they did not have access to in 2007. Furthermore, perhaps the system used by Bond and Lovley [31] did involve a type of electron transfer mediation. As discussed in previous chapters, hydrogen produced at the counter electrode can be transferred to the working electrode where it can be consumed by the EAB. In this case, hydrogen can act like an electron transfer mediator. Picioreanu et al. later expanded on their model and included multiple species by incorporating the International Water Association's aerobic digestion model (ADM1) [32] into the diffusion-based model [33]. Furthermore, Picioreanu et al. added electromigration using the Nernst–Planck equation and pH considerations [34]. Collectively, these models show that both EET mechanisms can be used to describe current production in MFCs. They have also helped researchers optimize power and address MFC design issues. Picioreanu's work has shown us that EAB model development is an ongoing process, and it will take time for modelers to construct appropriate models that work for a variety of conditions. However, these models did not investigate the combination, interaction, and relationship of diffusion- and conduction-based EET. This is necessary to determine which mechanism dominates under various experimental and natural conditions. Under certain conditions, the dominant electron transfer mechanisms can be different, and this can be verified by mathematical models. Furthermore, such interactions are critical for mixed-species biofilms, where a variety of species that use different EET strategies may coexist [35]. The fact that mixed-species biofilms generally produce higher current densities than most of the pure culture biofilms warrants the investigation of interactions between EET mechanisms. We believe that mathematical modeling can help us critically investigate these interactions and aid in experimental design.

In this chapter, we explore the possible connections between diffusion- and conduction-based EET and the conditions that affect the usage of each EET strategy by a biofilm. We should note that conduction-based EET can be either electron hopping or metallic-like conduction, as described by two separate chapters in this book. Specifically, we develop a model that includes four different cases of EET (Fig. 9.1): (1) diffusion-based EET; (2) conduction-based EET; (3) isolated-dual EET; and (4) interacting-dual EET. Isolated-dual EET considers electron transfer through a combination of diffusion- and conduction-based EET mechanisms. In this case, a biofilm may use both pathways; however, the pathways are independent, noninteracting, and thus isolated from each other. Interacting-dual EET also considers electron transfer through a combination of diffusion- and conduction-based EET mechanisms; however, in this case, the extracellular biofilm matrix acts as an extension of the electrode and can accept/donate electrons from/to a soluble mediator while conducting electrons from the cells to the electrode. This is in contrast to isolated-dual EET, in which the biofilm matrix cannot interact with mediators. By considering the presence of redox mediators in biofilms, this approach is reasonable. The model we present in this chapter is generic, intended for use with any biofilm-forming electrochemically active organism, applicable to a wide variety of conditions, and inclusive of both steady-state and dynamic solutions. It provides the ability to simulate both cyclic voltammetry (CV) and squarewave voltammetry (SWV) of a biofilm using experimentally observed parameters. This chapter includes a detailed description of how the model was implemented using a MATLAB® client connection to run finite element method software COMSOL Multiphysics®. Where needed, we have presented sample code to help the user see the actual commands used to build the model. As an example application, an *S. oneidensis* biofilm was simulated using experimentally generated parameters and compared to a simulated *G. sulfurreducens* biofilm. We believe that mathematical models are excellent tools for addressing science questions and for testing and refining hypotheses. Therefore, we used our model to address a set of questions, shown in italic in the following:

- *How does the distribution of the EET mechanisms affect theoretical current production?*
  This question addresses the contribution of each mechanism to total EET. Although this question is partially answered in the literature, readdressing

this question allowed us to compare our predictions with previous literature studies. Interestingly enough, we obtained identical results, which illustrated the reproducibility of models using different simulation packages and verified our model.

- *When would the simultaneous use of both diffusion- and conduction-based EET be beneficial for a biofilm?*
  As it has been experimentally demonstrated that some biofilms can use both mechanisms, we determined theoretically whether this generates any benefit for biofilm activity or the total electron transfer rate. Having a solution to address this question can allow future users of the model to design a reactor that utilizes EABs to generate higher electron transfer rates.
- *How does biofilm spatial activity affect the EET mechanism?*
  We asked this question to determine whether a biofilm with higher activity can produce higher current and whether nonuniform activity is connected with the EET mechanism. We expected that somehow microbial activity and EET mechanism should be optimized to maximize electron transfer rates.
- *What are the theoretical maximum biofilm thicknesses for different biofilm conductivities?*
  The current literature on this topic is conflicting. There are studies claiming that thickness is limited by conductivity and other studies claiming that thickness is not limited. The answer to this question can tell us what the limiting factors controlling biofilm thickness are. We should note that initially researchers claimed that certain biofilms do not grow more than a few tens of microns due to conductivity limitations. Remarkably, in the subsequent years, researchers managed to grow several-hundred-micron-thick biofilms. These results motivated us to ask this critical question.
- *How does the effective diffusion coefficient profile affect diffusion-based EET?*
  Often, effective diffusion coefficient values are assumed in the literature to be constant. Recently, it has been shown that this is not correct. This question addresses how critical the selected values of the diffusion coefficients are for the predicted current. During our literature search, we found that unrealistic effective diffusion coefficient values were used to successfully predict experimental data. Therefore, we extended this question to address the importance of using correct effective diffusion coefficients.
- *Can voltammetry determine the contributions of EET mechanisms used by the biofilm?*
  The answer to this question is critically needed to determine what information idealized CV or SWV may provide. We noted that there are many publications presenting CV as a method to determine electron transfer mechanisms. This question is also important to explore the power and limitations of CV as a technique to investigate EET mechanism in a conclusive way.
- *Which parameters are critical for determining the EET mechanisms and theoretical current production?*
  There are many parameters used in our model. By changing their values, we tested how they contribute to the overall current. Finding the key parameters

will help us in designing experiments and inform which variables to control or measure. Measuring parameters that do not critically control electron transfer parameters may not be useful.

- *What is the difference between the local biofilm potential and the redox potential?*
  Recently, our group discovered the local biofilm potential, which is different from the redox potential. The answer to this question will allow us to compare them, as one is based on conduction and defined for the extracellular matrix (local biofilm potential), whereas the other is defined for the liquid phase (redox potential). Local biofilm potential is a parameter that can be measured by electronically connecting the biofilm to the tip of a microelectrode. In this way, potential can be measured at a given point in a biofilm.
- *What benefit may a biofilm gain by allowing mediators to interact with the conductive biofilm matrix?*
  If the EPS that comprise the biofilm matrix are conductive and this matrix interacts with electron transfer mediators, this interaction should provide an advantage to biofilm growth. The answer to this question can help determine the extent of this advantage. There is strong experimental evidence that EPS contain redox mediators, which can transfer electrons. These mediators can be adsorbed to the EPS surface and transfer electrons by conduction as described in earlier chapters. This is an approach we introduced to model EET in EABs.

9.2 GENERAL MODEL FORMULATION

The proposed model simulates an EAB residing on an impenetrable electrode, consisting of a homogeneous distribution of electrochemically active species that can oxidize a substrate to release high-energy electrons. *G. sulfurreducens* biofilms generally grow more homogeneously and fits our proposed model assumptions. However, *S. oneidensis* biofilms generally grow more heterogeneously. As we described earlier, homogeneous biofilms can be represented by average values using stratified biofilm models (Lewandowski and Beyenal, 2014). Electrons from the microbial oxidation of substrate are transferred to the electrode through conduction-based and/or diffusion-based EET mechanisms. The simulated mechanisms will be controlled by the model input parameters. As most of the experimental data in the literature present voltammetric techniques, we include CV and SWV in our simulation, relying on first principles and experimentally observed input parameters when possible. This chapter presents the simplest cases considering the interaction of the two EET mechanisms and shows how to build this model simulation using selected software. We believe that the user can extend the model for more advanced conditions.

9.2.1 Biomass

The biofilm in our model is composed of a distribution of biomass (cells and EPS, the biofilm matrix). We assume that the biomass is homogenous in three dimensions and does not change thickness or composition over time. This is a reasonable assumption as long as the biofilm has reached a pseudo-steady state and only short time frames are considered [36]. Our model does not predict biofilm growth but predicts electron transfer rates for given conditions. Our model also explores the effect of biomass heterogeneity on current production and EET mechanisms by incorporating variable diffusion coefficients and density profiles. The biofilm has defined thickness, L (m); cell density, X (g m^{-3}); and isotropic conductivity, κ (S m^{-1}).

9.2.2 Biofilm Metabolism

The biofilm subsists on the oxidation of an organic substrate, S (mM), which is delivered to the biofilm matrix via diffusive mass transport. The substrate is supplied at a constant concentration to a large, well-mixed anodic chamber. This allowed us to assume that the bulk concentration of substrate is constant because of the size of the chamber and the relatively slow consumption rate of substrate by the biofilm. The concentration at the biofilm surface is equal to the bulk concentration because of mixing in the anodic chamber and because of simplification of the model. The substrate utilization rate is controlled by both the substrate (electron donor) concentration and the electron acceptor concentration, through multiplicative Monod substrate utilization equations [37, 38]. Equations 9.1 and 9.2 simply state that the biofilm can only metabolize in the presence of both an electron donor and an electron acceptor. The lack of either one prevents biofilm metabolic activity. In our model, we assume that there are two possible electron transfer pathways; thus, there are two substrate utilization equations. For diffusion-based EET, substrate utilization is given by:

$$q_{\mathrm{M}} = q_{\max}\left(\frac{S}{S+K_{\mathrm{S}}}\right)\left(\frac{M_{\mathrm{o}}}{M_{\mathrm{o}}+K_{\mathrm{M}}}\right) \tag{9.1}$$

where q_{M} (mol g^{-1} s^{-1}) is the specific substrate utilization rate via diffusion-based EET, $q_{\max}$ (mol g^{-1} s^{-1}) is the maximum specific substrate utilization rate, S (mM) is the substrate concentration, K_{S} (mM) is the half-saturation constant of the substrate, M_{o} (mM) is the oxidized mediator concentration [M_{r} (mM) is the reduced mediator concentration], and K_{M} (mM) is the half-saturation constant of the mediator. For conduction-based EET, substrate utilization is given by the Nernst–Monod equation [27]:

$$q_{\mathrm{C}} = q_{\max}\left(\frac{S}{S+K_{\mathrm{S}}}\right)\left(\frac{1}{1+\exp\left(-\frac{F}{RT}\left(E-E_{K_{\mathrm{A}}}\right)\right)}\right) \tag{9.2}$$

where q_{C} (mol g^{-1} s^{-1}) is the specific substrate utilization rate via conduction-based EET, F (s A mol^{-1}) is the Faraday constant, $\mathcal{R}$ (J K^{-1} mol^{-1}) is the universal gas constant, T (K) is the temperature, E ($\mathrm{V_{SHE}}$) is the local biofilm potential, and $E_{K_{\mathrm{A}}}$ ($\mathrm{V_{SHE}}$) is the half-maximum rate potential. The electron acceptor term, which is a function of E, is known as the Nernst–Monod term [27]. We should note that electron acceptor and its use in Equation 9.2 is different from the classical Monod description because it comprises an electrode surface rather than a soluble molecule. The maximum specific substrate utilization rate is considered to be the same, regardless of whether the EET is diffusion- or conduction-based, as it is assumed that the maximum specific substrate utilization rate is not affected by the EET mechanism used by the microorganism. In our model, diffusion- and conduction-based EET can be used simultaneously; thus, piecewise functions are needed to define a modified q_M^* and q_C^* if $q_M + q_C$ is larger than q_{max}.

$$q_M^* \equiv f(q_C, q_M, q_{max}) = \begin{cases} q_M, & \text{if } q_M + q_C \le q_{max}\left(\frac{S}{S+K_S}\right) \\ q_{max}\left(\frac{S}{S+K_S}\right)\frac{q_M}{q_M+q_C}, & \text{if } q_M + q_C > q_{max}\left(\frac{S}{S+K_S}\right) \end{cases} \tag{9.3}$$

$$q_C^* \equiv f(q_C, q_M, q_{max}) = \begin{cases} q_C, & \text{if } q_M + q_C \le q_{max}\left(\frac{S}{S+K_S}\right) \\ q_{max}\left(\frac{S}{S+K_S}\right)\frac{q_C}{q_M+q_C}, & \text{if } q_M + q_C > q_{max}\left(\frac{S}{S+K_S}\right) \end{cases} \tag{9.4}$$

Note that, by definition, multiplicative Monod substrate utilization allows for the limitation of metabolism when the electron acceptor is limiting. For EABs simulated in this model, the terminal electron acceptor is always the electrode; however, diffusion- and conduction-based EET allow for two possible access paths for the electrode. Whether EET is 100% diffusion based, 100% conduction based, or a mixture of the two, the maximum substrate utilization rate is the same. The total specific substrate utilization rate, q (mol g^{-1} s^{-1}), is given by:

$$q = q_M^* + q_C^* \le q_{max} \tag{9.5}$$

Equations 9.3 and 9.4 ensure q will never exceed q_{max}. Furthermore, the ratio q_M^*/q_C^* will not deviate from q_M/q_C, which is determined by the relative availability of mediators or a conduction pathway, even if $q_M + q_C$ does exceed q_{max}. Thus, when EET is a mixture of diffusion-based and conduction-based mechanisms, the EET mechanism percentages will not be a function of substrate (electron donor) concentration. The overall substrate reaction rate due to metabolism, R_S (mol m^{-3} s^{-1}), is given by:

$$R_S = Xq \tag{9.6}$$

9.2.3 Diffusion-Based Extracellular Electron Transfer

In the biofilm, the oxidized mediator, M_o, is reduced by cells at a rate proportional to the utilization of substrate, q_M^*. It is speculated that redox mediators, such as flavins, can interact with outer membrane *c*-type cytochrome proteins, for example, MtrC and OmcA, receiving electrons from the periplasmic cytochrome MtrA and transferring them to terminal electron acceptors outside of the cell. In this case, it is assumed that the mediators are not fixed in location and that the distances between mediators and the terminal electron acceptor are too great (greater than tens of Ångströms) for efficient electron hopping. Thus, for this mechanism, electrons are carried by mediators that move via diffusion in the biofilm and undergo subsequent cycles of oxidation and reduction reactions. Simply, mediators have been described as shuttles, carrying electrons to change their physical location and allowing them to interact with distant electron acceptors (e.g., electrodes). The following equations summarize the steps of diffusion-based EET:

Step 1:	$S(\text{bulk}) \rightleftarrows S(\text{biofilm})$	Transport
Step 2:	$S + \text{Cell}_o \rightarrow \text{Cell}_r + P$	Reaction
Step 3:	$\text{Cell}_r + M_o \rightarrow \text{Cell}_o + M_r$	Reaction
Step 4:	$M_r(\text{biofilm}) \rightleftarrows M_r(\text{electrode})$	Transport
Step 5:	$M_r \rightleftarrows M_o + ne^-$	Reaction

Kim et al. [39], Velasquez-Orta et al. [40], and Lanthier et al. [41] all found that the number of recovered electrons in actual systems is significantly lower than that expected from substrate oxidation with 100% coulombic efficiency. We also observed that this was true for our experimental data. Therefore, a correction factor, f, is defined to represent the fraction of electrons obtainable for external electrical energy production. The formation of the reduced form of the mediator, M_r, is given by:

$$R_M = \frac{yf}{n} X q_M^* \tag{9.7}$$

where R_M (mol m^{-3} s^{-1}) is the reduction rate of M_o in the biofilm due to metabolism, y (unitless) is the electron equivalence of the substrate, f (unitless) is the fraction of electrons recoverable for current, and n (unitless) is the number of electrons transferred by a redox mediator. Note that an uppercase "R" in the provided equations indicates that the reaction is taking place in the biofilm matrix (i.e., is a volumetric reaction), whereas a lowercase "r" indicates that the reaction is taking place at the electrode surface. The biofilm is grown on a polarized electrode, where M_r and M_o can be oxidized and reduced, respectively, depending on the polarization potential. By convention, *positive current* is produced when the electrode oxidizes M_r, releasing electrons to the electrode and ultimately the electrochemical cell circuit. *Negative current* is produced when M_o is reduced. While the reaction at the electrode is reversible, it is assumed that M_r does not have the ability to deliver electrons back into the cell. The reaction of the mediator at the surface is governed by Butler–Volmer kinetics. The oxidation rate of M_r is given by:

$$r_o = M_r k_0 e^{\left[(1-\alpha)\frac{nF}{RT}(\varepsilon - E^{0'})\right]} \tag{9.8}$$

where r_o (mol m^{-2} s^{-1}) is the oxidation rate of M_r at the electrode surface, M_r (mM) is the reduced mediator concentration, k_0 (m s^{-1}) is the standard heterogeneous rate constant for the mediator redox reaction, α (unitless) is the transfer coefficient for the mediator redox reaction, ε (V_{SHE}) is the polarized electrode potential, and $E^{0'}$ (V_{SHE}) is the standard redox potential for the mediator redox reaction. The reduction rate of M_o, r_r (mol m^{-2} s^{-1}), is given by:

$$r_r = M_o k_0 e^{\left[-\alpha\frac{nF}{RT}\left(\varepsilon - E^{0'}\right)\right]} \tag{9.9}$$

The reduction of M_o results in the formation of M_r; thus, the net generation rate of e^- at the electrode surface, r_M (mol m^{-2} s^{-1}), is given by:

$$r_M = n(r_o - r_r) \tag{9.10}$$

9.2.4 Conduction-Based Extracellular Electron Transfer

There is currently a debate as to exactly how conduction-based EET occurs in EABs, whether by electron superexchange or by metallic-like conduction [24, 42, 43]. We should note that there are two chapters in this book describing how to perform measurements related to the proposed mechanisms. We are hoping that when the measurements are replicated by other groups and new data are generated, it can be integrated into mathematical models to improve our understanding of electron transfer mechanisms. In both electron superexchange and metallic-like conduction, (1) electrons are conducted through the biofilm matrix, (2) electron transfer occurs on shorter time scales than molecular diffusion, and (3) current must be conserved. The steps of conduction-based EET can be summarized as follows:

Step 1:	$S(\text{bulk}) \rightleftarrows S(\text{biofilm})$	Transport
Step 2:	$S + \text{Cell}_o \rightarrow \text{Cell}_r + P$	Reaction
Step 3:	$\text{Cell}_r \rightarrow \text{Cell}_o + \text{Matrix}_r$	Reaction
Step 4:	$\text{Matrix}_r \rightarrow \text{Matrix}_o + e^-$	Transport

Electrons originate in the biofilm in proportion to substrate utilization, q_C^*, and propagate in a single direction toward the electrode. The rate at which e^- is generated in the biofilm through metabolism, R_C (mol m^{-3} s^{-1}), and becomes available for conduction-based EET is given by:

$$R_C = yfXq_C^* \tag{9.11}$$

Similarly to the assumption for diffusion-based EET, electrons cannot be sent back to the cells. (The microbial respiratory electron transport chain cannot work in reverse.) We believe that this assumption can be modified to model cathodic EABs that oxidize substrates extracellularly and may have applications in high-value chemical production (e.g., via electrosynthesis).

9.2.5 Mass Balances

The continuity equation for the substrate within the biofilm, on the basis of our assumptions, can be written as follows:

$$\frac{\partial S}{\partial t} = \frac{\partial}{\partial x}\left(D_{eS}\frac{\partial S}{\partial x}\right) - R_S \tag{9.12}$$

where D_{eS} (m^2 s^{-1}) is the effective diffusion coefficient of substrate in the biofilm. Diffusion through a biofilm is impeded by high tortuosity in the matrix and interactions with both cells and EPS; therefore, an *effective* diffusion coefficient is used. For solutes at low concentrations, where the chemical activity of the solute is close to unity (i.e., as in an ideal solution) [44], the effective diffusion coefficient of any solute to which Fick's laws are applicable can be obtained from:

$$D_{ei} = \frac{D_e}{D_{aq}} D_i \tag{9.13}$$

where D_{ei} ($m^2\ s^{-1}$) is the effective Fickian diffusion coefficient of solute i, D_i ($m^2\ s^{-1}$) is the bulk liquid Fickian diffusion coefficient of solute i, D_e ($m^2\ s^{-1}$) is the effective self-diffusion coefficient of water, and D_{aq} ($m^2\ s^{-1}$) is the bulk liquid self-diffusion coefficient of water. Note that the ratio D_e/D_{aq} is called the relative effective diffusion coefficient, D_r (unitless). Equation 9.13 is used for the solutes in this model. Mass balances of the oxidized and reduced mediator, parallel to the continuity equation given previously, are given by:

$$\frac{\partial M_o}{\partial t} = \frac{\partial}{\partial x}\left(D_{eM}\frac{\partial M_o}{\partial x}\right) - R_M \tag{9.14}$$

$$\frac{\partial M_r}{\partial t} = \frac{\partial}{\partial x}\left(D_{eM}\frac{\partial M_r}{\partial x}\right) + R_M \tag{9.15}$$

It is assumed that the effective diffusion coefficients of the oxidized and reduced forms of the mediator are identical. As the molecular weight change is relatively small for oxidized and reduced forms, the change in the effective diffusion coefficients will be ignorable. Full justification for this assumption is given as follows in the Example Model Application section. Thus, if the initial concentration of mediators is uniform, then $\frac{\partial M_o}{\partial t} + \frac{\partial M_r}{\partial t} = 0$.

9.2.6 Current and Potential

The overall current density produced by diffusion-based EET, j_M ($A\ m^{-2}$), is given by a form of Faraday's law:

$$j_M = F \cdot r_M \tag{9.16}$$

The current density produced by conduction-based EET, j_C ($A\ m^{-2}$), is a result of the cumulative electrons produced in the biofilm that are available through substrate utilization via conduction-based EET. Performing an electron balance on the biofilm yields:

$$\frac{\partial j_C}{\partial x} = F \cdot R_C \tag{9.17}$$

It is assumed that the time required for electron transfer through the biofilm matrix is negligible compared to the time scale of the diffusion of dilute solutes in the biofilm; electrons are conducted instantaneously. For simulations considering the isolated-dual and interacting-dual EET cases, the total current density, j ($A\ m^{-2}$), is the total current derived from both underlying EET mechanisms (diffusion- and conduction-based):

$$j = j_M + j_C \tag{9.18}$$

Therefore, for the isolated-dual and interacting-dual EET cases, the fraction of current provided by the underlying diffusion-based EET, d (unitless), and conduction-based EET, c (unitless), can be quantified as:

$$d = \frac{j_{\mathrm{M}}}{j} \equiv 1 - c = 1 - \frac{j_{\mathrm{C}}}{j} \tag{9.19}$$

The phrase "percentage of EET mechanism" is herein used to refer to the percentage of *current* collected at the electrode by means of a particular mechanism. Coulombic efficiency (unitless), the ratio of the rate at which electrons are delivered to the electrode to the rate at which electrons are taken from the substrate, is calculated as:

$$\text{Coulombic efficiency} = \frac{j/F}{\int_0^L y R_{\mathrm{S}} dx} \tag{9.20}$$

Two different potentials are of primary interest in EABs. Firstly, because the biofilm matrix is conductive, a local biofilm potential, E (V_{SHE}), can be calculated. Ohm's law describes the current–voltage relationship for metallic-like conduction or for electron hopping, assuming that the distance between adjacent redox molecules is small [24]. The conductivity, κ, is only associated with electron movement within the biofilm matrix; it does not consider any ions or redox couples that are in solution outside of the matrix:

$$\frac{\partial j_{\mathrm{C}}}{\partial x} = -\frac{\partial}{\partial x}\left(\kappa \frac{\partial E}{\partial x}\right) \tag{9.21}$$

In the future, a different current–voltage relationship may be used if a consensus is formed regarding the true nature of conduction-based EET. Note that electron hopping has also been simulated as a diffusional process, using an effective electron diffusion coefficient to dictate the rate of electron flux for a given potential gradient [23]. The second potential of interest, the redox potential, E_{redox} (V_{SHE}), which is associated with the soluble redox-active compounds in the biofilm interstitial space, can be calculated by applying the Nernst equation to the single redox couple, the mediator:

$$E_{\mathrm{redox}} = E^{0'} - \frac{RT}{nF} \ln\left(\frac{M_{\mathrm{r}}}{M_{\mathrm{o}}}\right) \tag{9.22}$$

Note that the redox potential is generally a function of both mediator concentrations and pH [45, 46]. This is based on the assumption that the mediator undergoes a proton-coupled redox reaction:

$$M_{\mathrm{o}} + ne^- + m\mathrm{H}^+ \rightleftarrows M_{\mathrm{r}} \tag{9.23}$$

where m is the number of protons transferred per mediator redox reaction. (Note that if $n \neq m$, other products, likely including buffering components, need to be included in this equation to ensure a net charge balance.) However, because the standard reduction potential for this reaction is given in the biochemical standard state, and because it is assumed that the pH in the biofilm will be neutral and unchanging, a pH correction term is not included in Equation 9.22. Further justification for this is provided in Section 9.4.

A succinct summary of the entire system of equations with the initial and boundary conditions is provided in Table 9.1. A schematic representation of the model is shown in Figure 9.2.

9.2.7 Interacting-Dual Extracellular Electron Transfer: Mediator Interactions with a Conductive Biofilm Matrix

The aforementioned model formulation represents the "standard case." However, we also wanted to explore the hypothesis that mediators are able to interact with, that is, exchange electrons with, a conductive biofilm matrix. This means that reduced mediators may transfer electrons to the matrix, and likewise, oxidized mediators may accept electrons from the conductive matrix. In this case, dubbed interacting-dual EET, the matrix acts as an extension of the electrode and electron exchange can occur between mediators and the matrix, just as it can between mediators and the electrode surface. Mediators may not have to travel the entire distance between the reducing cell and the oxidizing electrode surface to transfer electrons. We wanted to explore this idea because of the mounting evidence of mediator and cytochrome interactions and the critical role cytochromes play in EAB EET [47–51]. In addition, currently, we have strong experimental evidence that this mechanism is actually involved in electron transfer processes.

We classify any electrons that pass from a mediator into the conductive matrix and then proceed to the electrode as arriving via conduction-based EET, while any electrons that pass from the conductive matrix to a mediator that is subsequently oxidized at the electrode surface as arriving via diffusion-based EET. In simpler terms, the final form of transport that the electron encounters before being accepted by the electrode is the EET form attributed to it, regardless of the EET mode(s) it experienced along its journey. This distinction is required to quantify the percentage of each EET mode. The steps for interacting-dual electron transfer consist of steps from both diffusion- and conduction-based EET. Note that the steps shown here are only an example, as electrons may transfer between mediators and the conductive matrix several times before reaching the electrode.

Step 1:	$S(\text{bulk}) \rightleftarrows S(\text{biofilm})$	Transport
Step 2:	$S + \text{Cell}_o \rightarrow \text{Cell}_r + P$	Reaction
Step 3:	$\text{Cell}_r + M_o \rightarrow \text{Cell}_o + M_r$	Reaction
Step 4:	$M_r(\text{biofilm}) \rightleftarrows M_r(\text{matrix})$	Transport
Step 5:	$M_r + \text{Matrix}_o \rightleftarrows M_o + \text{Matrix}_r$	Reaction
Step 6:	$\text{Matrix}_r \rightleftarrows \text{Matrix}_o + e^-$	Transport

TABLE 9.1 The Model Consists of a Set of Four Second-Order Differential Equations.

Substrate mass balance		
$\frac{\partial S}{\partial t} = \frac{\partial}{\partial x}\left(D_{eS}\frac{\partial S}{\partial x}\right) - R_S$		
Initial condition:	$S(t = 0) = S(\text{bulk})$	Biofilm is saturated
Boundary condition 1:	$\frac{\partial S}{\partial x}(x = 0) = 0$	Neumann BC (No flux at electrode)
Boundary condition 2:	$S(x = L) = S(\text{bulk})$	Dirichlet BC (Defined concentration)
Oxidized mediator mass balance		
$\frac{\partial M_o}{\partial t} = \frac{\partial}{\partial x}\left(D_{eM}\frac{\partial M_o}{\partial x}\right) - R_M$		
Initial condition:	$M_o(t = 0) = 0$	All mediators are reduced in biofilm
Boundary condition 1:	$\frac{\partial M_o}{\partial x}(x = 0) = -\frac{1}{D_{eM}}(r_o - r_r)$	Neumann BC (Butler–Volmer flux at electrode)
Boundary condition 2:	$M_o(x = L) = M_o(\text{bulk})$	Dirichlet BC (defined concentration)
Reduced mediator mass balance		
$\frac{\partial M_r}{\partial t} = \frac{\partial}{\partial x}\left(D_{eM}\frac{\partial M_r}{\partial x}\right) + R_M$		
Initial condition:	$M_r(t = 0) = M_o(bulk)$	Equivalent concentration to bulk
Boundary condition 1:	$\frac{\partial M_r}{\partial x}(x = 0) = -\frac{1}{D_{eM}}(r_r - r_o)$	Neumann BC (Butler–Volmer flux at electrode)
Boundary condition 2:	$M_r(x = L) = 0$	Dirichlet BC (defined concentration)

Electron balance		
$\frac{\partial j_C}{\partial x} = -\frac{\partial}{\partial x}\left(\kappa \frac{\partial E}{\partial x}\right)$		
Initial condition:	$E(t = 0) = \varepsilon$	Biofilm has same potential as electrode
Boundary condition 1:	$E(x = 0) = \varepsilon$	Dirichlet BC (defined potential at electrode)
Boundary condition 2:	$\frac{\partial E}{\partial x}(x = L) = 0$	Neumann BC (no electrons out of top of biofilm)

The following are the relevant algebraic and differential expressions used in the four second-order differential equations:

Substrate utilization

Specific substrate utilization via diffusion-based EET	$q_M = q_{max}\left(\frac{S}{S + K_S}\right)\left(\frac{M_o}{M_o + K_M}\right)$
Specific substrate utilization via conduction-based EET	$q_C = q_{max}\left(\frac{S}{S + K_S}\right)\left(\frac{1}{1 + \exp\left(-\frac{F}{RT}\left(E - E_{K_A}\right)\right)}\right)$
Modified specific substrate utilization via diffusion-based EET	$q_M^* \equiv f(q_C, q_M, q_{max}) = \begin{cases} q_M, & \textit{if } q_M + q_C \le q_{max}\left(\frac{S}{S + K_S}\right) \\ q_{max}\left(\frac{S}{S + K_S}\right)\frac{q_M}{q_M + q_C}, & \textit{if } q_M + q_C > q_{max}\left(\frac{S}{S + K_S}\right) \end{cases}$
Modified specific substrate utilization via conduction-based EET	$q_C^* \equiv f(q_C, q_M, q_{max}) = \begin{cases} q_C, & \textit{if } q_M + q_C \le q_{max}\left(\frac{S}{S + K_S}\right) \\ q_{max}\left(\frac{S}{S + K_S}\right)\frac{q_C}{q_M + q_C}, & \textit{if } q_M + q_C > q_{max}\left(\frac{S}{S + K_S}\right) \end{cases}$

(continued)

TABLE 9.1 (*Continued*)

Reactions in the biofilm volume	
Overall substrate consumption rate:	$R_S = Xq = X(q_M^* + q_C^*)$
Reduction rate of M_o in the biofilm:	$R_M = \frac{yf}{n} X q_M^*$
Generation rate of e^- in the biofilm:	$R_C = yfXq_C^*$
Reactions at the electrode surface	
Oxidation rate of M_r at the electrode surface	$r_o = M_r k_0 e^{\left[(1-\alpha)\frac{nF}{RT}(\varepsilon - E^{0'})\right]}$
Reduction rate of M_o at the electrode surface	$r_r = M_o k_0 e^{\left[-\alpha \frac{nF}{RT}(\varepsilon - E^{0'})\right]}$
Generation rate of e^- at the electrode surface	$r_M = n(r_o - r_r)$
Current	
Current produced by diffusion-based EET	$j_M = F \cdot r_M$
Current produced by conduction-based EET	$\frac{\partial j_C}{\partial x} = F \cdot R_C$ with $j_C(x = L) = 0$
Total current	$j = j_M + j_C$

The four dependent variables are substrate, S; oxidized mediator, M_o; reduced mediator, M_r; and biofilm matrix potential, E. The following are all of the differential equations, initial conditions, and boundary conditions. Each boundary condition is specified as either a Dirichlet (first-type) or a Neumann (second-type) boundary condition. The electrode surface is at $x = 0$, and the top of the biofilm is at $x = L$.

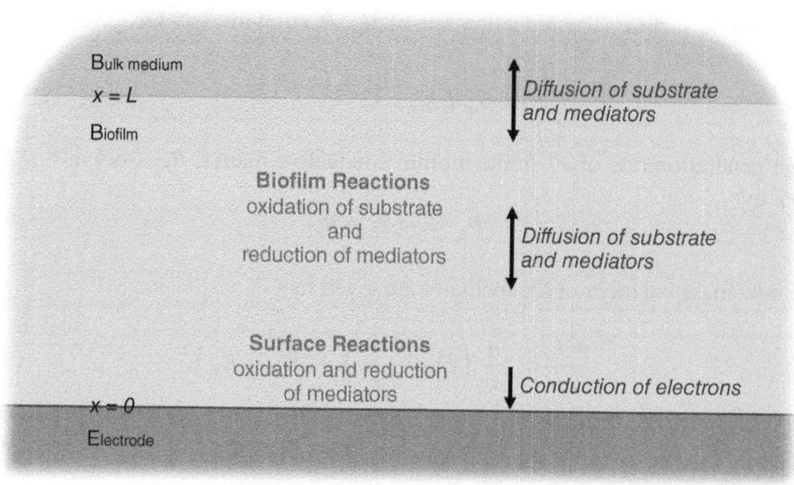

Figure 9.2 *Schematic representation of the model standard case.* The model considers a single domain, the biofilm, with two interfaces, the bulk medium at the top of the biofilm and the solid conducting electrode at the base of the biofilm.

A new variable is defined, the specific electrochemically active surface area per unit volume, S_a (m^{-1}), which is a measure of the active surface area available per volume of conductive biofilm matrix with which the mediator species can interact. This is based on the idea that even if the biofilm is highly conductive, generally, only specific sites will be electrochemically active and allow interaction with electron mediators. This is unlike metal conductors, whose entire surface is generally available for electron transfer. An electrochemically active area may include *c*-type cytochromes, which have been found on nanowires, bound in EPS, and on the outer membrane of *S. oneidensis* cells [50, 52–54]. By making this assumption, we provide a tool to researchers to check if this is acceptable.

The following equations and modifications are relevant only to the cases in which mediators are allowed to interact with the matrix. Firstly, the biofilm is now able to oxidize and reduce the mediators on the basis of the local biofilm potential, as defined by Equation 9.21. This leads to a new set of redox reactions for the mediators in the biofilm. Similarly to the redox reactions occurring at the surface of the electrode, it is assumed that the redox reactions occurring in the conductive biofilm matrix will follow Butler–Volmer kinetics. The oxidation rate of M_r in the biofilm due to interaction with the conductive matrix, R_o (mol m^{-3} s^{-1}), is given by:

$$R_o = S_a M_r k_0^b e^{\left[(1-\alpha_b)\frac{nF}{RT}(E-E^{0'})\right]} \tag{9.24}$$

A new standard heterogeneous rate constant, k_0^b (m s^{-1}), and transfer coefficient, α_b (unitless), are used because it is not currently known whether these values will be the same for redox reactions occurring on the conductive biofilm matrix as for those on the electrode surface. The reduction rate of M_o, R_r (mol m^{-3} s^{-1}), is given by:

$$R_r = S_a M_o k_0^b e^{\left[-\alpha_b \frac{nF}{RT}\left(E-E^{0'}\right)\right]} \tag{9.25}$$

A net generation rate of e^- in the biofilm conductive matrix, R_{S_a} (mol m^{-3} s^{-1}), is given by:

$$R_{S_a} = n(R_o - R_r) \tag{9.26}$$

The new mass balances of the mediator are given by:

$$\frac{\partial M_o}{\partial t} = \frac{\partial}{\partial x}\left(D_{eM}\frac{\partial M_o}{\partial x}\right) - (R_M - R_{S_a}) \tag{9.27}$$

$$\frac{\partial M_r}{\partial t} = \frac{\partial}{\partial x}\left(D_{eM}\frac{\partial M_r}{\partial x}\right) + (R_M - R_{S_a}) \tag{9.28}$$

Equations 9.27 and 9.28 replace Equations 9.14 and 9.15, respectively. Performing an electron balance on the biofilm yields:

$$\frac{\partial j_C}{\partial x} = F(R_C + R_{S_a}) \tag{9.29}$$

This equation replaces Equation 9.17.

9.3 MODEL IMPLEMENTATION

In this section, we provide the necessary details for implementation of the model. We assume that the reader is a graduate student or advanced undergraduate and has a working knowledge of MATLAB® and COMSOL®. Both software packages should be installed before attempting to build the model. Our model was executed using a MATLAB client connection (MathWorks, Natick, MA, USA) to run finite element method software COMSOL Multiphysics (COMSOL Inc., Burlington, MA, USA). This is similar to a previous SWV model developed by our group to simulate a voltammetric flavin microelectrode for use in biofilms [55]. A MATLAB m-file was initially generated (described in the following) using the COMSOL chemical engineering module diffusion application mode for a surface reaction. The most straightforward method to do this initial step is to use the COMSOL Model Wizard to select the appropriate dimensions (one-dimensional in our case), add the desired physics such as Transport of Diluted Species, and then select the study type, which will be Stationary for steady-state modeling and Time Dependent for simulating nonsteady state conditions. After the Model Wizard is finished, the m-file is generated by selecting "Save as Model M-file" under the File menu. More physics modules and study types can be added to the m-file at a later time manually, so it is not critical to set up all model traits in the COMSOL Model Builder prior to saving the m-file. For the model presented in this chapter, the scaffold m-file was structured to encompass all model parameters, geometry, finite element meshing, solvers (both Stationary and Time Dependent), and post-processing visualization. A step-by-step description of the code required to do this is given as follows. Note that to run the m-file in MATLAB, MATLAB must be opened using the "COMSOL with MATLAB" executable. If this executable is not available, check that the COMSOL installation was setup to link with the appropriate MATLAB installation directory. The target for the "COMSOL with MATLAB" executable should be: "**[dir]*comsolserver.exe***" MATLAB, where [dir] is the directory of the COMSOL installation, such as C:\\Program Files\COMSOL\bin\win64\.

To initiate each model, we use a set of commands to ensure that variables are reset during each simulation, to clear old figures, and to set our output format. As good coding practice, other variables will be cleared throughout the m-file to ensure that they are reset for each simulation.

```
clear all      % Removes all variables from the workspace
flclear fem    % Clears Comsol fem variable to reset model
clc            % Clear line command window
close all      % Close all figures
format shortE  % Sets command window output format
format compact % Sets command window output format
```

The fem element will comprise all constants, global expressions, geometries, application modes, custom partial differential equations (PDEs), meshes, and solutions. Each of these is discussed in order.

Any variables or constants that will not be used in COMSOL functions can be declared and manipulated using standard MATLAB commands. However, for the variables and constants that will be used by both MATLAB and COMSOL functions, the variables must be passed to the COMSOL server using the fem.const command. The command takes the following form: `fem.const = {'comsolvar1',matlabvar1,'comsolvar2',matlabvar2,...}`, where comsolvar is the variable name to be used by COMSOL, and matlabvar is the variable name as defined in MATLAB. Typically, we use the same variable name for COMSOL and MATLAB to avoid confusion. The following is an example for five variables that will be used in COMSOL:

```
R = 8.314472;        % Universal gas constant (J/K*mol)
T = 303.15;          % Temperature (K)
F = 96485.3399;      % Faraday Constant (s*A/mol)
f = F/(R*T*1000);    % f constant (1/mV)
Eka = -155;          % Half maximum rate potential (mV vs. SHE)
qmax = 0.002         % Maximum specific growth rate (mmol/(g*s))
Ks = 13.2;           % Half saturation constant for substrate (mmol/L)
Km = 0.0001;         % Half saturation constant for mediator (mmol/L)
% Send constants for use in Comsol functions
fem.const = {'Eka',Eka,'f',f,'qmax',qmax,'Ks',Ks,'Km',Km};
```

Expressions for use by COMSOL functions are defined using the fem.globalexpr command. Anytime a variable, including within expressions, is to be evaluated by a COMSOL function, it must be declared in MATLAB using single quotes. If an expression includes both COMSOL defined and MATLAB defined variables, the entire expression must be in single quotes, and the MATLAB fined variables must be passed using the fem.const function described previously. For an example of two expressions, the Nernst–Monod term (nm) and the specific rate of substrate utilization (qmr):

```
% Define expressions for use by Comsol
fem.globalexpr = {...
      % Nernst-Monod term (unitless)
      'nm', '(1+exp(-f*(E-Eka)))^-1'...
      % Specific rate of S utilization (mmol/(g*s))
      'qmr','qmax*(S/(S+Ks))*(Mo/(Mo+Km))'...
      };
```

For the Nernst–Monod term expression, the f and Eka variables were originally defined in MATLAB and needed to be first sent to COMSOL using the fem.const command as shown previously. The variable E is a solution component (i.e., dependent variable) of a generic PDE application mode that is described in detail later. It is only defined in COMSOL, and thus, it was not necessary to declare it in MATLAB or send it via the fem.const command. Likewise, for the specific rate of substrate utilization expression, q_{max}, K_s, and K_m were defined in MATLAB, whereas S and Mo are the solution components of the diffusion application mode. Note that in MATLAB and COMSOL, we do not use subscripts; instead, we use regular fonts. Because the expression contains both COMSOL defined and MATLAB defined variables, the expression must be in single quotes.

The one-dimensional biofilm geometry is set using the struct commands. However, several COMSOL specific functions must be used to build the geometry, which are then analyzed and finalized using the geomcsg command. Our geometry consists of a biofilm with height L:

```
% Biofilm pseudo steady-state thickness (m)
L = 100*10^-6;
% Define geometry for biofilm
g1 = solid1([0,L]);
% Create a cell array of geometry objects
s.objs={g1};
% Defines a tag for the geometry objects
s.tags={'g1'};
% Converts the geometric object to a Comsol structure
fem.draw=struct('s',s);
% Analyzes the geometry structure and finalizes geometry
fem.geom=geomcsg(fem);
```

Defining a COMSOL application mode consists of three parts: the application mode selection with dependent variable declaration, the boundary settings, and the subdomain settings. In our case, the application mode is diffusion of dilute species from the chemical engineering module:

```
% Diffusion Application Mode
% S: Substrate (lactate), Mo: Oxidized mediator, Mr: Reduced mediator
```

```
species = {'S','Mo','Mr'};       % Chemical species list (mmol/L)
clear appl                       % Clear the application mode
appl.mode.class = 'Diffusion';   % Application mode selection
appl.dim = species;              % Names of solution components
appl.module = 'CHEM';            % Using the chemical eng. module
appl.gporder = 4;                % Order of numerical quadrature
appl.cporder = 2;                % Constraints discretization order
appl.assignsuffix = '_chdi';     % Variable suffix
```

The boundary settings must be selected for each of the three species at both boundary locations.

```
% Boundary Settings
% Boundary 1: electrode surface
% Boundary 2: biofilm/bulk interface
% Bulk concentrations (mmol/L) for S, Mo, and Mr
BulkConc = {25;0.001;0};
% Reaction rates on electrode (mol/(m^2*s)) for S, Mo, and Mr
SurfReact = {0;'rbv_Mo';'rbv_Mr'};
clear bnd                  % Clear the boundary conditions
bnd.c0 = {0,BulkConc};     % Concentration boundary condition
bnd.N = {SurfReact,0};     % Flux boundary condition
% Boundary condition type for each species at each boundary
bnd.type = {{'N0';'N';'N'},{'C';'C';'C'}};
bnd.ind = [1, 2];          % Index value of each boundary
% Apply the boundary settings to the application mode
appl.bnd = bnd;
```

The concentration and surface reactions are declared as MATLAB variables and then passed to COMSOL, similarly to declaring initial conditions. The reaction rates rbv_Mo and rbv_Mr were defined as COMSOL global expressions and are thus written in single quotes. The boundary types are defined by COMSOL as follows: N0 for no flux, N for Neumann (Butler–Volmer flux at the electrode surface), and C for concentration. Only the concentration boundary condition applies at the biofilm/bulk liquid interface; thus, the first index value for `bnd.c0` is zero. Likewise, the reaction of the mediators occurs only at the electrode boundary; thus, the second index value for `bnd.N` is zero.

The final step to defining a COMSOL application mode is to set the subdomain settings. Similarly to how BulkConc and SurfReact were defined in MATLAB for the boundary conditions, DiffCoef, BflmReact, and InitConc would also all need to be defined as three-element arrays in MATLAB, with each element corresponding to the dependent variable of interest (S, M_o, and M_r).

```
% Subdomain Settings (biofilm)
clear equ                  % Clear the subdomain equations
equ.D = {DiffCoef};        % Diffusion coefficients in biofilm
equ.R = {BflmReact};       % Reaction rates in biofilm
equ.init = {InitConc};     % Initial concentrations in biofilm
equ.ind = 1;               % Index value for subdomain (biofilm)
% Apply the subdomain equations to the application mode
appl.equ = equ;
```

The application is now completely defined and can be added to the fem element:

```
% Apply the application mode to the comsol fem element
fem.appl{1} = appl;
```

It is often desired to add a differential equation in COMSOL that is not available through one of their built-in application modes. For example, in our model, we wanted to add a custom PDE to calculate the redox potential drop through the biofilm via Ohm's law; Equation 9.21. To achieve this, we use the generic PDE application mode. As in the built-in application modes, defining a custom PDE application mode consists of the application mode selection, boundary settings, and subdomain settings. However, in the generic PDE application mode, we must also define the PDE itself. For the coefficient form, which is used in this case, the PDE takes the form of:

$$d_{\mathrm{a}}\frac{\partial u}{\partial t} + \nabla \cdot (-c\nabla u - \alpha u + \gamma) + \beta \cdot \nabla u + au = f \tag{9.30}$$

where u is the dependent variable of interest, t is time, and all other coefficients are used to customize the equation. For more details regarding the PDE forms available, please see the COMSOL documentation in the Modeling Guide under PDE Modes for Equation-Based Modeling, particularly The Scalar Coefficient Form Equation section. For our case, the equation form simplifies to:

$$\nabla \cdot (-\nabla u) = f \tag{9.31}$$

The entire declaration of the custom PDE in MATLAB is given as follows:

```
% Generic PDE Application Mode - Redox Potential Profile
clear appl            % Clear the application mode
% Application mode is generic PDE (coefficient form)
appl.mode.class = 'FlPDEC';
appl.dim = {'E','E_t'};  % Array with names of the solution compo-
nents
appl.sshape = 2;          % Order of geometry approximation
appl.assignsuffix ='_c'; % Variable suffix
% Boundary Settings
% Boundary 1: electrode surface
% Boundary 2: biofilm/bulk interface
clear bnd             % Clear the boundary conditions
bnd.r = {'Ea',0};  % Dirichlet boundary condition (E=Ea at 0)
bnd.g = {0,'0'};   % Neumann boundary condition (dEdx = 0 at L)
% Boundary condition type for redox potential at each boundary
bnd.type = {'dir','neu'};
bnd.ind = [1, 2];  % Index value of each boundary
% Apply the boundary settings to the application mode
appl.bnd = bnd;
% Subdomain Settings (biofilm)
clear equ                          % Clear the subdomain equations
equ.da = 0;                        % Coefficient da
equ.c = 1;                         % Coefficient c
equ.al = 0;                        % Coefficient alpha
```

```
equ.ga = 0;                          % Coefficient gamma
equ.a = 0;                           % Coefficient a
equ.be = 0;                          % Coefficient beta
equ.f = '-10000*dJdx/(kbio)';        % Coefficient f (mV/m^2)
equ.ind = 1;                         % Index value for subdomain (biofilm)
% Apply the subdomain equations to the application mode
appl.equ = equ;
% Apply the application mode to the comsol fem variable
fem.appl{2} = appl;
```

The final step before solving the model is to provide mesh for the geometry. This is accomplished using the meshinit and meshextend functions. It is often beneficial for postprocessing to obtain the node coordinates from COMSOL, which can be done using the xmeshinfo command.

```
% Create a mesh with a defined mesh size
fem.mesh = meshinit(fem,'hmax',meshsize);
% Finalize the mesh in the fem element
fem.xmesh = meshextend(fem);
% Obtains the mesh node coordinates from Comsol
nodes = xmeshinfo(fem,'out','nodes');
xgrid = 0:meshsize:L;                             % Mesh node matrix
```

The solvers available in COMSOL are extremely customizable and capable of handling a wide array of PDEs, geometries, and initial/boundary conditions. Here, we present an example steady-state solver setup, with multiple optional features enabled.

```
[fem.sol,fem.solcompdof]=femstatic(fem, ...
                    'solcomp',{'S','Mr','Mo','E'},...
                    'outcomp',{'S','Mr','Mo','E'},...
                    'out',{'sol','solcompdof'},...
                    'blocksize','auto',...
                    'ntol',10^(-1*12),...
                    'Hnlin','on',...
                    'Maxiter',1000,...
                    'Initstep',0.01,...
                    'Rstep',1.1);
```

Each of the desired dependent variables must be listed. The solcomp array provides the names of variables that the solver is to solve for, and the outcomp array designates which variables will be saved for output. Ultimately, the variables listed in the outcomp array will be available in the fem.sol element and can be accessed for postprocessing and visualization. The options listed in the example include the relative tolerance (ntol), the highly nonlinear problem indicator (Hnlin), the maximum number of Newton iterations (Maxiter), the initial damping factor for the step length (Initstep), and the damping factor update factor (Rstep). Depending on the type of problem being solved, these options may be necessary for a solution to be reached. However, for simpler problems, most of these options can be omitted, as the default values will be sufficient. For time-dependent solvers, the femtime function is used in place of femstatic. The femtime function also has a different set of option available than femstatic. We highly recommend that users resort to the COMSOL manuals, as a full description of the solvers and available options has been given.

For postprocessing the model solution in MATLAB, postinterp can be used to deliver COMSOL variables back to the MATLAB workspace where they can be plotted and manipulated using typical MATLAB functions. The postinterp function has the following syntax: `[matlabvar1,matlabvar2,…] = postinterp (fem,'comsolvar1','comsolvar2',…,grid,options)`, where comsolvar is the variable name to be used by COMSOL, and matlabvar is the variable name as defined in MATLAB.

```
xgrid = 0:meshsize:L;       % Mesh node matrix
% Send comsol variables back to matlab for post-processing
[S,Mo,Mr,E] = postinterp(fem,'S','Mo','Mr','E',xgrid,'solnum','all');
```

Here, we have provided a short guide on implementing a model using a MATLAB client connection to run finite element method software COMSOL Multiphysics. Developing a model from scratch in MATLAB, or simply altering a COMSOL built model in an m-file, greatly expands the ease and capability of modeling with COMSOL. An m-file model can be as complex as any built solely in the COMSOL GUI. The COMSOL manuals are an ample resource and should be referenced generously.

For the present chapter, SWV and CV simulations were executed using independent custom m-files that controlled the dynamic electrode polarization potential, ε. Diagrams of the voltammetric wave forms are provided in Figure 9.3. To avoid jump discontinuities when squarewave simulations were run, which would prevent model convergence, the voltammetric signal changes between peak and trough potentials were approximated by a differentiable, and thus continuous, spline function consisting of two quarter-sine waves connected by a straight line, as developed by Nguyen et al. [55] and shown in Figure 9.4. This was chosen over a Fourier series expansion to avoid Gibbs ringing artifacts in the model.

The one-dimensional FEM mesh includes 10,000 elements, spaced 10 nm apart for the standard case. During SWV and CV simulations, the maximum time step was set to 100 μs. Relative and absolute tolerances were set to 10^{-11}. These FEM settings were tested to ensure that increasing the number of elements and decreasing the maximum time step and tolerances did not significantly change the solution. It was concluded that these settings were more than rigorous enough to handle the rapid changes in the SWV waveform.

A one-at-a-time sensitivity analysis was performed to measure the sensitivity of steady-state current production, j, and the fraction of current due to diffusion-based EET, d, to the model input parameters. The sensitivity was evaluated by calculating the resulting average elasticity given a ±25% change in each parameter. Elasticity, δ (unitless), also called relative sensitivity, measures the proportional effect of a change in a parameter value, p, on a model output. We use the sensitivity analysis to determine the relative importance of the selected parameter on the electron transfer rate:

$$\delta_{j,p} = \frac{p}{j}\frac{\partial j}{\partial p} \equiv \frac{\partial \log j}{\partial \log p} \tag{9.32}$$

$$\delta_{d,p} = \frac{p}{d}\frac{\partial d}{\partial p} \equiv \frac{\partial \log d}{\partial \log p} \tag{9.33}$$

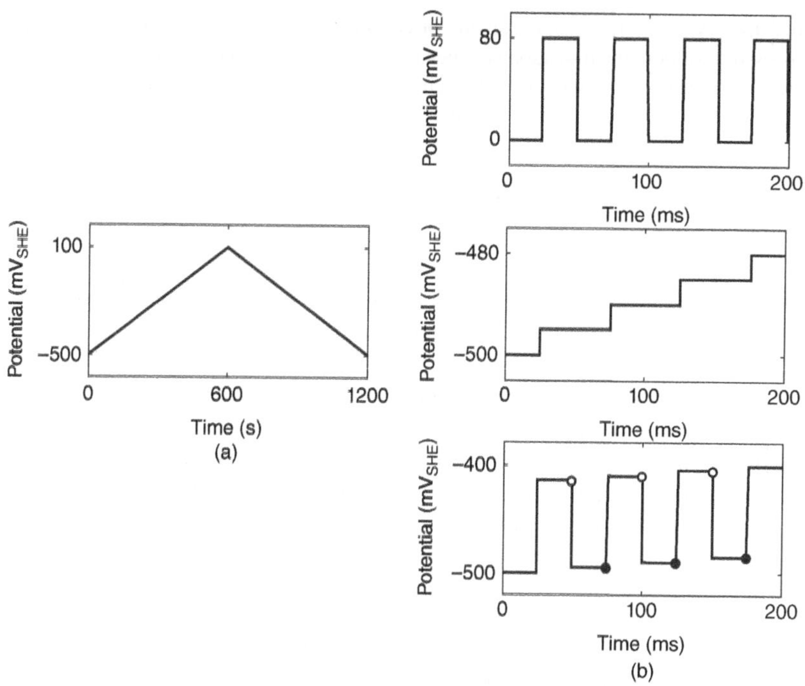

Figure 9.3 Wave forms for cyclic and squarewave voltammetry. (a) Cyclic voltammetry waveform with a 1-mV s^{-1} scan rate, a minimum potential of −500 mV$_{SHE}$, and a maximum potential of 100 mV$_{SHE}$. These parameters are identical to those used to generate the 1-mV s^{-1} scan rate voltammograms in this chapter. (b) The development of the potential waveform for squarewave voltammetry. At the top is shown a squarewave potential, and in the middle is shown a staircase potential. These two waveforms combined produce the final potential waveform for squarewave voltammetry, shown at the bottom. The open circles show the time when current is measured for the end of peak potential, and the filled circles show the time when current is measured for the end of trough potential. The final squarewave voltammogram is composed of the difference between the current measured at the end of the peak potential and that measured at the end of the trough potential. The waveform has a starting potential of −500 mV$_{SHE}$, a final potential of 100 mV$_{SHE}$, a potential step height of 5 mV, a pulse potential of 40 mV, and a 20-Hz scan frequency. These parameters are identical to those used to generate all squarewave voltammograms in this chapter.

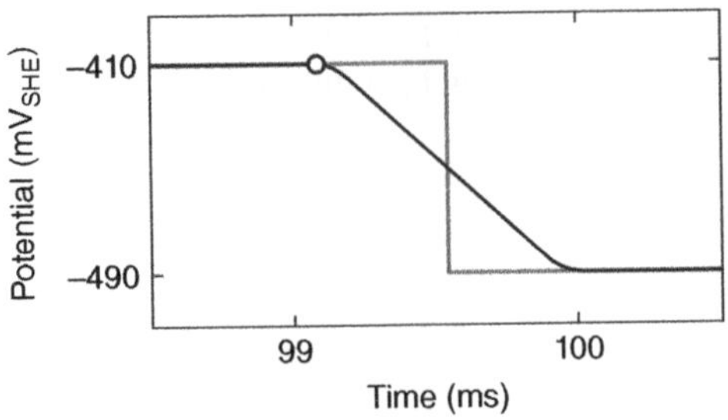

Figure 9.4 A spline function consisting of two quarter-sine waves connected by a straight line is used to smooth the transition between high and low potentials during simulated square-wave voltammetry (black line). The open circle shows the time when current is measured for the end of peak potential. The gray line shows the ideal unsmoothed transition.

9.4 EXAMPLE MODEL APPLICATION: A SHEWANELLA ONEIDENSIS BIOFILM

For the standard case, a 4-day-old *S. oneidensis* MR-1 biofilm growing on a polarized graphite electrode was simulated. A 4-day-old biofilm was chosen on the basis of the availability of literature data, which we used in our model and are discussed in detail in the following paragraphs. In the simulation, the electrode was the only available terminal electron acceptor, as no soluble electron-accepting chemicals were included and the system was therefore anaerobic. All parameter values discussed here for the standard case are summarized in Table 9.2. The polarization potential of the electrode was set at +300 mV_{SHE}. This value was chosen to ensure that current production would not be limited by an insufficiently positive electron acceptor potential. (Note that while electrochemical reactions are not theoretically limited at this potential via Butler–Volmer kinetics, the biomass of *Shewanellaceae* biofilms have been shown to increase at more positive potentials [74].) The temperature was set at 30 °C, which is common for the growth of *S. oneidensis* MR-1 [21, 22, 26, 41, 58, 59]. Most parameters were chosen assuming a 4-day-old biofilm, similar to that of the studies by Baron et al. [58] and Renslow et al. [59]. On the basis of these studies, a biofilm thickness of 100 μm and a lactate concentration of 25 mM were chosen. It was assumed that the impact of pH was negligible. This is based on microelectrode measurements by our group showing that inside a 100- to 200-μm-thick *S. oneidensis* biofilm respiring on a +600 mV_{SHE} polarized electrode, pH varied by only 0.08 units [46]. Furthermore, when the phosphate buffer strength was changed from 10 to 100 mM, the pH variance was found to be zero. Small pH variances for low-current, well-buffered systems have also been predicted by mathematical models [34, 75].

Renslow et al. [59] determined that the average relative effective diffusion coefficient, D_r, across a 100-μm-thick 4-day-old biofilm was 0.89. Although the biofilm in that study was grown on a glass substratum as opposed to a polarized electrode, the D_r was found to be nearly identical to that of a 14-day-old *S. oneidensis* biofilm that was grown on a polarized electrode (D_r of 0.88) [76]. The biofilm density can

TABLE 9.2 Summary of Model Parameter Values for the Standard Case.

Symbol	Description	Value	Units	References
Constants and system parameters				
ε	Polarized electrode potential	+300	mV_{SHE}	[56]
F	Faraday constant	96485.3399	$s\,A\,mol^{-1}$	[57]
R	Molar gas constant	8.314472	$J\,K^{-1}\,mol^{-1}$	[57]
T	Temperature	303.15	K	[21, 22, 26, 58, 59]
Computation parameters				
	Number of finite elements	10,000	unitless	
	Finite element size	0.01	μm	
Biofilm parameters				
D_r	Relative effective diffusion coefficient	0.89	unitless	[59]
f	Fraction of electrons recoverable for current	0.6	unitless	[41, 60, 61]
L	Biofilm thickness	100	μm	[59]
q_{max}	Maximum specific lactate utilization rate	0.002	$mmol\,g_{protein}^{-1}\,s^{-1}$	[62]
X	Biofilm density	1700	$g_{protein}\,m^{-3}$	[59, 63, 64]
Substrate parameters				
S	Bulk lactate concentration	25	mM	[59]
D_{eS}	Effective diffusion coefficient for lactate	1.09×10^{-9}	$m^2\,s^{-1}$	[59, 65–68]
K_S	Half-saturation constant for lactate	13.2	mM	[62]
y	Electron equivalence of lactate	4	$mmol\text{-}e^{-}\,mmol\text{-}La^{-1}$	[41, 69, 70]

(continued)

TABLE 9.2 (*Continued*)

Symbol	Description	Value	Units	References
Diffusion-based EET parameters				
M_o(bulk)	Bulk FMN concentration	1	μM	[26, 41]
M_r(bulk)	Bulk $FMNH_2$ concentration	0	μM	
D_{eM}	Effective diffusion coefficient for FMN/$FMNH_2$	0.43×10^{-9}	$m^2\ s^{-1}$	[59, 65–68]
$E^{0\prime}$	Standard redox potential for the FMN/$FMNH_2$ redox reaction	−217	mV_{SHE}	[71]
k_0	Standard heterogeneous rate constant for the FMN/$FMNH_2$ redox reaction	1.6×10^{-5}	$m\ s^{-1}$	[72]
n	Electrons transferred per FMN/$FMNH_2$ redox reaction	2	unitless	[72]
K_M	Half-saturation constant for FMN	0.1	μM	[30]
α	Transfer coefficient for the FMN/$FMNH_2$ redox reaction	0.5	unitless	[73]
Conduction-based EET parameters				
E_{KA}	Half-maximum rate potential	−155	mV_{SHE}	[56]
κ	Biofilm conductivity	0.5	$mS\ cm^{-1}$	[56]

TABLE 9.3 Details of Several Anaerobic *S. oneidensis* MR-1-Based MFCs with Lactate as the Electron Donor. Coulombic Efficiencies Are Based on the Partial Oxidation of Lactate to Acetate.

Study	Temperature (°C)	Lactate Concentration (mM)	Electrode Material	Coulombic Efficiency (%)
Bretschger et al. [60]	22	5	Graphite felt	54[a]
Lanthier et al. [41]	30	10	Graphite sticks	56
Newton et al. [61]	25	10	Graphite felt	48[a]
Watson and Logan [80]	23	18	Graphite fibers	16[b]
Watson and Logan [80]	23	18	Graphite paper	25[b]
Watson and Logan [80]	23	18	Graphite paper	24[b]
Watson and Logan [80]	23	18	Graphite brush	20[b]
Qian et al. [81]	RT	20	Gold	2.8 micro-MFC
Rosenbaum et al. [82]	30	20	Graphite paper	11
Velasquez-Orta et al. [40]	–	60	Graphite plate	6

[a]Indicates coulombic efficiencies that were corrected to be based on the partial oxidation of lactate to acetate rather than on the complete oxidation of lactate to CO_2. Calculations of coulombic efficiency are described in Logan et al. [83].
[b]Values read from a figure.
RT: room temperature.

be approximated from the following equation, which is modified from the studies by Renslow et al. [59] and Fan et al. [63]. It assumes a 50% weight percent protein composition [64, 77]. (This is similar to the 55% weight percent protein used by Marcus et al. [27], which was derived from the study by Whitman et al. [78]):

$$D_r = 1 - \frac{2.46X^{0.92}}{19,500 + X^{0.99}} \tag{9.34}$$

As defined by Equation 9.34, for a D_r of 0.89, X is 1700 g-protein m^{-3}. The maximum specific substrate utilization rate, 0.002 mmol g^{-1} s^{-1}, and the half-saturation constant for lactate, 13.2 mM, were selected on the basis of the studies by Tang et al. [62] and Zeng and Zhang [79]; however, other values are explored later in this chapter. Table 9.3 shows the coulombic efficiencies of several anaerobic *S. oneidensis* MR-1-based MFCs. The highest coulombic efficiencies are 48–56%. From these data, the maximum fraction of electrons obtainable for current, f, is set at 0.6.

In the simulation, cells gain energy through the partial oxidation of lactate to acetate [41, 69, 70]. This results in an electron equivalence value of 4 mmol-e^- mmol-lactate^{-1}:

$$\text{lactate}^- + 2H_2O \rightleftarrows \text{acetate}^- + HCO_3^- + 5H^+ + 4e^- \tag{9.35}$$

Note that the partial oxidation of lactate to acetate yields 66.7% fewer electrons than the complete oxidation to carbon dioxide and that while *S. oneidensis* MR-1 appears to be incapable of oxidizing lactate to completion under normal anaerobic conditions, other strains of *S. oneidensis* appear capable of the complete oxidation of lactate [60].

Recently, actively secreted flavins have been identified as the electron mediators in *S. oneidensis* EET to solids [21, 26, 40]. In the majority of studies, *S. oneidensis* under anaerobic conditions produces flavin mononucleotide (FMN) at the highest percent, followed by riboflavin (RF), and then by a very low percent of flavin adenine dinucleotide (FAD) [19, 20, 26, 40, 84, 85]. We should note that our research group could not find any FMN produced by *S. oneidensis* biofilms under anaerobic conditions. We found that FMN is produced only when the *S. oneidensis* are grown under aerobic conditions. However, our technique could not detect below 50 nM of FMN. We believe that these controversial findings will be another discussion in the future of EAB research. In this chapter, we wanted to follow current published literature values and provide a tool to test experimental results, rather than developing a model on the basis of our experimental results only. Table 9.4 shows a summary of the properties of each of these flavins. Because of its prevalence, FMN was chosen as the electron mediator for the standard case. Each flavin, including FMN, can undergo a two-step oxidation. When fully oxidized, or in the flavoquinone form (FMN), it can undergo a single-electron reduction to the flavosemiquinone form (FMNH) and then a second single-electron reduction to the flavohydroquinone form ($FMNH_2$). This is shown for FMN:

$$\mathrm{FMN} + \mathrm{H}^+ + \mathrm{e}^- \rightleftarrows \mathrm{FMNH} \tag{9.36}$$

$$\mathrm{FMNH} + \mathrm{H}^+ + \mathrm{e}^- \rightleftarrows \mathrm{FMNH_2} \tag{9.37}$$

It has been found that flavosemiquinones generally exist only in a low concentration (about 2% [86], although they were previously believed to be in much higher concentrations [71]), as they are rapidly converted through the fast reaction of 2 flavosemiquinones to 1 flavoquinone and 1 flavohydroquinone [72]. The full reduction of flavoquinones to flavohydroquinones is very rapid and can be considered a single, two-electron reduction step as shown:

$$\mathrm{FMN} + 2\mathrm{H}^+ + 2e^- \rightleftarrows \mathrm{FMNH_2} \tag{9.38}$$

For our model, we assume that the redox reactions occur in a single, two-electron transfer step, which occurs at a single redox potential. Note that rapid experiments that look at the mechanisms of each electron step report separate standard redox potentials for the flavoquinone–flavosemiquinone and flavosemiquinone–flavohydroquinone reactions for all three flavins [87]. In our model, we neglect the loss of flavin

TABLE 9.4 Properties of the Three Common Electron Mediator Flavins at 30 °C.

Flavin	Molar Mass (g mol^{-1})	Diffusion Coefficient (m^2 s^{-1})	Standard Redox Potential (V_{SHE})
RF	376.36	0.52×10^{-9}	−0.211
FMN	456.34	0.48×10^{-9}	−0.217
FAD	783.53	0.36×10^{-9}	−0.221

mediators due to adsorption to the electrode surface. Adsorbed flavins are believed to increase the electron transfer rate of other flavins, and, in fact, flavins may only be able to transfer electrons to graphite electrodes through contact with adsorbed flavins [72]. We assume that the graphite electrode has steady-state flavin coverage and that soluble flavin concentrations are not affected by adsorptive/desorptive processes. We believe that the adsorptive/desorptive processes would be dominant during initial biofilm development rather than in an established biofilm at pseudo-steady state. Common values for the standard redox potentials of flavins have been −208 mV versus the standard hydrogen electrode (SHE) for RF [88] and −219 mV_{SHE} for both FAD and FMN [89]. In this chapter, however, we chose to use values listed by Ksenzhek and Petrova [71], which have been updated to reflect more modern experiments. For all of the values in their study, we corrected the potential for temperature if it was different from 30 °C using the Nernst equilibrium equation ($nK = nFE^0/RT$), the van't Hoff equation, and enthalpy change values for the flavin reductions given by Beaudette and Langerman [90] and Watt and Burns [91]. By averaging the values in their table, we obtain the redox potentials listed in Table 9.4. For the standard-case simulation, the mediator (FMN) has a standard redox potential of −217 mV_{SHE}.

For the reduction and oxidation of FMN on the graphite electrode, a standard heterogeneous rate constant, k_0, of 1.6×10^{-3} cm s^{-1} was assumed on the basis of the value obtained by Verhagen and Hagen for FAD on a glassy carbon electrode [72]. (FAD is related to FMN, having the same redox-active isoalloxazine group as FMN.) This heterogeneous rate constant value is within the normal range given by Bard and Faulkner [73] and similar to the standard heterogeneous rate found for cytochrome c on a glassy carbon electrode [92]. Furthermore, it is assumed that the redox transfer coefficient is 0.5, which implies a symmetric energy barrier for the forward and reverse FMN redox reactions [73]. Later in this chapter, we explore the effect of different redox transfer coefficients, in part because Verhagen and Hagen found that for the redox of FAD on glassy carbon, the transfer coefficient was about 0.34 [72].

The standard-case simulation assumed an endogenous concentration of FMN only. Velasquez-Orta et al. [40] studied *S. oneidensis* in an MFC setup and found endogenous concentrations of extracellular FMN up to 0.1 μM and negligible extracellular RF and FAD in the anodic chamber after 4 days. On the other hand, von Canstein et al. [26] found that planktonic *S. oneidensis* cultures produced 2 μmol $g_{protein}^{-1}$ FMN, but negligible RF and FAD in 24 h. If the standard-case biofilm produced this amount, the concentration of FMN inside the biofilm would reach 3.4 μM. These numbers highlight the current dilemma in diffusion-based EET research, where the concentration of mediators inside the biofilm is likely to be much higher than that detected in the surrounding bulk liquid. For the standard-case simulation, we assumed a concentration of 1 μM FMN, which represents a conservative estimate based on current understanding. This is similar to the concentration our group found near the base of an *S. oneidensis* biofilm using a voltammetric flavin microelectrode [55]. Again, we should note that this is the concentration when biofilms were grown aerobically. Later in this chapter, the addition of exogenous mediators is explored through a parametric study to determine the effect on current and percentage of the EET mechanism. Baron et al. [58] and Velasquez-Orta et al. [40] added up to 5 μM of flavins, and Covington et al. [84] added up to 10 μM of flavins. In this chapter, we explore concentrations up to 30 μM. The half-saturation constant for FMN, K_m, is 0.1 μM, which is 10% of the initial bulk FMN concentration, analogous to the value used by Picioreanu et al. [30].

The diffusion coefficients of all species in this model were calculated using the Hayduk and Laudie method at 30 °C [67]. This method is based on the viscosity of the solvent and the Le Bas molar volume of the solute [65]. Calculations were performed using an online calculator [66] developed by the US Environmental Protection Agency for modelers with methods taken from Tucker and Nelken [68]. The diffusion coefficients of RF, FMN, and FAD are given in Table 9.4. On the basis of measurements by Visser and Hoek [93], Verhagen and Hagen [72] reported the diffusion coefficient of FAD at 30 °C to be 0.38×10^{-9} m^2 s^{-1}, revealing that the calculated diffusion coefficient of FAD was off by 5%, which is less than the expected 18% uncertainty for the Hayduk and Laudie method [94]. The calculated diffusion coefficient of lactate is 1.22×10^{-9} m^2 s^{-1}, which is only 4% off the value determined experimentally by Øyaas et al. [95]. These results support the appropriateness and accuracy of the Hayduk and Laudie method for obtaining diffusion coefficient values for modeling. The diffusion coefficients of FMN and $FMNH_2$ were assumed to be the same, because the diffusion coefficients of the oxidized and reduced forms were within 2% of each other. The diffusion coefficient was calculated to be 0.48×10^{-9} m^2 s^{-1} at 30 °C, and thus the effective diffusion coefficient of FMN was 0.43×10^{-9} m^2 s^{-1}. The calculated diffusion coefficient of lactate was 1.22×10^{-9} m^2 s^{-1}, and the effective diffusion coefficient was 1.09×10^{-9} m^2 s^{-1}.

The final set of parameters listed in Table 9.2 relates to the conduction-based EET mechanism. Torres et al. [56] showed that the macroscopic conductivity must be greater than or equal to 0.5 mS cm^{-1} using a model developed with the Nernst–Monod equation. For the standard case, we used 0.5 mS cm^{-1} because this conductivity is sufficient to prevent deviation from the Nernst–Monod ideal shape. Note that recent conductivity measurements of individual *S. oneidensis* nanowires were approximately 1000 mS cm^{-1} [96]. Table 9.5 shows the conductivity of selected materials for comparison. Later in this chapter, we show how the EET mechanism and current production are affected by different biofilm conductivities, which were in the range of the values given in Table 9.5. Torres et al. [56] found the measured half-maximum rate potential of a mixed-species wastewater biofilm with predominantly *G. sulfurreducens* to be −155 mV$_{SHE}$. Furthermore, Marcus et al. [27] originally predicted the half-maximum rate potential to be about 100 mV more positive than the open circuit anode potential. The anode open circuit potentials of *S. oneidensis*-based MFCs are generally about −260 mV$_{SHE}$ [60, 97]. From these studies, we assume that the E_{KA} of *S. oneidensis* will also be about −155 mV$_{SHE}$.

Currently, it is not known how much surface area of an *S. oneidensis* biofilm is available for electrochemical interactions with flavin mediators. However, it is known that outer membrane cytochromes are critical for EET [47, 51, 98], and flavin molecules have been confirmed to interact with these proteins [48, 51]. To estimate the surface area available for electron transfer between the matrix and the flavin

TABLE 9.5 Conductivities of Several Materials Compared to the Conductivity Used in the Standard Case *Shewanella oneidensis* Biofilm Model.

Material	Conductivity (mS cm^{-1})
Silver	630,000,000
Iron	100,000,000
Graphite	20,000
Individual *S. oneidensis* nanowire [96]	1,000
This chapter: standard case *S. oneidensis* biofilm	0.5
Silicon	0.0156
Rubber	0.0000000000001

Reference (except where otherwise specified): Serway, Raymond A. *Principles of Physics*. Fort Worth: Saunders College Pub., 1998.

mediators (S_a), we relied on atomic force microscopy and antibody recognition force microscopy of the outer membrane cytochromes (OmcA and MtrC) performed by Lower et al. [50, 99]. They estimated that *S. oneidensis* MR-1 has between 4×10^{15} and 7×10^{15} cytochromes/m^2 of surface area. The estimated surface area of a single rod-shaped *S. oneidensis* cell is 6.48 μm^2, calculated using average measurements of cell length (3.38 μm) and width (0.61 μm) at 22 °C [100]. Assuming a middling cytochrome coverage of 5.5×10^{15} m^{-2} and an average cytochrome diameter of 6.5 nm [101], the cytochrome surface area coverage percentage is about 18%, which is similar to the 16% estimated by Lower et al. [50]. Assuming a single cell density of 1 g cm^{-3} (yielding a cell mass of 4.64×10^{-13} g-protein, comparable to that of Neidhardt et al. [102]) and a biofilm density of 1700 g-protein m^{-3}, the estimated total cell surface area per unit volume is 24,000 m^2 m^{-3}. This yields an estimated S_a of 4300 m^{-1}. A reasonable range would be between 1900 and 8300 m^{-1}, based on the variability of cell sizes, cytochrome surface coverage, and cytochrome sizes. However, the true S_a may be much smaller (even zero) because of steric hindrance of redox-active sites or the inability of flavin–cytochrome reactions to translate electrons to/from the conductive matrix. Note that the aforementioned calculations are based on outer membrane cytochromes located on the cell surface. However, it has been shown that the *c*-type cytochromes of *Shewanella* species can be secreted onto the EPS and may not be associated with the cell surface [103]. Cytochromes located on EPS may not be available for electron transfer (which will likely be another future debate!). Future studies may further refine our estimation of S_a as more knowledge is gained on the role of cytochromes in the EPS and their function in EET. In the standard case, S_a is not utilized. However, when the possible interactions between diffusion- and conduction-based EET are explored for interacting-dual EET, S_a is set to 4300 m^{-1}, unless otherwise noted. It is also assumed that the redox transfer coefficient and the standard heterogeneous rate constant of the FMN/$FMNH_2$ redox reactions are the same at the biofilm conductive matrix as they are at the electrode surface (i.e., $\alpha_b = \alpha$ and $k_0^b = k_0$).

9.5 MODEL SIMULATION RESULTS AND DISCUSSION

To ensure that our model is robust and capable of handling a wide array of input parameters, a complete one-at-a-time parameters analysis was performed. This is a standard technique that all modelers should follow. Failure of model convergence can indicate problems associated with coding, model design, or model assumptions. The parameter analysis is presented in Figure 9.5. These parameter analysis results demonstrate that the model could indeed converge to a steady-state solution for a wide range of parameters. The initial conditions were also tested to ensure that our solution was not dependent on the initial conditions of the model. This is presented in Figure 9.6. Four initial conditions were tested, representing four extremes, and it was found that all the conditions yielded the identical final solutions (measured as current, percentage EET, and mediator depth profiles) and that they all converged smoothly and rapidly. This verified that the codes and developed model were satisfactory and did not include any errors that lead to the failure of the tested conditions.

9.5.1 Using Modeling as a Tool to Address Science Questions

Here, we apply our model to answer questions that are critical for understanding EET in *S. oneidensis*. Furthermore, by varying the model parameter values from the standard-case parameter values listed in Table 9.2, we compare and contrast features of another model EAB, *G. sulfurreducens*, with those of *S. oneidensis*. We chose the standard case from commonly observed experimental or theoretically accepted parameter values, as described in Section 9.4. Throughout this chapter, white-filled circles in figures represent the results at the given locations for the standard case, provided in Table 9.2.

We used the model described in this chapter to address selected science questions. We believe that models are most useful if they are used to address science questions and test hypotheses rather than curve fitting to experimental data. The following section describes some of the science questions we thought were worth exploring. We believe that the reader can extend these questions and apply the model to areas relevant to their research.

9.5.1.1 How Does the Fractional Distribution of Extracellular Electron Transfer Mechanisms Affect Theoretical Current Production? Figure 9.7 shows the steady-state current production and coulombic efficiency as a function of the percentage of diffusion-based EET (without allowing the EET mechanisms to interact, i.e., isolated-dual EET). In the standard case, the steady-state current density is 0.046 A m^{-2}, with approximately 13% of the current coming from diffusion-based EET (~87% coming from conduction-based EET) and a coulombic efficiency of 53%. This is similar to current measured in actual *S. oneidensis* systems [46, 61, 76, 104]. Figure 9.7 shows that as the percentage of diffusion-based EET decreases, the current increases to a maximum of 0.052 A m^{-2} at 100% conduction-based EET. The current reaches a minimum at 0.008 A m^{-2} at 100% diffusion-based EET. The coulombic efficiency also increases as more of the current is collected by conduction-based EET. Note that the highest coulombic efficiency

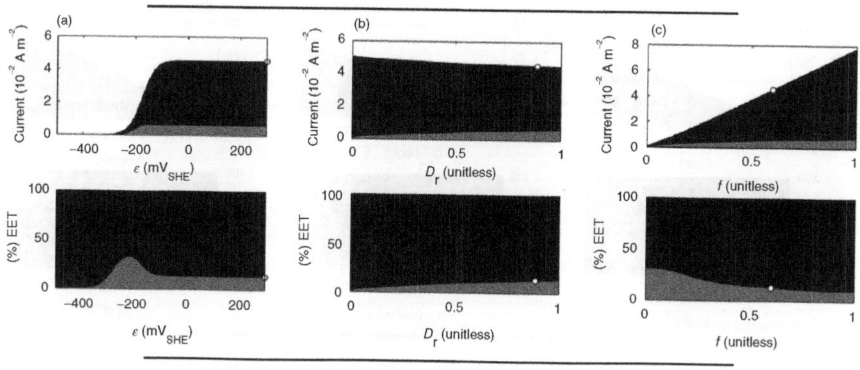

Figure 9.5 *Parameter Analysis.* Each panel (a–s) shows the effect of changing a single parameter value on anodic current production (top) and the percentage of the EET mechanism by which anodic current is generated (bottom). Gray represents the portion due to diffusion-based EET, and black represents the portion due to conduction-based EET. The colored areas are stacked, meaning that the interface of white and black in the current plots conveys the total current (sum of diffusion- and conduction-based EET current). Consequently, the percentage of EET total shown in each bottom plot is always 100%. The white-filled circles are provided to illustrate the parameter value and corresponding current and percent of EET for the model standard case.

The following parameter analyses are shown:

(a) Polarized electrode potential	ε
(b) Relative effective diffusion coefficient	D_r
(c) Fraction of electrons recoverable for current	f
(d) Biofilm thickness	L
(e) Maximum specific lactate utilization rate	q_{max}
(f) Biofilm density	X
(g) Bulk lactate concentration	S
(h) Effective diffusion coefficient for lactate	D_{eS}
(i) Half-saturation constant for lactate	K_S
(j) Electron equivalence of lactate	y
(k) Bulk FMN concentration	M
(l) Effective diffusion coefficient for FMN/$FMNH_2$	D_{eM}
(m) Standard redox potential for the FMN/$FMNH_2$ redox reaction	$E^{0'}$
(n) Standard heterogeneous rate constant for the FMN/$FMNH_2$ redox reaction	k_0
(o) Electrons transferred per FMN/$FMNH_2$ redox reaction	n
(p) Half-saturation constant for FMN	K_M
(q) Transfer coefficient for the FMN/$FMNH_2$ redox reaction	α
(r) Half-maximum rate potential	E_{KA}
(s) Biofilm conductivity	κ

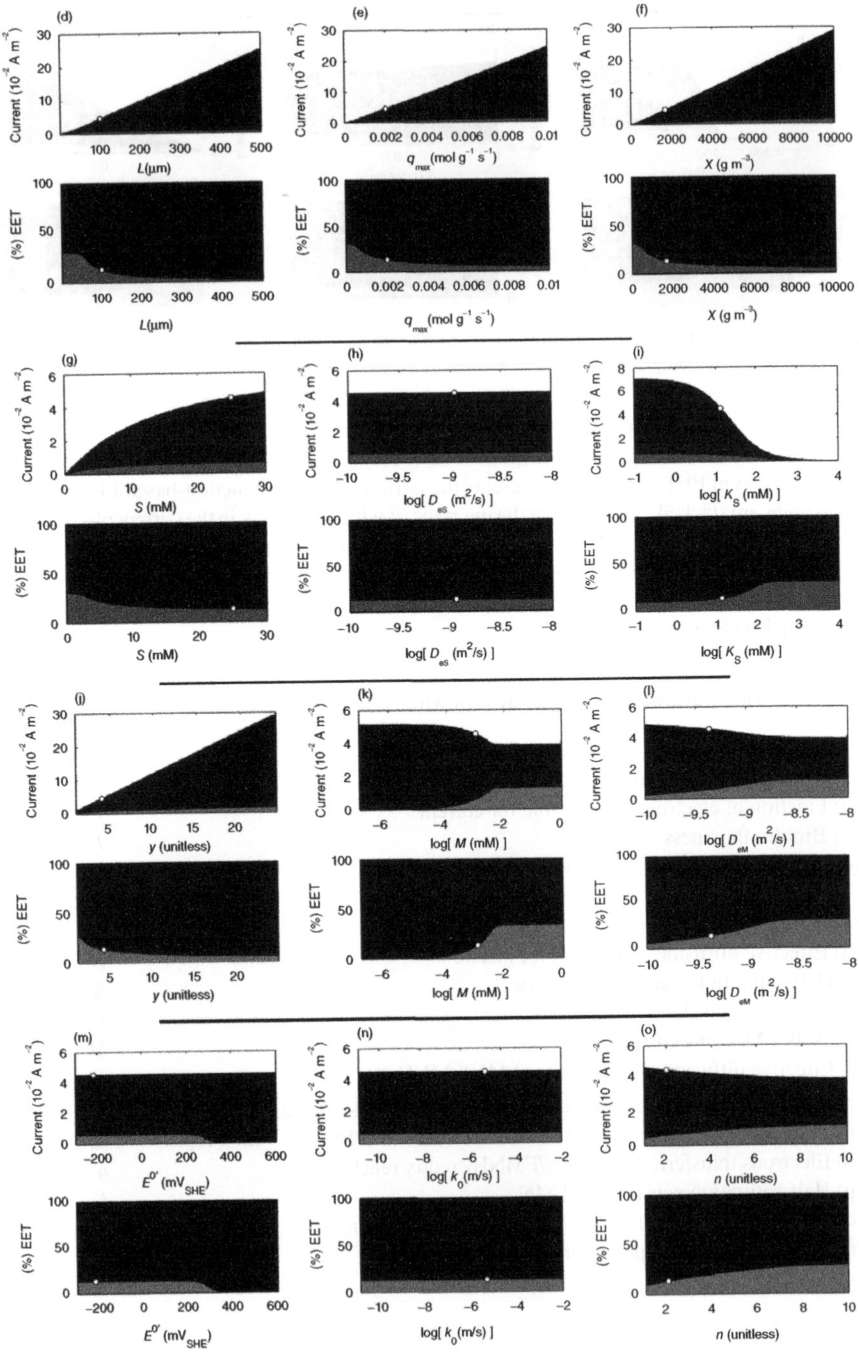

Figure 9.5 (*Continued*)

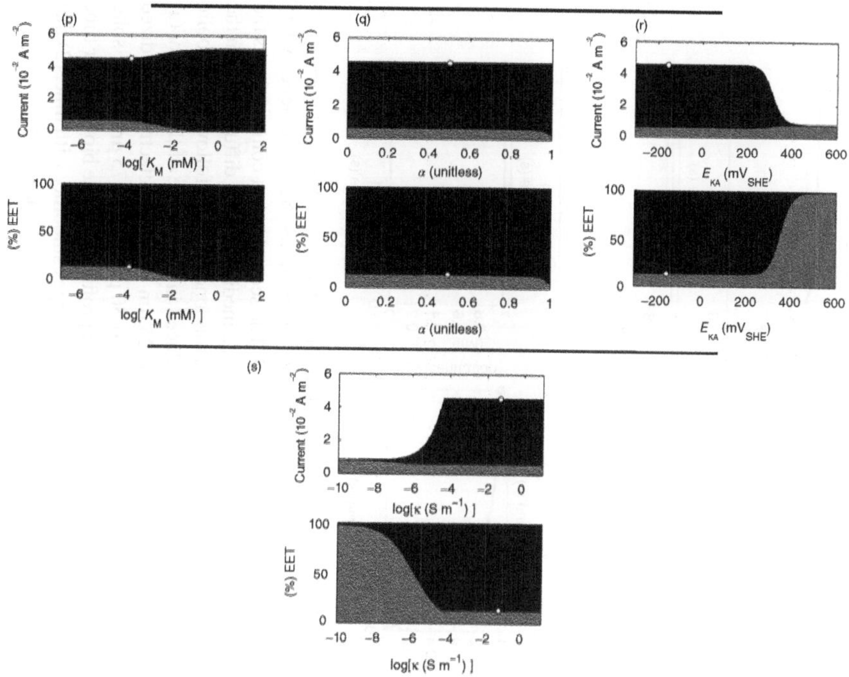

Figure 9.5 (*Continued*)

that the model can generate is 60% because that is the fraction of electrons that are recoverable for current (see Table 9.2). In practical terms, these results demonstrate that no electrons are lost along the conductive EET pathway. Compared to the conduction-based EET pathway, diffusion-based EET cannot sustain high current production. This is due to several causes: (1) the slow rate of electron transfer via diffusion (a mediator flux to the electrode of $\sim 4 \times 10^{-8}$ mol m^{-2} s^{-1} compared to an equivalent flux for conduction-based EET of $\sim 3 \times 10^{-7}$ mol m^{-2} s^{-1}); (2) the loss of current through the diffusion of reduced mediators out of the top of the biofilm (mediator flux loss of $\sim 4 \times 10^{-8}$ mol m^{-2} s^{-1}); and (3) the reduction of nearly all of the mediators in the middle of the biofilm, resulting in a low biofilm activity (low q_M) and uneven activity by depth. This is discussed further in the spatial activity section later in this chapter. In actual systems, diffusion-based EET may be further adversely affected by the loss of flavins due to adsorption to the electrode and the biofilm matrix [26].

9.5.1.2 Comparing Electron Transfer Rates for Low/High Biofilm Activity or Conductivity Three cases are defined in Table 9.6: a low-conductivity case and two high-activity cases (distinguished as the "high-activity case" and the "2× high-activity case"). These cases were designed to determine whether diffusion-based EET could generate higher current if the biofilm conductivity or activity were changed. In the low-conductivity case, the conductivity of the

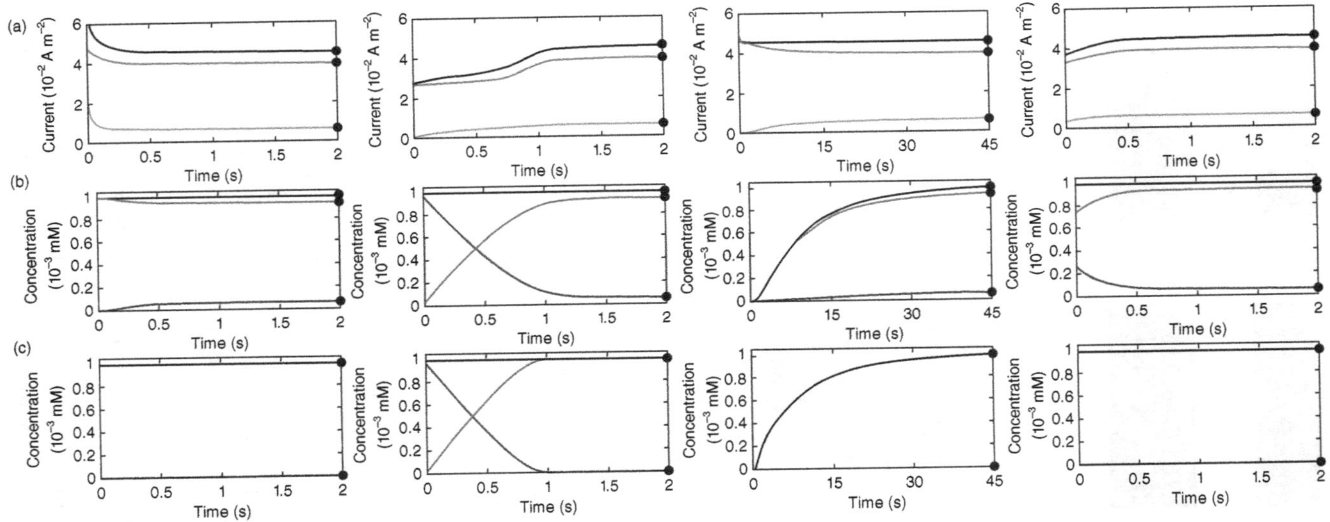

Figure 9.6 Convergence of a steady-state solution for multiple initial conditions. The left column shows the convergence of several parameters for the model as described in the chapter and in the Model System of Equations provided in Table 9.1. The next column shows the convergence when the initial conditions are altered so that the mediator is 100% oxidized in the biofilm as opposed to 100% reduced. The second to last column shows the convergence when the initial conditions are altered so that the mediator concentration is zero inside the biofilm (i.e., mediators must diffuse in from the bulk over time). The last column shows the convergence when the initial conditions are altered so that the mediator concentration follows a parabolic profile, with all mediators oxidized at the biofilm-electrode and biofilm-solution interfaces and all mediators reduced in the middle of the biofilm. (a) Current via diffusion-based EET, grey; current via conduction-based EET, grey; and total current, black. (b) At 25 μm above the electrode: oxidized mediator concentration, grey; reduced mediator concentration, grey; and total mediator concentration, black. (c) At 50 μm. (d) At 75 μm. Black solid dots mark the true steady-state solutions. (e) 3D plots showing convergence of the oxidized mediator concentration profile inside the biofilm over time (a 3D representation of the data presented in b, c, and d plots). These results confirm (1) that the model solution converges smoothly with time, (2) that the final solution is independent of the initial condition of the mediators, and (3) that the total mediator concentration is truly conserved.

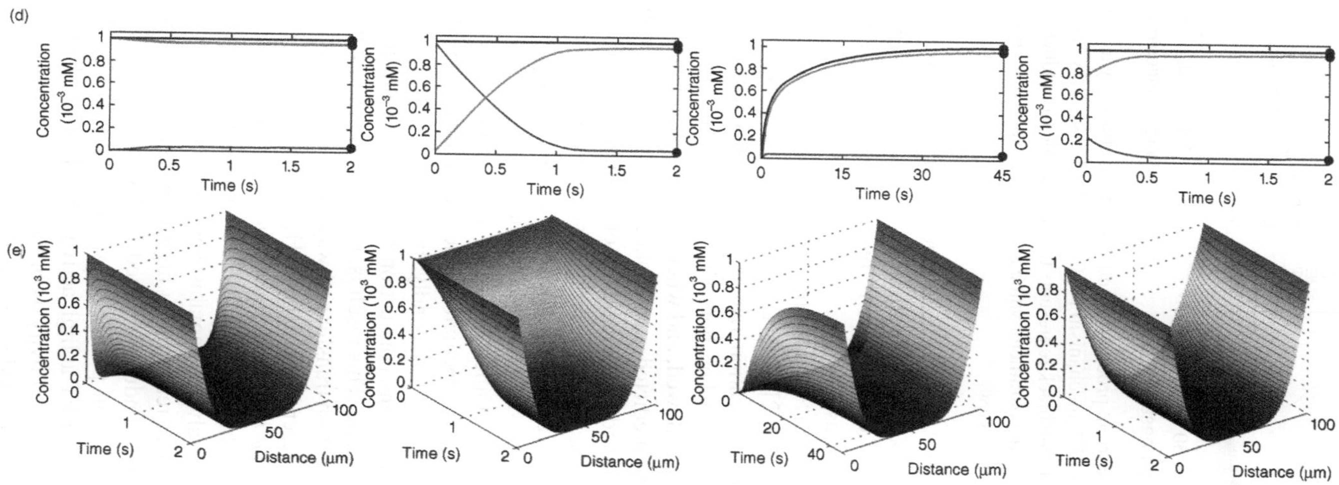

Figure 9.6 (*Continued*)

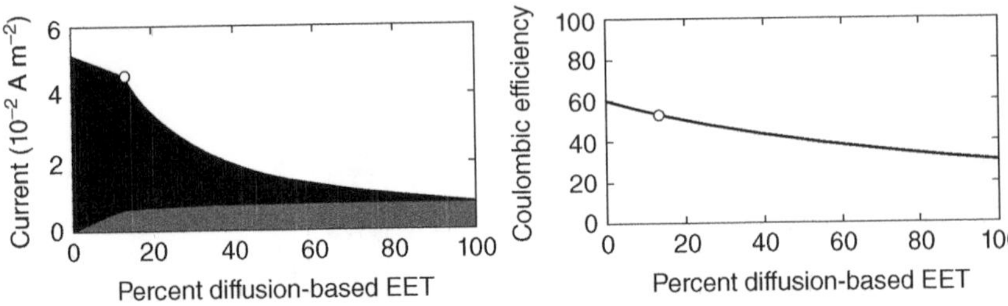

Figure 9.7 Steady-state current production and coulombic efficiency as a function of changing only the percentage of EET mechanism for the standard case. Gray represents the current due to diffusion-based EET, and black represents the current due to conduction-based EET. The white-filled circles represent the parameter value and the corresponding current and percentage of EET for the model standard case.

biofilm is lowered to 10^{-5} mS cm^{-1}, identical to that used by Marcus et al. for their low-conductivity case [27]. The high-activity case transforms the model to behave similarly to a *G. sulfurreducens* biofilm. For this, we assume a biofilm density similar to that measured in a *G. sulfurreducens* biofilm respiring on an electrode [76], 120,000 g m^{-3}, and a much higher maximum substrate utilization rate, 0.014 mmol g^{-1} s^{-1} [62]. The 2× high-activity case is identical, except that the maximum substrate utilization rate is set to 0.028 mmol g^{-1} s^{-1}. For each case, we explore the results when both EET mechanisms are allowed, when only conduction-based EET is allowed, and when only diffusion-based EET is allowed.

Table 9.6 shows the predicted steady-state current, the local biofilm potential drop inside the biofilm, and the percentage of mediators in the biofilm that are in the reduced form. The data reveal several reasons diffusion-based EET is not well suited to high current production. Firstly, in all cases, the majority of the mediators is in a reduced form. This shows that a mediator concentration of 1 μM severely limits diffusion-based EET. Other studies in the literature are consistent with this result because increasing the mediator concentration in the system can increase current production [40, 58, 84]. Secondly, in all cases except the low-conductivity case, having both diffusion- and conduction-based EET results in *lower* current than conduction-based EET alone. This result is not intuitive because it is anticipated that allowing for a second electron transfer pathway would increase current production by decreasing the electron acceptor bottleneck. Hence the question: *When would having both EET mechanisms be beneficial for the biofilm*? This question is addressed in the following section. A third trend in Table 9.6 data is that current produced solely by diffusion-based EET always remains very low (well under 1 A m^{-2}), while conduction-based EET can reach much higher current values. Specifically looking at the high-activity case, conduction-based EET is capable of significantly higher current comparable to current produced experimentally in *G. sulfurreducens* [45]. For 100% diffusion-based EET at high activity, nearly all of the mediators are in the reduced form. Increasing the activity as in the 2× high-activity case only increased the percentage of reduced mediators by 0.2%. This demonstrates the intrinsic limitation of diffusion-based EET: even if the biofilm is capable of higher activity,

TABLE 9.6 Steady-State Current Production, Local Biofilm Potential Drop, and Percentage of Mediators in the Reduced Form in the Biofilm for the Standard Case, Low-Conductivity Case, and High-Activity Cases.

Parameter values	Standard Case	Low-Conductivity Case	High-Activity Case	2× High-Activity Case
	See Table 9.2 $\kappa = 0.5$ mS cm^{-1}	$\kappa = 10^{-5}$ mS cm^{-1}	$q_{max} = 0.014$ mmol g^{-1} s^{-1} $X = 120{,}000$ g m^{-3}	$q_{max} = 0.028$ mmol g^{-1} s^{-1} $X = 120{,}000$ g m^{-3}
Both EET mechanisms	**0.046 A m^{-2}** <1-mV drop 80.1% mediators reduced	**0.024 A m^{-2}** 530-mV drop 82.4% mediators reduced	**24.09 A m^{-2}** 24-mV drop 99.1% mediators reduced	**45.62 A m^{-2}** 47-mV drop 99.2% mediators reduced
100% Conduction-based EET	**0.052 A m^{-2}** <1-mV drop	**0.022 A m^{-2}** 533-mV drop	**24.22 A m^{-2}** 25-mV drop	**45.81 A m^{-2}** 47-mV drop
100% Diffusion-based EET	**0.008 A m^{-2}** 84.5% mediators reduced	**0.008 A m^{-2}** 84.5% mediators reduced	**0.18 A m^{-2}** 99.2% mediators reduced	**0.25 A m^{-2}** 99.4% mediators reduced

Results are shown for when both EET mechanisms are allowed, for diffusion-based EET only, and for conduction-based EET only. The top row displays the parameter values that are used for each case, if they are different from the standard case as given in Table 9.2.

the low concentration and slow diffusion of mediators restrict the amount of current. Similarly to the predictions made by Torres et al. [18], these results confirm that diffusion-based EET cannot account for most of the current produced in common MFCs. It is clear that for most of higher-current-producing EABs, some form of conduction-based EET must be present.

In summary, our simulation results show that diffusion-based EET cannot sustain high current (>1 A m^{-2}). This is due to both the slow rate of transfer via diffusion and the loss of current caused by the diffusion of reduced mediators out of the biofilm into the bulk solution. If the cells were losing a biochemical that requires significant energy to produce (such as FMN), we wonder why they should continue its production. In our experiments, we noticed that FMN is generally produced by very young biofilms. It is likely that when high cell density is used for inoculation, the active cells in the inoculation may produce FMN, but later when the cells attach to the electrode and start growing as a biofilm, FMN production may be stopped. We believe that the fate of flavin needs more research under well-controlled conditions. Conduction-based EET is required for higher current production, such as that observed for *G. sulfurreducens* biofilms. Finally, we conclude that the use of any diffusion-based EET is likely to decrease the efficiency of electron transfer for *S. oneidensis*. This most likely explains why higher coulombic efficiencies are observed in batch systems. In batch systems, mediators are allowed to accumulate in the supernatant, decreasing the number of electrons that is lost from the system. When the concentration of the mediator in the biofilm decreases below the bulk concentration, the reduced mediators can diffuse back to the biofilm and transfer electrons. This will increase coulombic efficiency. In continuously fed systems, the mediators are washed out of the reactor. The same situation is observed for *G. sulfurreducens* biofilms. The coulombic efficiency is higher for batch biofilms, but low for flow through systems. Currently, we do not have an explanation for this, but we expect that some unexplored processes involving electron transfer in *G. sulfurreducens* biofilms may help explain this phenomenon.

9.5.1.3 When Would the Simultaneous Use of Both Extracellular Electron Transfer Mechanisms Be Beneficial for the Biofilm? From the results presented in Table 9.6, it is reasonable to determine whether a biofilm may benefit from using multiple EET mechanisms. From the standard-case results, it appears that 100% conduction-based EET will always yield the highest current and thus support the highest level of metabolism. However, there are cases when having both modes of EET available can enable higher levels of metabolism than either mode alone. The low-conductivity case shown in Table 9.6 presents one such example. When conductivity is low, the potential drop (533 mV) inside the biofilm begins to restrict conduction-based EET. Adding mediators provides a secondary electron acceptor and reduces the load on the conductive biofilm matrix. However, this addition is small, providing only a 9% increase in current over conduction-based EET alone.

A second example in which the utilization of both EET modes provides a metabolic benefit is when the electrode potential is very low, for example, when the electrode has a potential equal to the half-maximum rate potential. This condition is akin to the environment experienced by cells located in anoxic subsurface sediments or wetland soils, where the terminal electron acceptors (e.g., Fe^{3+}, SO_4^{2+}, and

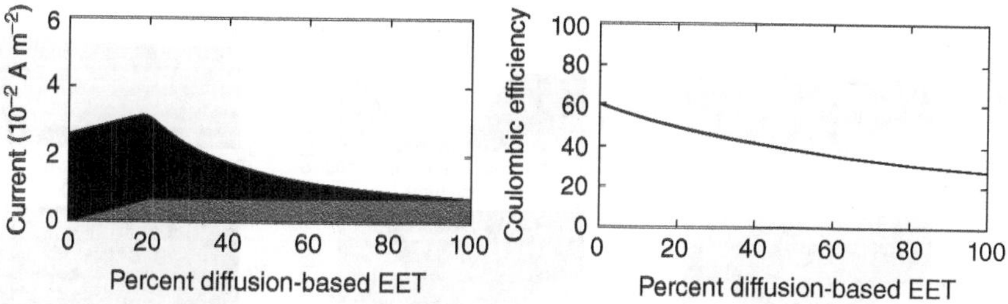

Figure 9.8 Steady-state current production and coulombic efficiency as a function of the percentage of the EET mechanism for a case in which the electrode potential is equal to the half-maximum rate potential. Gray represents the current due to diffusion-based EET, and black represents the current due to conduction-based EET.

CO_2) may have low standard reduction potentials [105]. Figure 9.8 shows current and coulombic efficiency as a function of the percentage of diffusion-based EET when the electrode potential is equal to the half-maximum rate potential. The peak current occurs when there is approximately 20% diffusion-based EET, giving a 21% increase in current over conduction-based EET alone. Note that this is only possible because the half-maximum rate potential for conduction-based EET is more positive than the standard reduction potential of FMN.

Because there is evidence of both diffusion- and conduction-based EET in *S. oneidensis*, we hypothesize that there must be a metabolic advantage for having both mechanisms. The model demonstrates that if the biofilm conductivity is low, if the biofilm half-maximum rate potential and mediator standard reduction potential are different, and/or if the terminal electron acceptor potential is very low, then there may be an advantage to having both mechanisms available concurrently. Later in this chapter, it is shown that having both diffusion- and conduction-based EET mechanisms further improves the current production if the mechanisms are allowed to interact (i.e., interacting-dual EET).

In summary, our simulation results show that the availability of diffusion-based EET does not contribute to increased biofilm metabolism unless conduction-based EET is restricted or unavailable. As a specific case, when the electrode potential is equal to the half-maximum rate potential, the total current can increase when some diffusion-based EET is allowed, although the overall coulombic efficiency drops. Finally, under natural conditions, having both mechanisms available could provide a metabolic advantage for *S. oneidensis*.

9.5.1.4 How Does Biofilm Spatial Activity Affect the Extracellular Electron Transfer Mechanism? Current production, while generally attributed to the biofilm as whole, is a product of electrons generated throughout the biofilm and is likely not a spatially homogenous phenomenon. It is important to understand where the electrons come from within the biofilm to optimize these systems. Furthermore, the spatial activity inside the biofilm can provide more clues as to why *S. oneidensis* may have more than one EET mechanism available simultaneously. One possibility is that cells in the biofilm may produce electrons, but the biofilm may not have the ability to deliver them to the electrode. In this case, having active cells in the

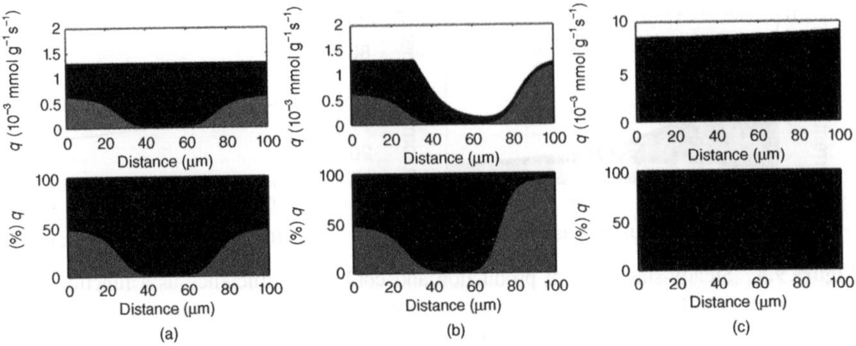

Figure 9.9 Biofilm activity by depth as measured by the substrate utilization rate, q, and percentage of q obtained using diffusion-based or conduction-based EET. Gray represents the portion due to diffusion-based EET, and black represents the portion due to conduction-based EET. (a) standard case, (b) low-conductivity case, (c) high-activity case.

biofilm does not produce any advantage for current generation. However, having a case such as this raises another question: if the biofilm has the capacity to produce more electrons, what are the possible paths for these electrons? Figure 9.9 shows the metabolic activity through the biofilm thickness, as measured by q (q_M and q_C shown separately as gray and black, respectively). Also shown are the percentages of q_M and q_C over the biofilm thickness. These represent the spatially resolved EET mechanism(s) utilized to accept electrons within the biofilm. Because the mediators may accept electrons and then diffuse out of the top of the biofilm, the values presented in Figure 9.9 do not necessarily reflect the EET mechanism percentages for electrons that contribute to current at the electrode. For the standard case, electrons are produced evenly throughout the biofilm. As discussed previously, diffusion-based EET is not utilized in the middle of the biofilm, solely because all the mediators are in the reduced form and cannot accept more electrons. Diffusion-based EET is maximized near the base of the biofilm, where the mediators are oxidized by the electrode, and near the top of the biofilm, where oxidized mediators diffuse into the biofilm. For the low-conductivity case, the base of the biofilm looks similar to the standard case; however, near the middle and top of the biofilm, q_M falls off rapidly. This is due to the local biofilm potential drop caused by the low conductivity. In this case, the top of the biofilm has a local biofilm potential of only -230 mV_{SHE}, preventing the transfer of electrons through conduction; however, diffusion-based EET allows the top of the biofilm to stay active. Note that most of the electrons produced near the top of the biofilm are subsequently lost to the bulk liquid via mediator diffusion, resulting in a coulombic efficiency of only 45%. While this state is counterproductive for extracting energy from EABs for practical applications, for biofilm survival, having both mechanisms provides a way for cells near the top to survive. A primary finding from these results is that the EET mechanism depends on the local microenvironments.

For the high-activity case, diffusion-based EET is negligible. This is because conduction is the dominant EET mechanism capable of transferring the large quantity of electrons associated with high activity. One peculiar feature of high activity is that more electrons are generated near the top of the biofilm, although this region is far from the electron-accepting surface. This is possible because the conductivity of the biofilm is high enough in this case not to be a limitation; thus, activity and electron production are highest near the top of the biofilm, where the substrate concentration is also the highest. This ability is highlighted by the simulation of a thick (750-μm) biofilm using the 2× high-activity parameter values and conduction-based EET alone (Fig. 9.10). Here, we see that the top of the biofilm is very active, but the base has negligible activity. The substrate concentration depth profile shows that the electron donor is restricted from the base of the biofilm (Fig. 9.10). Recently, our group obtained supporting experimental results for a several-hundred-micron-thick *G. sulfurreducens* biofilm grown in an electrochemical-nuclear magnetic resonance microimaging reactor [106]. In that study, the top of the biofilm was shown to be active, while the base was inactive because of the lack of substrate caused by mass transfer limitations. A profile such as this is only possible if the biofilm conductivity is sufficiently high to prevent a prohibitive local biofilm potential drop. Thus, if the conductivity is high enough, a biofilm grown on a solid electron-accepting surface will not have its thickness restricted by electrical resistance. It has been hypothesized in the literature that biofilms relying on conduction-based EET, composed of organisms such as *G. sulfurreducens*, have limited thickness because of the electrical resistance across the biofilm matrix [27, 107, 108]. However, as shown in Figure 9.10,

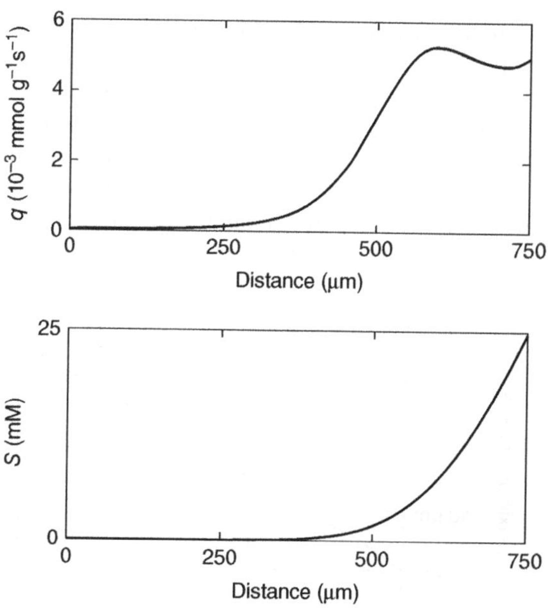

Figure 9.10 Biofilm activity by depth as measured by the substrate utilization rate, q, and the substrate concentration depth profile of a thick, highly active biofilm utilizing conduction-based EET alone.

cells at the top of high-conductivity biofilms may be metabolically active and use the biofilm matrix as an electrical conduit, even across metabolically inactive cells. We note that perhaps there is a critical biofilm conductivity that will prevent the biofilms from growing. This would support some of the claims in the literature. However, by considering the conductivity data presented in the literature, this is not likely for *G. sulfurreducens*, unless the biofilm conductivity changes by the biofilm thickness and with time, which requires further research to ascertain.

In summary, our simulation results show that the EET mechanism depends on local microenvironments. Biofilm metabolic activity controls which EET mechanism(s) will be utilized. Mediators can allow regions distal to an electron-accepting surface to be active when conductivity is low (10^{-5} mS cm^{-1}). When conductivity is high, the top of a conductive biofilm can be active, even when the base is limited by electron donor availability.

9.5.1.5 What Are the Theoretical Maximum Biofilm Thicknesses for Different Conductivities? While biofilms may grow thick enough to restrict cells at their base from receiving nutrients, for practical applications, it is often desirable to have a biofilm that is entirely active across its thickness. To determine the maximum theoretical biofilm thickness, the model is restricted to cases where the biofilm is not substrate limited and the entire biofilm is active. We define the limiting biofilm thickness to be the distance at which the Nernst–Monod term drops to below 0.01 (the point in the biofilm where the biofilm activity is restricted to just 1% of q_{max} by potential drop limitations). Figure 9.11 shows the results. For the standard case, the maximum biofilm thickness increases to several millimeter as the conductivity increases. This is because the biofilm activity is low for the standard case and does not produce a significant local biofilm potential drop. The low activity of *S. oneidensis* may be one of the reasons why it is generally capable of growing much thicker biofilms than *G. sulfurreducens*. On the other hand, the high-activity case predicts lower maximum biofilm thicknesses. In a previous study, we grew electrode-respiring *G. sulfurreducens* biofilms that were over 500-μm thick [106], revealing that the biofilm conductivity may be >0.01 mS cm^{-1}. In the literature, *G. sulfurreducens* biofilms have been shown to grow

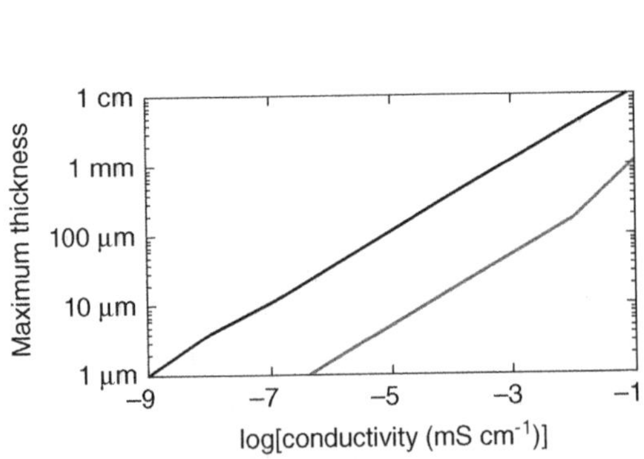

Figure 9.11 Maximum biofilm thickness as a function of conductivity, for conduction-based EET biofilms only. (Black) standard case, (gray) high-activity case.

across 1 cm gaps, indicating that conductivities of >0.1 mS cm^{-1} are likely. This is similar to the predicted minimum conductivity of 0.5 mS cm^{-1} given by Torres et al. [56], which is necessary to avoid non-Nernstian low-scan CV voltammograms. From these results, and the accumulating evidence in the literature, biofilm thickness does not appear to be limited by low conductivity for most *G. sulfurreducens* biofilms, the majority of which is <100-μm thick.

In summary, our simulation results show that the thickness of *G. sulfurreducens* biofilms is not limited by conductivity under the tested conditions.

9.5.1.6 How Does the Diffusion Coefficient Profile Affect Diffusion-Based Extracellular Electron Transfer? Historically, mathematical models of biofilm metabolism have used a constant, that is, flat, effective diffusion coefficient profile through the depth of the biofilm [23, 27, 30, 33, 34, 75, 109, 110]. This was explored in detail in a recent study by our group [76]. In that study, we showed that the assumed effective diffusion coefficient profile for the substrate could affect the prediction of current output for a conduction-based EET biofilm. We hypothesized that this effect would be more pronounced for diffusion-based EET, in which the effective diffusion coefficient would influence both the substrate diffusion and the mediators. This hypothesis is tested by examining three assumptions for the relative effective diffusion coefficient profile: (1) a constant profile, (2) a linearly decreasing profile, and (3) an empirical profile. The empirical profile was obtained using pulsed-field gradient nuclear magnetic resonance of a 100-μm-thick *S. oneidensis* biofilm (the biofilm on which the standard-case parameters were based) [59]. The linear profile was generated by making a linear fit between the empirical value at the top and that at the base of the biofilm. The constant profile was the average of the entire empirical profile. The profiles for these three cases are shown in Figure 9.12.

Table 9.7 shows the effects of different relative effective diffusion coefficient profiles on steady-state current and coulombic efficiency for a diffusion-based EET biofilm. Because D_r is indicative of the biofilm density, as seen in Equation 9.34 and discussed previously, the results are shown with and without a correction for biofilm density based on the diffusion coefficient profile. With no correction for

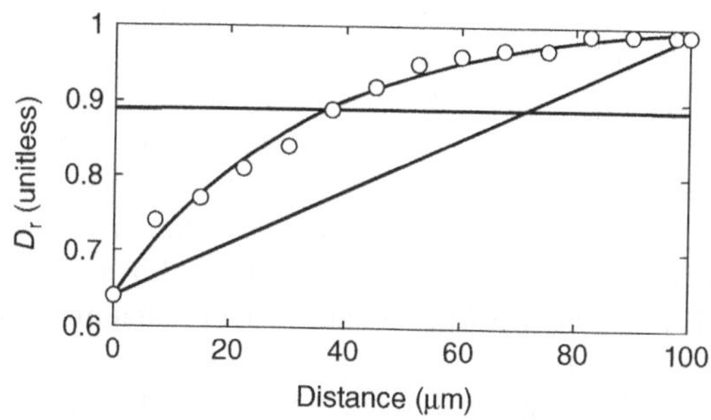

Figure 9.12 Three relative effective diffusion coefficient profiles. A constant profile, a linearly decreasing profile, and an empirical profile, based on data obtained experimentally using pulsed-field gradient nuclear magnetic resonance (data shown as white-filled circles).

TABLE 9.7 Effects of Different Relative Effective Diffusion Coefficient Profiles on Steady-State Current and Coulombic Efficiency for a Diffusion-Based EET Biofilm.

Profile	Current ($mA\ m^{-2}$)	Coulombic Efficiency (%)
No correction for biofilm density		
Constant	8.06	30
Linear	6.94	27
Empirical	7.12	27
With correction for biofilm density		
Constant	8.06	30
Linear	14.70	45
Empirical	14.44	49

biofilm density, current densities ranged from 6.94 to 8.06 mA m^{-2}. This variation in current densities is accounted for by the variation in coulombic efficiency and the variation in activity across the biofilm depth. When the biofilm density is corrected, the current varies by over 80%. The empirical profile yields a current density of 14.44 mA m^{-2} and a coulombic efficiency of 49% compared to 8.06 mA m^{-2} and 30% efficiency for a constant profile. These results show the significance of using empirical diffusion coefficients to predict current. With the correction for density, there is a higher biofilm activity very close to the electrode, resulting in fewer reduced mediators being lost to the bulk. In this case, the top of the biofilm acts as a protective diffusion barrier to slow the loss of mediators, keeping the mediators near the most active portion of the biofilm and the electrode. Perhaps, variable relative effective diffusion coefficient and density profiles, as measured in many biofilm systems, are a life strategy of EABs to minimize mediator loss and maximize redox cycling rates. In summary, our simulation results show that diffusion coefficient assumptions affect predicted current production.

9.5.1.7 Can Voltammetry Determine the Percentage Distribution of the Extracellular Electron Transfer Mechanisms Used by the Biofilm? Although both diffusion- and conduction-based EET mechanisms have been observed in *S. oneidensis* biofilms, there have not been direct measurements of both mechanisms occurring simultaneously. Each mechanism has only been studied independently. One of our goals was to test whether simulations of common voltammetric techniques could be used to determine the percentage of electrons delivered by each EET mechanism in isolated-dual EET. If so, this would enable the use of voltammetry coupled to model fitting to distinguish the EET percentages in actual systems. While CV of EABs has been simulated before [23], in this chapter, we use experimentally observed parameters for this, which is different from previous literature work. The use of experimental data is critical if we are ever to extract critical information from actual CVs using voltammetry models. Figure 9.13 shows the simulated CV at three different

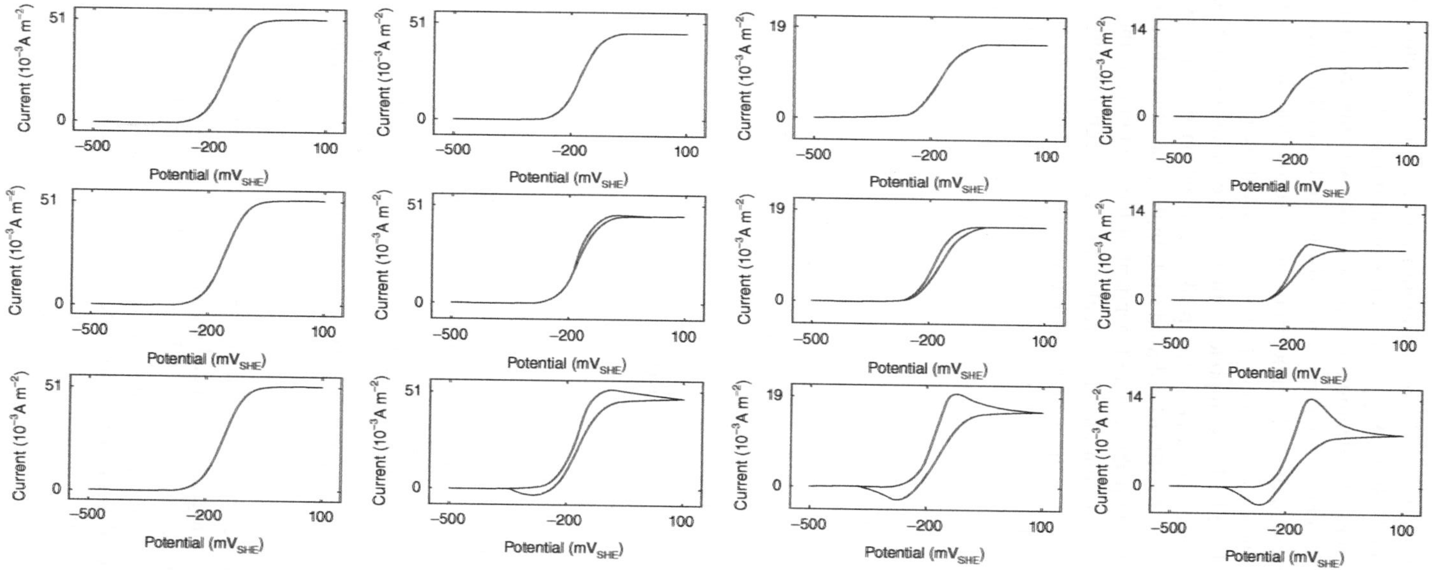

Figure 9.13 Simulated cyclic voltammograms. Each row represents a different scan rate: (Top) 1 mV s^{-1}, (middle) 200 mV s^{-1}, (bottom) 1000 mV s^{-1}. Each column represents a different percentage of the EET mechanism: (Column 1) 100% conduction-based EET only, (column 2) standard case (~13% diffusion-based EET), (column 3) 50/50% conduction-/diffusion-based EET, and (column 4) 100% diffusion-based EET only.

scan rates (1, 200, and 1000 mV s^{-1}) for different EET percentages, with a starting and minimum potential of −500 mV_{SHE} and a maximum potential of +100 mV_{SHE}. Because our model is an idealized simulation and lacks background currents, higher scan rates are required to observe the transient diffusion or Cottrell behavior. At the slow scan rates, typical sigmoid voltammograms are observed for all EET percentages. As the scan rate increases, Faradaic peaks become apparent for those systems that have a diffusion-based EET component. In the case of diffusion-based EET only, at 1000 mV s^{-1}, we see large redox peaks centered on the standard redox potential of flavins, typical for diffusion-based EET. In the case of conduction-based EET only, the voltammogram shape is independent of the scan rate, as there are no diffusion limitations and current can flow freely in the conductive biofilm matrix. We were unable to easily identify features in the voltammograms that would allow back-calculation of the percentage of electrons delivered by each EET mechanism. For example, the half-limiting-current slopes for the slow scan CVs (normalized by the limiting current) were not indicative of the percentage of the EET mechanism (9.6 mV^{-1}, 9.7 mV^{-1}, 8.8 mV^{-1}, and 11.8 mV^{-1}, for 100%, 87%, 50%, and 0% conduction-based EET, respectively). The ratio between maximum peak current and limiting current is linearly correlated with the percentage of the EET mechanism (e.g., at a 1000-mV s^{-1} scan rate, peak current to limiting current is 1, 1.13, 1.25, and 1.75 for 100%, 87%, 50%, and 0% conduction-based EET, respectively). However, to use this method to determine the percentages of the EET mechanisms, at least two calibration voltammograms representing biofilms with known EET mechanism percentages would need to be obtained. Currently, there are no methods to do this.

Unfortunately, the simulation of SWV did not offer a better solution for determining the percentage of the EET mechanism. The SWV parameters used were a starting potential of −500 mV_{SHE}, a final potential of 100 mV_{SHE}, a potential step height of 5 mV, a pulse potential of 40 mV, and a 20-Hz scan frequency. These are identical to the parameters used in our previous study, which used SWV to detect flavin concentrations inside *S. oneidensis* biofilms [55]. Figure 9.14 shows the simulated SWV results. The shapes of the differential-current voltammograms are nearly identical, regardless of the EET mechanisms. We measured the half-height peak widths to determine whether they corresponded to percentages of EET; however, the trend was not exploitable (124.09 mV, 112.28 mV, 114.52 mV, and 103.38 mV for 100%, 87%, 50%, and 0% conduction-based EET, respectively). At least for this idealized model, CV and SWV are limited and are not good tools for determining the percentage contribution of each EET mechanism for biofilms that utilize more than one EET mechanism. Furthermore, our idealized model cannot currently replicate some of the complex voltammograms produced in experimental systems. In summary, our simulation results show that idealized CV and SWV are currently not good tools for determining the specific percentage of the EET mechanism. A more sophisticated model, capable of simulating experimental voltammograms, will need to be developed so that the limits of these techniques can be explored, and their true ability to discern the percentage distribution of the various EET mechanisms can be tested.

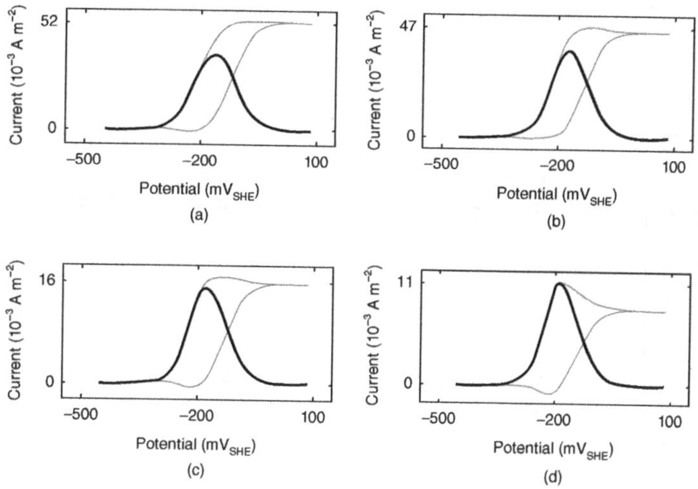

Figure 9.14 Simulated squarewave forward and backward (gray) and differential current (black). From (a) to (d): 100% conduction-based EET only, standard case (~13% diffusion-based EET), 50/50% conduction-/diffusion-based EET, and 100% diffusion-based EET only.

9.5.1.8 Which Parameters Are Critical for Determining the Extracellular Electron Transfer Mechanisms and Theoretical Current Production? Although the CV and SWV simulations were not capable of determining the contributions of the various EET mechanisms, it is possible to determine which parameters are critical. To get a better idea of the impact of each parameter on current and the contributions of the various EET mechanisms, sensitivity analysis was performed by measuring elasticity. Furthermore, a complete parameter analysis was performed to cover larger parameter ranges. In principle, a model with a huge number of parameters makes it difficult to study the real underlying physical phenomena of the simulated system. Given enough parameters, a model can be forced to fit any experimental data. However, parameter and sensitivity analysis can be used to determine which parameters are critical and which are trivial. This knowledge is critical for experimenters. Having such knowledge in advance can help design better experiments. Table 9.8 shows the results of the sensitivity analysis. The most critical parameters for the total current were biofilm thickness, fraction of electrons recoverable for current, biofilm density, maximum specific lactate utilization rate, and electron equivalence of lactate. These parameters control how much biomass is present and how efficient that biomass is at converting substrate into electrons. It is important to remember that this sensitivity analysis is based on a variation in parameters of ±25% from the standard case. Therefore, the parameters in the sensitivity analysis were weighted to favor conduction-based EET, which accounts for 87% of current in the standard case. In this case, the electron donor is not limiting, because the potential drop is not significant across the 100-μm biofilm thickness. Parameters of secondary importance include the bulk lactate concentration and the half-saturation constant for lactate. These are only

TABLE 9.8 Sensitivity Analysis as Measured by Elasticity [see Equations 9.32 and 9.33].

Symbol	Description	Total Current, j	Percentage of Diffusion-Based EET, d
System parameters			
ε	Polarized electrode potential	0	0
Biofilm parameters			
D_r	Relative effective diffusion coefficient	−0.07	0.57
f	Fraction of electrons recoverable for current	1.07	−0.59
L	Biofilm thickness	1.13	−1.23
q_{max}	Maximum specific lactate utilization rate	1.07	−0.59
X	Biofilm density	1.07	−0.59
Substrate parameters			
S	Bulk lactate concentration	0.38	−0.21
D_{eS}	Effective diffusion coefficient for lactate	0	0
K_S	Half-saturation constant for lactate	−0.37	0
y	Electron equivalence of lactate	1.07	−0.59
Diffusion-based EET parameters			
M_o(bulk)	Bulk FMN concentration	−0.07	0.64
D_{eM}	Effective diffusion coefficient for FMN/$FMNH_2$	−0.07	0.57
$E^{0'}$	Standard redox potential for the FMN/$FMNH_2$ redox reaction	0	0
k_0	Standard heterogeneous rate constant for the FMN/$FMNH_2$ redox reaction	0	0
n	Electrons transferred per FMN/$FMNH_2$ redox reaction	−0.07	0.57
K_M	Half-saturation constant for FMN	0.01	−0.07
α	Transfer coefficient for the FMN/$FMNH_2$ redox reaction	0	0
Conduction-based EET parameters			
E_{KA}	Half-maximum rate potential	0	0
κ	Biofilm conductivity	0	0

Elasticity measures the proportional effect of a change in a parameter value. These relative sensitivities were evaluated by successively making a finite change of 25% in each independent variable and monitoring how total current, j, and percentage of current due to diffusion-based EET, d, responded.

secondary because the standard case operates at a substrate concentration that would typically be described as nonlimiting.

The most critical parameter affecting the percentage of EET is the biofilm thickness. This is expected because the current produced via diffusion-based EET consists entirely of electrons delivered to the electrode. Thus, for the standard case, in which the diffusion coefficient profile is constant, half of the electrons produced and subsequently accepted by mediators are lost to the bulk solution. (Thus, the coulombic efficiency for 100% diffusion-based EET, shown in Figure 9.7, is half of *f*, i.e., 30%.) Another critical parameter for determining the percentage of diffusion-based EET is the mediator concentration. Currently, the intrabiofilm mediator concentration is unknown in continuous flow systems. In a recent study by our group, we measured only up to 0.7 μM flavin in an *S. oneidensis* biofilm using a flavin microelectrode, and this was localized to within approximately 50 μm of the base of a 300-μm-thick biofilm, where it was confirmed that no O_2 was present [55]. Elsewhere in that biofilm, flavin concentrations were below detection. Our model confirms that such low concentrations of mediators cannot account for the high currents measured in many systems. Thus, we believe that the best measure of the dominating EET mechanism in EABs is the steady-state current production. If the current rises higher than 1 A m^{-2}, then conduction-based EET is likely to be the dominant mechanism, contingent on confirmation that any possible mediator concentration is low. For this reason, we also see that any parameter that increases the electron-producing activity per volume of the biofilm (such as q_{max}, X, f, or y) will decrease the percentage contribution of diffusion-based EET. The mediator concentration is too low to handle any further increase in electron production by the biofilm. As shown in Table 9.6, most of the mediators are already in the reduced form and unable to accept more electrons even in the standard case, which has low activity.

The full parameter analysis (available earlier in the chapter; Fig. 9.5) gives a better look at how each parameter controls current and percentage of EET. We tested a large range for each parameter to allow for all possible realistic values and to highlight the regions where the parameter values have a large impact on the final solution. As in the sensitivity analysis, it is quickly apparent from the parameter analysis that the percentage of diffusion-based EET is maximized when the biofilm activity and amount of biomass are low. This may be one reason that diffusion-based EET is observed in *S. oneidensis*, but not in *G. sulfurreducens*, which is denser and has a higher activity. It is interesting to see that even if the diffusion coefficient of the mediator is increased by a factor of over 10, the current stays relatively low. Only by changing the conductivity or the half-maximum rate potential value is it possible to achieve 100% diffusion-based EET. Therefore, the three most revealing measurements for determining the contributions of the different EET mechanisms in EABs are the steady-state current, biofilm conductivity, and mediator concentration.

9.5.1.9 What Is the Difference Between the Local Biofilm Potential and Redox Potential? Recently, we introduced the concept of the local biofilm potential and compared it to redox potential [45]. In that study, we showed that there was a difference between the potential measured with a redox microelectrode and that measured with an electrode that was electronically connected to the biofilm matrix. Briefly, redox potential is determined by the soluble redox species with which an electrode (i.e., a platinized platinum microelectrode) is able to equilibrate. In most systems, this results in a mixed potential, which is dependent on multiple redox couples. Note that not all redox couples are capable of interacting (i.e., exchanging electrons) with an electrode [105]. The redox potential is governed by the Nernst equation (Equation 9.22) and is intimately associated with diffusion-based EET. The local biofilm potential is only measurable after an electrode is electronically connected with the biofilm matrix [45]. It is currently unknown exactly what this potential comprises. Here, we simulate it as the potential drop associated with resistance through the biofilm matrix, governed by Ohm's laws (Equation 9.21); it is therefore intimately associated with conduction-based EET. Although the model in this chapter does not yet fully represent all of the complexities of an actual system, it is capable of demonstrating that the local biofilm potential and redox potential represent two distinct and critically important functions in EABs. Figure 9.15 shows the profiles of local biofilm potential and redox potential for the standard case and for the low-conductivity case. For the standard case, the conductivity is high enough to prevent any significant potential drop for the local biofilm potential. The redox potential declines rapidly from the electrode potential to between −200 and −350 mV_{SHE} through the biofilm.

The slopes of the local biofilm potential and the redox potential are indicative of the net flux of electrons in the system for conduction-based and diffusion-based EET, respectively. When the slope is negative, there is a local net flux of electrons toward the electrode; alternatively, when the slope is positive, there is a local net flux of electrons toward the bulk. For the standard case, the slope of the local biofilm potential is negative (electrons flow toward the polarized electrode), although this is not visible at the scale shown. The negative slope of the local biofilm potential becomes clear in the low-conductivity case. In this figure, we see a local biofilm potential very similar to those predicted originally by Marcus et al. [27]. The cause of the variable biofilm activity in Figure 9.9 is now apparent, as the local biofilm potential drops to values slightly more negative than the half-maximum rate potential value, thus prohibiting significant electron conduction from regions distal to the electrode. The positive slope of the redox potential at regions further than 50 μm from the electrode

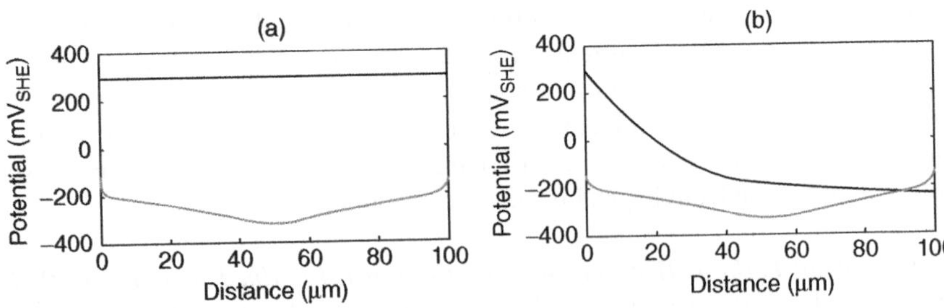

Figure 9.15 Local biofilm potential (black) and redox potential (gray) for the standard case (a) and the low-conductivity case (b).

visually shows the loss of mediators (and thus electrons) into the bulk solution. A reduced flavin molecule at a distance >50 μm from the bottom of the biofilm has a higher probability of being lost to the bulk solution than it has of interacting with the electrode. The redox potential profile in Figure 9.15, with a negative slope near the electrode and a positive slope near the top of the biofilm, was also predicted by Picioreanu et al. [34]. This type of profile is indicative of redox-active molecules acting as mediators for electron transfer. However, experimentally measured redox profiles in *S. oneidensis* biofilms do not have this profile [46]. Our group observed a redox profile in an *S. oneidensis* biofilm that was continually positive (increasing potential toward the top of the biofilm), regardless of the polarization potential of the electrode substratum. In view of our modeling results, this experimental profile suggests that the biofilm was not respiring using diffusion-based EET to the electrode. This also supports the hypothesis that FMN is not likely involved in electron transfer in steady-state *S. oneidensis* biofilms.

In summary, our simulation results show that a large decrease in local biofilm potential is indicative of low biofilm conductivity. Local biofilm potential and redox potential are two distinct measurements, and it cannot be assumed that they will have the same value.

9.5.1.10 What Benefit May a Biofilm Gain by Allowing Mediators to Interact with the Conductive Biofilm Matrix? Currently, it is not known whether any biofilm uses both EET mechanisms simultaneously. Our simulations have theoretically confirmed that there may be a metabolic advantage for biofilms that utilize both mechanisms. We hypothesize that if the two EET mechanisms interact (interacting-dual EET), this may provide a further metabolic advantage. To test this, the term specific electrochemically active surface area (S_a) was introduced. This is the surface area within the biofilm matrix on which mediators may be reduced or oxidized through the depth of the biofilm. Figure 9.16 shows a parameter analysis plot for S_a and biofilm activity by depth when S_a is set to 4300 m^{-1}. The parameter analysis plot shows that high S_a values allow for higher steady-state current and that the electrons delivered via diffusion-based EET approach 0% as S_a increases. This is due to the definition of the percentage of EET, where the final mode in which an electron arrives at the electrode is attributed to that EET mechanism, even if it originated from a different mechanism. These results show that as S_a increases, electrons produced via metabolism ultimately enter the conductive biofilm matrix. The cause for the increase in current is demonstrated in the rightmost figures in Figure 9.16. Throughout the biofilm thickness, the number of electrons derived from the diffusion-based EET mechanism (gray) remains constant. This is in contrast to the results in Figure 9.9, which show that, without interaction, diffusion-based EET drops to zero in the center of the biofilm. Adding the possibility of interaction enables two things: (1) q_M stays maximized because of the continuous oxidation of local mediators by the conductive biofilm matrix, and (2) fewer electrons are lost to the bulk because mediators are rapidly reduced locally without having to diffuse long distances (i.e., to the electrode). Thus, having interaction between diffusion- and conduction-based EET (interacting-dual EET) allows for a further metabolic advantage over having two independent mechanisms.

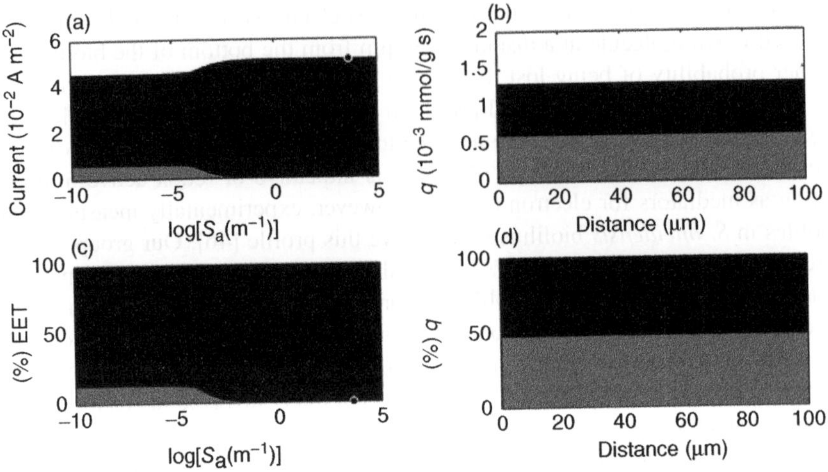

Figure 9.16 (a, c) Effect of parameter S_a on steady-state current and percentage EET. The black-filled circles illustrate the S_a value (4300 m^{-1}) used for all subsequent interaction figures. (b, d) Biofilm activity by depth as measured by substrate utilization rate, q, and percentage of q obtained from diffusion-based or conduction-based EET. Gray represents the portion due to diffusion-based EET, and black represents the portion due to conduction-based EET.

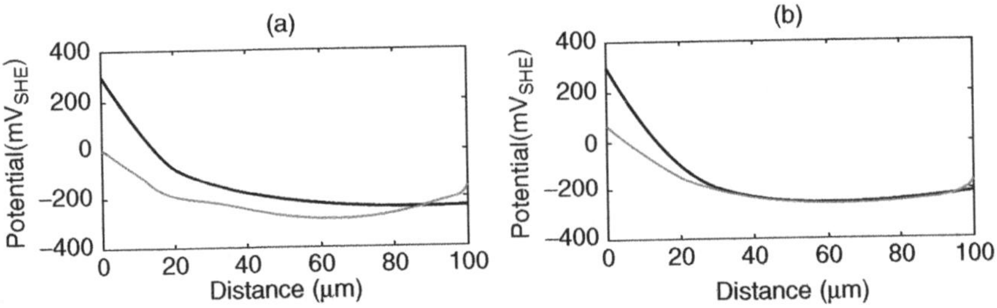

Figure 9.17 Local biofilm potential (black) and redox potential (gray) for (a) the low-conductivity case with interactions allowed and (b) the low-conductivity case with the electrochemically active surface area increased 100 times.

Figure 9.17 demonstrates the effect of S_a on the local biofilm potential and redox potential profiles. Unlike the case in which interactions are not allowed (see rightmost plot in Fig. 9.15), the redox potential profile is dependent on the local biofilm potential if interactions between the two mechanisms are allowed. This can be observed in the exaggerated case in which S_a is increased 100-fold. The local biofilm potential and redox potential are nearly identical from the middle of the biofilm to the top. This is because the local biofilm potential controls the rate of the mediator reaction at the biofilm matrix surfaces. The mediator concentration defines the redox potential.

In summary, our simulation results show that interaction between mediators and the conductive biofilm matrix could support higher biofilm metabolic activity. Interaction between mediators and the conductive biofilm matrix may be advantageous for the biofilm because it reduces the requirement for high mediator concentrations, limits mediator losses, and allows for higher biofilm densities. Closely related local biofilm potential and redox potential profiles could be indicative of a strong interaction between mediators and the conductive biofilm matrix.

9.6 CONCLUSIONS

We have developed a mathematical model simulating four different EET cases for EABs: (1) diffusion-based EET; (2) conduction-based EET; (3) isolated-dual EET; and (4) interacting-dual EET. Our simulations clearly demonstrate that differentiating these or other mechanisms will critically depend on a variety of measured parameters values. That is one of the reasons why this book is focused on providing protocols to measure critical parameters in EABs. We believe that by the time researchers advance and replicate experimental measurements under different conditions, new tools will be provided for the modelers to fully address the questions surrounding electron transfer mechanisms.

From our model, we concluded that

- The availability of diffusion-based EET does not contribute to increased biofilm metabolism unless conduction-based EET is restricted or unavailable;
- Under environmental, limited nutrient, or stressful conditions, having both mechanisms available may provide a metabolic advantage for *S. oneidensis*;
- When conductivity is high, the upper biofilm layer near the aqueous phase boundary of a conductive biofilm can be active, even when the base is limited by a low electron donor concentration;
- On the basis of our model, simulated CV and SWV are unable to determine the specific percentages of the EET mechanism with the current version of the model;
- The most critical parameters for the total current from an *S. oneidensis* biofilm are biofilm thickness, fraction of electrons recoverable for current, biofilm density, maximum specific lactate utilization rate, and electron equivalence of lactate;
- The most critical parameters for determining the EET mechanism are mediator concentration and current production;
- Interaction between mediators and the conductive biofilm matrix can support higher biofilm metabolic activity;
- Interaction between mediators and the conductive biofilm matrix may be advantageous for the biofilm because it reduces the requirement for high mediator concentrations, limits mediator losses, and allows for higher biofilm densities.

Nomenclature

α	Transfer coefficient for the mediator redox reaction at the electrode surface (unitless)
α_b	Transfer coefficient for the mediator redox reaction at the conductive biofilm matrix (unitless)
$Cell_o$	Illustrative of a cell capable of oxidizing substrate
$Cell_r$	Illustrative of a cell that has stored electrons from oxidizing substrate
c	Fraction of current due to conduction-based EET (unitless)
d	Fraction of current due to diffusion-based EET (unitless)
D_{aq}	Bulk liquid self-diffusion coefficients of water ($m^2\ s^{-1}$)
D_e	Effective self-diffusion coefficient of water ($m^2\ s^{-1}$)
D_{ei}	Effective Fickian diffusion coefficient of solute i ($m^2\ s^{-1}$)
D_{eM}	Effective diffusion coefficient of mediator ($m^2\ s^{-1}$)
D_{eS}	Effective diffusion coefficient of substrate ($m^2\ s^{-1}$)
D_i	Bulk liquid Fickian diffusion coefficient of solute i ($m^2\ s^{-1}$)
D_r	Relative effective diffusion coefficient (unitless)
E	Local biofilm potential (V_{SHE})
$E^{0'}$	Standard redox potential for the mediator redox reaction at the biochemical standard state (V_{SHE})
E_{KA}	Half-maximum rate potential (V_{SHE})
E_{redox}	Redox potential (V_{SHE})
ε	Polarized electrode potential (V_{SHE})
F	Faraday constant ($s\ A\ mol^{-1}$)
f	Fraction of electrons recoverable for current (unitless)
j	Total current produced by the biofilm ($A\ m^{-2}$)
j_C	Current produced by conduction-based EET ($A\ m^{-2}$)
j_M	Current produced by diffusion-based EET ($A\ m^{-2}$)
k_0	Standard heterogeneous rate constant for the mediator redox reaction at the electrode surface ($m\ s^{-1}$)
k_0^b	Standard heterogeneous rate constant for the mediator redox reaction at the conductive biofilm matrix ($m\ s^{-1}$)
K_M	Half-saturation constant for the mediator (mM)
K_S	Half-saturation constant for the substrate (mM)
κ	Biofilm conductivity ($S\ m^{-1}$)
L	Biofilm thickness (m)
M_o	Oxidized mediator concentration (mM)
M_r	Reduced mediator concentration (mM)
$Matrix_o$	Illustrative of a matrix capable of oxidizing substrate
$Matrix_r$	Illustrative of a matrix capable of reducing substrate
m	Protons transferred per mediator redox reaction (unitless)
n	Electrons transferred per mediator redox reaction (unitless)
P	Product concentration (mM)
q	Total specific substrate utilization rate ($mol\ g^{-1}\ s^{-1}$)
q_C	Specific substrate utilization rate via conduction-based EET ($mol\ g^{-1}\ s^{-1}$)

q_C^*	Modified specific substrate utilization rate via conduction-based EET (mol g^{-1} s^{-1})
q_M	Specific substrate utilization rate via diffusion-based EET (mol g^{-1} s^{-1})
q_M^*	Modified specific substrate utilization rate via diffusion-based EET (mol g^{-1} s^{-1})
q_{max}	Maximum specific substrate utilization rate (mol g^{-1} s^{-1})
$\mathcal{R}$	Molar gas constant (J K^{-1} mol^{-1})
R_C	Generation rate of e^- in the biofilm due to metabolism (mol m^{-3} s^{-1})
r_M	Net generation rate of e^- at the electrode surface (mol m^{-2} s^{-1})
R_M	Reduction rate of M_o in the biofilm due to metabolism (mol m^{-3} s^{-1})
R_o	Oxidation rate of M_r in the biofilm due to interaction with the conductive matrix (mol m^{-3} s^{-1})
r_o	Oxidation rate of M_r at the electrode surface (mol m^{-2} s^{-1})
R_r	Reduction rate of M_r in the biofilm due to interaction with the conductive matrix (mol m^{-3} s^{-1})
r_r	Reduction rate of M_o at the electrode surface (mol m^{-2} s^{-1})
R_S	Overall substrate consumption rate due to metabolism (mol m^{-3} s^{-1})
R_{Sa}	Generation rate of e^- in the biofilm due to mediator interaction with the conductive matrix (mol m^{-3} s^{-1})
S	Substrate concentration (mM)
S_a	Specific electrochemically active surface area (m^{-1})
T	Temperature (K)
t	Time (s)
X	Biofilm density (g m^{-3})
y	Electron equivalence of lactate (unitless)
x	Distance from electrode surface (m)
$\delta_{j,p}$	Elasticity for parameter p and model output j (unitless)
$\delta_{d,p}$	Elasticity for parameter p and model output d (unitless)

ACKNOWLEDGMENTS

This research was supported by the US Office of Naval Research (ONR), grant #N00014-09-1-0090, NSF-CAREER award #0954186 and by the US Department of Energy (DOE), Office of Biological and Environmental Research (BER), as part of BER's Subsurface Biogeochemistry Research Program (SBR). This contribution originates, in part, from the SBR Scientific Focus Area (SFA) at the Pacific Northwest National Laboratory (PNNL). A portion of this chapter was written under the Laboratory Directed Research and Development Program at Pacific Northwest National Laboratory, a multiprogram national laboratory operated by Battelle for the US Department of Energy. Ryan S. Renslow is grateful for the support of the Linus Pauling Distinguished Postdoctoral Fellowship program. Furthermore, both Ryan Renslow and Jerome Babauta acknowledge the NIH Protein Biotechnology Training Grant (T32-GM008336). Haluk Beyenal acknowledges additional support from

Fundamental and Applied Chemical and Biological Catalysts to Minimize Climate Change, Create a Sustainable Energy Future, and Provide a Safer Food Supply with a project #WNP00807. This chapter is built on our previously published paper: R. Renslow, J. Babauta, A. Kuprat, J Schenk, C. Ivory, J. Fredrickson, and H. Beyenal, *Physical Chemistry Chemical Physics* 2013, **15**, 19262–19283 – Reproduced by permission of the PCCP Owner Societies.

REFERENCES

1. Borole AP, Reguera G, Ringeisen B, Wang ZW, Feng YJ, Kim BH. Energy Environ Sci 2011;4:4813–4834.
2. Logan BE. Nat Rev Microbiol 2009;7:375–381.
3. Lovley DR. Nat Rev Microbiol 2006;4:497–508.
4. Lovley DR. Curr Opin Biotechnol 2008;19:564–571.
5. Donovan C, Dewan A, Heo D, Beyenal H. Environ Sci Technol 2008;42:8591–8596.
6. Zhang F, Tian L, He Z. J Power Sources 2011;196:9568–9573.
7. Geelhoed JS, Stams AJM. Environ Sci Technol 2011;45:815–820.
8. Liu H, Grot S, Logan BE. Environ Sci Technol 2005;39:4317–4320.
9. Du Z, Li H, Gu T. Biotechnol Adv 2007;25:464–482.
10. Liu H, Ramnarayanan R, Logan BE. Environ Sci Technol 2004;38:2281–2285.
11. Finneran KT, Anderson RT, Nevin KP, Lovley DR. Soil Sediment Contam 2002;11:339–357.
12. Lovley DR. J Ind Microbiol Biotechnol 1997;18:75–81.
13. Lovley DR, Baedecker MJ, Lonergan DJ, Cozzarelli IM, Phillips EJP, Siegel DI. Nature 1989;339:297–300.
14. Beech WB, Sunner J. Curr Opin Biotechnol 2004;15:181–186.
15. Kan J, Chellamuthu P, Obraztsova A, Moore JE, Nealson KH. J Appl Microbiol 2011;111:329–337.
16. Gralnick JA, Newman DK. Mol Microbiol 2007;65:1–11.
17. Shi L, Squier TC, Zachara JM, Fredrickson JK. Mol Microbiol 2007;65:12–20.
18. Torres CI, Marcus AK, Lee HS, Parameswaran P, Krajmalnik-Brown R, Rittmann BE. Fems Microbiol Rev 2010;34:3–17.
19. Brutinel ED, Gralnick JA. Appl Microbiol Biotechnol 2012;93:41–48.
20. Kotloski NJ and Gralnick JA. mBio 2013;4.
21. Marsili E, Baron DB, Shikhare ID, Coursolle D, Gralnick JA, Bond DR. Proc Natl Acad Sci U S A 2008;105:3968–3973.
22. Gorby YA, Yanina S, McLean JS, Rosso KM, Moyles D, Dohnalkova A, Beveridge TJ, Chang IS, Kim BH, Kim KS, Culley DE, Reed SB, Romine MF, Saffarini DA, Hill EA, Shi L, Elias DA, Kennedy DW, Pinchuk G, Watanabe K, Ishii S, Logan B, Nealson KH, Fredrickson JK. Proc Natl Acad Sci U S A 2006;103:11358–11363.
23. Strycharz SM, Malanoski AP, Snider RM, Yi H, Lovley DR, Tender LM. Energy Environ Sci 2011;4:896–913.

24. Strycharz-Glaven SM, Snider RM, Guiseppi-Elie A, Tender LM. Energy Environ Sci 2011;4:4366–4379.
25. Renslow R, Babauta J, Kuprat A, Schenk J, Ivory C, Fredrickson J, Beyenal H. Modeling biofilms with dual extracellular electron transfer mechanisms. Phys Chem Chem Phys 2013;15:19262–19283.
26. von Canstein H, Ogawa J, Shimizu S, Lloyd JR. Appl Environ Microbiol 2008;74:615–623.
27. Marcus AK, Torres CI, Rittmann BE. Biotechnol Bioeng 2007;98:1171–1182.
28. Pinto RP, Srinivasan B, Manuel MF, Tartakovsky B. Bioresour Technol 2010;101:5256–5265.
29. Zeng YZ, Choo YF, Kim BH, Wu P. J. Power Sources 2010;195:79–89.
30. Picioreanu C, Head IM, Katuri KP, van Loosdrecht MCM, Scott K. Water Res 2007;41:2921–2940.
31. Bond DR, Lovley DR. Appl Environ Microbiol 2003;69:1548–1555.
32. Batstone DJ, Keller J, Angelidaki I, Kalyuzhnyi SV, Pavlostathis SG, Rozzi A, Sanders WTM, Siegrist H, Vavilin VA. Water Sci Technol 2002;45:65–73.
33. Picioreanu C, Katuri KP, Head IM, van Loosdrecht MCM, Scott K. Water Sci Technol 2008;57:965–971.
34. Picioreanu C, van Loosdrecht MCM, Curtis TP, Scott K. Bioelectrochemistry 2010;78:8–24.
35. Nevin KP, Richter H, Covalla SF, Johnson JP, Woodard TL, Orloff AL, Jia H, Zhang M, Lovley DR. Environ Microbiol 2008;10:2505–2514.
36. Kissel J, McCarty P, Street R. J Environ Eng 1984;110:393–411.
37. Bae W, Rittmann BE. Biotechnol Bioeng 1996;49:683–689.
38. Rittmann BE, Dovantzis K. Water Res 1983;17:1727–1734.
39. Kim JR, Premier GC, Hawkes FR, Rodriguez J, Dinsdale RM, Guwy AJ. Bioresour Technol 2010;101:1190–1198.
40. Velasquez-Orta SB, Head IM, Curtis TP, Scott K, Lloyd JR, von Canstein H. Appl Microbiol Biotechnol 2010;85:1373–1381.
41. Lanthier M, Gregory KB, Lovley DR. FEMS Microbiol Lett 2008;278:29–35.
42. Malvankar NS, Tuominen MT, Lovley DR. Energy Environ Sci 2012;5:6247–6249.
43. Strycharz-Glaven SM, Tender LM. Energy Environ Sci 2012;5:6250–6255.
44. IUPAC. *Compendium of Chemical Terminology (the "Gold Book")*. 2nd ed. Oxford: Blackwell Scientific Publications; 1997.
45. Babauta JT, Nguyen HD, Harrington TD, Renslow R, Beyenal H. Biotechnol Bioeng 2012;109:2651–2662.
46. Babauta JT, Hung Duc N, Beyenal H. Environ Sci Technol 2011;45:6654–6660.
47. Bouhenni RA, Vora GJ, Biffinger JC, Shirodkar S, Brockman K, Ray R, Wu P, Johnson BJ, Biddle EM, Marshall MJ, Fitzgerald LA, Little BJ, Fredrickson JK, Beliaev AS, Ringeisen BR, Saffarini DA. Electroanalysis 2010;22:856–864.
48. Johs A, Shi L, Droubay T, Ankner JF, Liang L. Biophys J 2010;98:3035–3043.
49. Lewandowski Z, Beyenal H. Fundamentals of biofilm research, 2nd ed. Boca Raton, FL: CRC Press-Taylor & Frances Group; 2014.

50. Lower BH, Yongsunthon R, Shi L, Wildling L, Gruber HJ, Wigginton NS, Reardon CL, Pinchuk GE, Droubay TC, Boily JF, Lower SK. Appl Environ Microbiol 2009;75:2931–2935.
51. Ross DE, Brantley SL, Tien M. Appl Environ Microbiol 2009;75:5218–5226.
52. Leang C, Qian X, Mester T, Lovley DR. Appl Environ Microbiol 2010;76:4080–4084.
53. Myers CR, Myers JM. Lett Appl Microbiol 2003;37:254–258.
54. Shi LA, Richardson DJ, Wang ZM, Kerisit SN, Rosso KM, Zachara JM, Fredrickson JK. Environ Microbiol Rep 2009;1:220–227.
55. Nguyen HD, Renslow R, Babauta J, Ahmed B, Beyenal H. Sens Actuators B – Chem 2012;161:929–937.
56. Torres CI, Marcus AK, Parameswaran P, Rittmann BE. Environ Sci Technol 2008;42:6593–6597.
57. Mohr PJ, Taylor BN, Newell DB. Rev Mod Phys 2008;80:633–730.
58. Baron D, LaBelle E, Coursolle D, Gralnick JA, Bond DR. J Biol Chem 2009;284:28865–28873.
59. Renslow RS, Majors PD, McLean JS, Fredrickson JK, Ahmed B, Beyenal H. Biotechnol Bioeng 2010;106:928–937.
60. Bretschger O, Cheung ACM, Mansfeld F, Nealson KH. Electroanalysis 2010;22:883–894.
61. Newton GJ, Mori S, Nakamura R, Hashimoto K, Watanabe K. Appl Environ Microbiol 2009;75:7674–7681.
62. Tang YJ, Meadows AL, Keasling JD. Biotechnol Bioeng 2007;96:125–133.
63. Fan LS, Leyvaramos R, Wisecarver KD, Zehner BJ. Biotechnol Bioeng 1990;35:279–286.
64. Nelson DL, Lehninger AL, Cox MM. *Lehninger Principles of Biochemistry*. New York: W. H. Freeman; 2008.
65. Bas GL. *The Molecular Volumes of Liquid Chemical Compounds from the Point of View of Kopp*. London: Longmans, Green & Co; 1915.
66. EPA. EPA on-line tools for site assessment calculation: estimated diffusion coefficients – extended chemical range. Available at http://www.epa.gov/athens/learn2model/part-two/onsite/estdiffusion-ext.html. Accessed 2010 Dec 16.
67. Hayduk W, Laudie H. AIChE J 1974;20:611–615.
68. Tucker WA, Nelken LH. In: Lyman WJ, Rosenblatt DH, Reehl WF, editors. *Handbook of Chemical Property Estimation Methods: Environmental Behavior of Organic Compounds*. Washington, DC: American Chemical Society; 1982.
69. Rosenbaum M, Cotta MA, Angenent LT. Biotechnol Bioeng 2010;105:880–888.
70. Lovley DR, Phillips EJP, Caccavo F. Appl Environ Microbiol 1992;58:3205–3206.
71. Ksenzhek OS, Petrova SA. Bioelectrochem Bioenerg 1983;11:105–127.
72. Verhagen MF, Hagen WR. J Electroanal Chem 1992;334:339–350.
73. Bard AJ, Faulkner LR. *Electrochemical Methods: Fundamentals and Applications*. New York: Wiley; 2001.
74. Carmona-Martinez AA, Harnisch F, Kuhlicke U, Neu TR, Schroder U. *Bioelectrochemistry*. Amsterdam, Netherlands: ; 2012.
75. Marcus AK, Torres CI, Rittmann BE. Electrochim Acta 2010;55:6964–6972.

76. Renslow RS, Babauta JT, Majors PD, Beyenal H. Energy Environ Sci 2013;6:595–607.
77. Rittmann B, McCarty P. Chapter 3: Microbial Kinetics. New York, NY, U S A: McGraw-Hill; 2000. p 768.
78. Whitman WB, Coleman DC, Wiebe WJ. Proc Natl Acad Sci U S A 1998;95:6578–6583.
79. Zeng H, Zhang TC. Water Res 2005;39:4941–4952.
80. Watson VJ, Logan BE. Biotechnol Bioeng 2010;105:489–498.
81. Qian F, Baum M, Gu Q, Morse DE. Lab Chip 2009;9:3076–3081.
82. Rosenbaum MA, Bar HY, Beg QK, Segre D, Booth J, Cotta MA, Angenent LT. Bioresour Technol 2011;102:2623–2628.
83. Logan BE, Hamelers B, Rozendal RA, Schrorder U, Keller J, Freguia S, Aelterman P, Verstraete W, Rabaey K. Environ Sci Technol 2006;40:5181–5192.
84. Covington ED, Gelbmann CB, Kotloski NJ, Gralnick JA. Mol Microbiol 2010;78:519–532.
85. Wang H, Hollywood K, Jarvis RM, Lloyd JR, Goodacre R. Appl Environ Microbiol 2010;76:6266–6276.
86. Müller F. Free Radical Biol Med 1987;3:215–230.
87. Anderson RF. Biochim Biophys Acta, Bioenerg 1983;722:158–162.
88. Michaelis L, Schubert MP, Smythe CV. J Biol Chem 1936;116:587–607.
89. Lowe HJ, Clark WM. J Biol Chem 1956;221:983–992.
90. Beaudette NV, Langerman N. Arch Biochem Biophys 1974;161:125–133.
91. Watt GD, Burns A. Biochem J 1975;152:33–37.
92. Hagen WR. Eur J Biochem 1989;182:523–530.
93. Visser AJWG, Hoek A v. Proc SPIE: Time-Resolved Laser Spectrosc Biochem 1988;909:61–68.
94. Perry RH, Green DW, Maloney JO, editors. *Perry's Chemical Engineers' Handbook.* New York: McGraw-Hill; 1997.
95. Øyaas J, Storrø I, Svendsen H, Levine DW. Biotechnol Bioeng 1995;47:492–500.
96. El-Naggar MY, Wanger G, Leung KM, Yuzvinsky TD, Southam G, Yang J, Lau WM, Nealson KH, Gorby YA. Proc Natl Acad Sci U S A 2010;107:18127–18131.
97. Manohar AK, Bretschger O, Nealson KH, Mansfeld F. Bioelectrochemistry 2008;72:149–154.
98. Lower BH, Yongsunthon R, Shi L, Wildling L, Gruber HJ, Wigginton NS, Reardon CL, Pinchuk GE, Droubay TC, Boily J-F, Lower SK. Appl Environ Microbiol 2009;75:2931–2935.
99. Lower BH, Shi L, Yongsunthon R, Droubay TC, McCready DE, Lower SK. J Bacteriol 2007;189:4944–4952.
100. Abboud R, Popa R, Souza-Egipsy V, Giometti CS, Tollaksen S, Mosher JJ, Findlay RH, Nealson KH. Appl Environ Microbiol 2005;71:811–816.
101. Wigginton NS, Rosso KM, Lower BH, Shi L, Hochella MF Jr. Geochim Cosmochim Acta 2007;71:543–555.
102. Neidhardt FC, Ingraham JL, Schaechter M. *Physiology of the Bacterial Cell: a Molecular Approach.* Sinauer Associates, Sunderland, 1990.
103. Cao B, Shi L, Brown RN, Xiong Y, Fredrickson JK, Romine MF, Marshall MJ, Lipton MS, Beyenal H. Environ Microbiol 2011;13:1018–1031.

104. Ringeisen BR, Henderson E, Wu PK, Pietron J, Ray R, Little B, Biffinger JC, Jones-Meehan JM. Environ Sci Technol 2006;40:2629–2634.

105. Delaune R, Reddy K. In: Hillel D, editor. *Encyclopedia of Soils in the Environment*. Academic Press; 2005. p 366–371.

106. Renslow RS, Babauta JT, Dohnalkova AC, Boyanov MI, Kemner KM, Majors PD, Fredrickson JK, Beyenal H. Energy Environ Sci 2013;6:1827–1836.

107. Jain A, Gazzola G, Panzera A, Zanoni M, Marsili E. Electrochim Acta 2011;56:10776–10785.

108. Liu Y, Kim H, Franklin R, Bond DR. Energy Environ Sci 2010;3:1782–1788.

109. Marcus AK, Torres CI, Rittmann BE. Bioresour Technol 2011;102:253–262.

110. Merkey BV, Chopp DL. Bull Math Biol 2012;74:834–857.

10

APPLICATIONS OF BIOELECTROCHEMICAL ENERGY HARVESTING IN THE MARINE ENVIRONMENT

Clare E. Reimers

College of Earth, Ocean and Atmospheric Sciences, Oregon State University, OR, USA

10.1 INTRODUCTION

A microbial fuel cell (MFC) is an engineered bioelectrochemical system, designed to convert biochemical to electrical energy, using bacteria to oxidize organic matter with the anode of a galvanic cell serving as an intermediate electron acceptor [1–3]. The MFC reaction commonly terminates with reduction of dissolved O_2, and the O_2 is delivered directly by water, or by air in contact with a polymeric proton exchange membrane pressed against the cathode material [4]. While initially developed for wastewater treatment, MFCs have found a unique and evolving niche as power supplies for unattended sensor systems distributed to measure physical, chemical, geological, and biological variables in the marine environment [5]. In such applications, MFCs are configured to scavenge reactants from the environment and to sustain their active "biocatalyst" through natural processes of bacterial division and colonization at the anode [6]. This chapter reviews engineering and environmental considerations that have influenced the development of marine MFC devices, provides examples of their construction and testing, and describes from experience some of the factors that challenge their long-term operation. It is hoped that this information will guide future development efforts and expand recognition of the utility of these systems.

Biofilms in Bioelectrochemical Systems: From Laboratory Practice to Data Interpretation, First Edition.
Edited by Haluk Beyenal and Jerome Babauta.

Characteristics of the marine environment that are well suited to MFCs are the ubiquity of a conductive electrolyte (seawater; average conductivity ~5.3 S m^{-1} at 25 °C), locales such as sediment–water interfaces, vents and seeps with natural redox gradients and concomitant communities of heterotrophic bacteria and archaea, and dissolved oxygen transported in a turbulent fluid. However, the marine environment is not uniform, or it is usually benign. The idea of using an MFC to produce low levels of power, for a very long time in a remote marine location, must be balanced against the forces that impact any marine instrument installation or which may cause the chemical environment to be unstable. Users need to consider that MFCs may have to withstand strong currents, macrofauna disturbance, biofouling, materials fatigue, or disruption by human activities such as bottom fishing. In deep-sea locations, these devices are under extreme pressure and may be difficult to place and impractical to recover. These realities and the power requirements of different sensors and communication devices will continue to shape how MFCs are designed for the marine environment [7]. These factors and relative costs also determine when and where MFCs have advantages over common marine battery packages as sources of electricity [8].

10.2 DESIGN OF UNDERWATER MICROBIAL FUEL CELL DEVICES

To generate bioenergy using an MFC in the environment, it must be possible to draw current while maintaining a redox potential difference on the order of 0.5 V between an anode and a cathode. (Herein, this potential difference is referred to as the *cell potential*, E_{cell}.) Some of the earliest studies of marine redox gradients emphasized the "built-in" potential difference across the sediment–water interface as a promising locale "to generate power by immersing one array of inert electrodes in the mud and exposing a second array to the overlying water" [9]. Sedimentary systems have also been likened to natural "geobatteries" [10, 11]. Thus, sediment or benthic MFCs have received the most attention of any form of MFC for marine applications [12, 13]. Benthic systems also make sense technically because their physical framework can form fixed nodes for environmental sensors. In such applications, MFCs become subsystems combined with power management and the application sensor and communication components (Fig. 10.1). Design efforts must consider all these elements. For example, components may be packaged in common housings and feedback controls built in to improve energy efficiency and data recovery.

10.2.1 Configuration Options for Benthic MFCs

Within 1000 km from shore, most sediment-covered areas of the ocean are anaerobic at horizons just centimeters beneath the sediment–water interface due to the degradation of organic matter derived from surface primary production and terrestrial biomass carried to continental margins by rivers and ground water [14]. Targeting these locations, and depending on sediment properties such as grain size and porosity, MFCs have been designed with anodes that are shallowly buried, or

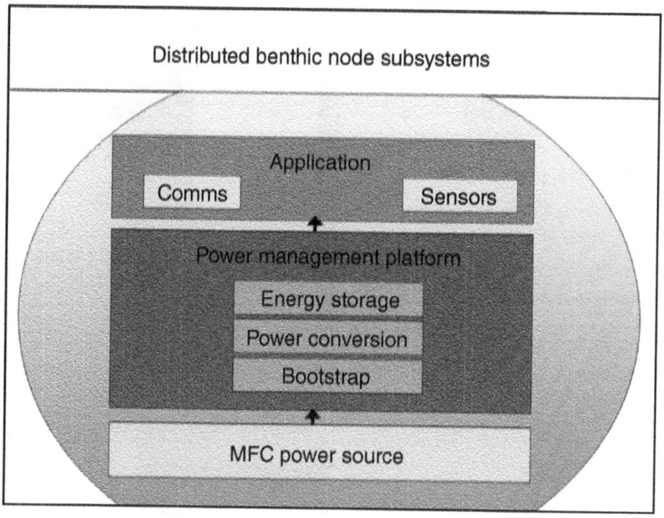

Figure 10.1 The components of an underwater benthic node with MFC power source. Subsystems are described in the text.

else placed on top of the sediment enclosed in benthic chambers (Fig. 10.2). With either design, electrodes are usually made from graphite or other high-surface-area, low-cost, carbon fiber materials that will promote charge transfer reactions, and the general goal is to produce enough power to sustain a particular application or energy store. Anode reactions will usually limit the power supplied, and this is attributed to the higher mass transfer resistance of the anode environment (compared to the cathode) and biological energy losses linked to underlying microbial metabolic pathways [15]. The communities of microorganisms that are effective in promoting electron transfer to an anode (acting as an extracellular electron acceptor) appear to be functionally complex, but ubiquitous in the marine environment, and readily become established biofilms on anode surfaces [16].

Advantages of the fully buried anode approach can be that (1) fuel in the form of organic detritus is continually renewed by sedimentation from above, and (2) anodes can be made to have large contact areas with anoxic sediment by wiring many individual electrodes in parallel [17], or using carbon cloth [18]. Disadvantages are that large forms are difficult to bury and to keep buried, especially in regions with energetic bottom currents or wave pumping. An account of an unsuccessful experiment illustrates how even partial anode exposure can lead to MFC failure.

In 2004, my research group developed a prototype benthic MFC design in which the anode was formed by four 50-cm-long by 4-cm-diameter solid G-10 graphite rods (with bottom ends milled into cone-shaped tips) wired in parallel and secured to a supporting polyvinyl chloride cylinder by titanium hardware (Fig. 10.3). The geometric surface area of each rod was 0.05 m^2. Other components were (1) a cathode made from a 1-m-long carbon fiber brush secured with a Ti bolt to a 14-cm-long by 4-cm-diameter section of G-10 graphite (the brush was taken from a commercially

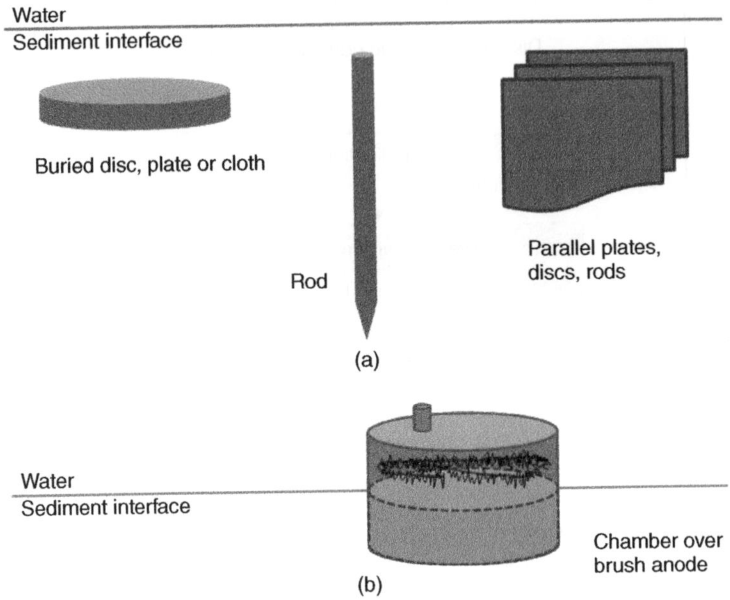

Figure 10.2 Shapes and modes of emplacement of benthic MFC anodes. (a) Buried and (b) chamber enclosed.

available seawater battery: SWB 1200, Kongsberg Maritime AS, Norway, where it also served as a cathode [19]); (2) a variable load designed as a single step potentiostat (NW Metasystems, Bainbridge Island, WA) housed with data loggers in a titanium case, as illustrated in Ref. [7]; and (3) an ORE Shallow Water acoustic release with pop-up flotation canister (Fig. 10.3a). Six of these MFCs were deployed in muddy sediments on the Oregon shelf using the weight stack of a frame-mounted gravity corer to push the anode rods into the sediment (Fig. 10.3a and b). Deployments were made at two locations with muddy sediments (43°56.0′; 124°19.0-.4′; water depth 110 m; 43°44.2-.6′; 124°16.8′; water depth 95 m); a video camera mounted on the corer showed that anodes were fully inserted at the time of deployment and that supporting structure of the fuel cell released from the corer weights as the corer frame was pulled from the bottom and recovered.

Figure 10.4 shows time-series of E_{cell} recorded by three of the six experimental cells that all fell disappointingly short of the preset potentiostatic load potentials of 0.5 V. When the cell potential of the MFC meets the setpoint, the function of the potentiostat is to actively regulate the potential to the setpoint by allowing current to flow. Each record shows an initial rise in E_{cell} when the anodes were fully buried and both anodes and cathodes were equilibrating with the environment. The potentials drop off due to anode exposure at different times, illustrating that a partially buried anode cannot support a potential difference. The factors that worked against success during these experiments are believed to have been the buoyancy of the recovery

(a) (b) (c)

Figure 10.3 A prototype benthic MFC. In (a), the MFC is ready for deployment, coupled to a gravity corer's weight stack by an aluminum releasing head. Mounted to the right side is a flotation canister designed to carry a connecting line to the sea surface after release by an acoustic command. Tie-wrapped to the supporting structure above the anode rods is a titanium pressure housing for a potentiostat and data loggers. The cathode extends to the left. In (b), the whole corer frame and MFC are in view while being launched. A small underwater video camera is mounted to the frame on the bottom right. An outward-looking digital still camera and strobe are mounted on the left side of the frame. Panel (c) shows one of these MFCs being recovered after 9 months on the seafloor of the Oregon shelf. The orange line was connected to the pop-up float. Sea anemones and barnacles are attached to parts of the framework, and the anode rods display a color change indicating that approximately one third of the length of each rod was not buried during most of the deployment. It is expected that not enough ballast and the buoyancy of the release float caused the MFC anodes to work their way out of the sediment during these experiments. (*See insert for color representation of this figure.*)

canisters and insufficient counter weight. Under the influence of surface wave pumping, forces on the canisters worked the anodes tens of centimeter out of the sediment (Fig. 10.3c). Thus, no energy was ever harvested during these trials. The design requirement of maintaining a fully anaerobic anode environment, and low power densities observed with large solid graphite anodes [20], led my research group to begin work on chambered MFC designs. These designs have the advantages that higher effective power densities (scaled to seafloor area) can be achieved by incorporating high-surface-area carbon fiber brush electrodes as anodes as well as cathodes [21]. Potting custom underwater cables to the titanium wire cores of carbon fiber brush electrodes is also easier and more reliable than wire connections to solid graphite. This is because the porosity of graphite eventually leads to seawater infiltration and

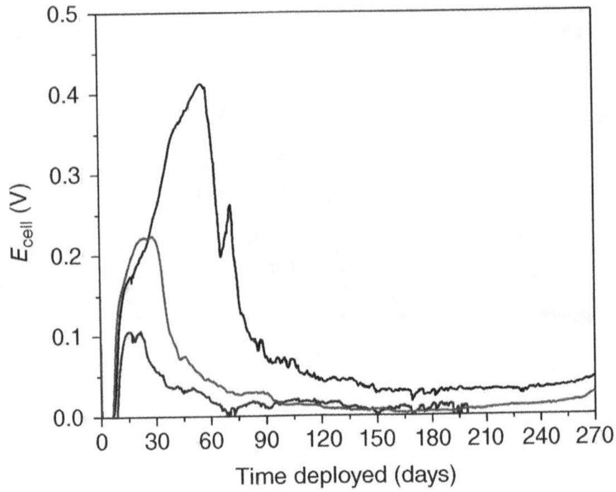

Figure 10.4 Records of cell potential from three in situ benthic MFCs where exposure of the anodes short-circuited the cells at different times. Figure 10.3 shows the design features of these prototype cells.

corrosion of wire contacts. Other advantages associated with chambered MFCs are their relative ease of deployment, the ability to enhance mass transport around the anode either mechanically or through natural advection [7], and the ability to add supplemental materials as priming electron donors [22].

Anode exposure can plague chambered MFCs, however, just as with other designs. One comical example comes from a field experiment in San Diego Bay, CA, during which an MFC deployed by my research group was thwarted by an octopus. This MFC was constructed from a PVC cylinder (0.5-m tall × 0.4-m inside diameter) fitted with a domed top and a one-way valve for venting. The anode was attached to anchor points inside the lid and assembled from four 1-m sections of carbon fiber brush electrode (Fig. 10.5a). The cathode was formed by two 1-m carbon brush sections attached to framework above the chamber. Divers facilitated deployment, and the chamber was pushed approximately 0.3 m into muddy sediments near a small-boat marina. After oxygen in the chamber was depleted by sediment respiration, a potential difference was developed between the anode in the chamber and the cathode in the oxic bay-water. Power levels rose to 32 mW. These levels were short-lived, however, and over 4 months, we observed the power drop out, rise to low levels for short periods, and drop out again and again [23]. A diver later determined that an octopus had taken up residence in the MFC chamber and was ventilating the semienclosed chamber volume probably each time it ventured out to hunt for prey. Now I include a "bioturbation barrier" in the chambered MFCs my group deploys (Fig. 10.5b). A pore size of 0.5 in. diameter has been used to effectively exclude large burrowers. The barrier pictured in Figure 10.5b was made from 0.25-in. thick perforated PVC sheeting.

Figure 10.5 (a) Bottom-up view of a 0.4-m inner diameter, chambered benthic MFC with carbon fiber brush anode attached to the domed top. (b) A top-down view of a later chamber with a perforated PVC bioturbation barrier fixed to the midsection of the chamber. (*See insert for color representation of this figure.*)

10.2.2 Energy Harvesting and Power Management

10.2.2.1 In Situ Monitoring of MFCs As long as MFCs are under development, there will be a need to monitor and test the limits of performance. Although many battery-powered devices regularly record battery voltage, MFC devices have additional characteristics that are of interest because these parameters reflect how the MFC interacts with the environment. As with other power sources, MFC performance will also affect the utility of dependent components of autonomous sensing systems intended for long-term operations (Fig. 10.1). For these reasons, polarization and power curves that reflect cell voltage and cell power as a function of current (or current density) are frequently obtained to track MFC functioning through time [2]. Many researchers use variable potentiostats able to impose a linear potential decrease (or increase) to study polarization behavior (e.g., [24]); however, these instruments are not designed to operate autonomously on low power. Instead, with minimal control and energy utilization, a series of decreasing resistances can be applied to mimic increasing external loads over regular timed intervals to study MFC power production. For example, in recent experiments with chambered benthic MFCs, daily power curves have been recorded using a microprocessor-based power management platform (PMP) to first isolate the MFC from other system components and then apply a series of six resistances each for 10 min [22, 25]. Using Ohm's law ($E_{cell} = IR_{ext}$), the voltage difference between the cathode and anode (E_{cell}) at the end of each 10-min interval can then be divided by the corresponding external resistance (R_{ext}) to give a measure of the MFC current output (I) as a function of voltage. It is commonplace that performance will vary in response to shifts in energy harvesting (long-term loads), environmental conditions such as temperature and salinity, chemical conditions in close proximity to the anode, and/or biofilm development on anodes and cathodes. Figure 10.6 shows example polarization and power curves from such an evolution over a 36-day test deployment of a chambered benthic MFC in Yaquina Bay, OR in 2011. A description of this MFC design and associated subsystems is given in Section 3.3 and Ref. [25]. The resistance series ranged from 10,000 to 5.8 Ω. The greatest maximum power point (MPP) was recorded 14 days after deployment, but the evolution of symmetry around the MPP of the power curves over time indicates continuing stabilization of electrode overpotentials. In other words, in this case, the internal resistance of the MFC was not constant. In MFC systems, the internal resistance stems from activation, metabolic and mass transport losses at the electrodes, added to fixed ohmic losses arising from the conductive components of the MFC itself [2]. Temperature, although variable (Fig. 10.6c), was not a major factor causing changes in power in this example. Longer-term experiments in littoral zone settings with seasonally variable temperature have shown improved MFC performance at higher temperatures, but these MFCs were also affected by unidentified aperiodic environmental conditions lasting over weeks [17].

10.2.2.2 Power Management Components The low voltage and low power levels produced by a benthic MFC cannot directly power most commercially off-the-shelf (COTS) oceanographic instruments that nonetheless represent target applications for environmental energy harvesting. Such devices usually draw current from power sources at 3–24 V DC [7, 17]. The energy harvested from the MFC must therefore be converted to higher voltage levels and stored in a way to meet the duty cycles of underwater devices. The goals in power management are to accomplish power conversion and energy storage with high efficiency and to also cycle certain components between idle and active states. This is why bootstrapping capabilities, power conversion, and energy storage are considered the rudimentary components of most power management systems (Fig. 10.1). By bootstrapping is meant both analog circuitry and software utilities (sometimes accomplished through a state machine) that can optimize energy harvesting, reduce energy consumption, and/or restart and protect system components. One example is the inclusion of a microprocessor-controlled dynamic gain charge pump that adjusts an MFC to always deliver current near the MPP [22]. As power curves will vary over an MFC life cycle (Fig. 10.6), this feature improves overall efficiency compared to discharging an MFC at a fixed potential that may fall on either side of the MPP. Other feedbacks include "failsafe modes" that shut down loads when stored energy gets low. If this is finely tuned, a power management system can preserve key functions such as data storage while limiting other system activities such as communications until energy stores are replenished. Ideally, bootstrapping would also mean having the PMP able to recover from a complete discharge to full operation, as long as the MFC starts supplying power again. That would require a specialized circuit (not yet available) that runs directly off the MFC's approximately 0.5 V and charges up a higher voltage reservoir that could then reactivate the microprocessor and other parts of the PMP.

Starting with a large energy reserve is another prudent strategy for ensuring a reliable renewable power source. Components that can serve to accumulate energy before

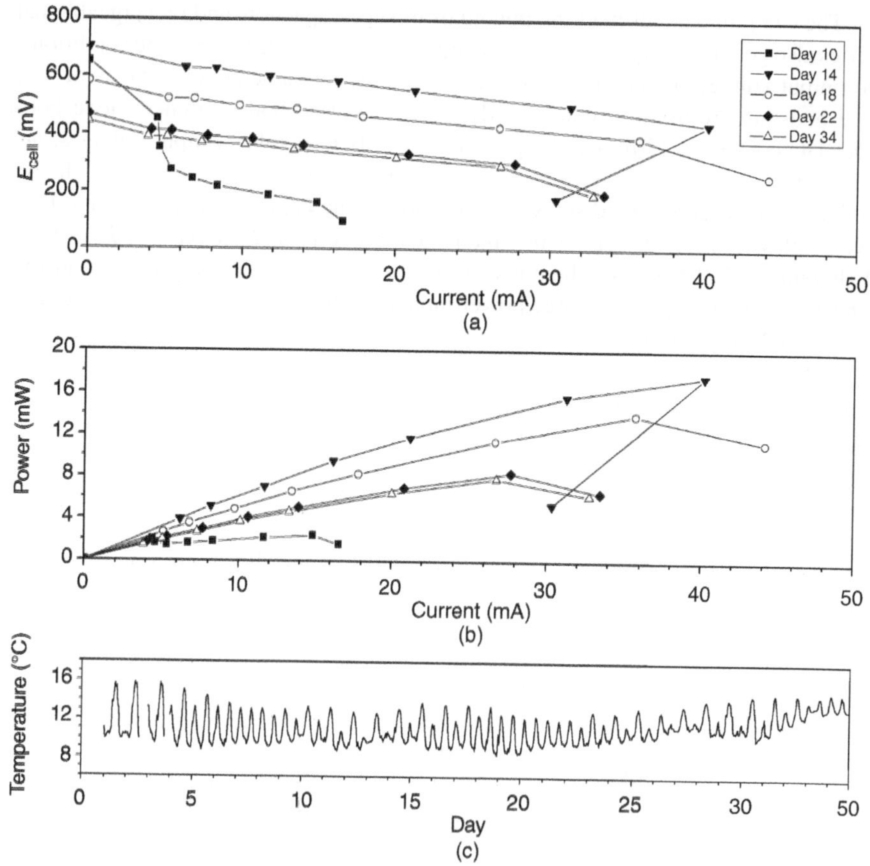

Figure 10.6 Polarization and power curves generated by a chambered benthic MFC after deployment in Yaquina Bay, OR on August 23, 2011. The temperature record reflects hourly readings in the water above the MFC from an *Aanderaa* optode powered by the MFC.

and during a benthic node deployment are banks of supercapacitors and rechargeable batteries [26], although in most PMPs, only rechargeable batteries offer high enough capacity and specific power (W kg^{-1}) to both store energy harvested by an MFC and to back up the MFC should its power delivery decline. Studies that have used single capacitors alone have necessitated that energy is not utilized at all or else very sparingly during intervals when the capacitor is collecting charge from the MFC [27]. This mode of energy storage and utilization feeds back to drive intermittent energy harvesting – which may improve MFC performance compared to continuous energy harvesting using a constant electrical resistance or potential [28]. The cause of improvement is attributed to reduction of the overall mass transfer resistance by diffusion of electron donors to the anode during periods when energy harvesting is relaxed [28, 29].

The charging and discharging potentials of supercapacitors and rechargeable batteries also govern voltage conversion requirements. For MFC power to be conditioned to output voltages compatible with oceanographic instruments, two stages of voltage boosting are usually required. My group has used both switched mode boost DC–DC converters [30] and charge pump circuits to regulate output voltages, and these continue to be optimized to have reduced power dissipation losses (i.e., higher efficiencies). Figure 10.7 displays a schematic of the present operation of the PMP we employ. This technology is the intellectual property of *Teledyne Benthos* (North Falmouth, MA, USA, www.benthos.com). The efficiencies of the primary and secondary charge pump with source power inputs of approximately 10 mW are at least 80% and 98%, respectively.

10.2.2.3 Power Source Uses The architecture of power management systems described previously has two motivations: (1) operating underwater devices indefinitely off energy scavenged from the marine environment and (2) controlling sensor network data gathering and communications requirements that are intermittent for many applications. As a result, a continuous power delivery of approximately 10 mW by an MFC may be enough to sustain essential microcontroller chips, real-time clock crystals and data buffers while also building energy stores for event-driven duties such as communicating detection of an environmental condition.

An illustrative example has been the target application of powering a moored submersible ultrasonic receiver (SUR-1; Sonotronics Inc. Tucson, AZ, USA) designed for detecting acoustic tags fixed to free-ranging marine mammals and fish (for an example study using this receiver, see Ref. [31]). This lightweight, 30- to 90-kHz frequency-scanning COTS hydrophone will log when a tagged animal is detected and can signal a Cellular Uplink Buoy after a certain preprogrammed number of detections. Its circuitry includes a microprocessor, flash memory, and real-time clock. When maintaining scanning and detection logging operations as originally designed, it draws 4–8 mA continuously from 3.6-V lithium D-cell batteries (or ~14–26 mW). Donovan et al. [32] showed by separating the power flow for the SUR

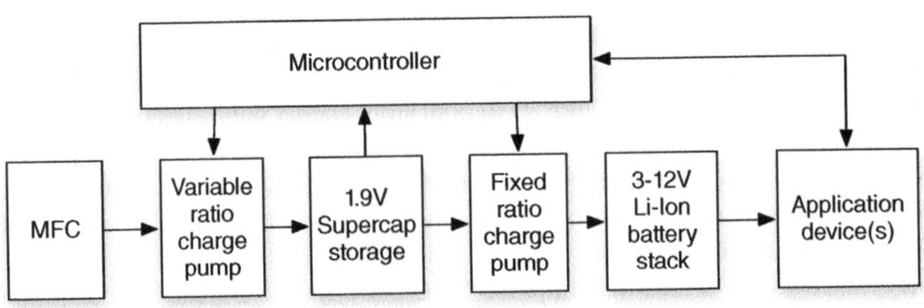

Figure 10.7 Schematic of the interactive components within a power management platform applied when utilizing energy from an underwater MFC.

into (1) essential clock and power management functions (requiring <2 mW continuously) and (2) active scanning and writing to flash memory functions (requiring on average 15.3 mW for intervals of 20 s intermittently) that the sensor could be fully powered by a simple sediment MFC supplying 3–10.4 mW. Over the course of a 6-week experiment, sensor scans were maintained on intervals from 1 to 7.13 min. The power management system used in this demonstration relied only on a 350 F capacitor for energy storage and had DC/DC conversion efficiencies of 70.3–81.4%. One take-home message from this and related work is that most commercial sensors are not optimized for ultra-low-power electrochemical energy sources [33]. However, through fairly simple modifications to existing devices, or new modular designs complemented by programmer-tunable power control, the ultimate goal of unattended instrument nodes monitoring the ocean for years to decades should be realizable.

10.2.3 Deployment and Recovery Considerations

One frequently overlooked consideration in designing MFCs for marine applications is how to emplace and recover them. These strategies link to MFC configuration options and how the MFC power source is wired to other benthic node subsystems. For shallow water applications, SCUBA divers have been relied on to deploy and recover most prototype systems, including burying anodes or water-jetting anode chambers into the sediment. Towed sleds that unroll carbon cloth anodes behind a sediment plow and then drop other parts of a benthic node have also been tried in shallow water [18]. Divers can influence the location of devices on a small scale, but diver-guided emplacement tasks are challenging as visibility is hampered by sediment resuspension. The deep-water alternative to SCUBA is using remote-operated vehicle (ROV) or human-occupied submersible platforms for deployments and recoveries. Pragmatically, however, the services of these mobile platforms and their support personnel are expensive and not always available.

The deployment approach that was the essence of the prototype benthic MFC described in Figure 10.3 is a wire lowered and released instrument package. In the marine geoscience community, this approach is used to deploy a variety of instruments such as ocean-bottom seismometers (OBSs) or acoustic Doppler current profilers (ADCPs) that are usually heavily weighted and may be covered with trawl-resistant shields. Acoustic releases are useful end-of-wire release devices because they can be triggered at any depth by sound pulses sent from the surface vessel. For water depths shallower than 200 m, acoustically triggered pop-up buoy releases may be attached to seafloor instruments for recovery, as was the method in the experiment illustrated in Figure 10.3. Scientists deploying benthic instruments in deeper water will often use ROVs to recover a lifting line (Fig. 10.8) [34]. These approaches should be customized more around MFC-powered marine instrument packages in the future. As the deployment duration of an OBS or ADCP is usually battery limited, these instrument packages represent two of the many promising target applications for bioelectrochemical energy harvesting.

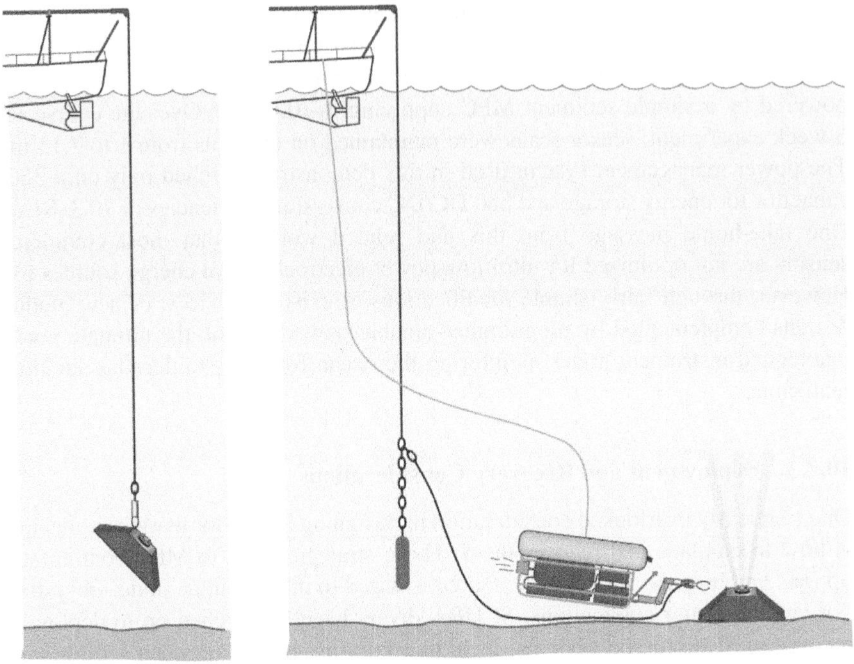

Figure 10.8 Schematic sketch of winch and wire-based deployment and recovery operations from a surface ship using an acoustic release for placement and a ROV for instrument recovery.

10.3 MARINE APPLICATIONS

The most promising marine applications for MFCs or other power sources that harvest low levels of energy from the environment are sensors and communication devices that have long-term missions and will function reliably for years without servicing. This means, they must be rugged, enclosed in well-designed and isolated, corrosion-resistant housings, and not compromised by inevitable biofouling. Acoustic devices (hydrophones, receivers, modems, seismometers, and Doppler current meters) meet these criteria. In deep water away from sunlight where algae will not grow on sensors, many chemical and optical sensors will also hold calibrations and function long term because microbial biofouling and macrofouling by organisms such as bivalves, bryozoans, polychaetes, and anemones can be controlled with copper or other biocide generation systems or coatings. Those designed to signal event detection such as sea-level changes due to tsunami waves, or oil and gas leakage, are strong candidates for power sources that draw energy from the environment. This motivates continuing research and development to increase the power levels that can be generated continuously by bioelectrical systems while also working to reduce the power requirements of autonomous sensors.

10.3.1 Environmental Variables of Importance to Optimizing Marine MFCs

Previous studies have demonstrated that benthic MFCs can be expected to generate continuous power densities ranging from 5 to 30 mW m^{-2} (with areas representing the footprint of anode chambers or buried graphite anodes) in a wide variety of seafloor environments [8, 13, 20, 23]. Under unique environmental conditions, when benthic MFCs have been deployed at a methane cold seep, peak power densities as high as 380 mW m^{-2} have been observed with a chambered design and as high as 1100 mW m^{-2} with a 91.4-cm-long rod-shaped graphite anode centered in the sediments of an active seep [7, 35]. These findings illustrate that the factor that most often limits power generation is the supply of reactive electron donors to the anode environment. Seeps are exceptional because dissolved electron donors in the form of organic acids, hydrocarbons, and hydrogen sulfide are in high concentration and channeled through stratigraphically controlled regions of the sediment column as a consequence of hydrostatic pressure differences and the buoyancy of gas-rich fluids [36]. In contrast, the supply of remineralized nutrients and dissolved organic compounds to an anode in contact with more typical marine sediments quickly becomes limited by molecular diffusion involving low ambient concentrations; these concentrations in turn are fed largely by slow fermentative reactions, mediated by microorganisms in adjacent sediment [16, 37].

Strategies to increase power delivery are discussed in the following section. Although the introduction of electrocatalysts and genetically engineered microorganisms with enhanced capabilities for power output has been suggested as approaches to investigate [38, 16], any improvements realized in marine MFCs will first require systems engineered to maintain a stable biofuel-enriched anode environment and an oxic cathode environment.

10.3.2 Approaches to Sustained Energy Harvesting

10.3.2.1 Biofuel Supplementation With the rate of electron transfer to an MFC anode being limited by the oxidation rate of naturally available fuels in most marine settings, the seemingly most direct path to improved sustainable performance is biofuel supplementation. This approach has been demonstrated with laboratory-scale sediment MFCs [39]. However, achieving a long-term passive introduction of exogenous substrates (i.e., without added power demands from pumps or injection devices) poses another engineering problem for in situ marine applications.

My group and collaborators have tried introducing into benthic anode chambers Hydrogen Release Compound (HRC-X, *Regenesis*, San Clemente, CA, USA), fresh macroalgae harvested from a local mudflat, and 1 -liter blocks of gelatin with an acetate concentration of 200 mM in each block. The HRC-X is a commercially available, viscous liquid electron donor material, which when hydrated is specifically designed to produce an extended, controlled release of lactic acid. The compound is used in soil and groundwater bioremediation to fuel anaerobic biodegradation processes, but during in situ MFC trials of 1–2 months duration, it produced no noteworthy enhancements of MFC performance. Fresh algae may be considered a form of batch mode supplementation of typical marine organic matter, and when introduced at the start of a deployment, it has helped to boost energy production in a chambered benthic MFC [22]. As discussed in the following section, recent MFC deployments in which blocks of acetate-enriched gelatin were added had no certain impact. A general concern is that supplements may largely enhance microbial metabolic processes that are not closely integrated with microbial anode reduction (electron transfer). This is why research into substrate-affected differences in power production from sediment MFCs will continue to be of value [5] and should be coupled with engineering approaches to achieve quasi-continuous supplement supply to an MFC.

10.3.2.2 Anode Cycling A promising approach to overcoming concentration losses that occur when the mass transfer of electron donors or products of electrode reactions limits current production [2] is duty cycling achieved by using multiple-anode MFCs [29]. Multiple-anode, marine, benthic MFCs can be of buried or chambered anode configurations and have an added advantage of redundancy. The design and operation factors that may contribute to duty cycling have a wide range of variables (e.g., number of anodes, cycling interval, E_{cell} when discharging, and means of control) but have had very little study. With microprocessor-controlled PMPs, the programming of duty cycling should be easily achievable. In laboratory experiments, an order of magnitude more current per day was harvested with a sediment MFC system of 15 anodes cycled at an interval of 15 s when compared to a single anode held at a 0.5-V E_{cell} continuously [29]. Current profiles in cycled systems exhibit approximately $t^{1/2}$ decay rates that can be explained by an expanding diffusion layer. In addition, microbial diversity analyses revealed no discernible difference in the community composition of anode biofilms among a range of cycling intervals [29]. This suggests that the communities of microorganisms that exploit MFC anodes as electron acceptors may be tolerant of a fairly wide range of redox potentials.

10.3.3 Case Study at an Ocean Observatory

A point emphasized throughout this chapter is that the success of specific marine MFC applications will depend on organic matter sources available to the anode environment and engineering able to maintain these sources in the context of ocean conditions during operations. Each new field trial increases the understanding of areas of promise as well as pitfalls. As a final illustration, the most ambitious project I have conducted to date is one in which two benthic MFCs were constructed using a cylindrical chambered design with a 0.28-m^2 footprint [25]. Each MFC was used to harvest power for the application of a *Teledyne Benthos* compact acoustic modem interfaced with an *Aanderaa* optode that measured either dissolved oxygen and temperature or conductivity and temperature. The canisters housing the compact acoustic modem electronics also contained an integrated PMP designed as in Figure 10.7 to manage and store the electrical energy generated by the MFC. The PMP stored energy both in a 2-V supercapacitor, and after the second stage of voltage conversion, in a 7.6-V stack of two Li-ion batteries (18,650 size, 2.6 AHr). The rechargeable batteries were independently switchable between the communication and sensor application devices (loads).

In November 2011, the modem/sensor/MFC packages were deployed from the deck of a surface vessel after first attaching a flotation pack, then lifting the packages with a crane and releasing them at the sea surface. Each package descended freely and slowly, under their own weight buoyed by the "elevator" floats that were later jettisoned by ROV. Four 1 -liter blocks of gelatin enriched with 200 mM acetate as a biofuel supplement were attached underneath each chamber in mesh bags (see Section 3.2.1). The chambers landed on the seafloor at locations on opposite sides of the Monterey Accelerated Research System (MARS) cabled observatory node in Monterey Bay, California, approximately 0.5 km away from the main observatory infrastructure, and at depths of 863 and 895 m. The same day, they were checked, pushed slightly deeper in the sediment, and photographed by ROV (Fig. 10.9). The MARS Observatory location is on a plateau adjacent to the Monterey Submarine Canyon, and sediment cores taken by the ROV showed it to have only an approximately 10- to 15-cm thick cover of unconsolidated sediment above a layer of consolidated clay. Nonetheless, these MFC-powered benthic nodes were left to record data from the sensors and the power supply (E_{cell}, capacitor voltage, and battery voltage) on an hourly basis and to monitor the overall fuel cell performance (via a polarization sweep) on a daily basis. Data packets were recovered by activating the compact modem and

Figure 10.9 A chambered MFC in place on the seafloor near the MARS Observatory. Each MFC also carried an osmotic-pump sampler (mounted on the left side of the picture) designed to draw small water samples from the chambers throughout the deployment. The sensor and modem (outside the photo) were suspended above the carbon brush cathode that is pictured above the chamber. Lead weights are visible, fixed to the underside of the chamber lid. Image captured from high resolution underwater video recorded from the ROV Ventana operated by Monterey Bay Aquarium Research Institute. (*See insert for color representation of this figure.*)

prompting acoustic transmissions using either a hydrophone lowered from a surface vessel or acoustic signals sent from a *Teledyne Benthos* Telesonar SM 75 modem cabled to and operated through the MARS Observatory. Communications with the two MFC systems were infrequent at first but increased after several months when it became clear that energy was being added to the rechargeable batteries of the PMP. The two MFC systems operated as designed for 158 and 210 days, respectively. After these time periods, the energy draw of approximately weekly acoustic communications exceeded the energy stored in each system's rechargeable battery pack, causing their PMPs to shut down, with no means of restart (a PMP feature to be corrected). The project team then waited for ROV support to become available for recoveries. The equipment was retrieved on November 13, 2012, after 1 year on the seafloor.

Figure 10.10 shows acoustically transmitted data from one system. Plotted are individual voltages of the two stacked Li-ion batteries that were charged by the MFC and the maximum power observed in daily polarizations versus deployment time. This record indicates that the MFC came up to maximum power slowly, although osmopump samples analyzed after recovery showed persistent acetate concentrations ≤2 mM in this system's anode chamber from near the beginning of the deployment. By day 150, MFC power had recharged the batteries, but then, under the higher loading of more frequent acoustic communications, supply levels fell below the system's minimum energy requirement (~5 mW). Sensor function was reliable up until the PMP shut down. Figure 10.11 shows sensor data from days 208 and 209 of the deployment as an example of acoustically reported measurements of environmental variability.

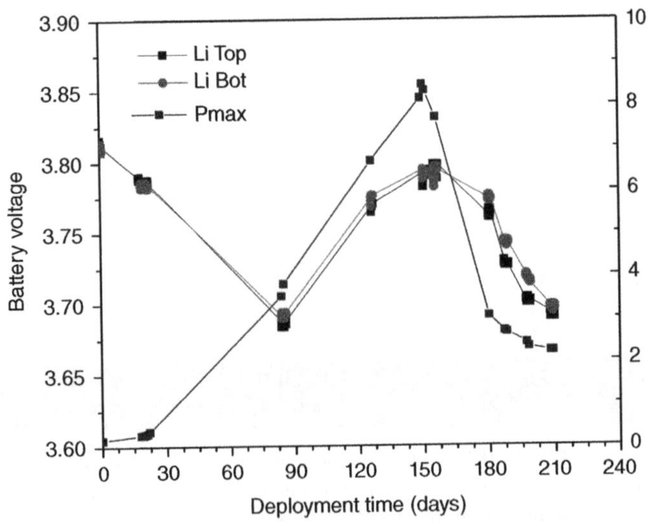

Figure 10.10 Battery voltages and maximum power levels at the MPP from one MFC over the course of its deployment near the MARS Observatory. These data were recovered acoustically.

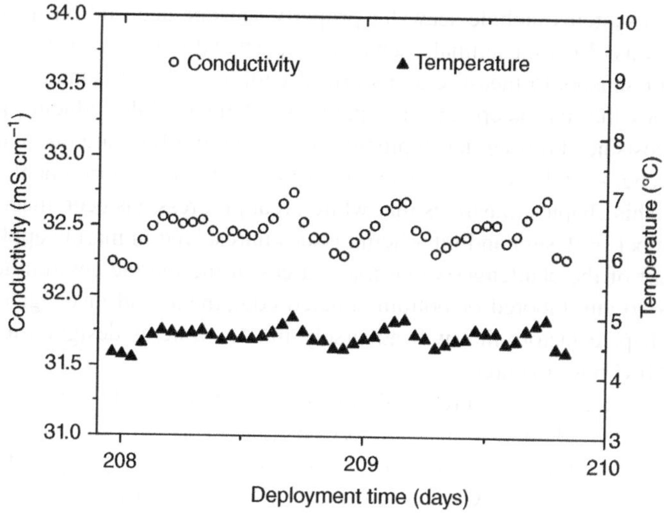

Figure 10.11 An example of hourly sensor data recovered acoustically from a benthic node powered by an MFC at 863-m water depth.

The energy generated from each MFC has been calculated and is reported in Ref. [25]. The long lag time in power production by the MFCs is attributed to the especially shallow penetration depth of the anode chambers into the sediment due to the underlying compact clay deposit seen in ROV push cores. This led to an excessive volume of seawater in the chamber, requiring more time and microbial activity to establish a favorable redox gradient and microbial community for energy harvesting. The acetate added appears to have had no, or else very little, effect in stimulating direct electron transfer from respiring microorganisms to the MFC anode. This lack of response combined with a sediment organic carbon content of $\leq 1\%$ (determined from core samples) likely limited the power output of the MFCs overall.

While the maximum power generated (<10 mW), total energy produced, and the operational length of time were all less than targeted for this experiment, the results are encouraging, given the general inhospitable nature of the deployment site in terms of low temperatures, low organic carbon content, and the physical makeup of the sediment. When future MFC system demonstrations are planned, more attention will be given to predeployment site characterizations. Chambered benthic MFCs will also be preloaded with an internal layer of organic-matter-enriched sediments to help prime environmental energy recovery as a new form of biofuel supplementation (see Section 3.2).

10.4 CONCLUSIONS

The impetus to monitor ocean conditions to supply data indicative of climate change, populate predictive models, or warn of ocean hazards is increasing. Modern observation systems include seawater property sensors, acoustic receivers to study the migrations of marine animals, sensors to detect and report subsurface oil and gas seepage, or sensors to measure and warn of earthquakes and tsunami waves. New technologies that can keep sensors operational longer while reducing monitoring program costs need further development. In this regard, MFCs are fascinating, open, bioelectrochemical devices that tie energy harvesting to the environment where they function. This chapter illustrates that while great progress has been made, there are still a number of design and interfacing issues unresolved in marine applications of MFCs. Part of the challenge is that many areas of the marine environment are not hospitable to any moored or bottom-situated equipment, and investigators need to better anticipate factors that will adversely impact systems designed to utilize an external MFC power source.

All underwater power sources fail more often than most field oceanographers and engineers like to admit. Testing systems over time scales of many months to years is also a slow development process. Only a few groups are currently engaged in developing operational marine MFCs, and so more are encouraged. I contend that once MFC and specific ocean observation devices become highly tailored to each other, MFCs will offer real practical systems advantages over other renewable power sources, or primary batteries in a number of applications. They are on the cusp of having a firm place in the marine technology toolbox.

ACKNOWLEDGMENTS

Work in the Reimers lab at Oregon State University to advance the application of benthic MFCs has been generously supported for over a decade by the US Office of Naval Research. The most recent ONR Award Number: N000141110402 supported the field experiments described in Section 3.3 and the writing of this chapter. The author thanks Michael Wolf (Teledyne Benthos) for his contributions to the design of power management platforms and Paul Schrader (OSU) for leading the field work at the MARS Observatory and comments on this chapter.

REFERENCES

1. Rabaey K, Rodriguez J, Blackall LL, Keller J, Gross P, Batstone D, Verstraete W, Nealson KH. Microbial ecology meets electrochemistry: electricity-driven and driving communities. ISME J 2007;1:9–18.
2. Logan BE, Hamelers B, Rozendal R, Schröder U, Keller J, Freguia S, Aelterman P, Verstraete W, Rabaey K. Microbial fuel cells: methodology and technology. Environ Sci Technol 2006;40:5181–5192.
3. Clauwaert P, Aelterman P, Pham TH, De Schamphelaire L, Carballa M, Rabaey K, Verstraete W. Minimizing losses in bio-electrochemical systems: the road to applications. Appl Microbiol Biotechnol 2008;79:901–913.

4. Liu H, Logan BE. Electricity generation using an air-cathode single chamber microbial fuel cell in the presence and absence of a proton exchange membrane. Environ Sci Technol 2004;38:4040–4046.

5. Logan BE, Regan JM. Microbial fuel cells – challenges and applications. Environ Sci Technol 2006;40:515–523.

6. Biffinger JC, Pietron J, Ray R, Little B, Ringeisen BR. A biofilm enhanced miniature microbial fuel cell using *Shewanella oneidensis* DSP10 and oxygen reduction cathodes. Bioelectronics 2007;22:1672–1679.

7. Nielsen ME, Reimers CE, White HK, Sharma S, Girguis PR. Sustainable energy from deep ocean cold seeps. Energy Environ Sci 2008;1:584–593.

8. Tender LM, Gray SA, Groveman E, Lowy DA, Kauffman P, Melhado J, Tyce RC, Flynn D, Petrecca R, Dobarro J. The first demonstration of a microbial fuel cell as a viable power supply: powering a meteorological buoy. J Power Sources 2008;179:571–575.

9. Whitfield M. The electrochemical characteristics of natural redox cells. Limnol Oceanogr 1972;17:383–393.

10. Bigalke J, Grabner EW. The Geobattery model: a contribution to large scale electrochemistry. Electrochim Acta 1997;42:3443–3452.

11. Naudet V, Revil A. A sandbox experiment to investigate bacteria-mediated redox processes on self-potential signals. Geophys Res Lett 2005;32:L11405. DOI: 10.1029/2005GL022735.

12. Reimers CE, Tender LM, Fertig S, Wang W. Harvesting energy from the marine sediment–water interface. Environ Sci Technol 2001;35:192–195.

13. Tender LM, Reimers CE, Stecher HA III, Holmes DE, Bond DR, Lovley DR, Lowry DA, Pilobello K, Fertig S. Harnessing microbially generated power on the seafloor. Nat Biotechnol 2002;20:821–825.

14. Jahnke RA. Constraining organic matter cycling with benthic fluxes. In: Boudreau BP, Jørgensen BB, editors. *The Benthic Boundary Layer-Transport Processes and Biogeochemistry*. New York: Oxford University Press; 2001. p 302–319.

15. Schröder U. Anodic electron transfer mechanisms in microbial fuel cells and their energy efficiency. Phys Chem Chem Phys 2007;9:2619–2629.

16. Lovley DR. The microbe electric: conversion of organic matter to electricity. Curr Opin Biotechnol 2008;19:564–571.

17. Bandyopadhyay PR, Thivierge DP, McNeilly FM, Fredette A. An electronic circuit for trickle charge harvesting from littoral microbial fuel cells. IEEE J Ocean Eng 2013;38:32–42.

18. Chadwick DB, Kagan J, Wotawa-Bergen AQ and Davis WC. Sled for benthic microbial fuel cell deployment with carbon fabric anodes. IEEE Oceans 2011 Conference Proceedings; 2011; p. 1–7.

19. Hasvold Ø, Henriksen H, Mevar E, Citi G, Johansen BØ, Kjønigsen T, Galetti R. Sea-water battery for subsea control systems. J Power Sources 1997;65:253–261.

20. Dewan A, Beyenal H, Lewandowski Z. Scaling up microbial fuel cells. Environ Sci Technol 2008;42:7643–7648.

21. Nielsen ME, Reimers CE, Stecher HA III. Enhanced power from chambered benthic microbial fuel cells. Environ Sci Technol 2007;41:7895–7900.

22. Gong Y, Radachowsky S, Wolf M, Nielsen M, Girguis P, Reimers CE. Benthic microbial fuel cell as direct power source for an acoustic modem and seawater oxygen/temperature sensor system. Environ Sci Technol 2011;45:5047–5053.

23. Wotawa-Bergen AQ, Chadwick DB, Richter KE, Tender LM, Reimers CE, Gong Y. Operational testing of sediment-based microbial fuel cells in the San Diego Bay. IEEE Oceans 2010 Conference Proceedings; 2010; p. 1–6. DOI: 10.1109/OCEANS.2010.5664612.

24. Aelterman P, Rabaey K, The Pham H, Boon N, Verstraete W. Continuous electricity generation at high voltages and currents using stacked microbial fuel cells. Environ Sci Technol 2006;40:3388–3394.

25. Schrader P, Reimers CE, Doolan C, Girguis P, Wolf M, Green D. Sensors and acoustic modems powered by benthic microbial fuel cells at the MARS Observatory. IEEE Oceans 2013 Conference Proceedings; 2013; p. 1–5.

26. Winter M, Brodd RJ. What are batteries, fuel cells and supercapacitors? Chem Rev 2004;104:4245–4269.

27. Shataram A, Beyenal H, Veluchamy RRA, Lewandowski Z. Wireless sensors powered by microbial fuel cells. Environ Sci Technol 2005;39:5037–5042.

28. Dewan A, Beyenal H, Lewandowski Z. Intermittent energy harvesting improves the performance of microbial fuel cells. Environ Sci Technol 2009;43:4600–4605.

29. Gardel EJ, Nielsen ME, Grisdela PT Jr, Girguis PR. Duty cycling influences current generation in multi-anode environmental microbial fuel cells. Environ Sci Technol 2012;46:5222–5229.

30. McBride LR, Girguis P, Reimers CE. Power storage and conversion from an ocean microbial energy source. IEEE Oceans 2006 Conference Proceedings; 2006; p. 1–5. DOI: 10.1109/OCEANS.2006.306847.

31. Cabanellas-Reboredo M, Alós J, Palmer M, March D, O'Dor R. Movement patterns of the European squid *Loligo vulgaris* during the inshore spawning season. Mar Ecol Prog Ser 2012;466:133–144.

32. Donovan C, Dewan A, Heo D, Lewandowski Z, Beyenal H. Sediment microbial fuel cell powering a submersible ultrasonic receiver: new approach to remote monitoring. J Power Sources 2013;233:79–85.

33. Hempstead M, Tripathi N, Mauro P, Wei G-Y, Brooks D. An ultra low power system architecture for wireless sensor network applications. Proceedings of the 32nd International Symposium on Computer Architecture (ISCA'05), Madison, WI; 2005.

34. Barclay AH, Gassier D, Webb SC, Koczynski T, Oletu V, Gaherty JB, Tolstoy M. A trawl-resistant ocean-bottom seismometer. Poster and Abstract S13C-2036 presented at 2010 Fall Meeting, AGU, San Francisco, CA, 2013 Dec. 13–17.

35. Reimers CE, Girguis P, Stecher HA III, Tender LM, Ryckelynck N, Whaling P. Microbial fuel cell energy from a ocean cold seep. Geobiology 2006;4:123–136.

36. Tryon MD, Brown KM, Torres ME. Fluid and chemical flux in and out of sediments hosting methane hydrate deposits on Hydrate Ridge, OR, II: hydrological processes. Earth Planet Sci Lett 2002;201:541–557.

37. Reimers CE, Alleau Y, Bauer JE, Delaney J, Girguis PR, Schrader PS, Stecher HA III. Redox effects on the microbial degradation of refractory organic matter in marine sediments. Geochim Cosmochim Acta 2013;121:582–598.

38. Rosenbaum M, Zhao F, Schröder U, Scholz F. Interfacing electrocatalysis and biocatalysis with tungsten carbide: a high-performance, noble-metal-free microbial fuel cell. Angew Chem 2006;118:1–4.

39. Nielsen ME, Wu DM, Girguis PR, Reimers CE. Influence of substrate on electron transfer mechanisms in chambered benthic microbial fuel cells. Environ Sci Technol 2008;43:8671–8677.

11

LARGE-SCALE BENTHIC MICROBIAL FUEL CELL CONSTRUCTION, DEPLOYMENT, AND OPERATION

JEFF KAGAN, LEWIS HSU, AND BART CHADWICK
Space and Naval Warfare Systems Center Pacific, San Diego, CA, 92152 USA

11.1 INTRODUCTION

Benthic microbial fuel cells (BMFCs) have been shown to generate sufficient energy for low-power applications [1–4]. The main challenges preventing use in a broader range of applications are their characteristic low power densities and scale-up problems. To generate higher power outputs, BMFCs with larger electrode surface areas will need to be designed, built, and deployed. Recent work has shown that making larger electrodes does not necessarily result in a proportional increase in power [5]. In our laboratory, we have found that in a system with an anode and a cathode that have the same surface area, power output is typically limited by the anode. In these anode-limited systems, current generated by the anodes does not increase linearly with an increase in surface area. Instead, current and power are observed to increase linearly with size up to a certain range and then only marginally increase above this size range.

In the rest of this chapter, when we refer to a BMFC, we refer to an anode-limited BMFC, as we encounter this scenario in most of our experiments. We posit that there is an optimally sized anode that will maximize BMFC power output, and minimize complexity, and that further increases in the size of the anode will not substantially increase the power [5]. Therefore, our current scale-up strategy is to deploy larger systems made of many smaller, optimally sized anodes to take advantage of the observed higher power density. These "segmented BMFCs," as they are referred

Biofilms in Bioelectrochemical Systems: From Laboratory Practice to Data Interpretation, First Edition.
Edited by Haluk Beyenal and Jerome Babauta.

to hereafter, are BMFC systems comprised of two or more single BMFCs capable of being deployed as a single system but operated as a large single system or as many BMFC modules depending on the energy harvesting electronics and strategy employed. The goal of this chapter is to describe the construction, deployment, and operation of a segmented, multianode BFMC with a selected example data set.

Before constructing a BMFC, it is necessary to consider the deployment environment and process because cost and effort limitations in burying or installing a BMFC are critical with large systems. Deployment methods for BMFCs generally fall under three categories: (1) autonomously deployed, (2) diver-deployed, and (3) boat-deployed. Autonomously deployed BMFCs are BMFCs that descend to the ocean bottom and complete installation and setup in the sediment without further human intervention [6]. These types of BMFCs require significant engineering and design work upfront to ensure proper placement of electrodes at a range of depths underwater. Practically, this can be considered the preferred case for the end users if engineering solutions exist, but most applications of this nature have been for very small, very-low-power systems. To date, there is limited research in developing autonomously deployed BMFC, so most reported studies involve deployment with diver and/or boat support.

Diver-deployed BMFCs are an option for shallow water deployments (typically <30–50 m) where depths are not prohibitive to diving and the size of the BMFC can be handled easily by one or two people in the water (i.e., BMFC systems requiring sediment displacement of volumes <0.5 m^3). BMFC systems larger than this require large teams of divers, which become cost prohibitive to deploy on the basis of labor costs for the deployment team to properly handle, place, and bury the anodes.

Boat-deployed systems are those that can be launched and installed by the crew on the boat. For our systems, an installation device is lowered to the sediment. Burial of the anode and placement of cathode and electronics package are started by towing the system with the boat [7]. Our experience has shown that deploying large BMFCs is most easily done with a boat-deployment system such as the systems developed in our laboratory. Therefore, in this chapter, we focus on boat-deployed systems.

In general, BMFCs consist of an anode and a cathode, where the anode is buried under the sediment, and the cathode is in the oxygenated water column (Fig. 11.1). Organics in the sediment serve as a source of renewable fuel for BMFC systems. The microorganisms in the sediment oxidize the organics [8] or hydrogen sulfide [9, 10] compounds present in the sediment. BMFC performance is generally thought to be dominated by oxidation of organic carbon (Equation 11.1) and hydrogen sulfide (Equation 11.2) to produce protons and electrons.

$$\text{Organic-C} + H_2O \rightarrow CO_2 + H^+ + e^- \quad (11.1)$$

$$HS^- \rightarrow S^o + H^+ + 2e^- \quad (11.2)$$

The electrons generated in these reactions are collected at the anode and transferred through the external circuit to the cathode (Fig. 11.1). Note that actual reactions are

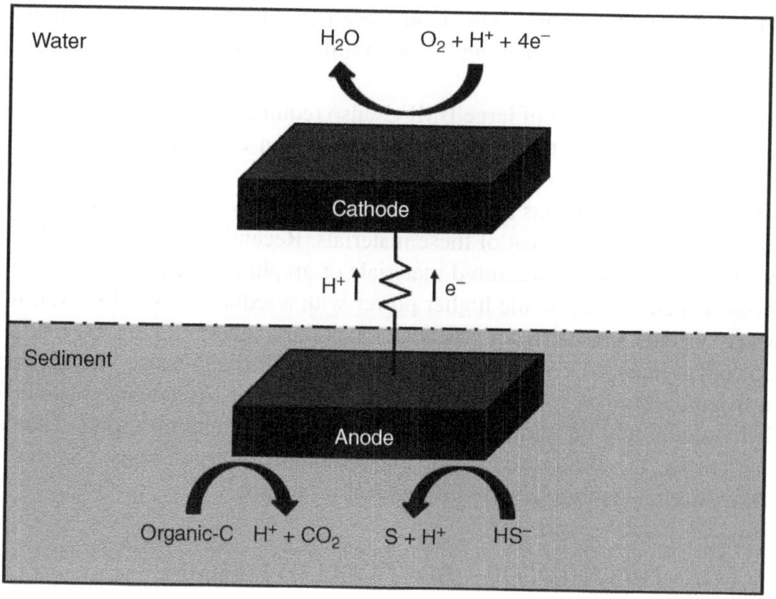

Figure 11.1 Schematic diagram of a benthic microbial fuel cell (BMFC). The anode was buried under the sediment, and the cathode floated underwater above the water–sediment interface. In this figure, electrons pass through a resistor. Instead of a resistor, a power management system can be used to increase potential and to operate a sensor.

more complex and include multiple reactions. As the reaction mechanisms are not our goal, we do not discuss them in this chapter.

Oxygen reduction reaction (ORR) is the primary source of cathodic current employed in BMFCs. The cathodic current could be abiotic or biotic in nature. Regardless of the exact mechanisms, the presence of oxygen is required to generate cathodic current for BMFCs. When oxygen is the only source of cathodic current, ORR leads to the production of hydrogen peroxide, a side reaction known to occur on graphite [11–13] and in BMFCs [14, 15]. On uncatlalyzed graphite surfaces, the ORR can proceed by a 2-electron reduction pathway to hydrogen peroxide (Equation 11.3), followed by another 2-electron reduction to water (Equation 11.4) [16].

$$O_2 + 2e^- + 2H^+ \leftrightarrow H_2O_2 \tag{11.3}$$

$$H_2O_2 + 2e^- + 2H^+ \leftrightarrow 2H_2O \tag{11.4}$$

From Equations (11.3) and (11.4), we expect that ORR at cathodes, in the absence of cathodic biofilms, will tend to accumulate hydrogen peroxide and decrease the pH near the cathode surface. Babauta et al. demonstrated that when graphite electrodes were used, Equation 11.3 controls ORR at the electrode when a large offset potential (relative to open circuit) is applied to the cathode [16]. However, when we operate

BMFCs in the laboratory and in the field, the observed cathodic overpotential is generally less than 100 mV, indicating the possibility of different reactions occurring on the cathodes.

Successful deployment of large BMFCs also requires careful consideration of the electrode materials. Electrodes should be inert when exposed to sea water. These systems require large anode surface areas, so the selected materials should be inexpensive. The use of precious metal catalysts on the anode or on the cathode surfaces is not possible due to the cost of these materials. Recently, several researchers have proposed the use of nanostructured materials or graphite doped with precious metal catalysts, which could provide higher power with a reduced cost. However, in open seawater systems, these are not practical. Once these electrodes are exposed to seawater, contamination by seawater and organisms will likely cause the electrodes to perform similarly to inexpensive carbon-based electrodes. As a result, most deployed BMFC systems rely on carbon-based electrodes with titanium wiring. Titanium is used primarily due to its resistance to corrosion and ability to make connections to graphite materials with relatively low contact resistance.

11.2 CONSTRUCTION OF A BOAT-DEPLOYED BMFC

The construction process and final form factor for large-scale BMFCs are highly dependent on the deployment method. Experience has shown that the nature of the site (e.g., water depth and sediment type), the proposed deployment methodology, and the desired power output should all be considered in determining the final BMFC configuration before any construction work was performed. For this chapter, we assume a typical large-scale BMFC with anode surface area of order 4 m^2, and we describe our typical conditions and boat-based deployment process, to provide context for the construction and monitoring methods. The brands of materials and tools are listed primarily for the benefit of the reader, and our goal is not to promote them. Note that the materials and methods in this case are simply what worked well for us. Someone following these procedures may use suitable alternatives and still achieve reasonable results.

11.2.1 Deployed Configuration

For a typical BMFC described in this chapter (Fig. 11.2), the anode is deployed about 15 cm below the sediment surface with a protective screen located about 7.5 cm below the sediment surface and a similar distance above the anode. The protective screen impedes burrowing ocean fauna from exposing the anode to oxygenated water, which would impact the system's performance. In areas where burrowing organisms are not prevalent, the protective screen can be replaced by a second anode, thus effectively doubling the size of the anode. The cathode chassis rests on the sediment surface with the cathodes held upright in the water column. For a typical test scenario, cables from both the anode and cathode are run up to a boat house for testing and data collection. For instrument power applications, the power is managed in a subsurface pod co-located with the cathode, and there are no surface cables.

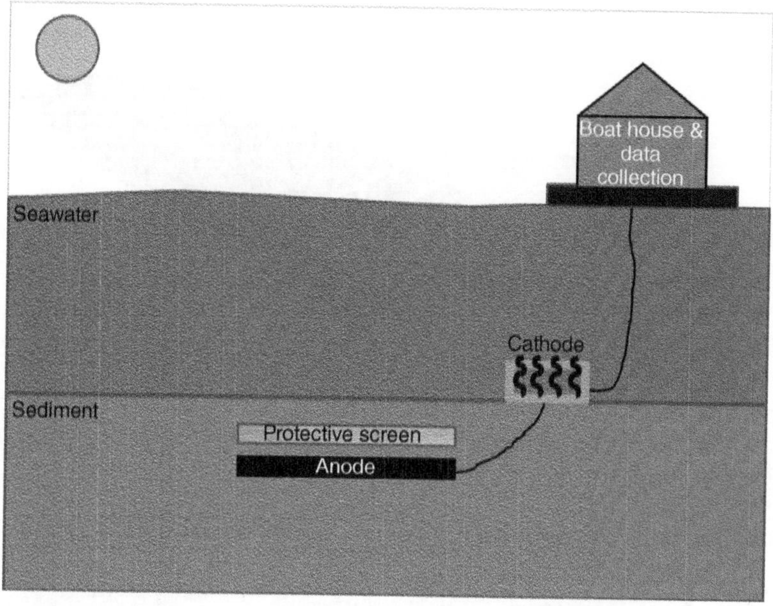

Figure 11.2 View of a deployed anode and cathode. In this chapter, we describe procedures to construct these BMFCs and describe how they are deployed. For our examples, our anode size is 4 m^2 carbon fabric.

11.2.2 BMFC Deployment System and Methodology

The boat-based deployment system consists of a stainless steel sled with a tow bracket and cathode/instrument bay on the top and two horizontal plow blades on the bottom (Fig. 11.3). The sled is about 1.2 m long by 1.0 m wide. The tow bracket is configured to pivot, so it can accommodate a range of tow angles from the boat. The cathode/instrument pod provides a location to house the cathode cage and any electronics that will be deployed as part of the system and allows these components to be ejected off the back of the sled once towing begins. The lower plow blade accommodates the fabric anode, which is rolled onto an axle and deploys into the sediment at a depth of 15 cm (below the sediment surface) through a slot in the back of the plow blade. The upper plow blade is comparable to the lower blade but deploys material at a burial depth of about 7.5 cm and can accommodate either the protective screen or a second anode layer. The back edges of the anode and the protective screen protrude through the slots in the plow blades and are secured in place by Velcro straps prior to the deployment. These straps are rigged with rope so that the straps will detach during the initial phase of deployment.

A schematic of the BMFC boat-based deployment process is shown in Figure 11.4. The anode and the protective screen are loaded into the plow blades, and the cathode cage is loaded into the deployment bay on the top of the sled. The sled is first lowered to the seafloor using a winch cable from the boat. The winch cable is connected to a

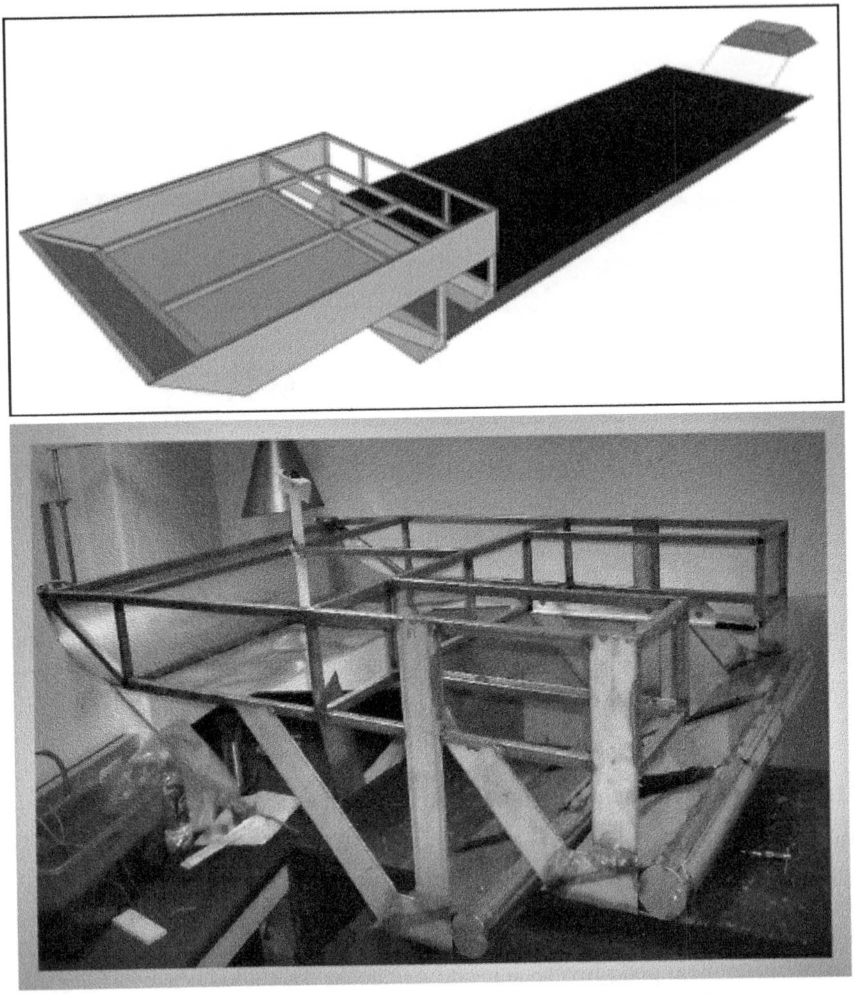

Figure 11.3 The boat-based deployment sled showing the general concept of operation (a), and the two-bladed prototype under construction (b). *(See insert for color representation of this figure.)*

rope bridal that acts to keep the sled level during deployment, as well as secure the cathode cage until the sled is on the seafloor. Care is taken to maintain proper orientation of the sled when it lands on the bottom using tag lines (not shown). The winch cable, bridal, and taglines are removed and the boat then backs away sufficiently such that there is about 6–8× the water depth of tow cable distance between the boat and the sled to insure that the sled is not pulled out of the bottom during towing. The boat then tows the sled at a slow speed (~0.5–1.0 m s^{-1}), and sediment begins flowing over the blades of the sled, driving the blades into the sediment until they are fully buried to the deployment depth. The initially displaced sediment passing over the

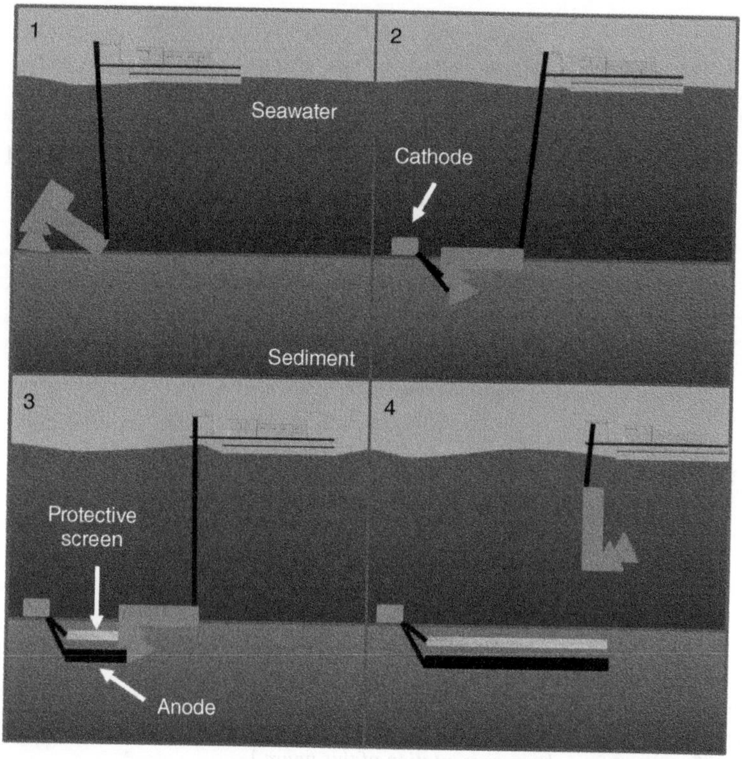

Figure 11.4 Step-by-step BMFC deployment: (1) BMFC sled is lowered to sediment; (2) cathode pod releases as boat begins pulling the BMFC sled; (3) the cathode pod acts as an anchor to initiate paying out the anode; (4) the BMFC is fully deployed and the BMFC sled is recovered.

plow blades creates pressure on the bottom plate of the cathode support structure and pushes it off the back of the sled. In this configuration, the cathode structure and the back edge of the anode (and protective screen) are tied together with short Dacron lines, so that the cathode acts as an anchor point to begin unfurling the anode and protective screen from the sled axles. The anode and protective screen are unrolled and covered as the sled continues to plow and displace more sediment. If the BMFC was deployed successfully, the sled is brought back onboard the boat, leaving behind the BMFC. The cathode support structure may include electronics to control/measure power generation and operate sensors if the BMFC is deployed in remote area where shore-side monitoring is not possible.

The full configured large-scale BMFC and deployment sled system is shown in Figure 11.5. The anode and protective screen are rolled onto axles. The anode is installed behind the lower blade, and the protective screen is installed in the upper blade. The cathode is installed in the compartment above the blades and is designed to disengage from the sled as the sled is towed in the sediment. The following sections

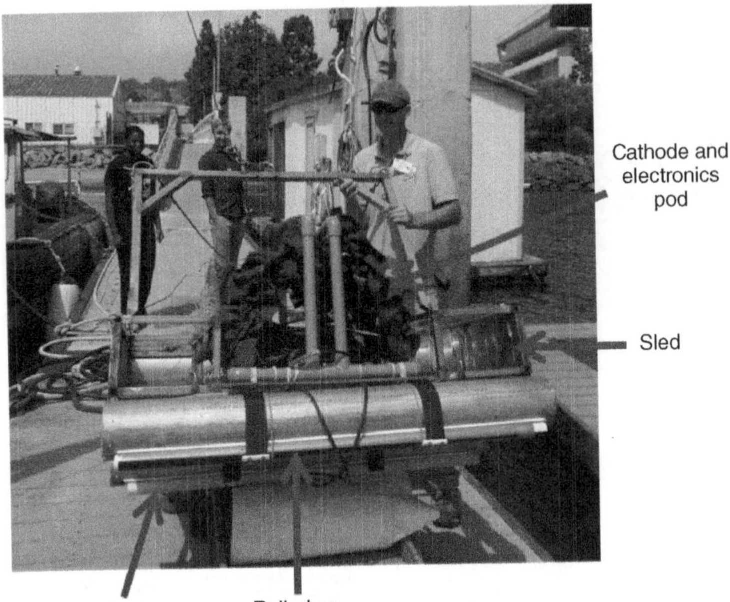

Figure 11.5 A sled loaded with the anode, cathode, protective screen, and cables. Electronics are used when BMFC is deployed in a remote area to record power generation or to operate sensors. *(See insert for color representation of this figure.)*

describe how to build the anode and cathode for deployment in a sled such as the one shown.

11.2.3 Anode Construction

In this example case, we describe the construction of a carbon fabric 1 m × 2 m segmented anode consisting of four 0.5 m × 1 m electrodes having a total surface area (including both sides) of about 4 m^2. Each segment is electrically isolated to allow the overall system to act as four independent BMFCs from one sled deployment. Four insulated copper flat wire current collectors run down the middle of each anode. Stainless steel staples are used to make a secure electrical connection between the carbon fabric and copper current collectors. The copper flat wire is connected to an underwater cable and integrated into a load and monitoring system. Table 11.1 lists the materials used to make the described segmented anode.

11.2.3.1 Constructing Each Segment The objective in this case is to make a four-segment anode from carbon fabric, with each segment measuring 0.5 m × 1 m. This segmented anode will be rolled onto an axle within the sled and deployed as

TABLE 11.1 The Materials Needed to Build Anode, Part Number, and Their Provider.

Material	Part Number	Manufacturer
Carbon fabric	PX30FBPW06 Panex 30	Zoltek; St. Louis, MO, USA
PVC 1 in. diameter	531194	JM Eagle; Los Angeles, CA, USA
Duct tape 5 cm (2 in.)	50-555	Roberts; Boca Raton, FL, USA
Gorilla tape 2.5 cm (1 in.)	61001	Gorilla Glue Company; Cincinnati, OH, USA
Gorilla tape 7.5 cm (3 in.)	6003001	Gorilla Glue Company; Cincinnati, OH, USA
5200 fast cure sealant	292592	3M; Maplewood, MN; USA
Copper flat wire	55921601	FlatWire Technologies Division; Carrollton, GA, USA
Stainless steel staples	508SS1	Arrow Fastener; Saddle Brook, NJ, USA
Underwater cables	Underwater pluggable	Teledyne Impulse; San Diego, CA, USA
Blue painters tape	2090-1J	3M; Maplewood, MN, USA
Easy mask masking paper	12102	Trimaco; Morrisville, NC 27560, USA
Super 77 spray adhesive	77-24	3M; Maplewood, MN; USA
The original mighty sealer	Z2H061458	The Sherwin-Williams Co.; Cleveland, OH, USA
Fiberglass screen	3000016	Phifer; Tuscaloosa, AL, USA
West System 105 Epoxy Resin	105-B	West System Inc; Bay City, MI, USA
West System 206 Hardener	206-B	West System Inc; Bay City, MI, USA

described previously. The following steps describe how to build a segmented anode (depicted in Fig. 11.6):

1. Measure and cut a piece of fabric that encompasses the total footprint of the system. In this example, we would like four segments that are 0.5 m × 1 m that will be separated by gaps of 2.5 cm (1 in.). Thus, the initial piece of carbon fabric that is needed approximately measures 2.1 m × 1 m.
2. Lay the carbon fabric panel on a flat, clean surface. Make sure that the fabric is aligned (i.e., the corners are 90° and opposite sides are parallel to each other) throughout the rest of this procedure. This is critical to making sure the anode retains its shape and rolls cleanly onto the sled axle.
3. Mark the separations between the segments with 7.5-cm Gorilla tape. In our example, this would result in three pieces of duct tape separating the four segments.
4. Place pieces of 7.5-cm Gorilla tape at the ends of the fabric. This will prevent fraying of the fabric as it is handled.

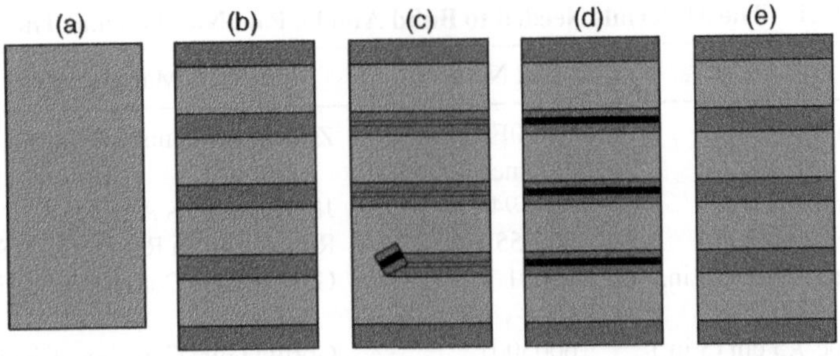

Figure 11.6 Step-by-step instructions for segmenting the anode. (a) Cut a suitably sized piece of carbon fabric cloth (Step 1); (b) tape ends and mark segment divisions with tape (Steps 3 and 4); (c) place the masked 7.5-cm Gorilla tape (Step 6) on the other side of the fabric, aligned with the tape marking the segment separations (Step 7); (d) cut out 2.5-cm section of tape and fabric, leaving the 7.5-cm masked Gorilla tape exposed (Step 10); (e) reinforce the anode by taping over the portion cut out to make the segment divisions (Step 14).

5. Flip the anode over so that the tape is now on the bottom, facing the floor.
6. We will need a 2.5-cm gap between the segments for adequate electrical isolation. To do this, we begin by masking the middle inch of the 7.5-cm Gorilla tape with the 2.5-cm Gorilla tape.
   a. Begin by cutting a piece of 7.5-cm Gorilla tape to match the width of the anode. In our case, this is 1 m.
   b. Cut a piece of 2.5-cm Gorilla tape that is the same length as the 7.5-cm tape measured in Step 6a.
   c. Place this 2.5-cm Gorilla tape piece in the center of the 7.5-cm Gorilla tape piece so that the adhesive sides of both pieces are facing each other. This should create a 1-m long piece of tape that is 7.5 cm wide, with the center inch of adhesive masked by the 2.5-cm tape.
7. Affix the 7.5-cm tape from Step 6 onto the carbon fabric on the side with no tape. Be sure to align the 7.5-cm tape, with the location of the tape already on the fabric marking separations in anode segments from Step 3. At this point, there should be tape on both sides of the fabric where the segments will be separated.
8. Flip the anode over and ensure that the fabric is properly aligned.
9. Apply pressure along all pieces of tape to ensure good adhesion.
10. There should be a raised area along the tape created by the 2.5-cm tape used in Step 6. With a sharp blade, carefully score the tape along both sides of this raised area, taking care to cut only the top layer of tape and the fabric and not the bottom layer of tape.
11. Remove the raised area of tape. The adhesive of the tape should result in the removal of the cut section of carbon fabric as well. At this point, only the masked tape from Step 5 will remain in the gaps between the anode segments.

12. Cut out stray fabric strands from separating gaps.
13. Test for electrical connections between panels with a multimeter or ohmmeter. If there are short circuits, check for and remove stray fabric strands that could be bridging the gap between anode segments. This step is critical to ensuring that each panel is electrically isolated.
14. Use the 7.5-cm Gorilla tape to tape over the cut tape removed in Step 10. This ensures mechanical strength between segments.

11.2.4 Installing Current Collectors

To collect current from the segmented anodes, copper flat wire is run to each anode segment and is electrically connected to the fabric with stainless steel staples. Installation of insulated copper flat wire is described as follows and depicted in Figure 11.7. The goal here is to create good, waterproof electrical contacts between the carbon fabric and the copper wire.

1. Cut a 2.5-m and a 1.5-m piece of copper flat wire. Each copper flat wire piece contains two separate copper conductors, so we have two separate pieces of insulated copper flat wire with a total of four conductors.
2. Lay both pieces of flat wire along the length of the fabric. These pieces should be centered around the middle of the fabric and separated by 7.5 cm. The objective in this case is to place the wires in a way that will allow the anode to roll up evenly on the sled axle. This is important and ensures that the anode will deploy smoothly.

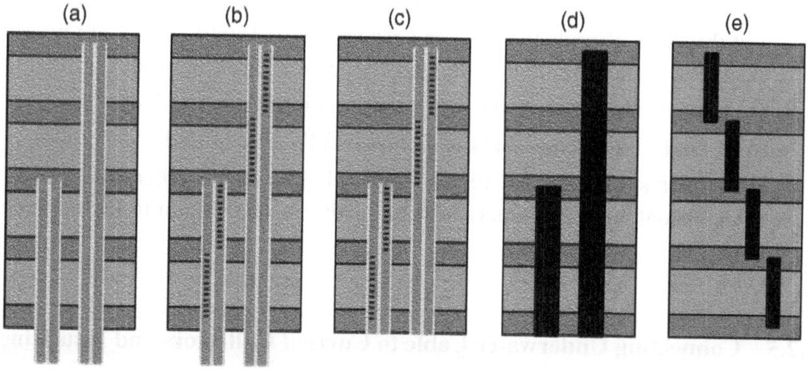

Figure 11.7 Step-by-step instructions for installing the copper current collectors. (a) Adhere copper flat wire to the anode (Step 4); (b) staple the copper flat wire to the anode, with one copper conductor dedicated for each carbon fabric segment (Step 6); (c) trim the copper flat wire flush to the anode (Step 7); (d) seal copper flat wire with waterproof sealant (Step 11–16); (e) seal the staples on the other side of the anode (Step 17).

3. Mark where the flat wires will be placed on the carbon fabric. Mask the surrounding carbon fabric with blue painters tape and masking paper to prevent excessive overspray from the spray adhesive.
4. Affix each piece of copper flat wire to the carbon fabric with 3M Super 77 spray adhesive. Allow the adhesive to cure/dry for at least 3 h.
5. Mark the locations where the staples will be placed on the copper flat wire. There will be one independent copper collector for each carbon fabric segment, with staples connecting the segment to the copper collector for each segment.
6. Staple each location along the copper flat wire made in Step 5. Staples should be aligned so that the carbon fabric can be rolled easily.
7. Now that the copper flat wire pieces are securely placed, trim any excess flat wire from the end of the anode. Note that this end of the flat wire will not be waterproofed.
8. Place the anode on the floor with the copper flat wire facing the floor. You will only see the staples on this side.
9. Strike each staple repeatedly with a rubber mallet, taking care not to damage the fabric. The goal in this case is to make sure that the staples are secure and make good contact with the copper.
10. Use a multimeter to confirm the resistance from the copper flatwire to the carbon fabric segment to which it is attached. We see typical resistance from carbon fabric to the copper wire about 0.7 Ω.
11. Seal and waterproof the ends of the copper wire that lay on the carbon fabric with 3M 5200 sealant.
12. Cover the area around the copper flat wire and staples with masking paper and painters tape to prevent overspray of the spray sealant when coating the copper flatwire.
13. Apply one light coat of the Mighty Sealer spray sealant to the copper flat wire.
14. Let this coat of sealant to dry and repeat Step 13 until staples are no longer visible to make sure that the stapled connections are waterproof.
15. Remove masking before the sealant is completely cured/dried.
16. Wait for 12 h to ensure that the sealant is fully cured/dried.
17. Flip the anode over and repeat Steps 11–15 for the other side. After the final sealant step, the anodes will be ready for connection to the underwater cable.

11.2.5 Connecting Underwater Cable to Current Collectors and Insulating Electrical Connections

In this example case, we planned to monitor power generation in a boat house next to the deployment, so we needed to wire BMFCs directly to the inside of the boat house. Step-by-step underwater cable connections and water-proofing instructions are given as follows and depicted in Figures 11.8 and 11.9 for this type of configuration.

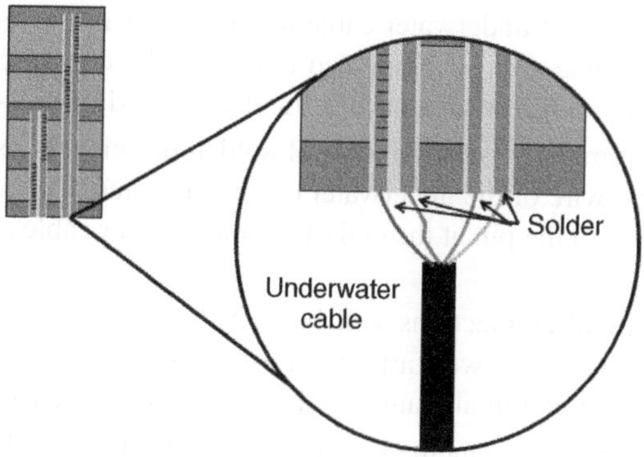

Figure 11.8 Soldering the anode to an underwater cable.

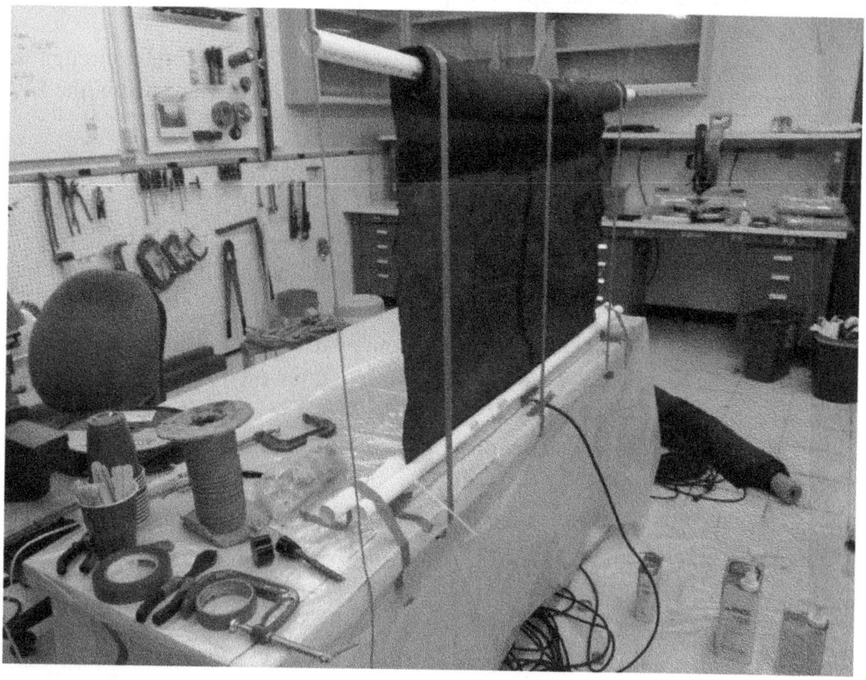

Figure 11.9 Setup for potting of the electrical connections with epoxy. *(See insert for color representation of this figure.)*

Alternatively, these connections can be run to an underwater pressure housing with associated electronics and instrumentation.

1. Clean the unsealed end of the copper flat wire to prepare it for soldering to the underwater cable conductors.

2. Solder a four-wire underwater cable to the copper flat wire. Each wire in the underwater cable will be connected to one of the copper conductors so that each copper conductor remains electrically isolated from the rest.
3. With a multimeter, test continuity of soldered connections, and ensure that each pin and wire of the underwater cable is isolated. The resistance from the carbon fabric to the pin at the end of the underwater cable should be less than 1 Ω.
4. These electrical connections will be potted in waterproof epoxy to prevent corrosion. To do this, we start with a 5-foot section of standard Schedule 40 PVC pipe (1 in. nominal diameter) that will serve as a casing.
5. Cut 12-mm groove along the length of the PVC. This is where the anode will enter this casing.
6. Cut out a notch from the middle of the PVC tube to accommodate the underwater cable. Once the epoxy is poured, the cable should pass through the potting/casing and emerge about 150° from the groove.
7. Carefully place the soldered end of the anode in the PVC tube through the opening made in Step 5. Be sure that all exposed wiring are within the PVC tube and that none of the individual wires is shorted.
8. Place the underwater cable in the notch cut out in Step 6.
9. Secure the anode, so it will extend straight out of the PVC pipe.
10. Seal the ends of the PVC pipe with blue painters tape.
11. Seal the area around the underwater cable and notch with hot glue and blue painters tape.
12. Secure the PVC pipe to a rigid, horizontal, flat structure with the slot oriented up so that the epoxy will flow evenly through the pipe and the heat from the epoxy will not deform the pipe while it sets up.
13. Mix the epoxy resin and hardener and carefully fill the PVC pipe cavity with epoxy.
14. Once the epoxy has set, the blue tape should be removed.
15. Allow at least 3 h to cure; 24 h is preferable.
16. Cut the excess PVC pipe/epoxy ends off with a hacksaw, so that there is only a 2-cm extension past the anode on both ends.

11.2.6 Sacrificial Screen Preparation for the Anode

Experience has shown that at the end of our sled deployment, the axle on the sled will spin faster than material can be deployed. When this occurs, the anode unravels or tears inside the sled and can jam, resulting in damage to the anode. As a preventative measure, we attach approximately 1 m of sacrificial fiberglass screen to the anode end that will deploy last. This is the end furthest from the underwater cable. The screen is perforated at the connection to the anode to ensure detachment of the screen from the anode if mechanical stress due to jamming is encountered. Step-by-step procedures for attaching the screen are given as follows:

1. Cut out a 0.5 m × 1 m section of fiberglass screen.
2. Lay out the anode and properly align it so that the corners are square and opposite sides are parallel.
3. Lay the screen piece at the end of the anode opposite the underwater cable.
4. Attach the screen along the edge of the anode with a single piece of 7.5-cm Gorilla tape.
5. Flip the anode and properly align it to remove distortions.
6. Affix a piece of 7.5-cm Gorilla tape at the same location, but on the other side of the anode, as the piece from Step 4.
7. Perforate the screen 3 cm from the Gorilla tape by cutting 3-cm incisions along the width of the screen. These incisions should be separated by 5 cm (Fig. 11.10).

11.2.7 Protective Screen Preparation

Previous BMFC experiments have seen performance impacts attributed to burrowing ocean fauna. As these organisms burrow toward the anode from the water column,

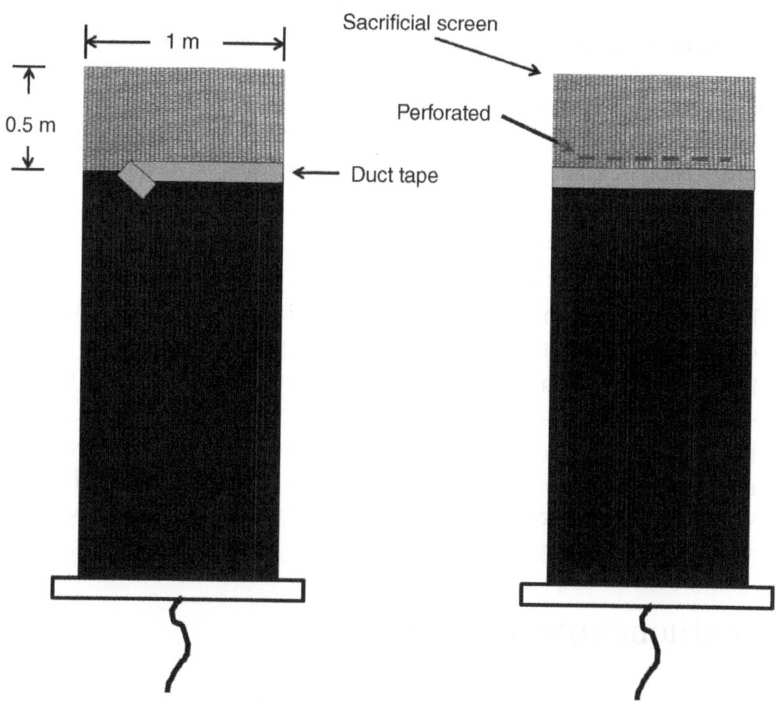

Figure 11.10 Schematic of the sacrificial screen preparation.

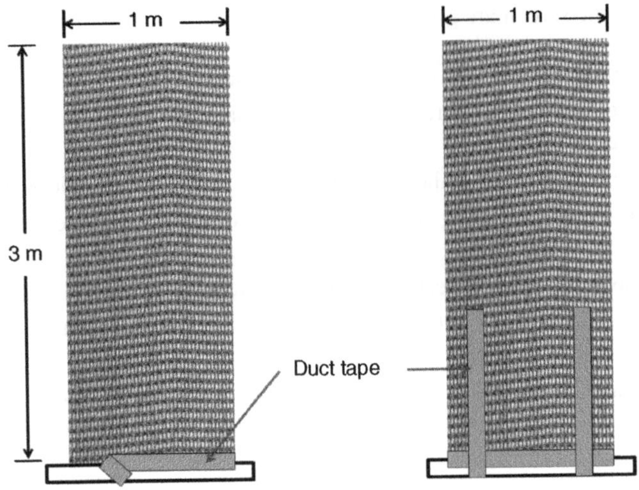

Figure 11.11 Schematic of the protective screen.

oxygenated water may reach the anode and eliminate the electrochemical gradient powering the BMFC. To impede burrowing, a protective screen is deployed parallel to and approximately 7.5 cm above the anode in the sediment. The procedure for constructing the protective screen is summarized in Figure 11.11, and step-by-step instructions for making the protective screen are given as follows:

1. Cut a 1 m × 3 m piece of fiberglass screen.
2. Tape along a 1-m side of the screen with the 5-cm duct tape. Only half of the tape should be attached to the screen. The other half will be used to attach the screen to a PVC pipe.
3. Cut a 1.2-m section of Schedule 40 PVC pipe (1 in. diameter).
4. Center the PVC pipe along the screen.
5. Roll the PVC onto the duct tape to make good attachment to the PVC and minimize any wrinkles in the tape.
6. Place a piece of 5-cm duct tape connecting screen to the PVC on the opposite side of the screen aligned with the first piece.
7. Place another two pieces of 5-cm duct tape along the length of the screen, starting at the PVC and ending 75 cm from the edge of the screen. These pieces should run perpendicular to the ones attached to the PVC and screen.

11.3 CATHODE CONSTRUCTION

The cathodes and support structure are built atop a flat plastic sheet outfitted with angle brackets on the underside. These angle brackets are located to provide traction

TABLE 11.2 List of Materials Used for Cathode Construction, Catalog Numbers, and Their Vendors.

Material	Part Number	Manufacturer
PVC 1 in. diameter	531194	JM Eagle; Los Angeles, CA, USA
PVC tee junction	401-010P5	Mueller Industries, Inc.; Memphis, TN, USA
PVC elbow junction	406-010P5	Mueller Industries, Inc.; Memphis, TN, USA
PVC glue and primer kit	302483	Oatey; Cleveland, OH, USA
Titanium wire	90455K53	McMaster-Carr; Elmhurst, IL, USA
Thread	UNKN	UNKN
Cable ties	295925T	Commercial Electric Products; Cleveland, OH, USA
Waterproof butt splices	309025	Marinco; Menomonee Falls; WI, USA
Underwater cable	IE2/IE4/IE6/IE8	Teledyne Impulse; San Diego, CA, USA
Marine Fast Cure 5200 adhesive sealant	292592	3M; Maplewood, MN, USA
PVC clamp	3192T52	McMaster-Carr; Elmhurst, IL, USA
PVC sheet 0.625 cm (1/4 in. thick)	8747K159	McMaster-Carr; Elmhurst, IL, USA
Aluminum angle iron	56840	Crown Bolt; Aliso Viejo, CA, USA
West System 105 Epoxy Resin	105-B	West System Inc; Bay City, MI, USA
West System 206 Hardener	206-B	West System Inc; Bay City, MI, USA
316 SS bolt	93190A542	McMaster-Carr; Elmhurst, IL, USA
316 SS nut	94804A029	McMaster-Carr; Elmhurst, IL, USA
316 SS washer	90107A013	McMaster-Carr; Elmhurst, IL, USA
316 SS lock washer	92147A029	McMaster-Carr; Elmhurst, IL, USA

to the sediment as it flows over the sled blades when the sled is plowing through the sediment. The support structure is meant to suspend carbon fabric cathodes in the water column and to provide an anchored location for cabling and electronics. For our purposes, the cathode consists of the base plate, cathode support frame, carbon fabric electrodes, and cathode conductors. Table 11.2 lists the materials used to construct the cathode.

11.3.1 Cathode Base Plate

The step-by-step construction of the cathode base plate is described as follows:

1. Cut the 0.625-cm PVC sheet to 59 cm × 90 cm.
2. Cut the aluminum angle bracket into 45-cm sections.

3. Arrange two angle bracket sections on the PVC sheet parallel to the 59-cm side and centered. These brackets should be spaced such that they are 10 cm from the 59-cm end.
4. Evenly space the brackets length wise, leaving 10 cm of space on both sides of the PVC plate.
5. Mark the location of the angle brackets, and drill 0.625-cm diameter mounting holes through the aluminum brackets and PVC sheet. Each bracket has two mounting holes.
6. Using stainless steel hardware, fix the angle brackets to the PVC sheet. Each hole will need one each of the bolt, nut, washer, and lock washer.
7. Label one of the 59-cm sides of the PVC to be the front. On the angle bracket mounted closest to this side, drill a couple of 0.625-cm holes in the angle bracket on the side that is not flush with the PVC. These holes should be centered 1.5 cm from the edge, one above the other. These holes will be used for line that will attach the anodes to this cathode support structure.

11.3.2 PVC Cathode Support

Step-by-step construction of the PVC cathode support is described as follows:

1. You will need the PVC joints listed as follows:
   a. 6–1 in. elbows
   b. 2–1 in. tees
2. From 1 in. PVC pipe and the PVC joints, construct the frame shown as follows. When working with PVC, it is easier to dry fit all the pieces together to ensure that all pieces are cut to the correct dimensions. If everything is suitable, apply PVC primer and cement to permanently bond each joint.
3. Drill 0.625-cm holes along the top of the PVC cathode support and along the sides. These holes should be spaced approximately 15 cm apart and are needed to flood the pipe when the support structure is submerged. This reduces buoyancy and ensures that the cathode assembly will firmly anchor itself in the ocean sediment (Fig. 11.12).

11.3.3 Carbon Fabric Electrodes Construction

Step-by-step construction of a single carbon fabric cathode is given as follows. You will need one cathode for each anode panel. So in our example, you will need a total of four cathodes.

1. Cut a 2 m × 0.15 m strip of carbon fabric that will serve as the main electrode piece.
2. Cut a 2 m length of titanium wire that will serve as the current collector.

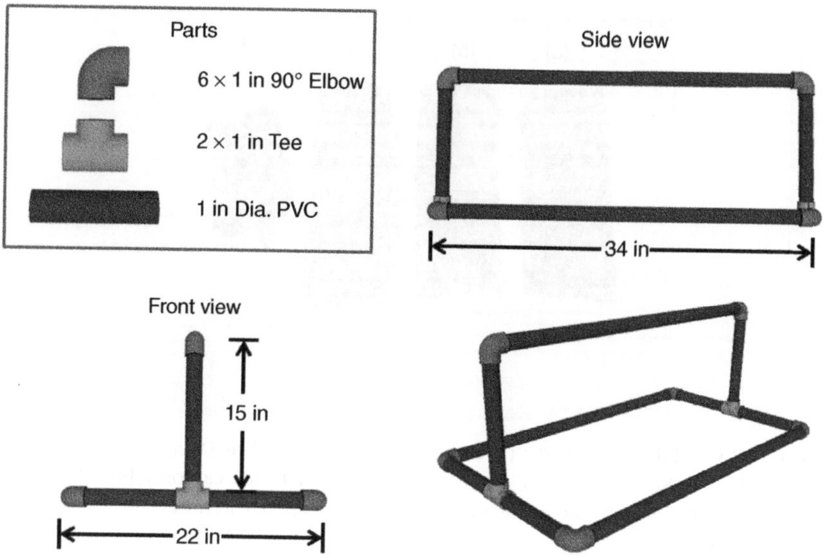

Figure 11.12 PVC cathode chassis.

3. Weave the titanium along the middle of the carbon fabric.
4. Using a sewing machine, sew the titanium wire to the carbon fabric to hold the fabric against the wire for good electrical contact.
5. Scrunch the carbon fabric to minimize its volume and provide pressure for electrical contact against the titanium wire. Bend the wire at the end of the carbon fabric to prevent movement.
6. You will need approximately 0.1 m of wire to make the connection to the underwater cable. Any excess should be trimmed off.
7. Unsheath 30 cm from the end of a four-pin, underwater cable.
8. Expose 1.5 cm of bare copper from one of the individual wires in the underwater cable.
9. Make the connection to the underwater cable with the waterproof butt splice connectors. Each cathode should be connected to an individual line in the underwater cable. Use a heat gun to ensure that the waterproof adhesive in the butt splice creates an adequate seal.
10. Use the 3M 5200 adhesive sealant to encase the butt splice and 2 cm of the wires extending from the butt splice. This is to add another layer of protection against corrosion and seawater.
11. Use epoxy to seal the part of the underwater cable where the unsheathed wires meet the waterproof jacketing. This will prevent unnecessary exposure of the jacketed wires from seawater (Figs. 11.13 and 11.14).

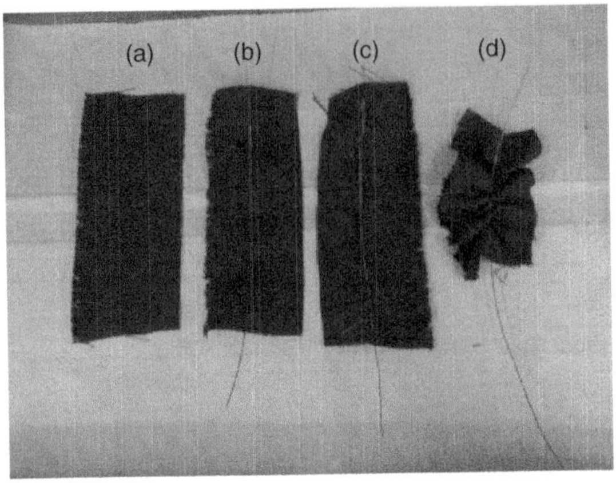

Figure 11.13 Making a cathode: (a) cut the strip of carbon fabric (Step 1); (b) weave in titanium conductor (Step 3); (c) sew the titanium conductor to the carbon fabric (Step 4); (d) scrunch the carbon fabric onto the titanium conductor (Step 5). *(See insert for color representation of this figure.)*

Figure 11.14 Example setup of sewing the carbon fabric to the titanium collector wire to form the cathode. *(See insert for color representation of this figure.)*

11.3.4 Assembling the Cathode

Step-by-step assembling of the different parts of the cathode is described as follows:

1. Attach the base plate and PVC support together with the PVC clamps. Two will be needed at each corner, for a total of eight. Three hundred sixteen stainless steel hardware (bolts, nuts, washers, and lock washers) will be required.
2. Attach each carbon fabric cathode electrode to the top of the PVC support with a cable tie around the titanium wire and PVC, so that the end attached to the underwater cable is pointed toward the base plate.
3. Fix the end of carbon fabric electrode to the bottom of the PVC support using a cable tie around the titanium wire. Some manipulation and bending of these electrodes may be necessary.
4. Coil the underwater cable and attach to the cathode base with a cable tie (Fig. 11.15).

11.3.5 Monitoring Power Generation from a Sediment Microbial Fuel Cell

In the example we gave here, the monitoring of deployed systems is accomplished through the use of data acquisition systems capable of automated data capture and storage. This data can then be used to calculate power generation from each panel for performance evaluation. The current collectors from each anode and cathode are measured against an Ag/AgCl reference electrode connected to a high-impedance digital multimeter (Model 2700; Keithley Instruments Inc.; Cleveland, OH, USA) with a multiplexer module (Model 7700; Keithley Instruments Inc.; Cleveland, OH, USA). Anode (V_{anode}) and cathode ($V_{cathode}$) potentials are used to calculate cell potential

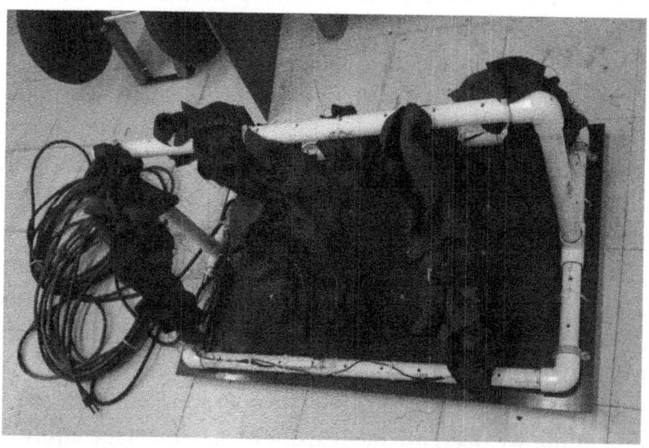

Figure 11.15 Attaching cathodes to the PVC chassis. *(See insert for color representation of this figure.)*

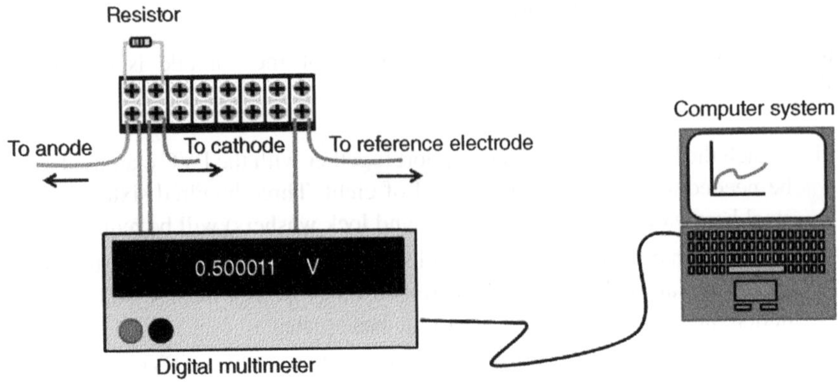

Figure 11.16 Example wiring setup to monitor the electrode potentials of one BMFC system.

(V_{cell}) according to the following equation (Fig. 11.16).

$$V_{cell} = V_{cathode} - V_{anode} \tag{11.5}$$

For our purposes, a Microsoft Excel add-in (ExcelLINX; Keithley Instruments Inc.; Cleveland, OH, USA) is used to communicate with and configure the digital multimeter. Criteria (e.g., sampling interval, log file location, and measurement speed) for data measurement and storage can be specified through this program. We generally read data from the multimeter every 5 min. Once the software and digital multimeter are configured, the system is ready to begin measuring and storing data. As data are accumulated, the amount of data may start to impact computer system performance. Therefore, the size of the data files should be kept manageable by starting new data log files periodically. The size of the data file critically depends on your computer configurations. As the BMFC is being monitored, real-time data can be plotted by simply creating a plot in Excel. This is used to evaluate the performance of the system and make adjustments to the system loads if needed.

During startup (1–7 days), the system is normally allowed to equilibrate under open circuit conditions (i.e., no external load). This allows the microbial community to settle on the electrode surfaces. Also, this prevents oxygen generation on the cathodes; therefore, a higher cathodic potential can be accomplished after drawing a current from the BMFC. By monitoring the electrode potentials, we are able to determine steady-state open circuit conditions before continuing with system conditioning. During the BMFC conditioning period, the BMFC is loaded with sequentially lower resistive loads with a goal of establishing a steady-state operation at a point of maximum power. In our evaluations, we start with a large resistor (about 100 kΩ) and begin decreasing the resistance as steady-state operating voltages are established.

As a rule of thumb, we decrease the resistor by a percentage equivalent to the percentage from our desired operating voltage. For example, if the cell potential with a 100-kΩ load is 0.8 V and we would like to operate our systems at 0.4 V, we would decrease the load by 50% to 50 kΩ. Once steady-state voltages are observed, power can be calculated and the resistor can be lowered to a different value. This process continues until lower resistor values begin to have minimal impact on increasing power production.

Once power production has been maximized and the system is at steady state, we generate a polarization curve with sequential load changes as described [17, 18]. Additional electrochemical tests such as linear scan voltammetry, cyclic voltammetry, and electrochemical impedance spectroscopy can be performed to further characterize each electrode. Samples can also be collected for analysis of the microbial community and microscopy.

11.3.5.1 Example Data Set Showing Power Generation Figure 11.17 shows a sample data set from startup to steady state. Typically, during the first several days following BMFC installation, the system is allowed to come to a stable potential under open circuit conditions or under high load. Here, we allow for a stable working potential to develop under a high resistive load of 1 MΩ. Initial monitoring of the initial potential difference is critical to determine if the BMFC is deployed correctly. Such as, if the anode was buried properly, which may be determined if the anode potential does not go to negative values (∼−400 $mV_{Ag/AgCl}$). Typical open circuit cell potential for marine BMFCs is approximately 0.8 V. We hypothesize that changing resistors slowly allows electrodes to accumulate a proper microbial community for higher electron transfer rates.

Once an acceptable, stable open circuit potential is observed, the external load is reduced methodically. We generally reduce the resistor by half its resistance for each change. A period of time is required to allow the system to reach a steady-state potential after each load change. This period of time may vary between systems. In this example, the equilibration time ranged from 6 days in the beginning of startup

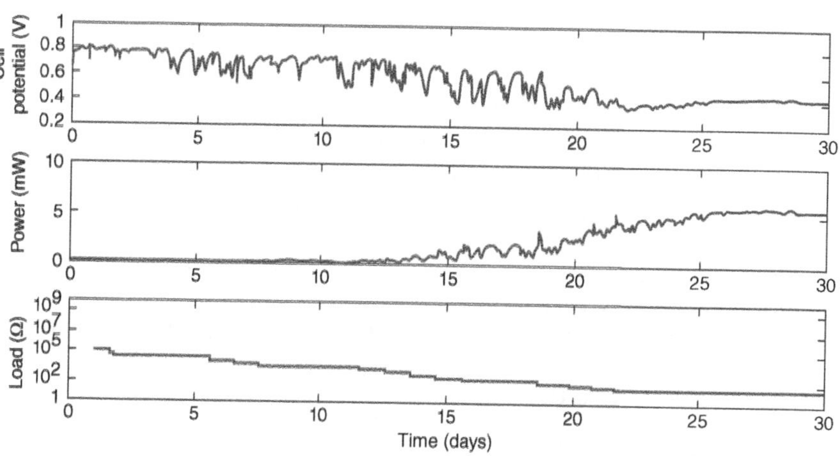

Figure 11.17 Sample data showing the initial startup period of a BMFC.

to 1 day toward the end of the startup. The resistor needs to be replaced after a new steady-state cell potential is established as exemplified in Figure 11.17.

11.4 CONCLUSIONS

Remaining challenges we have found with preparing and deploying these systems revolve mainly around ruggedizing the system to withstand the marine environment while minimizing internal losses. The marine environment is quite harsh and requires a great deal of care to protect electronics and those materials susceptible to corrosion. For example, one of the main areas of concern for our systems is the connections between the anode and the copper current connectors. The mating of carbon fabric with the copper with minimal contact resistance proves to be a persistent challenge. Other common sources of system failure are flooding of the electronics by seawater and partial unburial of the anode material during operation.

Deploying large surface area electrodes is the most challenging aspect of installing large-scale BMFC systems. We overcome this challenge by using fabric anodes and using our innovative technology – a boat-towed BMFC sled to bury the anodes. In this chapter, we described, in detail, the construction of a BMFC with a segmented anode electrode configuration. The procedures described here may be further customized for various types of applications and sediment types. Overall, we expect that the segmented BMFCs will lead to increased power generation by minimizing internal resistances associated with electrode material and decreasing system variability.

REFERENCES

1. Donovan C et al. Batteryless, wireless sensor powered by a sediment microbial fuel cell. Environ Sci Technol 2008;42(22):8591–8596.
2. Tender LM et al. The first demonstration of a microbial fuel cell as a viable power supply: Powering a meteorological buoy. J Power Sources 2008;179(2):571–575.
3. Donovan C et al. Power management system for a 2.5 W remote sensor powered by a sediment microbial fuel cell. J Power Sources 2011;196(3):1171–1177.
4. Donovan C et al. Sediment microbial fuel cell powering a submersible ultrasonic receiver: new approach to remote monitoring. J Power Sources 2013;233:79–85.
5. Hsu L et al. Scale up considerations for sediment microbial fuel cells. RSC Adv 2013;3(36):15947–15954.
6. Richter KE, Kagan J, George R. Autonomous, retrievable microbial fuel cell. 221st ECS Meeting, Seattle, WA; 2012.
7. Chadwick DB, Kagan J, Wotawa-Bergen AQ, Davis WC. "Sled for Benthic microbial fuel cell deployment with carbon fabric anodes," Proceedings of Oceans'11 MTS/IEEE Kona, sept 19–22, 2011, Kona, HI.
8. Aller RC. The sedimentary Mn cycle in long-Island sound – its role as intermediate oxidant and the influence of bioturbation, O_2, and C(Org) flux on diagenetic reaction balances. J Mar Res 1994;52(2):259–295.

9. Reimers CE et al. Microbial fuel cell energy from an ocean cold seep. Geobiology 2006;4(2):123–136.
10. Ateya BG, Al-Kharafi FM. Anodic oxidation of sulfide ions from chloride brines. Electrochem Commun 2002;4(3):231–238.
11. Morcos I, Yeager E. Kinetic studies of the oxygen—peroxide couple on pyrolytic graphite. Electrochim Acta 1970;15(6):953–975.
12. Paliteiro C, Hamnett A, Goodenough JB. The electroreduction of oxygen on pyrolytic graphite. J Electroanal Chem Interfacial Electrochem 1987;233(1–2):147–159.
13. Hossain MS, Tryk D, Yeager E. The electrochemistry of graphite and modified graphite surfaces: the reduction of O_2. Electrochim Acta 1989;34(12):1733–1737.
14. Vetter KJ. Electrochemical kinetics : theoretical and experimental aspects. New York: Academic Press; 1967.
15. Fu L et al. Synthesis of hydrogen peroxide in microbial fuel cell. J Chem Technol Biotechnol 2010;85(5):715–719.
16. Babauta JT et al. Microscale Gradients of Oxygen, Hydrogen Peroxide, and pH in Freshwater Cathodic Biofilms. Chemsuschem 2013;6(7):1252–1261.
17. Watson VJ, Logan BE. Analysis of polarization methods for elimination of power overshoot in microbial fuel cells. Electrochem Commun 2011;13(1):54–56.
18. Menicucci J et al. Procedure for determining maximum sustainable power generated by microbial fuel cells. Environ Sci Technol 2006;40(3):1062–1068.

INDEX

Biofilms in Bioelectrochemical Systems: From Laboratory Practice to Data Interpretation, First Edition.
Edited by Haluk Beyenal and Jerome Babauta.

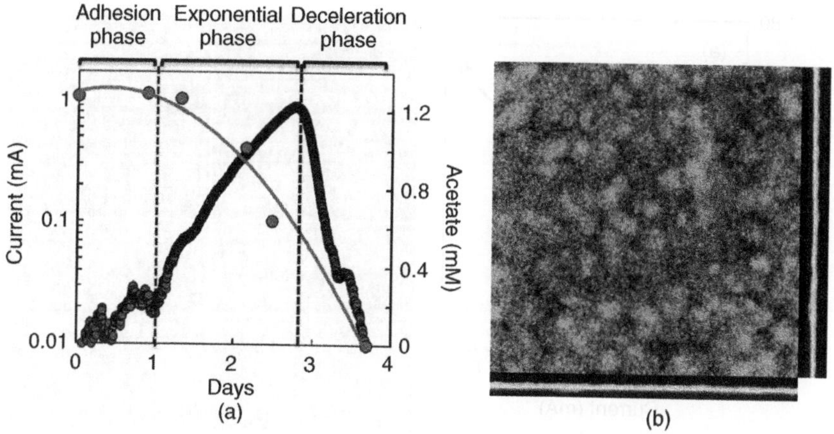

Figure 2.1 (a) Stages in the development of an anode biofilm by *G. sulfurreducens* coupled to current production (red symbols) and acetate utilization (shown as decrease in acetate concentration, blue symbols) in an MEC. After an initial adhesion phase, the attached cells grow exponentially on the anode electrode coupling the oxidation of the electron donor (acetate) to current production (exponential phase). Once the electron donor concentration decreases to growth-limiting levels, the biofilm cells enter stationary phase, the biofilm stops growing, and current production declines until all the residual acetate has been utilized (deceleration phase). (b) CLSM micrograph ($200 \times 200\ \mu m^2$ field) of an acetate-fed anode biofilm is ~10 μm thick examined at the end of the experiment (3.7 days). The anode biofilm was grown using the standard cultivation protocol described in Table 2.1 and incubation at 30 °C.

Biofilms in Bioelectrochemical Systems: From Laboratory Practice to Data Interpretation, First Edition.
Edited by Haluk Beyenal and Jerome Babauta.

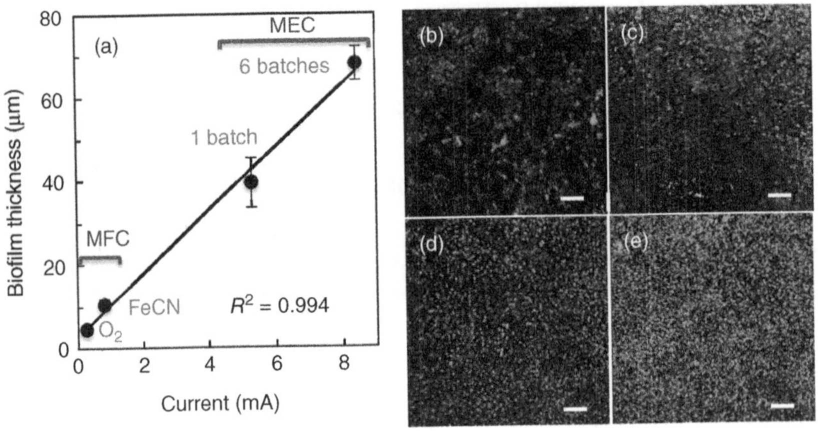

Figure 2.3 (a) Correlation between the growth of *G. sulfurreducens* on the anode electrode (quantified as biofilm thickness) and current generation (mA) in cathode-limited systems (MFCs) with oxygen (O_2) or ferricyanide (FeCN) as the catholyte and in cathode-unlimited systems (MECs) in which the anode electrode was poised at a potential of 300 mV versus an Ag/AgCl reference electrode using a potentiostat and fed with 10 mM acetate once (one batch) or six times (six batches). (b–e) CLSM micrographs of the anode biofilms from MFCs with O_2 (b) or FeCN (c) and from MECs fed once (d) or six times (e). Scale bar, 20 μm. All BES used fresh water medium and were incubated at 25 °C, as described in Reguera et al. [21].

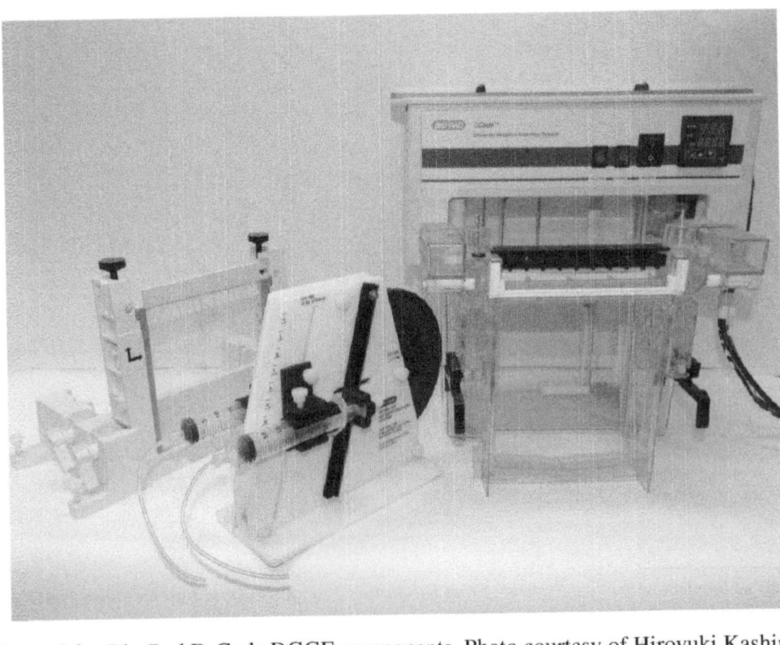

Figure 3.2 Bio-Rad D-Code DGGE components. Photo courtesy of Hiroyuki Kashima.

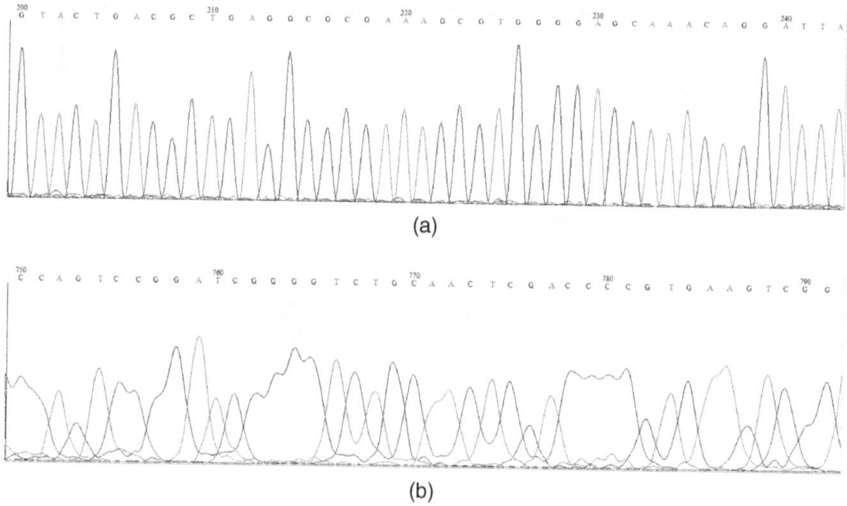

Figure 4.7 Examples of (a) "Good" quality and (b) "Bad" quality sequence data.

Figure 5.7 Picture of the fully assembled electrochemical cell with the yellow-colored solution containing ferricyanide.

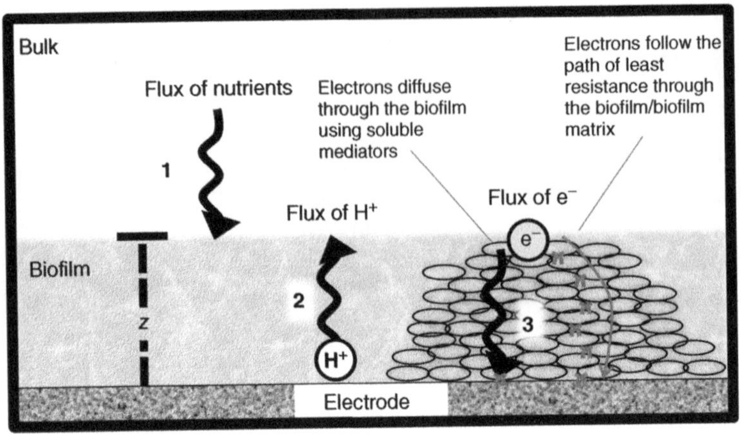

Figure 5.11 Biofilm cartoon showing: (1) the flux of nutrients, (2) flux of protons, and (3) flux of electrons inside a biofilm of finite thickness Z. This research was originally published in Ref. [26]. Copyright 2012 the Biochemical Society.

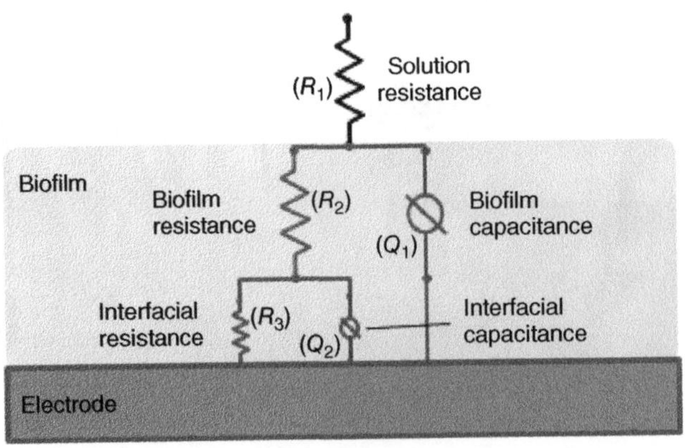

Figure 5.23 Equivalent circuit of *G. sulfurreducens* biofilm consisting of the film resistance, film capacitance, interfacial resistance, interfacial capacitance, and pseudocapacitance. Reproduced with permission from Ref. [50]. Copyright 2013 Wiley Periodicals, Inc.

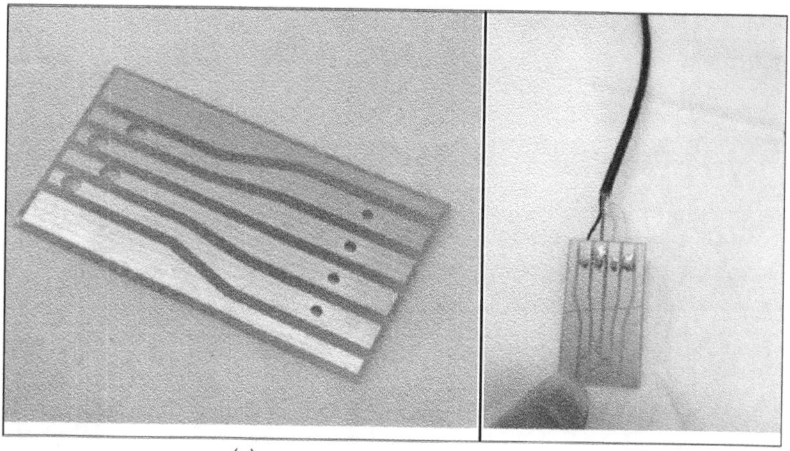

(a) (b)

Figure 6.13 (a) Circuit board used for making electrical contact to IDA chip. (b) Circuit board with electrical wires soldered in place.

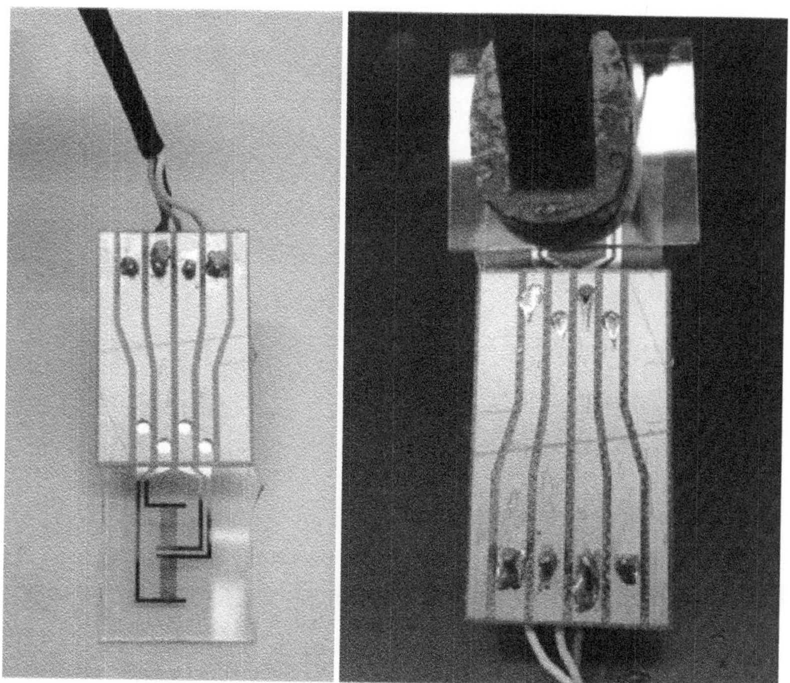

(a) (b)

Figure 6.14 (a) IDA chip glued to circuit board before being soldered together. (b) After being soldered together.

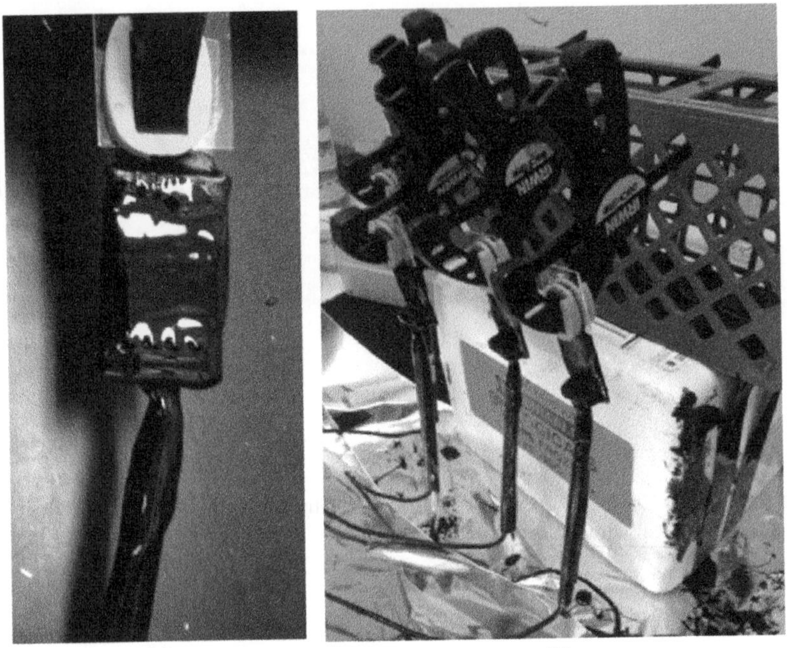

(a) (b)

Figure 6.16 Sealing IDA packaging with water-insulating splicing epoxy.

Figure 6.17 Sealed IDA in electrochemical cell.

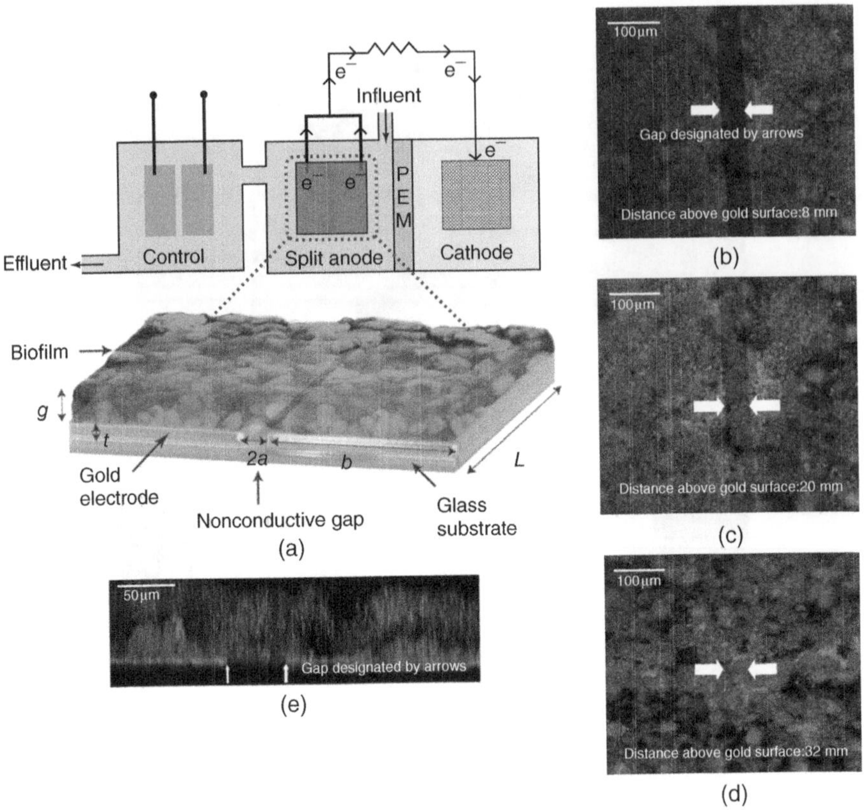

Figure 7.3 Strategy to measure in situ biofilm conductivity. (a) Schematic of microbial fuel cell with two gold electrodes serving as an anode, separated from cathode by a proton exchange membrane (PEM). The biofilm grows over the electrodes and the nonconducting gap between the two electrodes. No biofilm forms on the control electrode pair. Gap width $2a = 50$ μm, electrode width $b \approx 1.27$ cm, electrode length $L \approx 2.54$ cm, electrode thickness $t \approx 50$ nm, and biofilm height, g. (b–e) Representative fluorescent confocal scanning laser microscopy images of split-electrodes. Images were taken when microbial current was 0.25 mA and biofilm height was 36 ± 1.4 μm. (b–d) Top-down confocal image slices of biofilm spanning the nonconductive gap. X–Y image slices (parallel to the electrode surface). Scale bar 100 μm. (e) Cross-sectional image of biofilm spanning the nonconductive gap. X–Z image slice through the biofilm, in a direction perpendicular to the surface of the gold anode and across the 50-μm gap. Scale bar 50 μm. Figure adapted from Ref. [4] with permission.

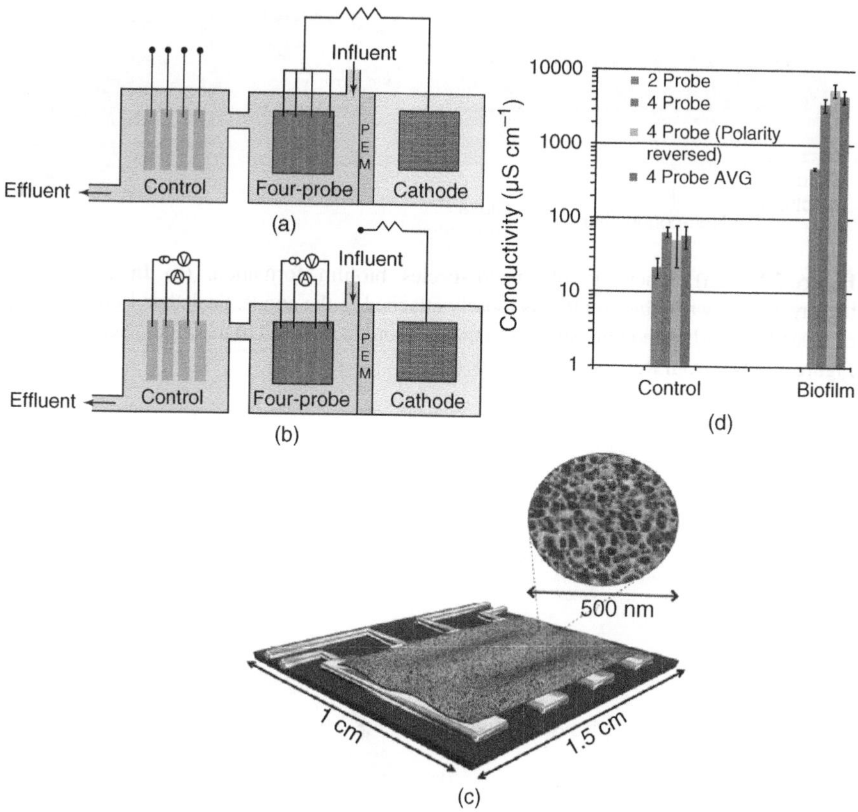

Figure 7.5 Four-probe measurements. (a) Schematic of microbial fuel cell with four gold electrodes, each separated by 50-μm nonconductive gap, serving as an anode (side-view). (b) Schematic of conductivity measurements using four electrodes (side-view). Current is injected through outer two electrodes, and voltage is measured across inner two electrodes. (c) Schematic of four-probe setup used for biofilm and pili measurements (top-view). Inset, AFM image of pili network placed on gold electrodes. (d) Comparison of conductivity measured using two-probe and four-probe methods. Error bars show SD of individual measurements for four biofilms of KN400. Figure adapted from Ref. [4] with permission.

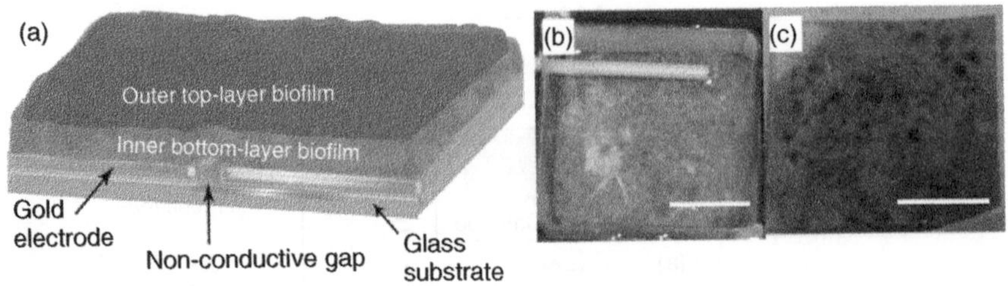

Figure 7.11 (a) Schematic of mixed-species biofilm formation. (b) Image of outer, top-layered brownish biofilm that is loosely attached to the anode. Scale bar 1 cm. (c) Image of inner, bottom-layered pinkish biofilm that is strongly attached to the anode. Scale bar 1 cm. Figure adapted from Ref. [5] with permission.

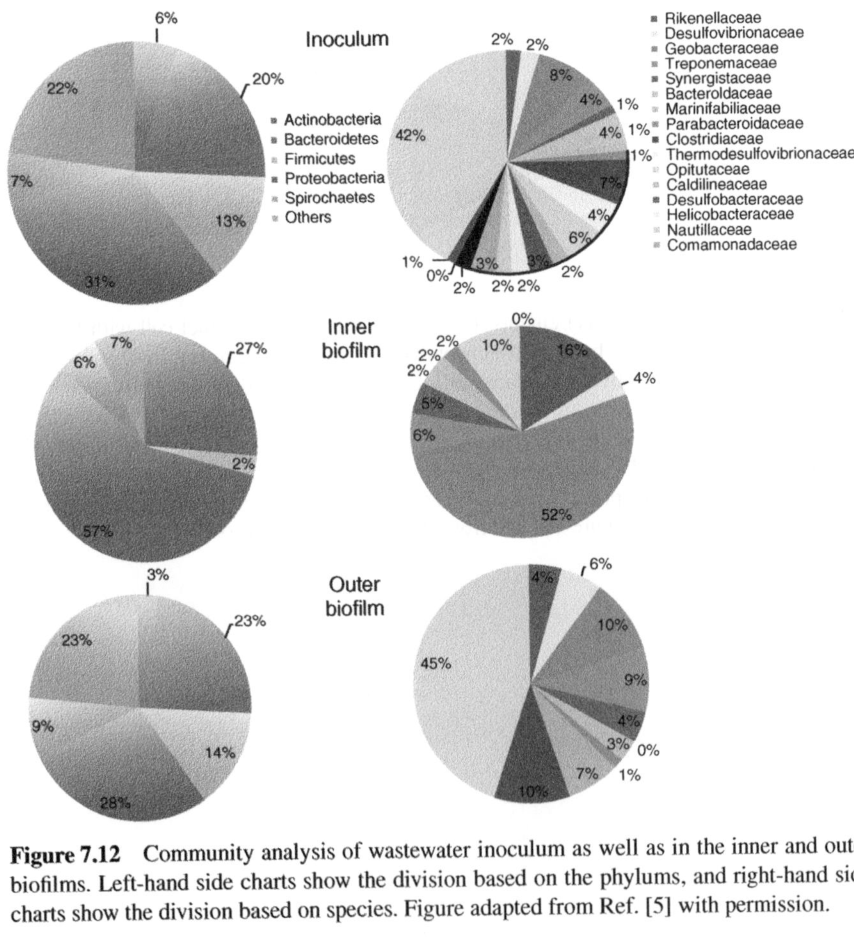

Figure 7.12 Community analysis of wastewater inoculum as well as in the inner and outer biofilms. Left-hand side charts show the division based on the phylums, and right-hand side charts show the division based on species. Figure adapted from Ref. [5] with permission.

(a) (b) (c)

Figure 10.3 A prototype benthic MFC. In (a), the MFC is ready for deployment, coupled to a gravity corer's weight stack by an aluminum releasing head. Mounted to the right side is a flotation canister designed to carry a connecting line to the sea surface after release by an acoustic command. Tie-wrapped to the supporting structure above the anode rods is a titanium pressure housing for a potentiostat and data loggers. The cathode extends to the left. In (b), the whole corer frame and MFC are in view while being launched. A small underwater video camera is mounted to the frame on the bottom right. An outward-looking digital still camera and strobe are mounted on the left side of the frame. Panel (c) shows one of these MFCs being recovered after 9 months on the seafloor of the Oregon shelf. The orange line was connected to the pop-up float. Sea anemones and barnacles are attached to parts of the framework, and the anode rods display a color change indicating that approximately one third of the length of each rod was not buried during most of the deployment. It is expected that not enough ballast and the buoyancy of the release float caused the MFC anodes to work their way out of the sediment during these experiments.

Figure 10.5 (a) Bottom-up view of a 0.4-m inner diameter, chambered benthic MFC with carbon fiber brush anode attached to the domed top. (b) A top-down view of a later chamber with a perforated PVC bioturbation barrier fixed to the midsection of the chamber.

Figure 10.9 A chambered MFC in place on the seafloor near the MARS Observatory. Each MFC also carried an osmotic-pump sampler (mounted on the left side of the picture) designed to draw small water samples from the chambers throughout the deployment. The sensor and modem (outside the photo) were suspended above the carbon brush cathode that is pictured above the chamber. Lead weights are visible, fixed to the underside of the chamber lid. Image captured from high resolution underwater video recorded from the ROV Ventana operated by Monterey Bay Aquarium Research Institute.

Figure 11.3 The boat-based deployment sled showing the general concept of operation (a), and the two-bladed prototype under construction (b).

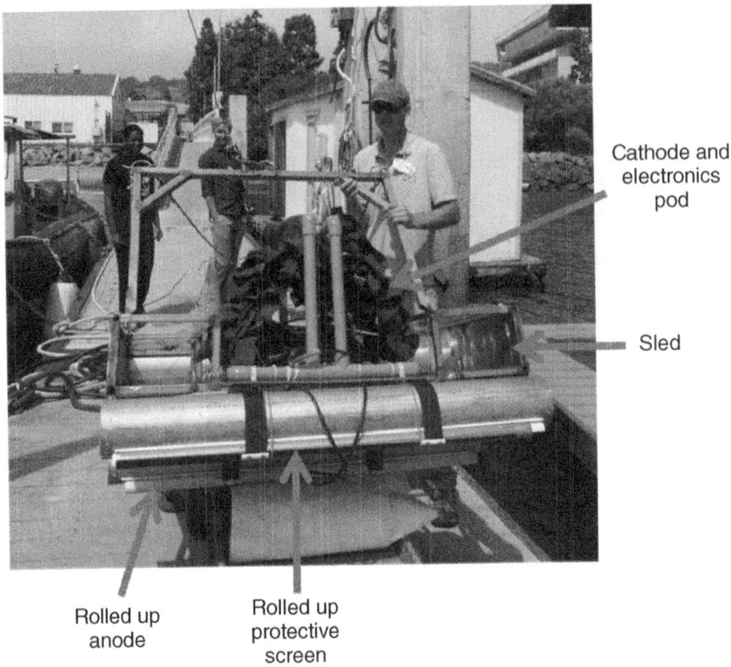

Figure 11.5 A sled loaded with the anode, cathode, protective screen, and cables. Electronics are used when BMFC is deployed in a remote area to record power generation or to operate sensors.

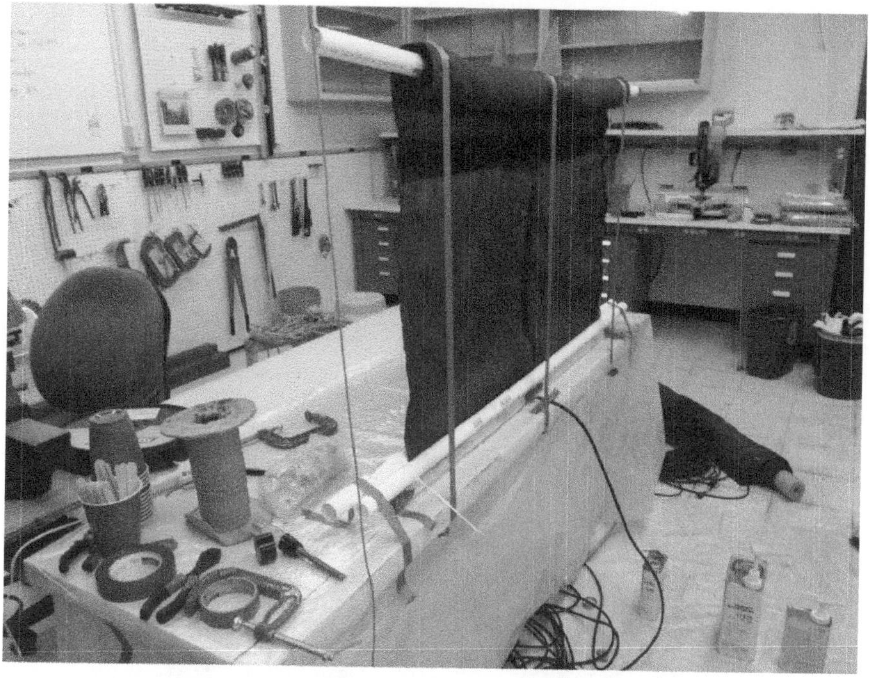

Figure 11.9 Setup for potting of the electrical connections with epoxy.

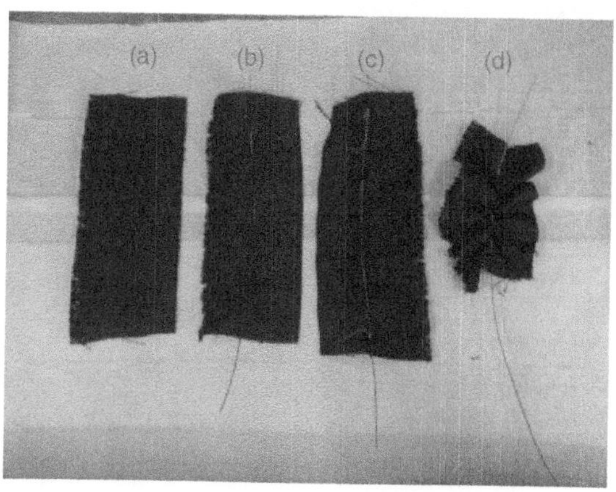

Figure 11.13 Making a cathode: (a) cut the strip of carbon fabric (Step 1); (b) weave in titanium conductor (Step 3); (c) sew the titanium conductor to the carbon fabric (Step 4); (d) scrunch the carbon fabric onto the titanium conductor (Step 5).

Figure 11.14 Example setup of sewing the carbon fabric to the titanium collector wire to form the cathode.

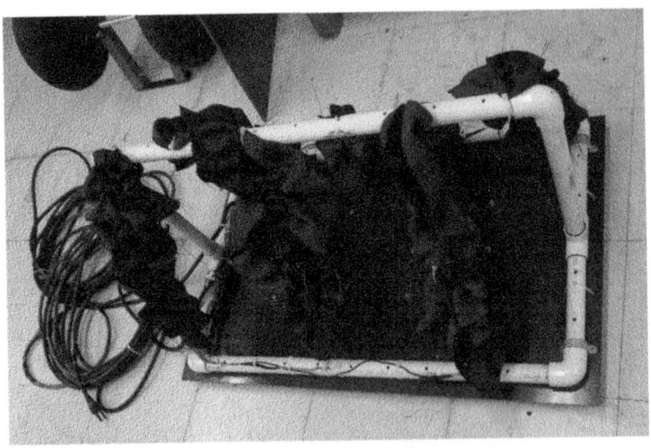

Figure 11.15 Attaching cathodes to the PVC chassis.

Printed and bound by CPI Group (UK) Ltd, Croydon, CR0 4YY
16/09/2026
14963423-0003